INCOMPRESSIBLE FLUID DYNAMICS

Incompressible Fluid Dynamics

P. A. DAVIDSON

University of Cambridge

OXFORD
UNIVERSITY PRESS

OXFORD

UNIVERSITY PRESS

Great Clarendon Street, Oxford, OX2 6DP,
United Kingdom

Oxford University Press is a department of the University of Oxford.
It furthers the University's objective of excellence in research, scholarship,
and education by publishing worldwide. Oxford is a registered trade mark of
Oxford University Press in the UK and in certain other countries

First Edition published in 2022

Impression: 1

Published in the United States of America by Oxford University Press
198 Madison Avenue, New York, NY 10016, United States of America

British Library Cataloguing in Publication Data
Data available

Library of Congress Control Number: 2021937986

ISBN 978–0–19–886909–2 (hbk.)
ISBN 978–0–19–886912–2 (pbk.)

Printed and bound by
CPI Group (UK) Ltd, Croydon, CR0 4YY

For Catherine, Sarah and James

PREFACE

This is a textbook for advanced undergraduate and postgraduate students of engineering, applied mathematics, and physics. It is not intended to be encyclopaedic in coverage and so difficult choices had to be made regards content. The topics addressed have been chosen largely on the grounds that they help establish the broad conceptual framework of the subject, expose key phenomena, and play an important role in the myriad of applications that exist in both nature and technology. Of course, the choice of material is ultimately a personal one, and so the author seeks the indulgence of those readers who find that their favourite topic is given less space than they would have liked.

In the preface to his 1946 text *Mechanics of Deformable Bodies*, Arnold Sommerfeld comments:

> I shall not detain myself with the mathematical foundations, but proceed as rapidly as possible to the physical problems themselves. My aim is to give the reader a vivid picture of the vast and varied material that comes within the scope of theory when a relatively elevated vantage-point is chosen.

This author fully endorses these sentiments, and so physical insight has been given priority over mathematical detail, as seems appropriate for a subject as physically rich as this. For example, when it comes to the classical theory of potential flow, it is probably less important for a student to master the intricacies of some fiendishly cunning 19th-century potential than to appreciate that, outside the field of water waves and a few choice problems in aerodynamics, an irrotational analysis will, most of the time, utterly fail to capture the real flow. It is also important for the student to understand that this failure is not merely an embarrassing inconvenience, but rather tells us something quite profound about the nature of fluid dynamics. In this respect it is, perhaps, appropriate to recall Rayleigh's whimsical, but telling, observation:

> The general equations of (inviscid) fluid motion were laid down in quite early days by Euler and Lagrange ... (but) some of the general propositions so arrived at were found to be in flagrant contradiction with observations, even in cases where at first sight it would not seem that viscosity was likely to be important. Thus a solid body, submerged to a sufficient depth, should experience no resistance to its motion through water. On this principle the screw of a submerged boat would be useless, but, on the other hand, its services would not be needed.   (1914, Scientific papers of Lord Rayleigh, p 237)

With Rayleigh's warning in mind, the viscous equations of motion and the associated concepts of boundary layers and turbulence are introduced prior to potential flow theory and its applications in aerodynamics and surface waves. In this way, it is hoped that the student will fully appreciate the fundamental limitations of potential flow as that subject is developed. The author has lived long enough to realize that this ordering of the material will not be to everyone's taste, but he makes no apology for this choice.

This caveat aside, the first half the book, in Chapters 1→7, follows a rather traditional route, covering topics met in most undergraduate courses in engineering and applied mathematics. This includes the inviscid equations of Euler and Bernoulli, the Navier–Stokes equation and some of its simpler exact solutions, laminar boundary layers and jets, potential flow theory with its various applications to aerodynamics, the theory of surface gravity waves, and flows with negligible inertia, such as suspensions, lubrication layers, and thin films. Throughout, a close link is maintained between theory and applications.

The second half of the book is more specialized and has one eye to the needs of postgraduate students in engineering, applied mathematics, and physics. Vortex dynamics, which is so essential to many natural phenomena, is developed in Chapter 8. This is followed by chapters on stratified fluids and flows subject to a strong background rotation, both topics being central to our understanding of atmospheric and oceanic flows. Instabilities and the transition to turbulence are then covered in Chapters 11 and 12, followed by two chapters on fully developed turbulence. The topic of turbulence is integral to most engineering courses, on the grounds that turbulence is both ubiquitous and important, but it is less common in texts aimed at students of applied mathematics, perhaps because the subject is infamously resistant to mathematical attack. However, to neglect such an important topic is to deny the central nature of everyday fluid mechanics, and so a gentle introduction to this difficult subject is provided in Chapters 13 and 14.

I would like to thank all of those at Oxford University Press who assisted in the preparation of this book, as well as colleagues who helped suggest the various scientists whose images appear at the start of each chapter. Sadly, those portraits are all of men. However, I am certain that the legacy of the current and future generations of fluid dynamicists will be much more evenly balanced in terms of gender. Finally, I must thank my long-suffering wife, Catherine, for her enduring patience.

Peter Davidson
Cambridge, 2021

CONTENTS

Prologue   xvii

**1   Elementary Definitions, Some Simple Kinematics, and the
     Dynamics of Ideal Fluids** 1
     1.1 Elementary Definitions   2
          1.1.1 What is the Mechanical Definition of a Fluid?   2
          1.1.2 Fluid Statics and One Definition of Pressure   3
          1.1.3 Different Categories of Fluid and of Fluid Flow   6
     1.2 Some Simple Kinematics   8
          1.2.1 Eulerian Versus Lagrangian Descriptions of Motion   8
          1.2.2 The Convective Derivative   9
          1.2.3 Mass Conservation and the Streamfunction   10
     1.3 The Dynamics of an Ideal (Inviscid) Fluid of Uniform Density   12
          1.3.1 Euler's Equation for a Fluid of Uniform Density   12
          1.3.2 Bernoulli's Equation and Mechanical Energy Conservation   14
          1.3.3 Some Applications of Bernoulli's Equation   15
          1.3.4 Inviscid Momentum Conservation in Integral Form   17
          1.3.5 Examples of the Use of Momentum Conservation to Calculate
                 Pressure Forces   18
          1.3.6 More on Momentum Conservation: Inviscid Flow Through
                 a Cascade of Blades   20
     1.4 Examples of the Failure of Ideal Fluid Mechanics   23
          1.4.1 The Borda–Carnot 'Head Loss' in a Sudden Pipe Expansion   24
          1.4.2 The Hydraulic Jump   25

2   Governing Equations and Flow Regimes for a Real Fluid 33
     2.1 Viscosity, Viscous Stresses, and the No-slip Boundary Condition   34
     2.2 More Kinematics: Characterizing the Deformation and Spin
          of Fluid Elements   36
          2.2.1 Two Things that Happen to a Fluid Element as it Slides down a Streamline 36
          2.2.2 The Rate-of-strain Tensor and the Deformation of Fluid Elements 38
          2.2.3 Vorticity: the Intrinsic Spin of Fluid Elements   38
     2.3 Dynamics at Last: the Stress Tensor and Cauchy's Equation of Motion 39
     2.4 The Navier–Stokes Equation   41
          2.4.1 Newton's Law of Viscosity   41
          2.4.2 The Navier–Stokes Equation and the Reynolds Number   43
          2.4.3 Navier–Stokes as an Evolution Equation for the Velocity Field 44
          2.4.4 The Viscous Dissipation of Mechanical Energy   45
     2.5 The Momentum Equation for Viscous Flow in Integral Form   46

2.6 The Role of Boundaries and Prandtl's Boundary-layer Equation 48
    2.6.1 The Need for Boundary Layers at High Reynolds Number 48
    2.6.2 Changes in Flow Regime as the Reynolds Number Increases 50
2.7 From Linear to Angular Velocity: Vorticity and its Evolution Equation 52
    2.7.1 The Biot–Savart Law Applied to Vorticity: an Analogy with
            Magnetostatics 52
    2.7.2 The Vorticity Evolution Equation 54
    2.7.3 Where does the Vorticity come from? 57
    2.7.4 Enstrophy and its Governing Equation 58
    2.7.5 A Glimpse at Potential (Vorticity Free) Flow and its Limitations 58
2.8 Summing up: Real versus Ideal Fluid Mechanics 59

3 Some Elementary Solutions of the Navier–Stokes Equation **65**
3.1 Some Simple Laminar Flows 66
    3.1.1 Planar Viscous Flow 66
    3.1.2 The Boundary Layer near a Two-dimensional Stagnation Point 68
3.2 The Diffusion of Vorticity from a Moving Surface 69
    3.2.1 The Impulsively Started Plate: Stokes' First Problem 69
    3.2.2 The Oscillating Plate: Stokes' Second Problem 72
3.3 The Navier–Stokes Equation in Cylindrical Polar Coordinates 73
    3.3.1 Moving from Cartesian to Cylindrical Polar Coordinates 73
    3.3.2 Hagen–Poiseuille Flow in a Pipe 75
    3.3.3 Rotating Couette Flow 78
    3.3.4 The Diffusion of a Long, Thin, Cylindrical Vortex 79
    3.3.5 A Thin Film on a Spinning Disc 81
    3.3.6 The Azimuthal–poloidal Decomposition of Axisymmetric Flows 83

4 Flows with Negligible Inertia **89**
4.1 Motion at Low Reynolds Number: Stokes Flow 90
    4.1.1 The Governing Equations at Low Reynolds Number 90
    4.1.2 Flow past a Sphere at Low Reynolds Number 91
    4.1.3 The Oseen Correction for Flow over a Sphere at Low Re 94
    4.1.4 The Uniqueness and Minimum Dissipation Theorems for Low-Re Flows 96
    4.1.5 Two-dimensional Flow in a Wedge at Low Reynolds Number 98
    4.1.6 Suspensions 99
    4.1.7 The Subtleties of Self-propulsion at Low Reynolds Number 102
4.2 Lubrication Theory 105
    4.2.1 The Approximations and Governing Equations of Lubrication Theory 105
    4.2.2 Reynolds' Analysis of the Slipper Bearing 107
    4.2.3 Sommerfeld's Analysis of the Journal Bearing 109
    4.2.4 Rayleigh's Analysis of the Stepped Bearing 111
4.3 Thin Films with a Free Surface 112
    4.3.1 Approximations and Governing Equations 113
    4.3.2 The Gravity-driven Spreading of a Circular Pool 114
    4.3.3 A Film on an Incline 115
    4.3.4 A Thin Film on a Rotating Disc (Reprise) 117

5 Laminar Flow at High Reynolds Number **123**
   5.1 Prandtl's Boundary Layer and a Revolution in Fluid Dynamics 124
   5.2 The Archetypal Boundary Layer: a Flat Plate Aligned with a
       Uniform Flow 126
   5.3 A Generalization of Prandtl's Boundary Layer to Other Physical Systems 128
       5.3.1 A Popular Model Problem and the Concept of Matched
             Asymptotic Expansions 128
       5.3.2 Prandtl's Generalization of the Boundary Layer: Another
             Model Problem 130
   5.4 The Effects of an Accelerating External Flow on Boundary-layer
       Development 132
       5.4.1 The Falkner–Skan Solutions for Flow over a Two-dimensional Wedge 132
       5.4.2 The Boundary Layer near the Forward Stagnation Point of a
             Circular Cylinder 134
   5.5 Jeffery–Hamel Flow in a Convergent or Divergent Channel 136
   5.6 Boundary-layer Separation and Pressure Drag 138
   5.7 Thermal Boundary Layers 141
       5.7.1 Forced Convection 141
       5.7.2 Free Convection 143
   5.8 Submerged Laminar Jets 144
       5.8.1 The Two-dimensional Jet 145
       5.8.2 The Axisymmetric Jet 147

6 Potential Flow Theory with Applications to Aerodynamics **153**
   6.1 Some Elementary Ideas in Potential Flow Theory 153
       6.1.1 The Physical Basis for, and Dangers of, Potential Flow Theory 153
       6.1.2 The Retrospective Application of Newton's Second Law:
             Bernoulli Revisited 156
       6.1.3 Some Simple Examples of Two-dimensional Potential Flow 157
       6.1.4 D'Alembert's Paradox 159
   6.2 The Kinematics of Two-dimensional Potential Flow 160
       6.2.1 The Complex Potential 160
       6.2.2 Some Elementary Examples of the Complex Potential 161
       6.2.3 Flow Normal to a Flat Plate of Finite Width 164
       6.2.4 A Not so Simple Example: the Intake to a Submerged Duct 166
       6.2.5 The Method of Images for Plane and Cylindrical Boundaries 167
   6.3 The Lift Force Exerted on a Body by a Uniform Incident Flow 170
       6.3.1 Two-dimensional Flow over a Cylinder with Circulation:
             an Illustrative Example 170
       6.3.2 Flow over a Planar Body of Arbitrary Shape: the Kutta–Joukowski
             Lift Theorem 172
       6.3.3 Kelvin's Circulation Theorem 173
       6.3.4 The Role of Boundary-layer Vorticity in Establishing Circulation
             round an Aerofoil 174
       6.3.5 The Lift Generated by a Slender Aerofoil 177

7   Surface Gravity Waves in Deep and Shallow Water **185**

   7.1   The Wave Equation and Dispersive versus Non-dispersive Waves   186

      7.1.1   The Wave Equation and d'Alembert's Solution   186

      7.1.2   Two Classes of Waves: Dispersive versus Non-dispersive Waves   187

   7.2   Two-dimensional Surface Gravity Waves of Small Amplitude   189

      7.2.1   Surface Gravity Waves on Water of Arbitrary Depth   189

      7.2.2   Shallow-water and Deep-water Waves   190

      7.2.3   Particle Paths, Stokes Drift, and Energy Density in Deep-water Waves   191

      7.2.4   Wave Drag in Deep Water   193

   7.3   The General Theory of Dispersive Waves   194

      7.3.1   Dispersion, Wave Packets, and the Group Velocity   194

      7.3.2   The Energy Flux in a Wave Packet   198

   7.4   The Dispersion of Small-amplitude Surface Gravity Waves   199

      7.4.1   The Group Velocity and Energy Density for Waves on Water

         of Arbitrary Depth   199

      7.4.2   Waves Approaching a Beach   201

      7.4.3   The Influence of Surface Tension on Dispersion   202

   7.5   Finite-amplitude Waves in Shallow Water   204

      7.5.1   The Inviscid Shallow-water Equations   204

      7.5.2   Finite-amplitude Waves and Non-linear Wave Steepening   205

      7.5.3   The Solitary Wave 1: Rayleigh's Solution   207

      7.5.4   Solitary Waves 2: The KdeV Equation   210

      7.5.5   More General Solutions of the KdeV Equation: Cnoidal Waves   212

      7.5.6   The Hydraulic Jump Revisited   212

8   Vortex Dynamics: Classical Theory and Illustrative Examples **221**

   8.1   Vorticity and its Evolution Equation (Revisited)   221

   8.2   Inviscid Vortex Dynamics   222

      8.2.1   The Classical Theories of Helmholtz and Kelvin   223

      8.2.2   Helicity and its Conservation   225

      8.2.3   Steady, Axisymmetric Flows and the Squire–Long Equation   227

      8.2.4   Viscous versus Inviscid Vortex Dynamics   230

   8.3   A Qualitative Overview of some Simple Isolated Vortices   232

      8.3.1   The Interaction of Line Vortices   232

      8.3.2   A Glimpse at Vortex Rings   234

      8.3.3   Vortices due to Boundary-layer Separation   236

      8.3.4   Columnar Vortices in the Atmosphere and Oceans   239

   8.4   Viscous Vortex Dynamics I: the Prandtl–Batchelor Theorem   241

      8.4.1   The Physical Origins of the Prandtl–Batchelor Theorem   242

      8.4.2   A Proof of the Theorem   243

   8.5   Viscous Vortex Dynamics II: Burgers' Vortex   245

      8.5.1   A Dilemma in Turbulence: Finite Energy Dissipation for

         Vanishing Viscosity   245

      8.5.2   Burgers' Axisymmetric Vortex   246

      8.5.3   The Robust Nature of Burgers' Vortex   247

8.6 More Axisymmetric Vortices (both Viscous and Inviscid) 248
    8.6.1 Hill's Spherical Vortex 248
    8.6.2 The Velocity Field and Kinetic Energy of a Thin Vortex Ring 249
8.7 Viscous Vortex Dynamics III: the Impulse of Localized Vorticity Fields 251
    8.7.1 The Far field of a Localized Vorticity Distribution 252
    8.7.2 The Spontaneous Redistribution of Momentum in Space 253
    8.7.3 Conservation of Linear Impulse and its Relationship
        to Linear Momentum 254
    8.7.4 Conservation of Angular Impulse and its Relationship
        to Angular Momentum 256
    8.7.5 Axisymmetric Examples of Impulse and Vortex Rings Revisited 257

9 Waves and Flow in a Stratified Fluid **265**
9.1 The Boussinesq Approximation and a Second Definition of the
    Froude Number 265
9.2 The Suppression of Vertical Motion: a Simple Scaling Analysis 268
9.3 The Phenomenon of Blocking 270
9.4 Lee Waves 271
    9.4.1 Linear Lee Waves in Two Dimensions 271
    9.4.2 Finite-amplitude Lee Waves in Two Dimensions 275
9.5 Internal Gravity Waves of Small Amplitude 277
    9.5.1 Linear Theory and Simple Examples 277
    9.5.2 The Reflection of Internal Gravity Waves 281
9.6 Generalized Vortex Dynamics: Bjerknes' Theorem and Ertel's
    Potential Vorticity 284

10 Waves and Flow in a Rotating Fluid **291**
10.1 Rayleigh's Stability Criterion for Inviscid, Swirling Flow 291
10.2 The Equations of Motion in a Rotating Frame of Reference 294
    10.2.1 The Coriolis Force and the Rossby Number 294
    10.2.2 Rapid Rotation: the Taylor–Proudman Theorem and Drifting
        Taylor Columns 295
10.3 Inertial Waves of Small Amplitude 298
    10.3.1 Their Dispersion Relationship, Group Velocity, and Spatial Structure 298
    10.3.2 The Formation of Transient Taylor Columns by Low-frequency Waves 300
    10.3.3 The Spontaneous Focussing of Inertial Waves and the Formation
        of Columnar Vortices 302
    10.3.4 Helicity Generation and Helicity Segregation by Inertial Waves 310
    10.3.5 Finite-amplitude Inertial Waves 313
10.4 Rossby Waves 314
10.5 Ekman Boundary Layers and Ekman Pumping 317
    10.5.1 Confined Swirling Flows: the Solutions of Kármán, Bödewadt,
        and Ekman 317
    10.5.2 Ekman Layers as a Mechanism for Energy Dissipation 320
10.6 Tropical Cyclones 320
    10.6.1 The Anatomy of a Tropical Cyclone 320
    10.6.2 A Simple Model of a 'Dry' Cyclone 322

11   Instability **333**

   11.1   The Centrifugal Instability   334

      11.1.1   Rayleigh's Inviscid Criterion for Axisymmetric Disturbances   334

      11.1.2   Two-dimensional Inviscid Disturbances (Rayleigh again)   335

      11.1.3   Viscous Instability and Taylor's Analysis   336

      11.1.4   The Experimental Evidence   339

   11.2   The Stability of a Fluid Heated from Below   341

      11.2.1   Rayleigh–Bénard Convection   341

      11.2.2   Rayleigh's Stability Analysis   342

      11.2.3   Slip Boundaries Top and Bottom: an Artificial but Informative Case   344

      11.2.4   No-slip Boundaries   346

   11.3   The Stability of Parallel Shear Flows   347

      11.3.1   Rayleigh's Inflection Point Theorem for Inviscid, Rectilinear Flow   347

      11.3.2   The Subtle Effects of Viscosity   349

   11.4   The Kelvin–Helmholtz Instability   350

      11.4.1   The Instability of an Inviscid Vortex Sheet   350

      11.4.2   The Inviscid Instability of a Layer of Vorticity of Finite Thickness   354

   11.5   The Stability of Continuously Stratified Shear Flow   355

      11.5.1   The Taylor–Goldstein Equation for Fluctuations in a Stratified
Shear Flow   356

      11.5.2   The Richardson Number Criterion for the Stability of a Stratified
Shear Flow   357

      11.5.3   An Interpretation of the Stability Criterion in terms of Energy   359

   11.6   The Kelvin–Arnold Variational Principle for Inviscid Flows   361

      11.6.1   A Statement of the Theorem   361

      11.6.2   A Derivation of the Theorem   361

      11.6.3   Some Simple Applications of the Theorem   365

   11.7   A Variational Principle for Inviscid Flows based on the Lagrangian   367

   11.8   The Stability of Pipe Flow: a Qualitative Discussion   371

12   The Transition to Turbulence and the Nature of Chaos **379**

   12.1   Some Common Themes in the Transition to Turbulence   379

   12.2   A Definition of Turbulence   382

   12.3   The Nature of Chaos: the Logistic Map as an Example   384

   12.4   Landau's Inspired (but Incomplete) Vision of the Transition
to Turbulence   386

13   An Introduction to Turbulence and to Kolmogorov's Theory **391**

   13.1   Elementary Properties of Turbulence: a Qualitative Overview   391

      13.1.1   The Need for a Statistical Approach and the Problem of Closure   392

      13.1.2   The Various Stages of Development of Freely Decaying Turbulence   394

      13.1.3   Richardson's Energy Cascade   398

      13.1.4   The Rate of Destruction of Energy and an Estimate
of Kolmogorov's Microscales   400

   13.2   A Digression into the Kinematics of Homogeneous Turbulence   402

      13.2.1   Two Useful Diagnostic Tools: Correlation Functions and
Structure Functions   402

13.2.2 The Simplifications of Isotropy and the Taylor Scale   406

13.2.3 Scale-by-scale Energy Distributions in Fourier Space: the Energy Spectrum   409

13.2.4 Relating Real-space and Spectral-space Estimates of the Energy Distribution   414

13.2.5 A Common Error in the Interpretation of Energy Spectra   415

13.3 Kolmogorov's Universal Equilibrium Theory of the Small Scales (K41)   416

13.3.1 Does Small-scale Turbulence have a Universal, Isotropic Structure at Large Re?   416

13.3.2 Kolmogorov's Universal Equilibrium Theory: the Two-thirds and Five-thirds Laws   417

13.3.3 The Kármán–Howarth Equation   420

13.3.4 Kolmogorov's Four-fifths Law   422

13.3.5 Obukhov's Constant Skewness Closure Model   423

13.4 Subsequent Refinements to K41   424

13.4.1 Landau's Objection to K41 Based on Large-scale Intermittency of the Dissipation   424

13.4.2 Kolmogorov's 1961 Refinement of K41 based on Inertial-range Intermittency   426

13.5 The Probability Distribution of the Velocity Field   428

13.5.1 The Skewness and Flatness Factors   428

13.5.2 The Flatness Factor as a Measure of Intermittency   431

13.5.3 The Skewness Factor as a Measure of Enstrophy Production   433

14 Turbulent Shear Flows and Simple Closure Models **441**

14.1 Reynolds Stresses, Energy Budgets, and the Concept of Eddy Viscosity   442

14.1.1 Reynolds Stresses and the Closure Problem (Reprise)   442

14.1.2 The Eddy Viscosity Model of Boussinesq, Taylor, and Prandtl   445

14.2 The Transfer of Energy from the Mean Flow to the Turbulence   447

14.3 Turbulent Jets   450

14.3.1 The Plane Jet   450

14.3.2 The Round Jet   454

14.4 Turbulent Flow near a Smooth Boundary: the Log-law of the Wall   456

14.4.1 The Log-law of the Wall in Channel Flow   456

14.4.2 The Log-law and Viscous Sublayer for Other Smooth-walled Flows   461

14.4.3 Inactive Motion: a Problem for the Universality of the Log-law?   462

14.4.4 Energy Balances and Structure Functions in the Log-law Layer   464

14.4.5 Coherent Structures and Near-wall Cycles   466

14.4.6 Turbulent Heat Transfer near a Surface and the Log-law for Temperature   469

14.5 The Influence of Surface Roughness and Stratification on Turbulent Shear Flow   472

14.5.1 The Log-law for Flow over a Rough Surface   472

14.5.2 The Atmospheric Boundary Layer, Stratification, and the Flux Richardson Number   472

14.5.3 Prandtl's Weak-shear Model of the Atmospheric Boundary Layer   474

14.5.4 The Monin–Obukhov Theory of the Atmospheric Boundary Layer   475

14.6 Closure Models for Turbulent Shear Flows: the k-ε Model
      as an Example   477
      14.6.1 The Basis of the k-ε Closure Model   477
      14.6.2 The k-ε Model applied to Some Simple Turbulent Flows   480

Appendices . . . .

1 *Dimensional Analysis*   485
2 *Vector Identities and Theorems*   492
3 *Navier–Stokes Equation in Cylindrical Polar Coordinates*   495
4 *The Fourier Transform*   496
5 *The Physical Properties of Some Common Fluids*   501

Index   503

PROLOGUE

The scope of fluid mechanics is vast, finding a multitude of applications in biology, engineering, meteorology, geophysics, and astrophysics. Moreover, these flows vary enormously in scale, with characteristic length-scales that range from 0.001 mm (swimming bacteria) to 10^{10} km (protoplanetary accretion discs) and velocities that vary from 0.1 mm/s (convection in the molten core of the Earth) to 100 km/s (the eruption of a solar flare). A few randomly chosen examples are shown in the table below, arranged more or less by the scale of the motion.

The Class of Flow	Typical Scale
The swimming of microorganisms such as bacteria and sperm	0.001 mm→0.1 mm
Lubrication layers in bearings	0.01 mm→0.1 mm
The rustling of leaves, the flight of insects, and the mating call of mice	1 mm→1 cm
The fluid dynamics of planes, trains, and automobiles	10 cm→30 m
Tidal vortices in the oceans and dust devils in deserts	1 m→10 m
Flow down the spillway of a dam	2 m→30 m
Vortex rings (smoke rings) produced by volcanic eruptions	50 m
Lee waves behind mountain ranges	1 km
Tropical cyclones	100 km→10^3 km
Convection cells in the molten core of the Earth	10^3 km
Large-scale ocean gyres	10^3 km→10^4 km
A protoplanetary accretion disc rotating around a young star	10^{10} km

We often think of fluid mechanics as belonging to the domain of engineering, and it is true that it is central to much of mechanical engineering (for example in lubrication theory, natural and forced convection, combustion, and power generation), and to much of civil engineering (rivers and canals, dams, surface gravity waves, coastal erosion). It also lies at the heart of chemical engineering and aerodynamics. However, there are probably just as many applications outside engineering. For example, in biology fluid dynamics finds applications in the study of the cardiovascular system, respiratory disorders, the swimming of fish and micro-organisms, and in the flight of insects and birds. Fluid mechanics also dominates our study of the atmosphere and the oceans, including topics such as urban dispersion, the dispersal of pollutants in the oceans, meteorology (including extreme events like tornadoes and tropical cyclones), and the large-scale atmospheric and oceanic flows that control the weather. Even astrophysicists cannot avoid the subject, as it is central to such topics as magnetic field generation within the convective interiors of planets and stars, the violent activity at the surface of the Sun (solar flares and coronal mass ejections), the solar wind, and the spiralling motion within those vast accretion discs that surround young and dying stars. It is hard not to be intrigued by a subject that pervades so many aspects of our lives.

Unfortunately, given its central importance in so many branches of science and technology, fluid dynamics is not an easy subject, and at the heart of that difficulty lies the fact that the governing equations are *non-linear*. It is all too easy to underestimate the importance of this statement. Most common *linear* partial differential equations (PDEs) are physically well behaved and so are relatively easy to solve analytically. Indeed, in a given situation, one can often divine the physical content of a linear PDE without having to solve it in any great detail, by invoking such concepts as diffusion length, Green's inversion integral (the Biot–Savart law), or the ideas of group velocity and wave dispersion. So, in subjects such as electrodynamics or elasticity, where the governing equations are linear, there are hundreds, if not thousands, of exact, non-trivial solutions, and there exist reference books whose function is to simply catalogue this cornucopia of exact solutions. In fluid dynamics, on the other hand, we know of no more than a couple of dozen exact, non-trivial, non-linear solutions of the governing equations. (I exclude here inviscid potential flows, whose linear governing equations really belong to the world of kinematics rather than dynamics, as we shall see.) One consequence of the difficulty of finding closed-form solutions in fluid mechanics is that great emphasis is placed on the role of conservation principles and conserved quantities. The hope, of course, is that these conservation principles will constrain the behaviour of a flow sufficiently for its essential features to be established.

There is a second consequence of non-linearity, over and above a dearth of closed-form solutions. Unless they are heavily damped, non-linear systems tend to exhibit non-uniqueness and hysteresis, and so it is with fluid mechanics. Worse still, if the damping is sufficiently weak, many non-linear systems develop chaotic behaviour, with all the complexities which that entails. In fluid mechanics that chaos manifests itself as *turbulence*, which is the natural state of nearly all flows in engineering and applied physics. While a raindrop running down a window pane may be laminar (non-turbulent), the wind in the street outside, the flow of water out of a tap, and even the flow of air in and out of our lungs, are all examples of turbulent motion. In short, turbulence is the norm and not the exception, and we are still trying to come to terms with the complexities of turbulence. Indeed, the great English applied mathematician Horace Lamb is reputed to have said:

> I am an old man now, and when I die and go to heaven there are two matters on which I hope for enlightenment. One is quantum electrodynamics and the other is the turbulent motion of fluids. About the former I am rather optimistic. (Attributed to Lamb by Sidney Goldstein, 1932.)

Little has changed. Turbulence remains to this day a profoundly difficult theoretical problem and we shall have much more to say about it in due course.

One reaction to the difficulty of making analytical progress in fluid mechanics has been a strong drive towards numerical simulations, or perhaps we should say *numerical experiments*. In many senses this has been a major success story, fuelled by the fact that computing power has risen so relentlessly. As with laboratory experiments, the careful use of numerical simulations as a source of information can provide valuable insights, although of course the

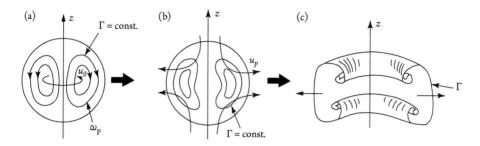

Figure 1 The self-induced, centrifugal bursting of an inviscid, axisymmetric, swirling blob of fluid. (a) The initial condition consists of azimuthal motion, u_θ, only. (b) A secondary flow, $\mathbf{u}_p$, develops which sweeps the angular momentum contours, Γ = const., radially outward. (c) The angular momentum contours form a thin sheet and the final asymptotic state takes the form of a mushroom-shaped vortex sheet which thins exponentially fast.

ability to compute a flow is no substitute for understanding it. Numerical experiments have proven to be particularly important in the difficult field of turbulence, and indeed they have become the research tool of choice for many theoreticians. However, the accurate simulation of even simple flows is not always as straight forward as one might think, as the following rather trivial example illustrates.

Consider an inviscid, axisymmetric flow whose initial velocity field is $u_r = u_z = 0$ and $u_\theta = \Omega r \exp(-\mathbf{x}^2/\delta^2)$ in cylindrical polar coordinates (r, θ, z). That is to say, the initial flow is simply a swirling bob of fluid centred on the origin and whose angular velocity falls off as a Gaussian on the scale of δ. It is convenient to introduce the notation $\Gamma = r u_\theta$ for the angular momentum density about the z-axis and $\mathbf{u}_p = (u_r, 0, u_z)$ for the secondary (poloidal) motion in the r–z plane. We now treat this as an initial-value problem and integrate forward in time, retaining axial symmetry. Given the smooth initial conditions, one might have expected that a numerical simulation of this simple flow would be straight forward, but it is not. It turns out that the swirling fluid wants to centrifuge itself radially outward, and it does this by creating a secondary poloidal velocity field, $\mathbf{u}_p$. It also happens that the lines of constant angular momentum move with the fluid, as if each atom wants to hold on to its initial value of Γ, and so the secondary flow, $\mathbf{u}_p$, sweeps the contours of constant Γ radially outward, as shown in Figure 1. So far there is nothing unusual. However, before long the angular momentum contours get swept up to form a thin, axisymmetric, mushroom-shaped vortex sheet, and that sheet then starts to thin exponentially fast, as discussed in Exercises 8.1 and 11.6.

Most numerical schemes now rapidly run into trouble, as it is extremely difficult to retain adequate spatial resolution of a vortex sheet that thins exponentially fast. Some schemes become numerically unstable as they lose resolution, while others artificially smear out the sheet, thus producing erroneous results. Yet other schemes try to retain both accuracy and stability by continuously refining the spatial resolution, but this quickly causes the time step required for numerical stability to become intolerably short. Clearly, none of these outcomes is satisfactory and that such a formidable numerical problem can emerge from benign initial conditions gives one cause to reflect.

There is a second consequence of the difficulty of making analytical progress in fluid dynamics, which is the need to make extensive use of *dimensional analysis*. Famously, dimensional analysis has been caricatured as a procedure by which one can establish the *scaling laws* which govern some physical process without any understanding of the phenomenon in question. Perhaps this is somewhat of an exaggeration, but it does at least highlight both the power of the method and the disconcerting thought that, in certain situations, scaling laws can be established with minimal physical understanding.

Let us consider a simple example. Suppose we are interested in surface gravity waves of wavelength λ propagating across deep water, and we have (somehow) decided that neither surface tension nor viscosity are important for this particular class of waves. We might then ask: how does the angular frequency of these waves, ϖ, depend on their wavelength? Detailed and nontrivial analysis (eventually) shows that the answer is $\varpi = \sqrt{2\pi g/\lambda}$, where g is the acceleration due to gravity, as discussed in Chapter 7. However, the scaling law $\varpi \sim \sqrt{g/\lambda}$ can be obtained much faster through simple dimensional considerations. The argument proceeds as follows. We ask what physical parameters ϖ might depend on, and the answer is that, since wave amplitude should not be relevant for a linear theory, and we have excluded surface tension and viscosity, the only relevant quantities left are λ and g. We then note that ϖ, λ, and g contain between them only two dimensions: length and time, and so we write $P = 3$ (for three parameters) and $D = 2$ (for two dimensions). The *Buckingham Pi theorem* then tells us that the number of dimensionless groups, G, that we can form from ϖ, λ, and g is $G = P - D = 1$, as discussed in Appendix 1. Next, we note by inspection that this dimensionless group can be written as $\Pi = \varpi \sqrt{\lambda/g}$. However, one dimensionless group can depend only on another dimensionless group, and in this case there are no other groups available to us. So Π is simply a constant and the scaling law $\varpi \sim \sqrt{g/\lambda}$ follows. Of course, this is a rather trivial example, and the application of dimensional analysis to more complex problems is rarely so straight forward. Nevertheless, it is the case that, in the hands of an expert, dimensional analysis provides a particularly powerful tool.

On that final note, perhaps it is time to bring this prologue to a close and invite the reader to immerse themselves in a topic full of surprises.

1

Elementary Definitions, Some Simple Kinematics, and the Dynamics of Ideal Fluids

Leonhard Euler (left, 1707–1783) and Augustin-Louis Cauchy (right, 1789–1857) were both outstanding mathematicians and mathematical-physicists who contributed to many fields, including the development of ideal fluid mechanics.

1.1 Elementary Definitions

1.1.1 What is the Mechanical Definition of a Fluid?

The answer to this question may seem trivially obvious, but it turns out to be worth pondering. The term *fluid* encompasses both liquids and gases, the former being characterized by the existence of a free surface and the latter by the ease with which it may be compressed. Somewhat surprisingly, the macroscopic dynamics of both liquids and gases can be accounted for by more or less the same theory, with only modest differences in emphasis.

To construct such a theory we adopt the *continuum approximation*, which assumes that matter is smeared continuously across space. This approximation rests on the large difference between the molecular scale (the distance between molecules) and the characteristic distance over which the *macroscopic* properties of a fluid, such as density or pressure, vary. So, for example, the density of a fluid is defined as the mass per unit volume measured over a scale which is large enough for all molecular fluctuations to be smoothed out, yet small enough for the density, ρ, to be considered a smoothly varying function of position. Likewise, the stresses exerted by one part of a fluid on another are considered to be a smooth function of position, being defined as the force per unit area transmitted across a small plane surface within the fluid, the surface being infinitesimally small on the macroscopic scale, yet large on the molecular scale. These may be normal stresses, arising from forces perpendicular to the surface in question, or shear stresses, arising from tangential forces.

The distinction between solids and fluids is, at first sight, rather obvious; *i.e.* solids exhibit rigidity, while fluids readily deform when acted upon by a force. However, there are subtleties in this distinction that are worth noting. For example, we cannot distinguish between solids and liquids if only normal stresses are in play. Rather, it is the way in which these two states respond to an imposed *shear stress* that distinguishes between the two. Suppose, for example, that we have two cylinders, each sealed by a movable piston. One cylinder is filled with oil and the other with a cylindrical block of rubber. We now pressurize the contents of the two cylinders using the pistons. Evidently both systems behave in exactly the same way: when a compressive stress is imposed by the pistons, both the oil and the rubber compress a little, and then return to a state of static equilibrium.

Now consider a different arrangement, consisting of two large flat metal plates which lie parallel to each other and are separated by a small gap, d. The bottom plate is fixed and the top one is free to slide, as shown in Figure 1.1. Suppose we have two such pairs of plates,

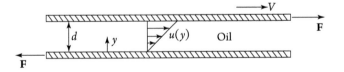

Figure 1.1 The gap d between two plates is filled with oil and tangential forces establish a shear stress, τ, within the oil. This stress causes layers of oil to slide over one another.

and in one case we fill the gap with oil and in the other the gap is filled with a rubber sheet that is bonded to the metal plates. We now apply equal and opposite tangential forces, **F**, to the two plates, which establish a shear stress, τ, in both the oil and the rubber. There is now a clear difference in behaviour. The rubber behaves exactly as before, giving a little when the stress is first applied, but then remaining in equilibrium thereafter. The oil, on the other hand, behaves differently. It sticks to the top and bottom plates and, for as long as the shear stress is applied, layers of oil slide over each other. Static equilibrium is re-established only when the stress is removed. This difference in behaviour provides the definition of a fluid: *a fluid, unlike a solid, continuously deforms under the action of a shear stress*. There is an important corollary to this, which is: *if a fluid is in static equilibrium, the shear stresses within it must be zero everywhere*.

Figure 1.1 also serves to introduce another important property of all fluids: that of viscosity. Suppose that y is the distance from the lower plate, V the speed of the top plate, and $u(y)$ the distribution of horizontal velocity of the fluid. It is observed that, for nearly all common fluids, the velocity gradient, $du/dy = V/d$, is proportional to the applied shear stress, τ. This is *Newton's law of viscosity* and it is written in the form

$$\tau = \mu \frac{du}{dy} = \rho\nu \frac{du}{dy}, \tag{1.1}$$

where the constant of proportionality, μ, is the *dynamic viscosity* of the fluid, and the related property, $\nu = \mu/\rho$, is known as the *kinematic viscosity*. Some fluids are clearly viscous, others less obviously so, but it is important to realize that *all* fluids (except superfluid helium) have *some* viscosity, even if they appear to be very 'thin'. Indeed, the kinetic theory of gases explicitly predicts that the kinematic viscosity is proportional to the product of the mean free path length and the mean thermal velocity of the molecules.

Actually, not all substances can be classified as simply as suggested above. For example, on short time scales asphalt behaves like a solid. You can walk on it without leaving footprints, and if struck by a hammer it shatters like glass. On the other hand, if subjected to forces for long periods of time (years), asphalt flows continuously like a fluid. Moreover, some thixotropic substances behave like an elastic solid if allowed to settle for long enough, yet flow like a fluid if subject to significant stresses. We shall not consider such complex fluids in this book.

1.1.2 Fluid Statics and One Definition of Pressure

Before discussing the dynamics of fluids, perhaps it is worth making a few comments about hydrostatics, if only to reinforce the notions of pressure and of stresses acting within a fluid. Let us start with *Pascal's law* for stationary fluids.

We have already seen that shear stresses are everywhere zero in a stationary fluid. Pascal's law follows directly from this and states that the magnitude of the normal stress acting at any given point is independent of direction. The proof is trivial. Let us use σ to denote a normal stress and reserve τ for shear stress. Consider a small wedge of fluid of mass m surrounding the point of interest, as shown in Figure 1.2. Let the wedge have sides δx, δy, and δz, and let

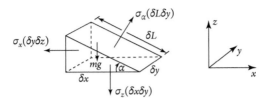

Figure 1.2 Forces acting on a wedge of static fluid arising from the normal stresses.

σ_x, σ_z, and σ_α be the normal stresses in the x and z directions and on the inclined surface. As there are no shear stresses, vertical equilibrium demands

$$[\sigma_\alpha(\delta L\,\delta y)]\cos\alpha = \sigma_\alpha(\delta x\,\delta y) = \sigma_z(\delta x\,\delta y) + \rho g(\tfrac{1}{2}\delta x\,\delta y\,\delta z), \qquad (1.2)$$

and as the size of the wedge tends to zero the weight mg drops out of (1.2) to give $\sigma_\alpha = \sigma_z$. Similarly, horizontal equilibrium requires $\sigma_\alpha = \sigma_x$. So, we conclude that because of the absence of shear stress, the normal stress, σ, is the same in all directions. In fluid dynamics the normal stresses are compressive, and so we define the fluid pressure to be $p = -\sigma$, and Pascal's law is sometimes paraphrased as saying that *the pressure at any given point is the same in all directions*.

Perhaps some comments are in order. First, the mechanical pressure as defined above is the same as the thermodynamic pressure. Second, sometimes it is convenient to pretend that a fluid has zero viscosity (a so-called *ideal fluid*), in which case there are no viscous stresses within the fluid when in motion. Once again the shear stresses vanish and the normal stress at any given point is the same in all directions. (Note that any inertial force vanishes to leading order in a local force balance like (1.2), just as the weight drops out of (1.2) as the volume of the fluid element goes to zero.) For a hypothetical inviscid fluid, then, the mechanical pressure is once again simply defined as $p = -\sigma$, whether that fluid is stationary or in motion. Conversely, if a *real* fluid is in motion there *will*, in general, be shear stresses acting within the fluid. The normal stresses at a given point then depend on direction and it is meaningless to define mechanical pressure in this way. We shall return to this point in Chapter 2, where we refine our definition of pressure.

Let us now return to hydrostatics and go to the next order in our force balance, gathering terms of order $\delta x\,\delta y\,\delta z$. We must now allow for spatial gradients in pressure if we want to balance the weight of a small fluid element. The easiest way to see this is to consider the cylindrical element of fluid shown in Figure 1.3, of cross-sectional area A, mass m, and height δz. Clearly the pressure at the bottom of the element must exceed that at the top in order to balance the weight of the fluid in between. Indeed, a vertical force balance requires

$$A\frac{\partial p}{\partial z}(\delta z) = -mg = -\rho A(\delta z)g,$$

from which

$$\frac{\partial p}{\partial z} = -\rho g. \qquad (1.3)$$

The horizontal gradients in pressure, by contrast, are zero, and so our net hydrostatic force balance is

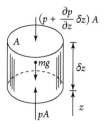

Figure 1.3 The vertical force balance for a cylindrical element in a stationary fluid.

$$\nabla p = \rho \mathbf{g}, \tag{1.4}$$

where $\mathbf{g} = -g\hat{\mathbf{e}}_z$ is the gravitational acceleration.

There is a more mathematical way of getting to the same result that is worth outlining, as we shall do something similar when it comes to dynamics. Consider a small volume, V, within the fluid which has surface S and is of arbitrary shape. The pressure force exerted by the surrounding fluid on a part of S, say dS, is then $-pd S\mathbf{n} = -pd\mathbf{S}$, where $\mathbf{n}$ is an outward pointing unit normal to S. Thus the total pressure force acting on the fluid within V is the surface integral of $-p$. This then converts into a volume integral using a variant of Gauss' theorem:

$$\text{net pressure force acting on } V = -\oint_S pd\mathbf{S} = -\int_V (\nabla p)dV. \tag{1.5}$$

We now ignore second-order derivatives in pressure on the grounds that V is small, so the net pressure force becomes $-(\nabla p)V$. Since this force, plus the weight of the fluid, $\rho V \mathbf{g}$, must sum to zero, we conclude that $\nabla p = \rho \mathbf{g}$, which brings us back to (1.4).

For the special case of a fluid of uniform density, (1.3) integrates to give $p = p_0 - \rho g z$, where p_0 is a reference pressure. When dealing with *liquids* this is normally rewritten as $p = \rho g\varsigma$, where ς is the depth below the free surface and p is now interpreted as gauge pressure, *i.e.* the pressure over and above atmospheric pressure. This reflects the fact that an upward force per unit area of $p = \rho g\varsigma$ is required to balance the weight $mg = (\rho\varsigma)g$ of an overlying column of fluid. Thus, for example, the pressure at the base of the Mariana trench, which is around 10 km deep, is 10^8 N/m^2, or 10^3 atmospheres. Note that the pressure force acting on a submerged body is now determined by the surface integral

$$\text{force on a submerged body} = -\oint pd\mathbf{S} = -\oint (\rho g\varsigma)d\mathbf{S}. \tag{1.6}$$

In practice, however, it is often easier to use Archimedes' principle to find the net pressure force on such a body. This states: *if a body is partially or wholly immersed in a fluid then it receives an upward pressure force equal to the weight of displaced fluid, with that force acting through the centre of gravity of the displaced fluid.* Readers may wish to prove this principle for themselves. If so, note that the proof requires no mathematics.

Let us close our discussion of hydrostatics with a whimsical paradox, devised by Den Hartog and based on Archimedes' principle. Consider Figure 1.4, which shows a box in

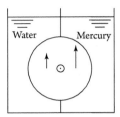

Figure 1.4 Den Hartog's perpetual-motion machine.

which sits a cylinder supported by a frictionless bearing and which is free to rotate. The left of the box is filled with water and the right with mercury, the two liquids being separated by a leak-proof partition. From Archimedes' principle there is an upward pressure force on the left of the cylinder equal to the weight of displaced water, and on the right a corresponding, but larger, pressure force equal to the weight of displaced mercury. The resulting force imbalance causes the cylinder to rotate, yielding a perpetual-motion machine! (If you decide to unravel this paradox, you might find it helpful to note that the fallacy does *not* lie in the application of Archimedes' principle.)

1.1.3 Different Categories of Fluid and of Fluid Flow

We now turn to dynamics. Before presenting a formal analysis of fluids in motion, it is, perhaps, useful to discuss the different approximations commonly employed in fluid dynamics, as well as the different regimes of behaviour normally encountered in practice. Perhaps the first important subdivision is between *incompressible* and *compressible* flows. It is tempting to associate the former with liquids and the latter with gases, but this is too simplistic. Compressibility cannot always be ignored in liquids (think of acoustic waves in the oceans), and, conversely, gasses in which the Mach number (the flow speed divided by the speed of sound) is less than 0.3 usually exhibit negligible variation in density. In this text we consider only incompressible flows.

The second subdivision is between *ideal* and *real fluids*. As mentioned above, it is sometimes convenient to ignore the viscous stresses in a fluid on the grounds that the viscosity, which is always finite, is sufficiently small for those stresses to be negligible. A hypothetical fluid with zero viscosity is called an ideal fluid. Real fluids *stick* to solid surfaces (Figure 1.1), whereas an ideal fluid is free to *slip* over such a surface. Moreover, viscous stresses convert mechanical energy into heat, whereas *ideal fluids conserve mechanical energy* (in the absence of external forces). So, an inviscid fluid can be a useful idealization, but one that invariably breaks down near solid surfaces in motion.

Real fluids tend to be classified as either *Newtonian*, if they obey the constitutive relationship (1.1), or *non-Newtonian*, if the relationship between shear stress and velocity gradient is more complicated. Nearly all common fluids are Newtonian, although non-Newtonian fluids are commonplace in chemical engineering and in some biological systems. For many fluids it is found that the applied shear stress is proportional to the resulting velocity gradient raised to some power, say $\tau \sim (du/dy)^n$. If $n < 1$, the fluid is described as *shear thinning*, which is typical of polymer solutions. Fluids for which $n > 1$ are described as *shear thickening*, and of course Newtonian fluids correspond to $n = 1$. In yet other fluids, a finite shear stress is required to initiate the flow, and these are called *Bingham plastics*. In this text, however, it seems appropriate to restrict the discussion to Newtonian fluids.

Finally, we come to the crucial distinction between *laminar* and *turbulent* flows. A flow is said to be steady if the velocity at every point is constant, so the flow pattern does not change with time. Laminar flows, which may be steady or unsteady, are perfectly reproducible in every respect and are typical of a flow in which the viscous stresses, and viscous damping, are large. Turbulent flow, on the other hand, is characterized by the fact that the velocity field has a component that is chaotic in both space and time. This is a manifestation of the fact that the governing equations are non-linear. (Chaos is the hallmark of a weakly damped, non-linear system.) Consider the case of flow in a pipe of diameter d. A convenient measure of the influence of viscosity is given by the *dimensionless Reynolds number*, $\mathrm{Re} = V d / \nu$, where V is the mean velocity in the pipe. For Re less than around 2000, the motion is always laminar (non-chaotic). However, if Re exceeds $\sim$2000, turbulence is readily triggered by even tiny disturbances. The motion in the pipe is then characterized by the superposition of a time-averaged component, which is perfectly reproducible and looks a little bit like the laminar flow, and a chaotic component of motion, whose precise details vary from one realization of an experiment to the next. This is typical of nearly all geometries: when a suitably defined Reynolds number is less than a few hundred, the flow is usually laminar, but for Re greater than a few thousand, the motion is almost invariably turbulent. Since the viscosity of most common fluids is small, the majority of flows in nature and technology are turbulent. Fully developed (*i.e.* strong) turbulence turns out to be highly dissipative.

These various categories of flow are listed in Table 1.1. In this text we limit ourselves to incompressible Newtonian fluids.

Table 1.1 Different categories of fluid and of fluid flow.

Incompressible	*Compressible*
Usually appropriate for liquids and for gas flows where the Mach number is less than $\sim$0.3.	Essential for acoustic waves in liquids and for gas flows where the Mach number exceeds $\sim$0.3.
Ideal (inviscid and non-dissipative)	*Real (viscous and dissipative)*
Such fluids do not exist, but this can be a useful approximation for high-Re flows away from boundaries in the absence of strong turbulence.	All real fluids are viscous and dissipative. This is important, even when $\mathrm{Re} \gg 1$, close to boundaries or in the presence of strong turbulence.
Newtonian	*Non-Newtonian*
Most common fluids are Newtonian.	These arise in biology and chemical engineering.
Laminar	*Turbulent*
Non-chaotic flow which occurs if Re is low-to-moderate. Uncommon in nature or technology.	Flow which is chaotic in time and space. Occurs if $\mathrm{Re} \gg 1$ and is common in nature and technology.

1.2 Some Simple Kinematics

1.2.1 Eulerian Versus Lagrangian Descriptions of Motion

In elementary mechanics the motion of a particle is usually described by specifying the position of the particle, $\mathbf{x}_p$, as a function of time, $\mathbf{x}_p(t)$. The velocity and acceleration of the particle are then defined in terms of the time derivatives of $\mathbf{x}_p(t)$. Such an approach, in which one tracks individual material particles, is referred to as a *Lagrangian* description of motion, and such time derivatives are known as Lagrangian time derivatives. Unfortunately, the Lagrangian formalism is cumbersome to implement in fluid mechanics, where there is an infinite number of particles to label.

Rather, in fluid mechanics, the various properties of the flow, such as the velocity distribution, $\mathbf{u} = (u_x, u_y, u_z)$, or the pressure distribution, p, are usually specified as functions of position and time, $\mathbf{x}$ and t. Thus we talk of the *velocity field*, $\mathbf{u}(\mathbf{x}, t)$, and *pressure field*, $p(\mathbf{x}, t)$. This is analogous to magnetism, where the spatial and temporal properties of a magnetic field, $\mathbf{B}$, are written as $\mathbf{B}(\mathbf{x}, t)$. Such an approach, in which $\mathbf{x}$ and t are *independent* variables, is known as an *Eulerian* description of the motion, and we shall adopt this formalism throughout the book.

A particularly convenient way of visualizing a steady flow, $\mathbf{u}(\mathbf{x})$, or the snapshot of an unsteady flow, $\mathbf{u}(\mathbf{x}, t = t_0)$, is to use *streamlines*. These are lines drawn in space which are everywhere parallel to $\mathbf{u}$ (Figure 1.5), just as magnetic field lines are everywhere parallel to $\mathbf{B}$. Such lines cannot cross, unless the velocity falls to zero, and tend to bunch together where the flow speed is large. Streamlines necessarily represent the trajectories of individual fluid particles (called *path-lines*) if the flow is steady. However, they do not, in general, coincide with particle trajectories if the flow is unsteady, as the streamline pattern then changes from one moment to the next, and so a given fluid particle will find itself on different streamlines as time progresses (see Exercise 1.1). If $\mathbf{u}(\mathbf{x}, t_0)$ is a known function of $\mathbf{x}$, then the equations for the corresponding family of streamlines can, at least in principle, be found by integrating

$$\frac{dx}{u_x(\mathbf{x}, t_0)} = \frac{dy}{u_y(\mathbf{x}, t_0)} = \frac{dz}{u_z(\mathbf{x}, t_0)}. \tag{1.7}$$

This expression follows from the fact that, by definition, a short portion of a streamline at point $\mathbf{x}$, $d\mathbf{x} = (dx, dy, dz)$, is parallel to $\mathbf{u}(\mathbf{x}, t_0)$.

A related concept is that of a *stream-tube*, which is the fluid analogue of a magnetic flux tube. This is an imaginary tube drawn in space such that $\mathbf{u}$ is everywhere parallel to the edges of the tube. Such tubes can be constructed from an aggregate of streamlines which all pass

Figure 1.5 Streamlines for flow over a wing. Note the stagnation point at the leading edge

through a closed curve within the fluid. If the flow is steady, then the fluid flows along a stream-tube as if it were a real tube, with impermeable boundaries.

There is, unfortunately, one potential difficulty with the Eulerian formalism. Frequently we want to know the rate of change of some fluid property, say temperature or velocity, of a *given particle* as it passes through the flow field. For example, suppose we have a steady flow, $\mathbf{u}(\mathbf{x})$, in which there is a steady temperature field, $T(\mathbf{x})$. Then we might want to know the rate of change of temperature of a particular fluid particle as it slides down a streamline, perhaps because we want to apply the laws of thermodynamics to that particle. (Note that the temperature of the particle changes with time despite the fact that the temperature field is steady, because the particle passes through a sequence of different points, each of which is at a different temperature.) Alternatively, we might want to know the rate of change of velocity of a fluid particle as it slides along its streamline, so that we can apply Newton's second law to it. So how can we extract such Lagrangian time derivatives from an Eulerian description of the flow? This is our next topic.

1.2.2 The Convective Derivative

The acceleration of a fluid element is not $\partial \mathbf{u}/\partial t$, which is the rate of change of $\mathbf{u}$ at a fixed point in space, through which a succession of fluid particles will pass, but rather the rate of change of $\mathbf{u}$ following that fluid particle. Such a derivative is written as $D\mathbf{u}/Dt$, a notation that was introduced by Stokes in 1845. Likewise, DT/Dt is the rate of change of temperature following a given element of fluid.

The operator $D(\cdot)/Dt$ is called the *convective derivative*, and we can get an expression for it as follows. Consider the change in temperature resulting from small changes in time and position. Then $\delta T = (\partial T/\partial t)\,\delta t + (\partial T/\partial x)\,\delta x + \cdots$ where, because we want to follow a particular fluid particle, $\delta \mathbf{x}$ and δt are related by $\delta \mathbf{x} = \mathbf{u}\delta t$. So, following a fluid particle we have $\delta T = (\partial T/\partial t)\,\delta t + (\partial T/\partial x)\,u_x\delta t + \cdots$, or

$$\delta T = (\partial T/\partial t + \mathbf{u} \cdot \nabla T)\,\delta t,$$

from which

$$\frac{DT}{Dt} = \frac{\partial T}{\partial t} + \mathbf{u} \cdot \nabla T. \tag{1.8}$$

Likewise, the acceleration of a fluid element is given by

$$\boxed{\frac{D\mathbf{u}}{Dt} = \frac{\partial \mathbf{u}}{\partial t} + (\mathbf{u} \cdot \nabla)\mathbf{u}}. \tag{1.9}$$

In those cases where the flow is steady, it is often convenient to rewrite the acceleration of a fluid particle in terms of *intrinsic coordinates*. To that end we write $\mathbf{u} = V(s)\hat{\mathbf{e}}_t$, where $V = |u|$, s is a coordinate measured along the streamline, and $\hat{\mathbf{e}}_t$ is a unit vector tangential to the streamline. To complete this description we need to introduce a unit normal vector,

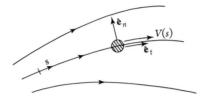

Figure 1.6 The intrinsic coordinates used in equation (1.10).

$\hat{\mathbf{e}}_n$, which is directed away from the local centre of curvature, as shown in Figure 1.6. The acceleration of a fluid particle, written in terms of $V(s)$ and the two unit vectors, is then

$$(\mathbf{u} \cdot \nabla)\mathbf{u} = V \frac{\partial}{\partial s}(V\hat{\mathbf{e}}_t) = V \frac{\partial V}{\partial s}\hat{\mathbf{e}}_t - \frac{V^2}{R}\hat{\mathbf{e}}_n, \tag{1.10}$$

where R is the radius of curvature of the streamline, defined in the usual way by $d\hat{\mathbf{e}}_t/ds = -\hat{\mathbf{e}}_n/R$. Equations (1.9) and (1.10), along with Newton's second law, provide the starting point for the derivation of the equations of motion of a fluid.

1.2.3 Mass Conservation and the Streamfunction

Let us continue with the theme of kinematics. The rate of flow of mass through an imaginary surface S drawn in a fluid is

$$\dot{m} = \int_S \rho\mathbf{u} \cdot d\mathbf{S}. \tag{1.11}$$

It follows that mass conservation applied to a *control volume*, V, fixed in space and with bounding surface, S, requires

$$\boxed{\frac{d}{dt}\int_V \rho dV = -\oint_S \rho\mathbf{u} \cdot d\mathbf{S}} \tag{1.12}$$

where, by convention, $d\mathbf{S}$ points outward. Appling Gauss' theorem to the surface integral in (1.12) allows us to rewrite this as

$$\int_V [\partial\rho/\partial t + \nabla \cdot (\rho\mathbf{u})]\, dV = 0,$$

and since this holds for any and all volumes, V, we conclude that mass conservation in differential form is

$$\boxed{\frac{\partial\rho}{\partial t} = -\nabla \cdot (\rho\mathbf{u})}. \tag{1.13}$$

This is known as the *continuity equation* and it is frequently rewritten in the more convenient form,

$$\frac{D\rho}{Dt} = -\rho \nabla \cdot \mathbf{u}.$$

Let us now restrict ourselves to incompressible fluids, whose defining property is $D\rho/Dt = 0$. Mass conservation then requires

$$\boxed{\nabla \cdot \mathbf{u} = 0}. \tag{1.14}$$

Evidently $\mathbf{u}$ is solenoidal in an incompressible fluid, and this is true even in those cases where the density is non-uniform.

Now, any solenoidal vector field can be written as the curl of another vector field, say $\mathbf{u} = \nabla \times \mathbf{A}$, where $\mathbf{A}$ is the *vector potential* for $\mathbf{u}$. This turns out to be particularly useful for two-dimensional flows, $\mathbf{u}(x, y) = (u_x, u_y, 0)$, where the simplest form of the vector potential is $\mathbf{A} = \psi(x, y)\hat{\mathbf{e}}_z$. In such cases we have

$$u_x = \frac{\partial \psi}{\partial y}, \quad u_y = -\frac{\partial \psi}{\partial x}, \tag{1.15}$$

and hence $\mathbf{u} \cdot \nabla \psi = 0$. Evidently the lines of constant ψ are parallel to $\mathbf{u}$ and so represent streamlines, the function ψ being known as the *streamfunction*. The obvious advantage of using the streamfunction in a two-dimensional flow is that a single scalar field replaces the two components of a vector field, whilst simultaneously ensuring that mass conservation is satisfied. As to the physical interpretation of ψ, readers may wish to confirm for themselves that, if the fluid density is uniform, the rate of flow of mass between two streamlines is equal to the difference in the values of ψ for the two streamlines, multiplied by the fluid density.

The equivalent streamlines in an axisymmetric flow, $\mathbf{u}(r, z) = (u_r, 0, u_z)$, expressed in terms of cylindrical polar coordinates (r, θ, z), are given by the stream-surfaces $\Psi = \text{constant}$, where the *Stokes streamfunction*, $\Psi(r, z)$, is defined by

$$\mathbf{u} = \nabla \times \mathbf{A}_\theta = \nabla \times \left[(\Psi(r, z)/r)\, \hat{\mathbf{e}}_\theta \right]. \tag{1.16}$$

In component form we have

$$\mathbf{u} = \left(-\frac{1}{r}\frac{\partial \Psi}{\partial z},\, 0,\, \frac{1}{r}\frac{\partial \Psi}{\partial r} \right), \tag{1.17}$$

from which we see that $\mathbf{u} \cdot \nabla \Psi = 0$, so that the surfaces of constant Ψ are indeed parallel to the streamlines. The physical interpretation of Ψ is that the mass flow rate between two stream-surfaces, Ψ_1 and Ψ_2, is given by $2\pi\rho(\Psi_2 - \Psi_1)$. This can be seen by integrating the axial velocity across the annular gap which separates Ψ_1 and Ψ_2:

$$\dot{m} = \rho \int_{r_1}^{r_2} 2\pi r u_z \, dr = 2\pi \rho \int_{r_1}^{r_2} (\partial \Psi / \partial r) \, dr = 2\pi \rho \, (\Psi_2 - \Psi_1). \qquad (1.18)$$

1.3 The Dynamics of an Ideal (Inviscid) Fluid of Uniform Density

1.3.1 Euler's Equation for a Fluid of Uniform Density

We now turn from kinematics to dynamics and derive the equation of motion of an ideal (inviscid) fluid of uniform density. Feynman et al. (1964), and indeed many other authors, rightly warn against the dangers of the idealized notion of an inviscid fluid, labelling it as 'the study of dry water'. We shall, in due course, dwell on those dangers. First, however, we explore some of the more useful results that stem from an inviscid analysis.

Consider a small blob of fluid of mass $\rho \delta V$. The net pressure force acting on this blob is, from (1.5),

$$\oint_S (-p) d\mathbf{S} = \int_{\delta V} (-\nabla p) \, dV = -(\nabla p) \, \delta V.$$

Since there are no shear stresses in an inviscid fluid, Newton's second law now gives us,

$$(\rho \delta V) \frac{D\mathbf{u}}{Dt} = -(\nabla p) \, \delta V + (\rho \delta V) \, \mathbf{g}, \qquad (1.19)$$

where we have allowed for the self-weight of the fluid element. Evidently our equation of motion is

$$\rho \frac{D\mathbf{u}}{Dt} = -\nabla p + \rho \mathbf{g}, \qquad (1.20)$$

which is known as *Euler's equation*. This is subject to the so-called *free-slip boundary condition*, $\mathbf{u} \cdot d\mathbf{S} = 0$, at any stationary, solid surface. In the interests of simplicity, we shall focus on fluids of uniform density throughout the remainder of this book (except in Chapter 9, where we consider stratified fluids), so our equation of motion simplifies to

$$\frac{D\mathbf{u}}{Dt} = -\nabla (p/\rho) + \mathbf{g}, \quad \mathbf{u} \cdot d\mathbf{S} = 0, \qquad (1.21)$$

which may be rewritten in a number of different ways. For example, it is usual to write $\mathbf{g} = -\nabla(gz)$ and then absorb this term into the pressure gradient, to give

$$\frac{D\mathbf{u}}{Dt} = -\nabla (p/\rho + gz). \qquad (1.22)$$

If self-weight is small this becomes

$$\boxed{\frac{D\mathbf{u}}{Dt} = -\nabla(p/\rho)} \text{ (version I)}. \tag{1.23}$$

Actually, (1.23) is often used in preference to (1.22) even when self-weight *is* significant, on the understanding that p now represents *the departure from a hydrostatic pressure distribution.* The reason is as follows. Even in cases where gravity is important, it is possible to show that, *in the absence of a free surface,* the gravitational term in (1.22) cannot influence the motion of a fluid of uniform density, and indeed its *only* role is to augment the pressure field with a hydrostatic contribution. In such cases it is usually more efficient to ignore gravity altogether, simply adding it in at the end of a calculation, if it is required. So it is common practice to use (1.22) if free surfaces need to be considered, but (1.23) if there are no free surfaces. Note that some authors adopt different symbols for the true pressure and the departure from a hydrostatic pressure distribution. However, this can quickly become cumbersome, and so we shall use the same symbol for both, on the grounds that it will be obvious what is meant in a given situation.

There is another common way of writing the Euler equation which rests on the vector identity

$$\nabla(u^2/2) = (\mathbf{u}\cdot\nabla)\mathbf{u} + \mathbf{u}\times(\nabla\times\mathbf{u}).$$

This allows us to rewrite (1.22) and (1.23) in the alternative form,

$$\boxed{\frac{\partial\mathbf{u}}{\partial t} = \mathbf{u}\times(\nabla\times\mathbf{u}) - \nabla H} \text{ (version II)}, \tag{1.24}$$

where $H = p/\rho + \mathbf{u}^2/2$ if weight is unimportant, or else

$$H = p/\rho + \mathbf{u}^2/2 + gz, \tag{1.25}$$

if weight is significant and p is the true pressure. This turns out to be a useful starting point for energy considerations and the quantity H is usually called *Bernoulli's function.*

A notable feature of both (1.23) and (1.24) is that, unlike Maxwell's equations of electrodynamics, or the equations of elasticity, they are non-linear. This makes them difficult to solve analytically, and also lies at the root of fluid turbulence, as noted above. It is, perhaps, no exaggeration to say that the non-linearity of the equations of motion for a fluid has helped keep generations of fluid dynamicists in gainful employment for the better part of two centuries, with turbulence often referred to as the last great unsolved problem in classical physics.

There is a third useful way of writing Euler's equation, and this is often used when dealing with *steady flows.* If we adopt intrinsic coordinates then (1.10) combines with (1.22) to give us, in a steady flow,

$$V\frac{\partial V}{\partial s}\hat{\mathbf{e}}_t - \frac{V^2}{R}\hat{\mathbf{e}}_n = -\nabla\left(p/\rho + gz\right).\qquad(1.26)$$

This provides particularly simple expressions for the pressure gradients parallel and perpendicular to a streamline. In the absence of gravity, and assuming steady motion, these are

$$\frac{\partial p}{\partial s} = -\rho V\frac{\partial V}{\partial s}, \qquad \frac{\partial p}{\partial n} = \rho\frac{V^2}{R}\quad\text{(version III)},\qquad(1.27)$$

where the normal coordinate n increases with distance from the local centre of curvature. The first of these equations is simply Newton's second law applied in the stream-wise direction to a cylindrical element of fluid of length ds and cross-sectional area A, i.e.

$$\text{streamwise pressure force} = -\left[\frac{\partial p}{\partial s}(ds)\right]A = \left[\rho A(ds)\right]V\frac{\partial V}{\partial s},$$

as shown in Figure 1.7. The second equation, by contrast, tells us that pressure increases as we move away from the centre of curvature of the streamlines. Thus, for example, the pressure at the centre of a tornado is subatmospheric. The force associated with this cross-stream pressure gradient is required to maintain the centripetal acceleration of the fluid particles, just as a mass rotated at the end of a rope induces tension in that rope.

If we include gravity in the first of the equations in (1.27), it can be rewritten as

$$\frac{\partial}{\partial s}\left[p + \frac{1}{2}\rho V^2 + \rho gz\right] = 0.\qquad(1.28)$$

This leads us to an important energy equation, known as *Bernoulli's equation.*

1.3.2 Bernoulli's Equation and Mechanical Energy Conservation

If a flow is *steady* then (1.24) tells us that $\mathbf{u}\cdot\nabla H = 0$. It follows that H is constant along a given streamline, although in general it will vary from streamline to streamline. The same result can be obtained directly from (1.28). In summary, then, in an *ideal, steady flow of uniform density,*

$$\boxed{H = p/\rho + \mathbf{u}^2/2 + gz = \text{const. (on a streamline)}}.\qquad(1.29)$$

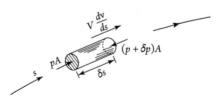

Figure 1.7 A fluid element slides along a streamline with speed V and acceleration $V\partial V/\partial s$.

Equation (1.29) is known as *Bernoulli's theorem*. For the *special case* of a steady flow in which $\nabla \times \mathbf{u} = 0$ at every point, (1.24) reduces to $\nabla H = 0$, and so (1.29) then generalizes to the statement that H is the same at all points in the flow. The cross-stream gradient in H, combined with (1.27), then yields

$$\frac{\partial p}{\partial n} = -\rho V \frac{\partial V}{\partial n} = \rho \frac{V^2}{R},$$

in the absence of gravity. This requires that $\partial V / \partial n = -V/R$, which is a particular kinematic property of flows in which $\nabla \times \mathbf{u} = 0$, as discussed in Exercise 1.10. (Flows in which $\nabla \times \mathbf{u} = 0$ are called *irrotational*.) However, we shall see later that it is unusual to have $\nabla \times \mathbf{u} = 0$ at all points, so we shall restrict ourselves to (1.29) for the time being.

We recognize the second and third contributions to H as the kinetic and potential energies of a particle of unit mass, which suggests, correctly as it turns out, that (1.29) is a statement that mechanical energy is conserved in the absence of frictional (*i.e.* viscous) forces. In a steady, incompressible flow, then, H is the mechanical energy per unit mass. But why does pressure constitute a form of mechanical energy in a fluid? The reason is that an element of fluid, but *not* an isolated particle, is subject to the conservative force $-\nabla (p/\rho)$, and p/ρ is the potential energy of that force. So, just as $gz_1 - gz_2$ is the work done per unit mass by the gravitational force in moving a particle from $\mathbf{x}_1$ to $\mathbf{x}_2$, the work done by the pressure force in moving a fluid element is $p_1/\rho - p_2/\rho$ (see Exercise 1.4).

The quantity H is often called Bernoulli's constant, although in the hydraulics literature it is known as the *total head* of the fluid, an obscure name that relates back to the use of manometers. In hydraulics the gravitational term is retained because weight is important and free surfaces common. In the aerodynamics literature, by contrast, the gravitational term is often omitted and Bernoulli's equation is rewritten as

$$p_0 = p + \rho \mathbf{u}^2/2 = \text{constant on a streamline.} \tag{1.30}$$

For a streamline that terminates at a *stagnation point*, such as at the leading edge of a wing, p_0 is the *stagnation pressure*.

1.3.3 Some Applications of Bernoulli's Equation

We now illustrate the utility of Bernoulli's equation by considering some simple devices employed to measure flow rates. Let us start with the *Pitot-static tube*, which is used in aerodynamics to measure air speed. This a thin, streamlined tube, aligned with the flow, and in which there are two small holes at which pressure is monitored. One hole is at the tip of the tube (pressure tapping 1) and the other further back (pressure tapping 2), as shown in Figure 1.8(a). Now suppose that the oncoming flow is uniform on the scale of the Pitot tube, with speed V, so that p_0 in (1.30) is the same on all streamlines and given by $p_0 = p_a + \rho V^2/2$, where p_a is atmospheric pressure. By symmetry, there is a stagnation point at the tip of the tube (see Figure 1.8 (b)), so the pressure there is $p_1 = p_a + \rho V^2/2$. Near pressure tapping 2, on the other hand, the streamlines are straight and parallel, and

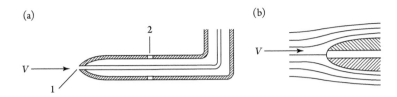

Figure 1.8 (a) A Pitot-static tube. (b) The stagnation point at the tip of the tube.

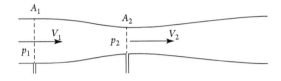

Figure 1.9 A Venturi tube.

so (1.27) tells us that there is no variation in pressure across the streamlines. It follows that the pressure measured at tapping 2 is $p_2 = p_a$, from which $p_1 - p_2 = \rho V^2/2$. Evidently, the speed V can be deduced from a measurement of this pressure difference.

In chemical engineering, by contrast, the flow rate in a pipe is often measured using a contraction, or *Venturi tube*, as shown in Figure 1.9. Suppose that upstream of the Venturi the pipe has a cross-sectional area of A_1 and that the speed of the fluid is uniform and equal to V_1. This cross-sectional area is slowly varied along the length of the Venturi tube, and we shall assume the flow remains uniform at each cross-section. At the throat of the Venturi the cross-sectional area and flow speed are A_2 and V_2. The pressure drop between cross-sections 1 and 2 is monitored and the question is, can we deduce the upstream speed in the pipe from a knowledge of $p_1 - p_2$? To that end, we first note that mass conservation yields $\rho A_1 V_1 = \rho A_2 V_2$. Moreover, Bernoulli's equation tells us that

$$p_1 - p_2 = \frac{1}{2}\rho \left(V_2^2 - V_1^2\right) = \frac{1}{2}\rho V_1^2 \left[(A_1/A_2)^2 - 1\right], \tag{1.31}$$

which can be rearranged to give V_1 in terms of $p_1 - p_2$, as required.

In civil engineering, the flow rate in a river or water channel is usually measured using a weir. Unlike the Pitot-static and Venturi tubes, whose operational principles are trivial, the behaviour of a weir is rather subtle. The simplest case to analyse is a *broad-crested weir*, an example of which is shown in Figure 1.10. In the vicinity of the weir crest, both the free surface of the water and the weir height are assumed to vary slowly, so that the flow speed at any one cross-section may be considered independent of height. If the depth and speed at any one cross-section are h and V, then the two-dimensional volumetric flow rate is $Q = Vh$, which is necessarily the same at all cross-sections. Moreover, Bernoulli's equation applied to the free surface tells us that

$$p_a/\rho + V^2/2 + gz_{\text{surface}} = p_a/\rho + gz_{\text{reservoir}}, \tag{1.32}$$

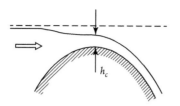

Figure 1.10 Flow over a broad-crested weir.

where $z_{reservoir}$ is the elevation of the water surface some distance upstream, where the speed V is negligible. We now introduce $d = z_{reservoir} - z_{surface}$ for the fall in water level below that of the reservoir, so that Bernoulli's equation reduces to $V^2 = 2gd$. The depth of the weir below the reservoir surface is then

$$h + d = \frac{Q}{V} + \frac{V^2}{2g}. \tag{1.33}$$

We now note that $d + h$ must exhibit a minimum at the weir crest. If we exclude the case where V itself is stationary at the crest, we may minimize $d + h$ as a function of V to yield

$$V_c = (gQ)^{1/3}, \quad h_c = Q/V_c = (Q^2/g)^{1/3}, \tag{1.34}$$

where the subscript c stands for either *crest* or *critical* (the latter for the reasons given below). Clearly, a measurement of h_c is sufficient to determine the flow rate Q.

Notice that (1.34) also gives $V_c = \sqrt{gh_c}$, which is interesting because the phase speed of shallow-water gravity waves relative to an undisturbed fluid is $\sqrt{gh}$. It would seem, therefore, that the flow speed equals the wave speed at the weir crest, and we shall see later that this is no coincidence. Shallow-water flow is said to be *subcritical* if $V < \sqrt{gh}$, *critical* if $V = \sqrt{gh}$ and *supercritical* if $V > \sqrt{gh}$, so the motion is subcritical upstream of the crest, critical at the crest, and supercritical downstream. Note that there is an analogy to flow of a compressible gas through a convergent–divergent nozzle, where a subsonic flow is accelerated up to the speed of sound at the throat of the nozzle, and is supersonic thereafter.

1.3.4 Inviscid Momentum Conservation in Integral Form

We now switch from energy conservation to momentum conservation. As with Bernoulli's equation, we shall restrict ourselves to the steady flow of an ideal fluid of uniform density. Euler's equation is then

$$\rho(\mathbf{u} \cdot \nabla)\mathbf{u} = -\nabla p + \rho \mathbf{g}, \tag{1.35}$$

which can be rewritten using $\nabla \cdot \mathbf{u} = 0$ in the form

$$\nabla \cdot (\rho u_i \mathbf{u}) = [-\nabla p + \rho \mathbf{g}]_i. \tag{1.36}$$

We now integrate this over a control volume V, fixed in space and with bounding surface S. Invoking Gauss' theorem, we have

$$\oint_S (\rho\mathbf{u})\mathbf{u} \cdot d\mathbf{S} = \oint_S (-p)d\mathbf{S} + \int_V \rho\mathbf{g}dV.$$ (1.37)

Noting that $\rho\mathbf{u} \cdot d\mathbf{S}$ is the mass flux through $d\mathbf{S}$, and that $\mathbf{u}$ is the momentum per unit mass, we recognize the integral on the left as the rate of flow of momentum out through the control surface S. Evidently (1.37) is simply an expression of momentum conservation. In words, it says that the net flux of momentum out through the control surface S is equal to the net force acting on the control volume, that force being the sum of the pressure force acting on S and the weight of the fluid within V. This is sometimes called the *momentum theorem*, and it can be generalized in an obvious way to unsteady flow and to include additional body forces. That is to say, if $\mathbf{f}$ represents the sum of all the body forces acting on the fluid per unit mass, then

$$\frac{d}{dt}\int_V \rho\mathbf{u}dV + \oint_S (\rho\mathbf{u})\mathbf{u} \cdot d\mathbf{S} = \oint_S (-p)d\mathbf{S} + \int_V \rho\mathbf{f}dV.$$ (1.38)

The inviscid momentum theorem is particularly useful when it comes to calculating pressure forces on bodies, as the following examples show.

1.3.5 Examples of the Use of Momentum Conservation to Calculate Pressure Forces

As our first example, consider a two-dimensional jet (a sheet of fluid) which impacts on a plate at an angle of α, as shown in Figure 1.11. We use the control volume indicated by the broken lines and adopt Cartesian coordinates where x is aligned with the plate. The speed and thickness of the incoming jet are V and b, and the fluid leaves the control volume at points 1 and 2 with a thickness of b_1 on the right and b_2 on the left. We wish to calculate the net pressure force exerted by the jet on the plate, and also the film thicknesses b_1 and b_2.

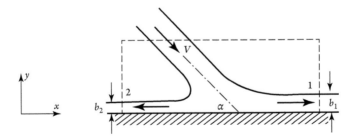

Figure 1.11 A two-dimensional jet impacts on a plate.

The first point to note is that the streamlines are straight and parallel at the entrance and exit of the control volume. The second equation in (1.27) then tells us that the pressure in the fluid at points 1 and 2 is atmospheric, just as it is in the incoming jet. Bernoulli's equation now requires that the speeds at locations 1 and 2 are the same as that in the jet, *i.e.* V. (We shall ignore weight.) Moreover, conservation of mass demands that what flows in must flow out, $bV = b_1 V + b_2 V$, from which we conclude that $b = b_1 + b_2$.

We now apply the momentum equation in the direction normal to the plate. This tells us the net pressure force in the y direction exerted *by the plate on the fluid*, F_y, satisfies

$$F_y = \oint_S (u_y)\rho\mathbf{u} \cdot d\mathbf{S} = (-V\sin\alpha)(-\rho V b) = \rho V^2 b \sin\alpha, \qquad (1.39)$$

where the first minus sign is because the y component of velocity in the incoming jet is $-V\sin\alpha$ and the second because the fluid is *entering* the control volume (*i.e.* $\mathbf{u} \cdot d\mathbf{S} < 0$). The pressure force exerted *on the plate* by the fluid is then equal and opposite to (1.39).

To get the film thicknesses, b_1 and b_2, we apply the momentum equation in the direction parallel to the plate and note that, because the fluid is assumed to be inviscid, there is no force exerted by the plate on the fluid in the x direction, $F_x = 0$. It follows that

$$F_x = \oint_S (u_x)\rho\mathbf{u} \cdot d\mathbf{S} = V(\rho V b_1) + (-V)(\rho V b_2) + V\cos\alpha(-\rho V b) = 0,$$

from which we obtain $b\cos\alpha = b_1 - b_2$. When combined with $b = b_1 + b_2$, this gives the film thicknesses as

$$b_1 = (b/2)(1 + \cos\alpha), \quad b_2 = (b/2)(1 - \cos\alpha). \qquad (1.40)$$

As our second example, we consider flow under a sluice gate, as shown in Figure 1.12. Once again, the control surface is indicated by broken lines. The flow enters and leaves the control volume at sections 1 and 2 and these are taken to be far enough upstream and downstream of the gate for the streamlines at the entrance and exit to be straight and parallel and the velocity independent of depth. The upstream depth and speed are h_1 and V_1, while

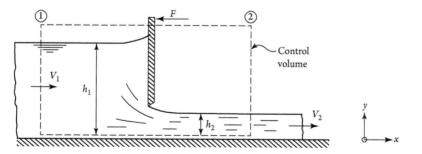

Figure 1.12 Flow under a sluice gate.

those downstream are h_2 and V_2. The gate is held in place by a force of magnitude F (per unit depth into the page) which is considered to be an external force acting on the control volume in the negative x direction. Our task is to find this force.

The momentum equation applied in the stream-wise direction, per unit depth into the page, is

$$-F + \int_0^{h_1} p_1 dy - \int_0^{h_2} p_2 dy = \oint_S (u_x)\rho \mathbf{u} \cdot d\mathbf{S} = V_2 (\rho h_2 V_2) + V_1 (-\rho h_1 V_1).$$

(1.41)

Clearly, we need to know the pressure distributions at sections 1 and 2. To that end, we note that the streamlines are straight and parallel at 1 and 2, and so the fluid is in hydrostatic equilibrium as far as a vertical force balance is concerned. It follows that there is a linear variation of pressure with depth at both sections, varying from zero at the surface to $\rho g h$ at the river bed. (We shall work with gauge pressure.) We conclude that the force required to hold the gate in place is

$$F = \frac{1}{2}\rho g h_1^2 - \frac{1}{2}\rho g h_2^2 + \rho h_1 V_1^2 - \rho h_2 V_2^2.$$

(1.42)

1.3.6 More on Momentum Conservation: Inviscid Flow Through a Cascade of Blades

A particularly interesting application of the inviscid momentum equation is the two-dimensional flow through a cascade of blades, as shown in Figure 1.13. Our discussion follows that of Prandtl (1952). The blades are regularly spaced with a separation of a and we use a control volume that encloses just one blade, as indicated by the broken lines. Crucially, the top and bottom surfaces of the control volume are identical streamlines, and so the pressure forces acting on the control volume through those two surfaces exactly cancel. Well upstream of the blades, at section 1, the velocity is uniform with a magnitude of V_1 and at an angle of β_1 to the horizontal. The flow is turned by the blades and exits the control volume with a new uniform velocity of magnitude V_2 and angle β_2. Each blade is held in place by a force $\mathbf{R}$(per unit depth into the page), which is treated as an external force applied to the control volume. We use Cartesian coordinates (x, y) and our task is to determine the force $\mathbf{R}$ required to hold each blade in place.

We start by noting that continuity of mass demands that the horizontal velocities at 1 and 2 are the same, with

$$\dot{m} = \rho a u_x = \rho a V_1 \cos \beta_1 = \rho a V_2 \cos \beta_2.$$

(1.43)

The momentum equation applied in the x direction then simplifies to

$$R_x + (p_1 - p_2)a = \dot{m} V_2 \cos \beta_2 - \dot{m} V_1 \cos \beta_1 = 0,$$

(1.44)

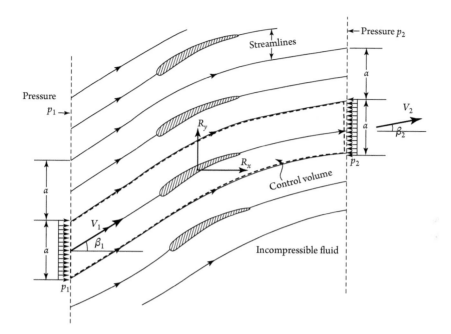

Figure 1.13 Flow through a cascade of blades. (Figure by G.D. Matthew.)

while that in the y direction is

$$R_y = \dot{m}V_2 \sin\beta_2 - \dot{m}V_1 \sin\beta_1 = \dot{m}(u_{y,2} - u_{y,1}).\tag{1.45}$$

Finally, we apply Bernoulli's equation from 1 to 2, noting that the horizontal velocity is the same at both locations. This yields

$$p_2 - p_1 = \frac{1}{2}\rho(V_1^2 - V_2^2) = \frac{1}{2}\rho(u_{y,1}^2 - u_{y,2}^2) = \rho\bar{u}_y(u_{y,1} - u_{y,2}),\tag{1.46}$$

where $\bar{u}_y = (u_{y,1} + u_{y,2})/2$ is the mean vertical velocity. We conclude that the components of force required to hold one blade in place are

$$R_x = \rho a\bar{u}_y(u_{y,1} - u_{y,2}), \quad R_y = -\rho a u_x(u_{y,1} - u_{y,2}).\tag{1.47}$$

Of course, the net pressure force exerted *by the flow on the blade*, which we shall denote **F**, is equal and opposite to **R**. So we write the components of **F** as

$$F_x = -\rho a\bar{u}_y(u_{y,1} - u_{y,2}), \quad F_y = \rho a u_x(u_{y,1} - u_{y,2}),\tag{1.48}$$

where $\bar{\mathbf{u}} = (\mathbf{u}_1 + \mathbf{u}_2)/2$ is the mean of the upstream and downstream velocities. Note that the magnitude of **F** is given by

$$F = \rho a \bar{V}(u_{y,1} - u_{y,2}), \tag{1.49}$$

where $\bar{V} = |\bar{\mathbf{u}}|$. Evidently, the larger the deflection of fluid momentum, the larger the back reaction on the blade.

Perhaps some comments are in order at this point. First, **F** is perpendicular to the mean velocity, $\bar{\mathbf{u}}$. Second, the force **F** is directed upward and to the left, which is not so surprising as the momentum of the flow is deflected downward and to the right. Third, let us introduce the closed line integral,

$$\Gamma = \oint \mathbf{u} \cdot d\mathbf{r}, \tag{1.50}$$

and evaluate this line integral in a *clockwise* sense around the perimeter of the control volume. (Such a closed line integral is known as the *circulation* in fluid dynamics.) The contributions to Γ from the top and bottom streamlines clearly cancel, and so we are left with the simple expression $\Gamma = a(u_{y,1} - u_{y,2})$. The magnitude of the pressure force acting on the blade can now be written as

$$F = \rho \bar{V} \Gamma. \tag{1.51}$$

These three observations are summarized in Figure 1.14.

It is interesting to speculate as to what happens as the separation of the blades becomes large, so that each blade begins to resemble a single aerofoil in a uniform incident flow. Suppose that the stream-wise width of the blades is W. Then it turns out that, as $a/W \rightarrow \infty$, the circulation Γ around each blade remains finite. Since $\Gamma = a(u_{y,1} - u_{y,2})$, this requires that the jump in vertical velocity falls to zero, and so the flows well upstream and downstream are identical. In this limit, then, the force on a single blade, or aerofoil, is

$$\boxed{F = \rho V_\infty \Gamma} \; (\Gamma \text{ evaluated in a clockwise sense}), \tag{1.52}$$

where V_∞ is the speed well upstream or downstream of the aerofoil. Such a situation is shown in Figure 1.15. Since **F** is perpendicular to the incident flow, it is an example of a *lift force*, there being no drag force in an inviscid fluid. It is now clear how the lift on an aerofoil arises. If the foil is cambered, or has a slight inclination to the oncoming flow, it deflects the momentum of the fluid downward, and the lift force is a direct reaction to this.

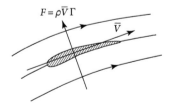

Figure 1.14 The pressure force on a single blade resulting from the turning of the flow.

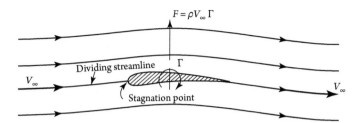

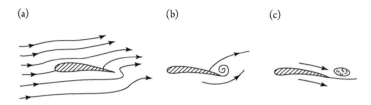

Figure 1.15 Flow over a two-dimensional aerofoil and the corresponding lift force.

Figure 1.16 (a) Starting flow without circulation. (b) Flow curls around the trailing edge to form a vortex. (c) The vortex is shed leaving behind a smooth flow with circulation.

The expression $F = \rho V_\infty \Gamma$ is an example of the famous *Kutta–Joukowski lift theorem* of aerodynamics, which says that the lift on a two-dimensional body in a perfect fluid is proportional to the incident speed and the magnitude of the circulation around that body. The reason why the flow over an aerofoil with camber, or a finite angle of attack (inclination), possesses circulation is not so obvious and is discussed in Chapter 6. As we shall see, it turns out that, without circulation, the air flow cannot pass smoothly over the rear of the aerofoil. This is the situation when the foil first starts to move through the air, as shown in Figure 1.16(a). The flow then curls around the trailing edge to form a vortex, known as the *starting vortex* (Figure 1.16(b)). That vortex is then shed, leaving behind a flow with circulation that passes smoothly over the trailing edge. We shall examine this process in more detail in Chapter 6, where we explain the relationship between the shed vortex and the subsequent circulation around the wing.

1.4 Examples of the Failure of Ideal Fluid Mechanics

So far, our inviscid equations of motion seem to have worked rather well. However, this is in large part because we have carefully chosen examples where the viscous dissipation of mechanical energy, which is always present, is not of central importance. Certainly, some hint that all is not well comes from the fact that our analysis of flow over an aerofoil predicts no drag force, which is clearly at variance with reality. As a prelude to our study of real fluids in Chapter 2, we now show how an inviscid analysis can rapidly run into trouble. To that end, we choose two examples in which violent turbulence is generated in the flow, then this turbulence dissipates mechanical energy. We start with an abrupt pipe expansion.

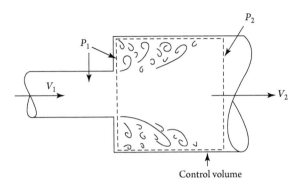

Figure 1.17 Flow behind the sudden expansion in a pipe.

1.4.1 The Borda–Carnot 'Head Loss' in a Sudden Pipe Expansion

Consider the pipe flow shown in Figure 1.17, where there is a sudden increase in cross-sectional area from A_1 to A_2. Upstream of the expansion the flow is almost uniform, with mean speed V_1 and mean pressure p_1. The Reynolds number is assumed to be large, and so immediately after the expansion the flow becomes violently turbulent, though steady-on-average. Finally, some distance downstream the flow settles down once again to a near uniform motion, this time with mean speed V_2 and mean pressure p_2.

It turns out that, in the vicinity of the expansion, the shear stresses exerted by the pipe wall on the fluid are negligible by comparison with the pressure forces acting on the fluid. Consequently, a good estimate of the pressure change across the expansion can be found by applying the inviscid momentum equation. We shall apply that equation to the control volume indicated by the broken lines. Of course, the flow is not steady, so strictly speaking we should use the unsteady version of the momentum equation, (1.38). However, from a time-averaged point of view the momentum in the control volume is constant, and so we are at liberty to use the steady version, which yields

$$p_1 A_2 - p_2 A_2 = \dot{m}(V_2 - V_1) = \rho A_2 V_2 (V_2 - V_1). \tag{1.53}$$

Notice that we have assumed that the upstream pressure, p_1, extends over the entire left-hand side of the control volume, which is found to be a good approximation in practice. From (1.53), the rise in pressure across the expansion is predicted to be

$$p_2 - p_1 = \rho V_2 (V_1 - V_2), \tag{1.54}$$

which turns out to be accurate to within a few percent. Let us now evaluate the mechanical energy per unit mass either side of the expansion in the form of the Bernoulli 'constant'. This gives us

$$H_1 - H_2 = (p_1 - p_2)/\rho + (V_1^2 - V_2^2)/2, \tag{1.55}$$

which combines with estimate (1.54) to yield

$$H_1 - H_2 = \frac{1}{2}(V_1 - V_2)^2. \tag{1.56}$$

Evidently, since H is not constant, Bernoulli's equation does not apply here. Of course, you might reasonably object that the flow is unsteady and so we have no right to expect it to apply, even in an ideal fluid. However, the real problem is not the unsteadiness. Rather, while there is zero dissipation of energy in an inviscid fluid (see Exercise 1.4), fully developed turbulence is *strongly* dissipative, and this is true for *all* values of the Reynolds number for which fully developed turbulence exists. In short, when the viscosity is small but finite, turbulence has the ability to generate strong dissipation within the fluid. This is true even in the limit of $\nu \to 0$, so long as ν is finite.

Note that (1.56) represents the loss of mechanical energy per unit mass of fluid passing down the pipe, and so the net rate of loss of energy in the expansion is

$$\text{rate of loss of energy} = \frac{1}{2}\dot{m}(V_1 - V_2)^2. \tag{1.57}$$

Of course, this reappears as internal energy (heat) in the fluid, and indeed (1.57) can be used to determine the rise in the temperature of the fluid across the expansion. This is the famous *Borda–Carnot head loss* equation, which has been used by engineers for over two centuries.

Equation (1.57) is intriguing in two respects. First, we have applied a *frictionless* version of Newton's second law in the form of (1.37), yet we seem to have acquired some thermodynamic information as a result; specifically, the irreversible rate of loss of mechanical energy. In this respect it is significant that we have supplemented Newton's second law with some empirical information, such as the pressure being uniform and equal to p_1 on the left of the control volume. Second, (1.57) tells us that the rate of loss of energy due to turbulence is independent of the value of the viscosity of the fluid, provided, of course, that the Reynolds number exceeds some threshold. (We need Re to be large enough for fully developed turbulence to be maintained.) This seems paradoxical as viscous stresses are responsible for this loss of energy. Moreover, it suggests that there is a fundamental distinction between setting $\nu = 0$ (zero dissipation) and taking the limit of $\nu \to 0$ (order unity dissipation). So, how can the viscous dissipation of energy be independent of the value of the very quantity that is responsible for that dissipation? This turns out to be one of the most fundamental properties of turbulence, and it has intrigued and perplexed many a famous scientist. We shall return to this paradox in later chapters.

1.4.2 The Hydraulic Jump

Let us now turn to a similar problem that occurs in open-channel flow. Recall from the discussion in §1.3.3 that the speed of shallow-water surface gravity waves (relative to undisturbed fluid) is $\sqrt{gh}$, where h is the water depth. Moreover, shallow-water flow is said to be subcritical if $V < \sqrt{gh}$, critical if $V = \sqrt{gh}$, and supercritical if $V > \sqrt{gh}$, where V is the speed of the fluid, assumed independent of depth. In the case of flow over a broad-crested weir we saw that the flow starts out as subcritical (slow and deep), becomes critical at the weir crest, and then supercritical (fast and shallow) thereafter. We also noted an analogy to the flow of a compressible gas through a convergent–divergent nozzle, which passes from subsonic to supersonic and is critical at the throat of the nozzle. The dimensionless group $Fr = V/\sqrt{gh}$ is called the *Froude number* in open-channel flow, and it is clearly the analogue of the Mach number (flow speed divided by the speed of sound) in a supersonic flow.

Now, just as a supersonic flow tends to quickly shock back down to a subsonic state through the formation of a shock wave, so supercritical flow in a water channel tends not to survive for long, but rather reverts to a subcritical flow through the formation of a *hydraulic jump*. Thus, for example, the supercritical flow that emerges from the sluice gate shown in Figure 1.12 will typically jump up to a slow and deep subcritical flow within a short distance of the gate. Hydraulic jumps are also inevitable at the base of dam spillways, and indeed they are often so violent that it is important to trigger such jumps in a region where they cannot do much structural damage.

A schematic of a hydraulic jump is shown in Figure 1.18. Just as strong turbulence appears downstream of a sudden pipe expansion, so strong turbulence is evident in a hydraulic jump (provided the Reynolds number is large enough). Upstream of the jump we have supercritical flow and downstream the flow is subcritical. The depth of fluid corresponding to critical flow is $h_c = \left(Q^2/g\right)^{1/3}$, where Q is the volumetric flow rate per unit depth into the page, $Q = Vh$. This is indicated by the horizontal broken line in Figure 1.18.

The two flows either side of the jump are known as *conjugate states*. Given the upstream conditions, we can estimate the conjugate depth using the inviscid momentum equation applied to a control volume that straddles the jump. Let us suppose that the inlet to the control volume (section 1) and the outlet (section 2) are sufficiently well removed from the jump for the streamlines there to be straight and parallel and the flow uniform. Then, applying the momentum equation, while neglecting friction on the river bed, yields

$$\int_0^{h_1} p_1 dy - \int_0^{h_2} p_2 dy = (\rho h_2 V_2) V_2 + (-\rho h_1 V_1) V_1. \tag{1.58}$$

Since the streamlines are straight and parallel at sections 1 and 2, the vertical pressure distribution must be hydrostatic, as discussed in §1.3.5. It follows that

$$\frac{1}{2}\rho g h_1^2 + \rho h_1 V_1^2 = \frac{1}{2}\rho g h_2^2 + \rho h_2 V_2^2, \tag{1.59}$$

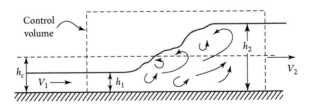

Figure 1.18 A hydraulic jump.

or, equivalently,

$$\frac{1}{2}\rho g h_1^2 + \frac{\rho Q^2}{h_1} = \frac{1}{2}\rho g h_2^2 + \frac{\rho Q^2}{h_2}. \tag{1.60}$$

Multiplying through by $h_1 h_2$ and then factoring out $\rho g(h_2 - h_1)$, we obtain

$$h_1 h_2 (h_1 + h_2) = \frac{2Q^2}{g}, \tag{1.61}$$

which may be solved for the downstream depth in terms of the upstream conditions:

$$2h_2/h_1 = \sqrt{1 + 8Q^2/gh_1^3} - 1 = \sqrt{1 + 8V_1^2/gh_1} - 1. \tag{1.62}$$

Equation (1.61) also yields simple expressions for the upstream and downstream Froude numbers,

$$\mathrm{Fr}_1^2 = \frac{Q^2}{gh_1^3} = \frac{h_2(h_1 + h_2)}{2h_1^2}, \quad \mathrm{Fr}_2^2 = \frac{Q^2}{gh_2^3} = \frac{h_1(h_1 + h_2)}{2h_2^2}, \tag{1.63}$$

to which we shall return shortly.

Turning now to energy, the Bernoulli 'constant' on the surface streamline (or equally the river bed) is $H = gh + V^2/2$, where we are working with gauge pressure. It follows that

$$H_1 - H_2 = g(h_1 - h_2) + \frac{Q^2}{2}\left(h_1^{-2} - h_2^{-2}\right),$$

and on substituting for Q using (1.61) we find the loss of mechanical energy per unit mass of fluid passing through the jump is

$$H_1 - H_2 = \frac{g}{4h_2 h_1}(h_2 - h_1)^3. \tag{1.64}$$

We conclude that the rate of dissipation of mechanical energy in the hydraulic jump, per unit depth into the page, is given by

$$\text{rate of loss of energy} = \frac{\rho g V_1}{4h_2}(h_2 - h_1)^3. \tag{1.65}$$

Since the mechanical energy must fall due to viscous dissipation in the turbulence, we conclude that $h_2 > h_1$, and it then follows from (1.63) that the upstream Froude number is greater than unity, while the downstream Froude number is less than unity. In short, a supercritical flow shocks down to a subcritical flow, as shown in Figure 1.18.

It is instructive to consider some numerical values. Suppose that the speed and depth of the flow at the base of a dam spillway are $V_1 = 20$ m/s and $h_1 = 1$ m, corresponding to a dam height of around 20 m and a Froude number of $Fr_1 = 6.39$. Then (1.62) tells us that $h_2 = 8.54$ m, and if the spillway is 15 m wide, (1.65) predicts that energy is dissipated in the jump at a rate of 37 Megawatts, which was the engine power of the *Titanic*. So much for ideal fluid mechanics with its zero dissipation of energy! Clearly, we must engage with the fact that *all* fluids are dissipative, which is our next task.

○ ○ ○

This concludes our introduction to kinematics and to inviscid dynamics. Readers seeking a more expansive introduction might consult Faber (1995), Feynman et al. (1964), or Prandtl (1952), the last of which offers a masterful, if occasionally idiosyncratic, overview.

··

EXERCISES

1.1 *Streamlines versus path-lines.* Consider the two-dimensional, unsteady flow $\mathbf{u} = (u_0 e^{-\alpha t}, u_0 e^{\alpha t})$, where u_0 is a constant. Show that the instantaneous streamlines are straight lines and find the path-line for a particle released from the origin at $t = 0$.

1.2 *Eulerian versus Lagrangian descriptions of motion.* Consider the two-dimensional, steady flow $\mathbf{u} = (\beta x, -\beta y)$ in the half-plane $y > 0$. Use (1.7) to show that the streamlines satisfy $xy = $ constant and (1.15) to show that $\psi = \beta xy$. Let $\mathbf{x}_p(t)$ locate a fluid particle in the flow which is located at $\mathbf{x}_0$ at $t = 0$. In a Lagrangian description $\mathbf{x}_p$ satisfies $d\mathbf{x}_p/dt = \mathbf{u}(\mathbf{x}_p)$. Confirm that $x_p = x_0 \exp(\beta t)$ and $y_p = y_0 \exp(-\beta t)$.

1.3 *Pressure gradients perpendicular to streamlines.* A bucket filled with water rotates with an angular velocity of Ω. The water rotates at the same rate and its upper surface adopts a curved profile. If r is the radius measured from the centre of the bucket, show that the depth of water, h, varies with r according to

$$h(r) = h(0) + \Omega^2 r^2 / 2g.$$

The whirlpool vortex that forms above a drain in a shallow tank of water has a tangential velocity that varies inversely with radius. What is the surface shape in that case?

1.4 *Euler's equation and energy conservation.* Show that Euler's equation (1.22) yields

$$\frac{d}{dt} \int_V \left(\mathbf{u}^2/2\right) \rho dV + \oint_S \left(\mathbf{u}^2/2 + gz\right) \rho \mathbf{u} \cdot d\mathbf{S} = \oint_S \mathbf{u} \cdot (-pd\mathbf{S}).$$

Thus, for an ideal fluid, kinetic energy is conserved in a closed domain of fixed shape. If there is flow across the boundary, S, the surface integral on the left is the flux of energy across S, while that on the right is the rate of working of the pressure forces acting on S. Evidently, in a steady flow,

$$\oint_S \left(p/\rho + \mathbf{u}^2/2 + gz\right) \rho \mathbf{u} \cdot d\mathbf{S} = \oint_S H\rho \mathbf{u} \cdot d\mathbf{S} = 0.$$

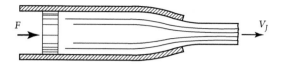

Figure 1.19 A piston pushes fluid through a contraction.

Deduce Bernoulli's equation by applying this to a stream-tube and show that the pressure term represents the work done by the pressure forces in moving a particle along the tube.

1.5 *An application of Bernoulli's equation.* A steady force, F, is applied to a piston, as shown in Figure 1.19. The speed and area of the piston and the jet are V_p, V_j, A_p, and A_j. The pressure on the left of the piston, and also that surrounding the jet, is atmospheric. Assuming that the flow is inviscid and quasi-steady, show that F and the rate of working of F are given by

$$F = \frac{1}{2}\rho V_j^2 A_p \left[1 - (A_j/A_p)^2\right], \qquad FV_p = \dot{m}\left[\frac{1}{2}V_j^2 - \frac{1}{2}V_p^2\right].$$

What is the physical interpretation of the second equation?

1.6 *Another application of Bernoulli's equation.* Two identical, open tanks of cross-sectional area A are filled with water up to the same height h relative to their bases. Both tanks have a small, circular drainage hole in their bases, of area a and with $a \ll A$. In tank 1 the hole opens up directly to the atmosphere, while in tank 2 a tube of length L and cross-sectional area a is connected to the hole and projects vertically downward from the base of the tank. The holes in the two tanks are unplugged at the same instant. Treating the flow as inviscid and quasi-steady, find the time taken to drain both tanks (to leading order in a/A) and show that the ratio of these two times is given by $\sqrt{1 + L/h} - \sqrt{L/h}$.

1.7 *Linear momentum.* Show that $u_i = \mathbf{u} \cdot \nabla x_i = \nabla \cdot (x_i \mathbf{u})$ in a fluid of uniform density. Hence show that, if the fluid is bounded by a stationary solid surface, $\int \rho \mathbf{u} dV = 0$.

1.8 *The angular momentum integral equation.* Confirm that $\mathbf{u} \cdot \nabla (\mathbf{x} \times \mathbf{u}) = \mathbf{x} \times (\mathbf{u} \cdot \nabla \mathbf{u})$ and hence use Euler's equation to show that, in the absence of gravity,

$$\rho\frac{\partial(\mathbf{x} \times \mathbf{u})}{\partial t} + \rho \mathbf{u} \cdot \nabla(\mathbf{x} \times \mathbf{u}) = \nabla \times (p\mathbf{x}).$$

Deduce the integral equation

$$\frac{d}{dt}\int_V \rho(\mathbf{x} \times \mathbf{u})dV + \oint_S (\rho\mathbf{x} \times \mathbf{u})\mathbf{u} \cdot d\mathbf{S} = -\oint_S p\mathbf{x} \times d\mathbf{S},$$

and give a physical interpretation of this expression in terms of angular momentum.

1.9   *Energy loss in a hydraulic jump.* From (1.59), $M = \rho g h^2/2 + \rho h V^2$ is conserved across a hydraulic jump. The energy, $H = gh + V^2/2$, is not. If h_c is the critical depth, show that

$$\frac{M}{\rho g h_c^2} = \frac{1}{2}\left(\frac{h}{h_c}\right)^2 + \frac{h_c}{h}, \qquad \frac{H}{g h_c} = \frac{1}{2}\left(\frac{h_c}{h}\right)^2 + \frac{h}{h_c}.$$

Plot $M/\rho g h_c^2$ as a function of h/h_c and show that, for $M/\rho g h_c^2 > 3/2$, there exist two possible solutions, one subcritical and the other supercritical. Now plot $H/g h_c$ on the same graph and show that, since $M/\rho g h_c^2$ is conserved across a jump, $H/g h_c$ must fall.

1.10 *The cross-stream invariance of H in irrotational motion.* Flows in which $\nabla \times \mathbf{u} = 0$ are called irrotational. In §1.3.2 we showed that Bernoulli's function H has the same value across all streamlines in an irrotational flow, and that this cross-stream invariance of H requires

$$\frac{\partial p}{\partial n} = -\rho V \frac{\partial V}{\partial n} = \rho \frac{V^2}{R},$$

where n is normal to the streamlines and R is the local radius of curvature of a streamline. It follows that $\partial V/\partial n = -V/R$ is a kinematic property of all irrotational flows. But why should this be so? First, show that we can rewrite this condition as $\partial(RV)/\partial n = 0$. Next, evaluate the circulation around a small quadrilateral formed from the intersection of:

Figure 1.20 The effect of surface tension on a stream of water. (Image by J.-M. Chomaz.)

(i)   two normal line segments which link two adjacent streamlines; and

(ii) the arcs formed by those streamlines.

Finally, show that a prerequisite for $\oint \mathbf{u} \cdot d\mathbf{r} = 0$, as required by $\nabla \times \mathbf{u} = 0$, is that $\partial(RV)/\partial n = 0$. (Hint: consult Prandtl (1952), page 59.)

1.11 *A party trick.* Hold two sheets of paper close together and blow between them. Instead of separating, they come together. Explain this using Bernoulli's equation.

1.12 *Surface tension.* The Young–Laplace law tells us that, if the free surface of a liquid is curved, the pressure jump across that surface is

$$\Delta p = \gamma \left(\frac{1}{R} + \frac{1}{R'} \right),$$

where R and R' are the principal radii of curvature and γ the surface tension coefficient. If $\gamma = 7.3 \times 10^{-2}\,\mathrm{Nm}^{-1}$ for water, estimate the internal pressure in the droplets shown in Figure 1.20.

....

REFERENCES

Faber, T. E., 1995, *Fluid dynamics for physicists*, Cambridge University Press.

Feynman, R.P., Leighton, R.B., & Sands, M., 1964, *The Feynman lectures on physics*, Vol. II, Addison-Wesley.

Prandtl, L., 1952, *Essentials of fluid dynamics*, Blackie & Son.

2

· · **·** · ·

Governing Equations and Flow Regimes
for a Real Fluid

The engineer Claude Navier and mathematical physicists George Stokes (left, 1819–1903) and Siméon Poisson (right, 1781–1840) all contributed to the viscous equations of motion. Stokes provided the first modern derivation, yet Navier and Poisson pre-empted many aspects of his findings. As Stokes noted in the introduction to his seminal 1845 paper: 'I afterwards found that Poisson had written a memoir on the same subject, and on referring to it I found that he had arrived at the same equations. The method which he employed was however so different from mine that I feel justified in laying the latter before this Society . . . The same equations have also been obtained by Navier in the case of an incompressible fluid, but his principles differ from mine still more than do Poisson's.'

2.1 Viscosity, Viscous Stresses, and the No-slip Boundary Condition

We now turn from ideal to real fluid mechanics, embracing the fact that all fluids (except superfluid helium) have a finite viscosity. As in Chapter 1, we shall assume that the fluid is incompressible so that $\nabla \cdot \mathbf{u} = 0$, and take the density to be uniform. We shall also restrict ourselves to Newtonian fluids which, in a one-dimensional shear flow, $u_x(y)$, obey Newton's law of viscosity (1.1), which we now write as

$$\tau_{xy} = \mu \frac{du_x}{dy} = \rho\nu \frac{du_x}{dy}. \tag{2.1}$$

As we shall see, for a fluid of uniform density it is the kinematic viscosity, ν, rather than the dynamic viscosity, μ, that appears naturally in the equations of motion, and so we shall work almost exclusively with ν.

The essential idea behind (2.1) is illustrated in Figure 2.1. Consider an initially rectangular element of fluid, $\delta x \delta y$, in a one-dimensional shear flow. In a short time interval, δt, the line element δy rotates by $\delta\gamma = (du_x/dy)\,\delta t$, and so the rate of shearing, or angular distortion rate, of the fluid element $\delta x \delta y$ is $d\gamma/dt = du_x/dy$. A shear stress acting in the x–y plane, τ_{xy}, is required to sustain this angular distortion, and in Newton's law of viscosity it is assumed that this shear stress is directly proportional to the angular distortion rate. Equation (2.1) then follows, with $\rho\nu$ being the constant of proportionality.

The introduction of viscosity has two distinct effects, both of which are critical. First, an imbalance in the viscous stresses acting on a fluid element gives rise to a net force on that element, which must be represented in the equation of motion of a fluid. So, Euler's equation needs to be generalized to incorporate such forces. We shall see that these viscous forces not only influence (or sometimes dominate) the flow pattern, but they also act to irreversibly convert mechanical energy into heat. The second, and equally important, consequence of viscosity is that it changes the boundary conditions at the interface of a fluid and a solid.

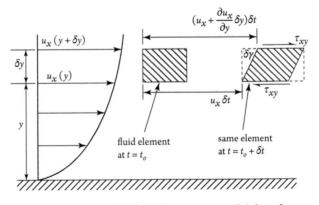

Figure 2.1 The distorsion of a rectangular fluid element in a parallel shear flow.

For an ideal fluid the boundary condition at a stationary, solid surface is simply $\mathbf{u} \cdot d\mathbf{S} = 0$, so that there is no normal component of velocity. However, for a viscous fluid adjacent to a solid surface it is observed that both the normal and tangential components of velocity vanish in a frame of reference moving with that surface. This is sometimes paraphrased by saying that real fluids *stick* to solid surfaces, and it is called the *no-slip boundary condition*. At a microscopic level, the no-slip condition arises because the atoms in the fluid ricochet off the stationary atoms in the solid, and so lose their forward momentum.

It is important to understand that the no-slip condition holds no matter how small the viscosity may be. This means that the behaviour of a real fluid with a finite, if tiny, viscosity can be radically different to that of an ideal fluid, whose motion is predicted by Euler's equation. Thus, there is a fundamental distinction between setting $\nu = 0$, as in an ideal fluid, and taking the limit of $\nu \to 0$, *i.e.* letting the viscosity get very small in a real fluid. (Actually, we met this distinction between $\nu = 0$ and $\nu \to 0$ in §1.4.1, where we discussed the turbulent dissipation of energy in a fluid of vanishingly small viscosity. However, in that particular case, the difference is *not* due to the no-slip condition.)

Figure 2.2 shows an example of the distinction between $\nu = 0$ and $\nu \to 0$ caused, at least in part, by the no-slip condition. A uniform flow of speed V approaches, and then passes over, a cylinder of diameter d. On the left is the motion of an ideal fluid, as predicted by Euler's equation. This flows smoothly over the cylinder and has left–right symmetry, this symmetry ensuring that there is no net pressure force acting on the cylinder. On the right, by way of contrast, is a schematic of the flow at large *Reynolds number*, $\mathrm{Re} = Vd/\nu \gg 1$. Upstream of the cylinder the flow is similar to that of a perfect fluid, except close to the surface of the cylinder where the tangential velocity is obliged to drop down to zero. The thin region in which this occurs is called the *boundary layer*, and we will have much to say about such layers in due course. Crucially, however, the flow downstream of the cylinder looks very different to that of an ideal fluid. In particular, the retarded flow within the boundary layer pulls away from the surface and spills out to form a *turbulent wake* behind the cylinder. Such a process is called *boundary-layer separation*. It turns out that this left–right asymmetry results in a lower pressure on the downstream face of the cylinder, giving rise to a drag force. This pressure force supplements the drag caused by the viscous shear stresses in the boundary layer.

Our main task in this chapter is to incorporate the viscous forces into the equation of motion for a fluid, and to explore some of the consequences of these forces and of imposing the no-slip condition. However, before doing this, we need to generalize Newton's law of viscosity from (2.1), which is restricted to a simple shear flow, to three dimensions. This, in

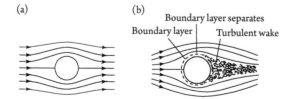

Figure 2.2 Flow over a cylinder. (a) A perfect fluid. (b) That of a real fluid at large Re.

turn, requires that we find a way of characterizing the deformation of fluid elements in three dimensions, since the shear stresses in a Newtonian fluid are directly proportional to the rate of deformation. So let us return, briefly, to kinematics.

2.2 More Kinematics: Characterizing the Deformation and Spin of Fluid Elements

2.2.1 Two Things that Happen to a Fluid Element as it Slides down a Streamline

Stokes (1845) and Helmholtz (1858) both start their seminal papers on modern fluid dynamics by noting that two distinct things happen to a blob of fluid as it slides along a streamline, over and above the obvious fact that the blob moves. First, the blob will tend to deform. For example, if it is initially spherical, then very quickly it deforms into an ellipsoid. Second, the blob as a whole may rotate or spin about its instantaneous centre. Both processes are important, but for different reasons. The rate of deformation is crucial because it sets the level of shear stress through Newton's law of viscosity, while the rate of rotation is important as it leads us to the topic of vortex dynamics and to the role of angular momentum conservation in fluid mechanics. So how can we categorize, and distinguish between, these two processes?

In order to get across some of the key ideas, let us start with the simple two-dimensional situation shown in Figure 2.3. Consider what happens to a small rectangular element of fluid, $\delta x \delta y$, as it is swept along by the flow. In a short time interval, δt, it will deform as shown. In particular, the anti-clockwise rotation of the short line element, δx, is $(\partial u_y / \partial x) \, \delta t$, while the clockwise rotation of the line element δy is $(\partial u_x / \partial y) \, \delta t$. We now define the *rate of strain* of this small fluid element in the x–y plane, denoted by S_{xy}, to be half of the total angular distortion rate,

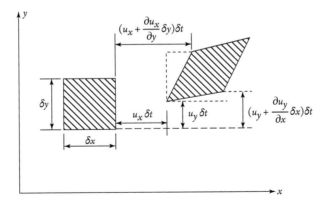

Figure 2.3 The distortion of a rectangular fluid element in a two-dimensional flow.

$$S_{xy} = \tfrac{1}{2}\left(\partial u_y/\partial x + \partial u_x/\partial y\right). \tag{2.2}$$

The two-dimensional generalization of (2.1) then states that the shear stress which causes this deformation is proportional to the angular distortion rate:

$$\tau_{xy} = 2\rho\nu S_{xy} = \rho\nu\left(\frac{\partial u_y}{\partial x} + \frac{\partial u_x}{\partial y}\right), \tag{2.3}$$

where τ_{xy} is the shear stress in the x–y plane acting on the sides of the fluid element.

By way of contrast, the *angular velocity* of the element, Ω, is defined as the average rate of rotation of the sides δx and δy. Taking rotation to be positive in the anti-clockwise direction, this gives

$$\Omega = (\partial u_y/\partial x - \partial u_x/\partial y)/2. \tag{2.4}$$

We now introduce an important quantity called the *vorticity*, defined as the vector field $\boldsymbol{\omega} = \nabla \times \mathbf{u}$. This allows us to rewrite (2.4) as $\Omega = \omega_z/2$, which is the first hint that:

the vorticity evaluated at any one location and any given time is equal to twice the angular velocity of a fluid blob passing through that point at that instant.

Let us now generalize these arguments to three dimensions. To that end, consider the relative motion of two adjacent fluid particles which are instantaneously located at points $\mathbf{x}$ and $\mathbf{x} + \delta\mathbf{x}$, having velocities $\mathbf{u}$ and $\mathbf{u} + \delta\mathbf{u}$. Evidently, $\delta\mathbf{u}$ is the velocity at $\mathbf{x} + \delta\mathbf{x}$ as measured in coordinates moving with the fluid at $\mathbf{x}$. Now $\delta u_i = (\partial u_i/\partial x_j)\,\delta x_j$ and it is natural to divide $\partial u_i/\partial x_j$ into symmetric and anti-symmetric parts. So we write

$$\delta u_i = \frac{\partial u_i}{\partial x_j}\delta x_j = \frac{1}{2}\left[\frac{\partial u_i}{\partial x_j} + \frac{\partial u_j}{\partial x_i}\right]\delta x_j + \frac{1}{2}\left[\frac{\partial u_i}{\partial x_j} - \frac{\partial u_j}{\partial x_i}\right]\delta x_j, \tag{2.5}$$

and introduce the *rate-of-strain tensor*, S_{ij}, defined by

$$S_{ij} = \frac{1}{2}\left[\frac{\partial u_i}{\partial x_j} + \frac{\partial u_j}{\partial x_i}\right], \tag{2.6}$$

which generalizes (2.2). A little algebra then shows that our expression for $\delta\mathbf{u}$ is

$$\delta\mathbf{u} = \delta\mathbf{u}^{(s)} + \delta\mathbf{u}^{(a)} = S_{ij}\delta x_j + \tfrac{1}{2}\boldsymbol{\omega} \times (\delta\mathbf{x}), \tag{2.7}$$

where $\boldsymbol{\omega} = \nabla \times \mathbf{u}$. As suggested above, the two terms on the right of (2.7) make distinct contributions to $\delta\mathbf{u}$, with S_{ij} associated with the deformation of fluid elements and $\boldsymbol{\omega}$ with the intrinsic spin of fluid blobs. We now explore these two processes one at a time.

2.2.2 The Rate-of-strain Tensor and the Deformation of Fluid Elements

Consider first the symmetric contribution to $\partial u_i / \partial x_j$. The tensor S_{ij} is symmetric and so it may be put into diagonal form through an appropriate orientation of the coordinate system. The resulting coordinate axes are called the *principal axes* of the tensor S_{ij}. Let us label these axes as 1, 2, and 3. If a, b, and c are the three *principal rates of strain* then $a = \partial u_1 / \partial x_1, b = \partial u_2 / \partial x_2$, and $c = \partial u_3 / \partial x_3$, with continuity demanding $a + b + c = 0$. In coordinates aligned with the principal axes we then have $\delta \mathbf{u}^{(s)} = (a\delta x_1, b\delta x_2, c\delta x_3)$ and so, if $\boldsymbol{\omega} = 0$, a short material line element oriented parallel to x_1 experiences the relative velocity field $\delta \mathbf{u} = (a\delta x_1, 0, 0)$. This element is therefore stretched or compressed at the rate $a\delta x_1$ while remaining parallel to x_1. Similarly, if $\boldsymbol{\omega} = 0$, line elements aligned with x_2 or x_3 stretch or contract at the rates $b\delta x_2$ or $c\delta x_3$ while remaining parallel to x_2 or x_3.

So we conclude that, provided $\boldsymbol{\omega} = 0$, an initially spherical blob of fluid deforms into an ellipsoid whose principal axes *do not rotate*. Thus S_{ij} is associated with the pure deformation of fluid blobs. Such deformations require stresses acting on the fluid, and so a stress tensor, τ_{ij}, is inevitably associated with S_{ij}. The essence of Newton's law of viscosity is that τ_{ij}, which consists of both the pressure stresses and the viscous stresses, is a linear function of S_{ij}. However, we shall defer our discussion of this relationship until §2.4.1, in part because we first need to describe the stress tensor in a little more detail.

2.2.3 Vorticity: the Intrinsic Spin of Fluid Elements

Consider now the anti-symmetric contribution to (2.7), $\delta \mathbf{u}^{(a)} = \frac{1}{2}\boldsymbol{\omega} \times (\delta \mathbf{x})$. Evidently, this represents rigid-body rotation about the point $\mathbf{x}$ with angular velocity $\boldsymbol{\omega}/2$. So, the relative velocity $\delta \mathbf{u}^{(a)}$ rotates fluid elements about $\mathbf{x}$ without causing any deformation of those elements. In short, the vorticity at location $\mathbf{x}$ and at time t is twice the angular velocity, $\boldsymbol{\Omega}$, of a fluid element passing through the point $\mathbf{x}$ at time t, the angular velocity being measured about the centre of that element. We conclude that vorticity is all about the intrinsic spin of fluid lumps as they slide along streamlines, as claimed above and as illustrated in Figure 2.4.

Perhaps some comments are in order at this point. First, vorticity is a crucial concept in fluid mechanics, as it allows us to tap into notions of angular momentum conservation. We shall return to this topic time and again in subsequent chapters. Second, the identity $\nabla \cdot (\nabla \times (\sim)) = 0$, combined with definition $\boldsymbol{\omega} = \nabla \times \mathbf{u}$, ensures that $\nabla \cdot \boldsymbol{\omega} = 0$, and

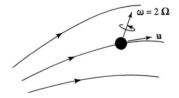

Figure 2.4 The vorticity, $\boldsymbol{\omega}$, at location $\mathbf{x}$ and at time t is twice the angular velocity, $\boldsymbol{\Omega}$, of a fluid element passing through the point $\mathbf{x}$ at time t.

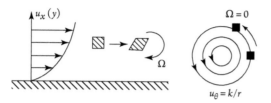

Figure 2.5 The fluid element on the left has vorticity, yet the streamlines are straight, while that on the right has no vorticity, yet the streamlines are circular.

this has nothing to do with incompressibility. Third, although we have said that ω is related to the spin of fluid particles as they slide along streamlines, the *local* vorticity is independent of any *global* rotation the flow may possess as a whole, as illustrated in Figure 2.5. The velocity field on the left is $\mathbf{u} = (u_x(y),0,0)$ in Cartesian coordinates. It is readily confirmed that this possesses vorticity, yet the streamlines are straight and parallel. On the other hand, the velocity field on the right is $\mathbf{u}(r) = (0,k/r,0)$ in (r,θ,z) coordinates. This has no vorticity, except for a singularity at $r = 0$, yet the streamlines are circular. In short, vorticity is all about the *intrinsic* spin of fluid elements.

2.3 Dynamics at Last: the Stress Tensor and Cauchy's Equation of Motion

We are now almost ready to establish the equation of motion for a viscous fluid. However, first we need to say a little more about the stress tensor acting on the fluid, τ_{ij}, which consists of *both* the pressure stresses and the viscous stresses. We start with some simple definitions. Consider a small element of fluid which is instantaneously cubic and aligned with the coordinate axes (x,y,z), shown in Figure 2.6. It has volume $\delta V = \delta x \delta y \delta z$ and is centred at location (x_0, y_0, z_0). Each surface element of our small cube will experience a force that is exerted on it by the adjacent fluid, and the magnitude of that force is proportional to its area. For example, the force $\delta \mathbf{F}$ exerted by the surrounding fluid on the surface element $\delta A = \delta x \delta y$, whose normal is $\hat{\mathbf{e}}_z$, is written as

$$\delta \mathbf{F}^{(\text{face } z)} = \left(\tau_{xz}\hat{\mathbf{e}}_x + \tau_{yz}\hat{\mathbf{e}}_y + \tau_{zz}\hat{\mathbf{e}}_z\right)\delta x \delta y. \tag{2.8}$$

Here, τ_{xz} and τ_{yz} are the shear stresses acting on the element of area, and τ_{zz} the normal stress, and these are all evaluated at the centre of that face, $(x_0, y_0, z_0 + \delta z/2)$. Note that the first subscript on τ_{ij} tells us about the direction of the associated component of force, while the second subscript identifies the normal to the surface. Now consider the stresses on the companion face below. Newton's third law applied across a horizontal plane demands that the stresses on the two faces are in opposite directions, being equal and opposite for $\delta z \to 0$. It follows that

$$\delta \mathbf{F}^{(\text{face}-z)} = -\left(\tau_{xz}\hat{\mathbf{e}}_x + \tau_{yz}\hat{\mathbf{e}}_y + \tau_{zz}\hat{\mathbf{e}}_z\right)\delta x \delta y, \tag{2.9}$$

where the stresses are now evaluated at $(x_0, y_0, z_0 - \delta z/2)$.

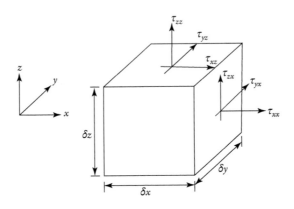

Figure 2.6 The stresses acting on a small rectangular element of fluid.

By extension, the force acting on the surface element $\delta y \delta z$, whose normal is $\hat{\mathbf{e}}_x$, is

$$\delta \mathbf{F}^{(\text{face } x)} = \left(\tau_{xx} \hat{\mathbf{e}}_x + \tau_{yx} \hat{\mathbf{e}}_y + \tau_{zx} \hat{\mathbf{e}}_z \right) \delta y \delta z. \qquad (2.10)$$

A torque balance on the cube shown in Figure 2.6 now requires that $\tau_{xz} = \tau_{zx}$, and a consideration of similar pairs of shear stresses shows that the stress tensor, τ_{ij}, must be symmetric. (Note that τ_{ij} is symmetric even in the presence of an angular acceleration of the fluid element, or of a volumetric body force, as discussed in Exercise 2.1.)

If these stresses are now allowed to vary in space then small differences in stress between companion faces can create a net force on the fluid element $\delta V = \delta x \delta y \delta z$. For example, small differences in the stresses top and bottom of the cube produce the net force

$$\delta \mathbf{F}^{(\text{face } z)} + \delta \mathbf{F}^{(\text{face} - z)} = \left(\frac{\partial \tau_{xz}}{\partial z} \hat{\mathbf{e}}_x + \frac{\partial \tau_{yz}}{\partial z} \hat{\mathbf{e}}_y + \frac{\partial \tau_{zz}}{\partial z} \hat{\mathbf{e}}_z \right) \delta x \delta y \delta z. \qquad (2.11)$$

A little algebra then shows that the net force arising from all six faces is

$$\delta F_i = \frac{\partial \tau_{ij}}{\partial x_j} \delta V. \qquad (2.12)$$

We are now finally in a position to generalize Euler's equation of motion by incorporating the force associated with the viscous stresses. Newton's second law applied to a moving blob of fluid of volume δV gives us,

$$(\rho \delta V) \frac{D\mathbf{u}}{Dt} = (\rho \delta V) \mathbf{g} + (\text{unbalanced fluid stresses}). \qquad (2.13)$$

When combined with (2.12) we obtain

$$\rho \frac{D\mathbf{u}}{Dt} = \frac{\partial \tau_{ij}}{\partial x_j} + \rho \mathbf{g}, \tag{2.14}$$

where the first term on the right represents both the pressure and viscous forces. This is called *Cauchy's equation*, and it is as far as Newton's second law will take us. If we are to make further progress we need to find a constitutive law that relates the stress tensor to velocity gradients in three dimensions, which is exactly what Newton's law of viscosity achieves.

2.4 The Navier–Stokes Equation

2.4.1 Newton's Law of Viscosity

In 1845 Stokes provided the first modern derivation of the three-dimensional constitutive law that relates viscous stresses to velocity gradients. We shall give only a schematic outline of that derivation here, and refer the reader to Batchelor (1967) for a full account.

The first problem we face is that, because of the presence of shear stress in a moving fluid, Pascal's law does not apply and the three normal stresses at a given point will not, in general, be the same. Evidently, we can no longer make the naïve assumption that the normal stresses are all equal to (minus) the pressure, and indeed we might ask what is meant by pressure in such a situation. Nevertheless, it is convenient to construct a scalar quantity for a moving fluid which is analogous to, or a generalization of, hydrostatic pressure. Since the trace of τ_{ij}, *i.e.* τ_{ii}, is independent of the orientation of the coordinate system, we take the judicious step of defining the *mechanical pressure* to be (minus) the average of the three normal stresses. Of course, this reduces to our conventional notion of pressure when there is no motion. From now on we will simply refer to this as 'the pressure' and label it as p.

Next, we assume that the stress tensor at any one point depends only on the *local* velocity gradients. Moreover, following Stokes, we hypothesise that:

(i) when $S_{ij} = 0$, so that there is no straining of the fluid, the stress tensor τ_{ij} reverts to the hydrostatic form, $\tau_{ij} = -p\delta_{ij}$;

(ii) the components of τ_{ij} are, at most, linear functions of the components of S_{ij};

(iii) there is no preferred direction in the relationship between τ_{ij} and S_{ij}.

Note that point (i) follows, in part, from Pascal's law, since a region of the fluid in which the velocity gradients are zero can be viewed as locally stationary through a change of frame of reference. However, (i) makes the stronger statement that $\tau_{ij} = -p\delta_{ij}$ holds even in the presence of velocity gradients, provided that those gradients take the form of pure vorticity (*i.e.* local rotation of the fluid elements without distortion). In effect, (i) amounts to the plausible assertion that, if there is no deformation of fluid elements, then there are no shear stresses in the fluid. Moreover (iii) simply says that we are restricted to fluids whose macroscopic behaviour is isotropic. So the key assumption is really (ii), which is a natural generalization of (2.1) and is analogous to Hook's law in elasticity.

Given assumptions (i) and (ii), we may write

$$\tau_{ij} = -p\delta_{ij} + \tau_{ij}^{\text{dev}}, \quad \tau_{ij}^{\text{dev}} = A_{ijmn}S_{mn}, \tag{2.15}$$

where τ_{ij}^{dev} is known as the *deviatoric stress* and the components of the tensor A_{ijmn} are constants, symmetric in i and j as well as m and n. Note that, in the light of our definition of p, the deviatoric stress satisfies $\tau_{ii}^{\text{dev}} = 0$. Now, if the fluid is isotropic, then the principal axes of τ_{ij}^{dev} and S_{ij} must be coincident. That is to say, if we are in the principal axes of stress, then a spherical blob of fluid will experience pure tension or compression along those axes and so the blob will be stretched or compressed along each axis, but not sheared. One form of the linear relationship (2.15) which satisfies this constraint is

$$\tau_{ij}^{\text{dev}} = k_1 S_{kk}\delta_{ij} + k_2 S_{ij}, \tag{2.16}$$

where k_1 and k_2 are constants. In fact, it turns out that this is the *only* form of (2.15) that enforces the constraint of coincident principal axes (see, for example, Batchelor, 1967). Note that, since S_{ij} is symmetric, (2.16) ensures the symmetry of τ_{ij}.

We are almost there. First, we note that $k_2 = 2\rho\nu$ for consistency with (2.3). Second, since $\tau_{ii}^{\text{dev}} = 0$, the trace of (2.16) requires $3k_1 + k_2 = 0$, or $k_1 = -(2/3)\rho\nu$. Finally, since $S_{ii} = \nabla \cdot \mathbf{u}$, we have

$$\tau_{ij}^{\text{dev}} = 2\rho\nu \left(S_{ij} - \tfrac{1}{3}\nabla \cdot \mathbf{u}\, \delta_{ij} \right). \tag{2.17}$$

We conclude that

$$\boxed{\tau_{ij} = -\left(p + \tfrac{2}{3}\rho\nu\nabla \cdot \mathbf{u} \right)\delta_{ij} + 2\rho\nu S_{ij}}, \tag{2.18}$$

which is Newton's law of viscosity, as derived by Stokes. For an incompressible fluid, (2.18) simplifies to

$$\boxed{\tau_{ij} = -p\delta_{ij} + 2\rho\nu S_{ij}}. \tag{2.19}$$

Perhaps some comments are in order. First, chemical engineering and biology apart, most common fluids are well approximated by (2.18), so we shall limit ourselves to Newtonian fluids throughout this book. Second, in a *compressible* fluid the mechanical pressure, which is an observable quantity, is not in general the same as the thermodynamic pressure, which in any event has a precise meaning only if the fluid is in equilibrium. Third, (2.19) was obtained by both Navier and Poisson some twenty years ahead of Stokes, but their derivations rested on various assumptions about the molecular origins of the viscous stresses, assumptions which turned out not to be generally valid. The significance of Stokes' derivation is that it is based purely on a macroscopic argument.

2.4.2 The Navier–Stokes Equation and the Reynolds Number

If we substitute Newton's law of viscosity into Cauchy's equation of motion, and invoke $\nabla \cdot \mathbf{u} = 0$, we obtain

$$\frac{D\mathbf{u}}{Dt} = \frac{\partial \mathbf{u}}{\partial t} + (\mathbf{u} \cdot \nabla)\mathbf{u} = -\nabla (p/\rho) + \mathbf{g} + \nu \nabla^2 \mathbf{u}, \tag{2.20}$$

which is known as the *Navier–Stokes equation*. This is the key equation of motion for an incompressible fluid. Note that the divergence of (2.20) yields

$$\nabla^2 (p/\rho) = -\nabla \cdot (\mathbf{u} \cdot \nabla \mathbf{u}). \tag{2.21}$$

In an infinite domain, this Poisson equation for pressure may be inverted to give p as an integral over all space of the instantaneous velocity field (see Appendix 2). So, as we shall see in §2.4.3, we may regard (2.20) as an evolution equation for the velocity field $\mathbf{u}$.

It is of interest to estimate the relative sizes of the inertial and viscous forces in (2.20). The viscous forces per unit mass are typically of the order of $f_\nu \sim \nu |\mathbf{u}|/\ell_\perp^2$, where $\ell_\perp$ is a characteristic length normal to the streamlines. (The reason for the choice of $\ell_\perp$ as a length scale will become clear.) The inertial forces per unit mass, on the other hand, are typically of the order of $f_{in} \sim |\mathbf{u}|^2 / \ell_\|$, where $\ell_\|$ is a characteristic length scale parallel to the streamlines (see (1.10)). The ratio of the inertial to viscous forces is then of the order of

$$\frac{\text{inertial forces}}{\text{viscous forces}} \sim \frac{u\ell_\perp^2}{\nu\ell_\|}. \tag{2.22}$$

If we do not distinguish between length scales, and approximate both $\ell_\perp$ and $\ell_\|$ by some geometric length scale, say ℓ, then this ratio becomes the *Reynolds number*, $\mathrm{Re} = u\ell/\nu$.

We now make an important observation: for nearly all flows of interest to the engineer, meteorologist, or geophysicist, the Reynolds number based on a characteristic geometric length scale is large. This reflects the fact that the kinematic viscosity of most common fluids is small, around 10^{-5} m^2/s for most common gases and 10^{-6} m^2/s for water and liquid metals. Some characteristic values of Re are listed in Table 2.1.

Table 2.1 Characteristic values of the Reynolds number in various flows.

Type of flow	Length scale used to define Re	Re
Raindrop sliding down a window	Drop size	1
Flow of water from a domestic tap	Diameter of tap	10^4
Flow at the base of a dam spillway	Depth of water	10^7
Flow over the wing of a commercial jet	Chord of wing	10^8
Convection in the liquid core of the Earth	Radius of the liquid core	10^9

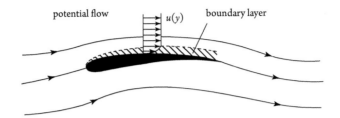

Figure 2.7 The boundary layer on an aerofoil.

The large value of Re in most of these flows tentatively suggests that viscous stresses might be ignored. However, as noted in Chapter 1, this is not the case for two reasons. First, the no-slip condition for a viscous fluid requires that the tangential velocity falls to zero adjacent to any stationary solid surface. Typically this occurs within a thin layer near the surface, called the boundary layer. Since it is the viscous stresses that cause the rapid reduction in tangential velocity, they must be of the same order of magnitude as the inertial forces within the boundary layer, and this is achieved by establishing a very high cross-stream gradient in velocity, so that the viscous forces remain order one, despite the smallness of ν. Consider, for example, the aerofoil shown in Figure 2.7. If we let $\ell_{\parallel} \sim \ell$ be the chord (width) of the wing, then (2.22) tells us that the ratio of this chord to the boundary-layer thickness, δ, is $\ell/\delta \sim \sqrt{u\ell/\nu}$. Thus, if the chord is measured in metres, the boundary-layer thickness will be measured in millimetres. We shall return to boundary layers in §2.6, and then again in Chapter 5.

There is, however, a second reason why viscous stresses remain important at large Reynolds numbers. As discussed in §1.1.3, flows at large Re tend to be turbulent. For confined flows, such as pipe flow in which Re > 2500, the entire flow tends to be turbulent, while in external flows the turbulence tends to be confined to certain regions of space. For example, for the foil shown in Figure 2.7, certain parts of the boundary layer may become turbulent, especially behind surface perturbations, or towards the rear of the foil where the boundary layer is thickest. The aerofoil may also be flying through patches of atmospheric turbulence, which can trigger turbulence across the entire boundary layer.

In fully developed turbulence the viscous stresses are always important, as they are responsible for the intense dissipation of energy which accompanies turbulence. The way this happens, despite the large nominal value of Re, is that very thin vortices develop within the turbulence, usually in the form of a tangle of intense *vortex tubes* (tubes of vorticity) whose typical diameter is only a tiny fraction of a millimetre. Such vortex tubes, which are embedded within larger eddies, are referred to as 'worms' because, at large Re, they look like a seething tangle of worms. In any event, the key point is that, as with a boundary layer, the small transverse scale of the worms gives rise to large velocity gradients, and hence the shear stresses are large despite the smallness of ν. This then leads to the intense dissipation of energy within the turbulence. We discuss turbulent worms in Chapter 13.

2.4.3 Navier–Stokes as an Evolution Equation for the Velocity Field

Note that the Navier–Stokes equation involves both the velocity and pressure fields. However, these are not independent, and indeed the pressure may be calculated directly from the instantaneous velocity field. There are different ways of seeing why this is so.

Let us, for simplicity, set gravity to zero. One way to see why p is fixed by the velocity field is to note that, since $\nabla \cdot \mathbf{u} = 0$, the divergence of (2.20) yields

$$\nabla^2 \left(p/\rho \right) = -\nabla \cdot \left(\mathbf{u} \cdot \nabla \mathbf{u} \right). \tag{2.23}$$

In an infinite domain, this may be inverted using Green's inversion formula (see Appendix 2) to give

$$p(\mathbf{x}) = \frac{\rho}{4\pi} \int \frac{\left[\nabla \cdot \left(\mathbf{u} \cdot \nabla \mathbf{u} \right) \right]'}{|\mathbf{x} - \mathbf{x}'|} \, d\mathbf{x}', \tag{2.24}$$

where the prime on the numerator in the integrand indicates that this term is evaluated at $\mathbf{x}'$. Evidently, the pressure here is a non-local function of the global velocity field. For a flow in which there are boundaries, the pressure field given by (2.24) will not, in general, satisfy the boundary conditions for $\partial p / \partial n$, the normal gradient of p, demanded by (2.20). In such cases an additional pressure field, satisfying $\nabla^2 p = 0$, must be calculated in order to satisfy this boundary condition. The total pressure is then the sum of that given by (2.24), which may be thought of as a particular integral of (2.23), plus the solution of $\nabla^2 p = 0$ which corrects for the boundary conditions. In any event, since the pressure may be calculated directly from the instantaneous velocity field, we may regard (2.20) as an evolution equation for $\mathbf{u}$ of the form $\partial \mathbf{u} / \partial t = \mathbf{f}(\mathbf{u}, p(\mathbf{u}))$. However, in order to march forward in time we must, at each time step, not only calculate a new velocity field from the old, but also update the pressure field using (2.24), or some other suitable procedure.

We close this section by noting a curious feature of (2.24). This equation demands that p is determined by the *instantaneous* velocity field. Thus, if the fluid is perturbed at one particular location, that local perturbation in velocity is felt *instantaneously* in the pressure field at *all* points in the fluid, no matter how distant. Of course, this cannot be true in practice, because information cannot be transmitted any faster than the speed of sound, and this is finite in any real fluid. The point is that the assumption of incompressibility is an idealization, and it is an idealization which demands an infinite speed of sound.

2.4.4 The Viscous Dissipation of Mechanical Energy

Let us now quantify the viscous dissipation of mechanical energy in a fluid. We can obtain a mechanical energy equation from Cauchy's equation of motion, (2.14), as follows. We start by noting that, because of the symmetry of τ_{ij},

$$u_i \frac{\partial \tau_{ij}}{\partial x_j} = \frac{\partial}{\partial x_j} \left(\tau_{ij} u_i \right) - \tau_{ij} \frac{\partial u_i}{\partial x_j} = \frac{\partial}{\partial x_j} \left(\tau_{ij} u_i \right) - \tau_{ij} S_{ij}. \tag{2.25}$$

Taking the dot product of $\mathbf{u}$ with Cauchy's equation, and ignoring gravity for simplicity, we obtain the energy equation

$$\frac{\partial}{\partial t} \left(\tfrac{1}{2} \rho \mathbf{u}^2 \right) = -\nabla \cdot \left(\left(\tfrac{1}{2} \rho \mathbf{u}^2 \right) \mathbf{u} \right) + \frac{\partial}{\partial x_j} \left(\tau_{ij} u_i \right) - \tau_{ij} S_{ij}. \tag{2.26}$$

On substituting for the stress tensor using Newton's law of viscosity, and dividing through by ρ, this becomes

$$\frac{\partial}{\partial t} \left(\tfrac{1}{2} \mathbf{u}^2 \right) = -\nabla \cdot \left((\tfrac{1}{2} \mathbf{u}^2) \mathbf{u} \right) - \nabla \cdot \left((p/\rho) \mathbf{u} \right) + \frac{\partial}{\partial x_j} \left(2\nu S_{ij} u_i \right) - 2\nu S_{ij} S_{ij}. \quad (2.27)$$

Finally, we integrate (2.27) over a control volume V fixed in space and with bounding surface S. This yields

$$\frac{d}{dt} \int_V \tfrac{1}{2} \mathbf{u}^2 dV = -\oint_S (\tfrac{1}{2} \mathbf{u}^2) \mathbf{u} \cdot d\mathbf{S} - \oint_S (p/\rho) \mathbf{u} \cdot d\mathbf{S} + \oint_S (2\nu S_{ij}) u_i dA_j - \int_V \varepsilon dV,$$

$$(2.28)$$

where $\varepsilon = 2\nu S_{ij} S_{ij}$. The surface integrals on the right represent, to within a factor of ρ, the material transport of kinetic energy across the bounding surface S and the rate of working of the pressure and viscous stresses on that surface. Conservation of energy now tells us that the integral of ε must represent the rate of loss of mechanical energy to heat. It follows that the rate of increase of internal energy per unit mass due to viscous dissipation is $\boxed{\varepsilon = 2\nu S_{ij} S_{ij}}$. This is usually just referred to as the *dissipation rate*. When written out in full, this is

$$\varepsilon = 2\nu S_{ij} S_{ij} = 2\nu \left(S_{11}^2 + S_{22}^2 + S_{33}^2 + 2S_{21}^2 + 2S_{31}^2 + 2S_{32}^2 \right), \quad (2.29)$$

and so the dissipation is zero only if all six components of S_{ij} are zero. Finally, it is readily confirmed that

$$\boxed{\varepsilon = 2\nu S_{ij} S_{ij} = \nu \omega^2 + \nabla \cdot (2\nu \mathbf{u} \cdot \nabla \mathbf{u})}. \quad (2.30)$$

Since the divergence often integrates to zero, $\nu \omega^2$ is often used as a proxy for dissipation.

2.5 The Momentum Equation for Viscous Flow in Integral Form

Let us now turn to the integral momentum equation for a real fluid, which is not so different to that of an ideal fluid. If we integrate Cauchy's equation of motion over a control volume, V, with bounding surface S, then we obtain

$$\frac{d}{dt} \int_V \rho \mathbf{u} dV + \oint_S (\rho \mathbf{u}) \mathbf{u} \cdot d\mathbf{A} = \oint_S \tau_{ij} dA_j + \int_V \rho \mathbf{f} dV,$$

where $\mathbf{f}$ represents the sum of all the body forces per unit mass acting on the fluid. Substituting for the stress tensor using Newton's law of viscosity then yields

$$\boxed{\frac{d}{dt} \int_V \rho \mathbf{u} dV + \oint_S (\rho \mathbf{u}) \mathbf{u} \cdot d\mathbf{A} = \oint_S (-p) d\mathbf{A} + \oint_S 2\rho \nu S_{ij} dA_j + \int_V \rho \mathbf{f} dV}. \quad (2.31)$$

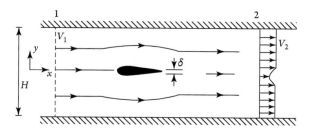

Figure 2.8 Two-dimensional flow over an aerofoil in a wind tunnel.

This generalizes (1.38) to incorporate the viscous stresses acting on the boundary, S. It is particularly useful for the calculation of the drag force acting on a body in a cross flow.

By way of a simple example, consider the two-dimensional flow over a symmetrical aerofoil in a wind-tunnel of cross-section H, as shown in Figure 2.8. Boundary layers form on either side of the foil and these spill out at the trailing edge to form a narrow wake of characteristic width δ, with $\delta \ll H$. The Reynolds number is taken to be large enough for viscous stresses to be confined to the boundary layers and wake, but not so large that the flow becomes turbulent. Suppose the flow well upstream of the aerofoil is uniform and equal to V_1, while that well downstream is $u_x(y) = V_2 - u'(y)$, where $u'(y)$ is the *velocity deficit* in the wake and V_2 is the downstream speed outside the wake, assumed uniform. Evidently, continuity requires

$$(V_2 - V_1)H = \int u'(y)dy, \tag{2.32}$$

from which $V_2 - V_1 \sim u'(\delta/H) \ll u'$.

Now suppose that there is a drag force acting on the aerofoil of F_D per unit depth into the page. (Since the foil is symmetric, there is no lift force.) Then a clamping force of F_D is required to hold the aerofoil in place, which points upstream in the negative x direction. Since the streamlines are more or less straight and parallel at sections 1 and 2, there is no cross-stream pressure gradient and the pressure is uniform at both sections. If we ignore any shear stress on the wind-tunnel walls, then the integral momentum equation applied to a control volume that spans the tunnel at sections 1 and 2 gives us

$$-F_D + p_1 H - p_2 H = \int \rho (V_2 - u')^2 \, dy - (\rho V_1 H) V_1. \tag{2.33}$$

Moreover, Bernoulli's equation applied to a streamline located outside the boundary layers and wake allows us to substitute for p_1 and p_2, resulting in

$$-F_D + \tfrac{1}{2}\rho (V_2^2 - V_1^2) H = \int \rho (V_2 - u')^2 \, dy - (\rho V_1 H) V_1. \tag{2.34}$$

This may be rearranged, with the help of (2.32), to give the simple expression

$$F_D = \rho \int (V_2 - u')\, u'\, dy + \tfrac{1}{2}\rho\, (V_2 - V_1)^2\, H, \qquad (2.35)$$

where the second term on the right is reminiscent of the Borda–Carnot head loss, (1.56). Finally, (2.32) tells us that we may neglect the last term on the right for $\delta \ll H$, and so our expression for the drag force takes on a particularly simple form:

$$\boxed{F_D = \rho \int (V_2 - u')\, u'\, dy}. \qquad (2.36)$$

The principal utility of this equation is that it allows the drag force F_D to be determined from a knowledge (or measurement) of the wake velocity profile (see Exercise 2.3).

2.6 The Role of Boundaries and Prandtl's Boundary-layer Equation

2.6.1 The Need for Boundary Layers at High Reynolds Number

Perhaps it is time to say a little more about boundary layers and boundary-layer separation. We shall study this in depth in Chapter 5, but in order to place the intervening chapters in context, it is helpful to establish some of the more elementary ideas now.

Ideal fluid dynamics had some early successes, starting with the pioneering work of Euler and Bernoulli and culminating in a seminal paper on vortex dynamics by Helmholtz (1858), about which we will have much to say in due course. Indeed, inviscid theory still finds its place in modern fluid dynamics, particularly in the study of water waves, both surface gravity waves and internal waves, and in vortex dynamics. However, by the time Stokes (1845) initiated the mathematical study of viscous flow, there was already a vast chasm between the applied mathematicians studying theoretical hydrodynamics (inviscid theory) and engineers who were trying to apply the ideas of hydraulics to real flows. Indeed, at one time the two subjects, theoretical hydrodynamics and hydraulics, seemed almost disconnected, as illustrated by Rayleigh's whimsical aside on d'Alembert's paradox. (See the quotation in the preface.)

This schism between theory and practice took a surprisingly long time to heal, in part because not everyone embraced the all-important no-slip boundary condition. It was a 29-year-old German engineer, Ludwig Prandtl, who, with astonishing clarity, showed exactly where the problems lay and how to circumvent them through the concepts of (i) the boundary layer and (ii) boundary-layer separation. Prandtl first presented his ideas at an international congress on mathematics in 1904, the proceedings of which appeared the following year. Although his paper is short, it was destined to revolutionize fluid dynamics. Curiously though, as with Maxwell and his equations of electrodynamics some 30 years earlier, the mathematical establishment took almost 20 years to appreciate the significance of what Prandtl had to say. (Note that Rayleigh's whimsical aside is dated 10 years after Prandtl first presented his ideas.)

Prandtl argued as follows. Consider the flow over a streamlined body at large Re, such as that in Figure 2.7. In order to satisfy the no-slip condition, the viscous forces must compete with inertia to reduce the velocity of the external flow down to zero. Since ν is small, this requires that the boundary-layer thickness, δ, is thin, so that the velocity gradients are large and the shear stresses significant. Since the ratio of the inertial to viscous forces is given by (2.22), this tells us that the boundary-layer thickness scales as $\delta \sim \sqrt{\nu \ell_{\|}/u}$, where the characteristic length scale, $\ell_{\|}$, might be taken as the chord of the wing, or in the case of a semi-infinite flat plate, the distance from the leading edge of the plate. Continuity now requires that the normal component of velocity is, at most, of order

$$u_n \sim \sqrt{\nu/u\ell_{\|}}\, u = \mathrm{Re}^{-1/2} u,$$

and so the acceleration of the fluid normal the surface is no larger than order $\mathrm{Re}^{-1/2}$ times that parallel to the surface. This, in turn, demands that, to leading order in the small parameter $\mathrm{Re}^{-1/2}$, we neglect normal gradients in pressure. In short, because there is no significant transverse component of acceleration, the pressure of the external flow at the edge of the boundary layer is *imposed* on the fluid within the boundary layer at the same streamwise location. Moreover, the extreme thinness of the boundary layer means that, to leading order in $\mathrm{Re}^{-1/2}$, we may treat the surface as locally flat. For a *laminar*, two-dimensional boundary layer, then, Prandtl gave the governing equation as

$$u_x \frac{\partial u_x}{\partial x} + u_y \frac{\partial u_x}{\partial y} = -\frac{1}{\rho}\frac{dp}{dx} + \nu \frac{\partial^2 u_x}{\partial y^2}, \tag{2.37}$$

where x and y are locally parallel and perpendicular to the surface.

Prandtl's next idea was to divide the flow into an external region, which may be treated as inviscid, and the boundary layer, where the shear stresses are large. In such a scheme one first solves the external problem subject to the free-slip boundary condition $\mathbf{u} \cdot d\mathbf{S} = 0$ on the surface. This then provides an outer boundary condition for the boundary layer, both in terms of the pressure gradient, dp/dx, within the boundary layer, and the tangential velocity at the top of the boundary layer. Given these boundary conditions we may then solve (2.37) to determine the laminar flow within the boundary layer.

By way of illustration, Prandtl considered the case of a thin, flat plate aligned with a uniform external flow, V. Here, the only possible choice for $\ell_{\|}$ is the distance from the leading edge of the plate, x, as there are no other length scales in the problem. In such a case, $\delta \sim \sqrt{\nu x/V}$ and the wall shear stress is of order $\tau_w \sim \rho\sqrt{\nu V^3/x}$. Given these scalings, (2.37) is readily solved to give the drag force on the plate (see Exercise 2.2).

Prandtl went on to point out that the situation for a bluff body is more complicated because of the phenomenon of boundary-layer separation. Suppose that, instead of an aerofoil, we consider flow over a cylinder at large Reynolds number. If the fluid were inviscid we would get a symmetric flow pattern like that shown in Figure 2.9(a). The pressures at the

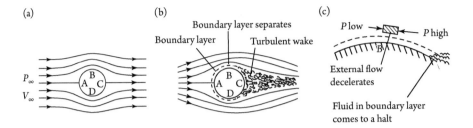

Figure 2.9 Flow over a cylinder. (a) Inviscid flow. (b) Real flow, Re $\gg$ 1. (c) Separation. (From Davidson, 2017.)

stagnation points A and C, ahead of and behind the cylinder, are then equal and given by $p_\infty + \frac{1}{2}\rho V_\infty^2$, where p_∞ and V_∞ are the upstream pressure and velocity. The real flow at large Re, by contrast, looks something like that shown in Figure 2.9(b). A boundary layer forms at the leading stagnation point and this remains thin as the fluid moves to the edges of the cylinder. However, near the edges of the cylinder (just upstream if the boundary layer is laminar, but downstream if it is turbulent) the fluid in the boundary layer is ejected into the external flow to form a turbulent wake.

This boundary-layer separation is caused by pressure forces. Outside the boundary layer the fluid, which tries to follow the inviscid flow pattern, starts to slow down as it passes over the outer edges of the cylinder (points B and D). This deceleration is caused by a positive pressure gradient which opposes the external flow. The same pressure gradient acts on the fluid in the boundary layer in accordance with (2.37), and so the boundary-layer fluid also begins to decelerate (Figure 2.9(c)). However, the fluid in the boundary layer has less kinetic energy than that in the external flow and rapidly comes to a halt, moving off and into the external flow to form a wake (see Exercise 2.4). Unlike the flat plate discussed above, which experiences only a viscous drag, the cylinder experiences both a viscous drag, because of the upstream boundary layer, and a pressure drag, because the high pressure at the forward stagnation point is not balanced by the pressure in the wake.

2.6.2 Changes in Flow Regime as the Reynolds Number Increases

The discussion above relates to flow at large Reynolds number. However, it is instructive to consider how the flow in a particular geometry varies as Re is increased from low to high. Consider, once again, a uniform flow V approaching a cylinder of diameter d, as shown in Figure 2.10. At low values of Re $= Vd/\nu$ we get a symmetric flow pattern. This is called *creeping flow*. Once Re reaches $\sim$10, steady vortices appear at the rear of the cylinder and by the time Re $\sim$ 100 these vortices peel off from the cylinder in a regular, periodic manner. This unsteady, but laminar, flow is called the *Kármán vortex street*. At yet higher values of Re, low levels of turbulence appear in the boundary layer, and this turbulence is carried off in the shed vortices. Finally, for values of Re in excess of 10^5, the flow at the rear of the cylinder loses its periodic structure, becoming fully turbulent.

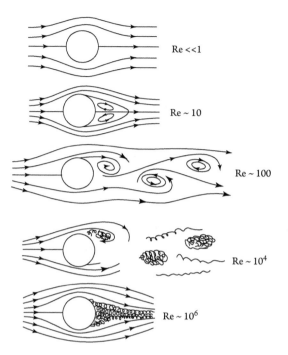

Figure 2.10 A uniform flow, V, approaches a cylinder at various values of Re.

Notice that upstream of the cylinder the fluid possesses no vorticity, whereas in the Kármán vortex street there is clearly some vorticity. Moreover, this vorticity seems to have come from the boundary layer. We shall return to this observation in §2.7.3.

It is also instructive to consider how flow in a pipe varies with Re $= Vd/v$, where d is now the pipe diameter and V the mean flow down the pipe. Famously, Reynolds used pipe flow to distinguish between laminar and turbulent states in a fluid, showing that it is the value of Re which controls the transition from one regime to another. He also showed that the critical value of Re at which turbulence first appears is very sensitive to disturbances at the entrance to the pipe, suggesting that the instability which initiates the turbulence might require a certain magnitude for the turbulence to take root. Thus, if no particular effort is taken to minimize disturbances, turbulence first appears at Re $\sim$ 2200, but when using a flared inlet to minimize disturbances, Reynolds was able to maintain laminar flow up to Re $\sim$ 13,000. Reynolds also examined what happens if turbulence is artificially created in the pipe and showed that, when Re $<$ 2000, that turbulence dies out.

It turns out that, as Reynolds suspected, the inlet conditions are very important. In the range Re $= 2200 \rightarrow 10^4$ the perturbations that initiate the turbulence are present at the pipe inlet, or else represent a finite-amplitude instability of the boundary layer near the inlet. For example, when we have a simple, straight inlet and quiescent inlet conditions, the flow is turbulent whenever Re exceeds a few thousand (Figure 2.11). The turbulence appears first as small patches in the annular boundary layer near the inlet. These patches then spread

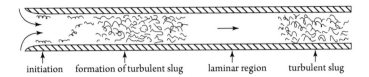

initiation   formation of turbulent slug   laminar region   turbulent slug

Figure 2.11 Turbulent slugs near the inlet of a pipe at $Re \sim 10^4$. (From Davidson, 2015.)

and merge until a slug of turbulence fills the pipe. Since Re exceeds $\sim$2000, such turbulent slugs can grow in length at the expense of the non-turbulent fluid between them, eventually merging to form fully developed turbulence.

2.7 From Linear to Angular Velocity: Vorticity and its Evolution Equation

We now turn to the important topic of vortex dynamics. We shall discuss this subject in detail in Chapter 8 and so our aim here is more modest. We seek merely to introduce the governing equations and provide some hints as to the underlying ideas. Let us start with a kinematic analogy between vorticity and magnetostatics, first noted by Helmholtz (1858).

2.7.1 The Biot–Savart Law Applied to Vorticity: an Analogy with Magnetostatics

As we shall see, it is typical of vortex dynamics that, at some particular instant, we know roughly what the vorticity distribution, $\boldsymbol{\omega}(\mathbf{x})$, is, and we would like to find the corresponding velocity field. However, the definition $\boldsymbol{\omega} = \nabla \times \mathbf{u}$ only allows us to calculate $\boldsymbol{\omega}$ from $\mathbf{u}$, and not $\mathbf{u}$ from $\boldsymbol{\omega}$. So it is natural to ask if we can invert, or perhaps we should say solve, $\nabla \times \mathbf{u} = \boldsymbol{\omega}$ subject to $\nabla \cdot \mathbf{u} = 0$ to give $\mathbf{u}$ as a function of $\boldsymbol{\omega}$. The question is not a foolish one, since a vector field is uniquely determined if its divergence and curl are specified and suitable boundary conditions given (see Appendix 2). It turns out that we can indeed perform this inversion, by borrowing a result from magnetostatics.

In magnetostatics the curl of a magnetic field $\mathbf{B}$ is given by *Ampère's law*, $\nabla \times \mathbf{B} = \mu_0 \mathbf{J}$, where μ_0 is the permeability of free space and $\mathbf{J}$ the current density vector. The magnetic field is also solenoidal, and so $\mathbf{B}$ is governed by the two equations $\nabla \cdot \mathbf{B} = 0$ and $\nabla \times \mathbf{B} = \mu_0 \mathbf{J}$. This might be compared with $\nabla \cdot \mathbf{u} = 0$ and $\nabla \times \mathbf{u} = \boldsymbol{\omega}$ for an incompressible fluid, and we see immediately that there is a kinematic analogy, in which $\mathbf{B} \leftrightarrow \mathbf{u}$ and $\mu_0 \mathbf{J} \leftrightarrow \boldsymbol{\omega}$. Now, in magnetostatics we typically know the distribution of $\mathbf{J}$ (the current flows in wires), and we would like to know the corresponding distribution of $\mathbf{B}$ associated with that current. So we are faced with a similar problem to vortex dynamics, in that we would like to invert $\nabla \times \mathbf{B} = \mu_0 \mathbf{J}$, subject to $\nabla \cdot \mathbf{B} = 0$. This is what the Biot–Savart law does. This law tells us that, in an infinite domain, the magnetic field corresponding to a given distribution of $\mathbf{J}$ may be calculated directly from

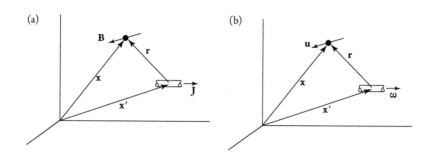

Figure 2.12 Coordinate system used for the Biot–Savart law in (a) (2.38) and (b) (2.39).

$$\mathbf{B}(\mathbf{x}) = \frac{\mu_0}{4\pi} \int \frac{\mathbf{J}(\mathbf{x}') \times \mathbf{r}}{|\mathbf{r}|^3} d\mathbf{x}', \quad \mathbf{r} = \mathbf{x} - \mathbf{x}', \tag{2.38}$$

where $\mathbf{x}'$ is a dummy variable which samples the current density at different points, as shown in Figure 2.12, and $\mathbf{r}$ takes us from the current $\mathbf{J}$ to the point where $\mathbf{B}$ is measured.

It follows immediately that, in an infinite domain, the velocity associated with a given distribution of $\boldsymbol{\omega}$ is

$$\mathbf{u}(\mathbf{x}, t) = \frac{1}{4\pi} \int \frac{\boldsymbol{\omega}(\mathbf{x}', t) \times \mathbf{r}}{r^3} d\mathbf{x}', \quad \mathbf{r} = \mathbf{x} - \mathbf{x}', \tag{2.39}$$

where $r = |\mathbf{r}|$. Perhaps some comments are in order. First, (2.39) not only ensures that $\nabla \times \mathbf{u} = \boldsymbol{\omega}$, it also enforces $\nabla \cdot \mathbf{u} = 0$, as shown in Exercise 2.6. Second, if $\boldsymbol{\omega}$ is localized in space, then $\mathbf{u}$ in (2.39) falls off with distance from the vorticity as a power law, with $\mathbf{u}(r \to \infty) = 0$. Third, we can add to (2.39) a curl-free velocity field which is established by a non-zero distribution of $\mathbf{u}$ at infinity and is governed by $\nabla \times \mathbf{u} = 0$ and $\nabla \cdot \mathbf{u} = 0$.

Mathematically, the Biot–Savart law follows almost directly from Green's inversion of Poisson's equation. The idea is the following. Since $\nabla \cdot \mathbf{u} = 0$, we can introduce a solenoidal vector potential, $\mathbf{A}$, for the velocity field. This is governed by $\nabla \cdot \mathbf{A} = 0$, $\nabla \times \mathbf{A} = \mathbf{u}$ and $\mathbf{A}(|\mathbf{x}| \to \infty) = 0$, which uniquely determines $\mathbf{A}$. The definition $\nabla \times \mathbf{u} = \boldsymbol{\omega}$ now converts to a Poisson equation for $\mathbf{A}$, $\nabla^2 \mathbf{A} = -\boldsymbol{\omega}$, which may be solved in an infinite domain using Green's inversion integral,

$$\mathbf{A}(\mathbf{x}) = \frac{1}{4\pi} \int \frac{\boldsymbol{\omega}(\mathbf{x}')}{|\mathbf{r}|} d\mathbf{x}', \quad \mathbf{r} = \mathbf{x} - \mathbf{x}'. \tag{2.40}$$

The link between (2.40) and the Biot–Savart law (2.39) now rests on the identity

$$\nabla \times (\boldsymbol{\omega}'/|\mathbf{r}|) = -\boldsymbol{\omega}' \times \nabla (1/|\mathbf{r}|) = \boldsymbol{\omega}' \times \mathbf{r} \Big/ |\mathbf{r}|^3, \tag{2.41}$$

where $\boldsymbol{\omega}' = \boldsymbol{\omega}(\mathbf{x}')$ and ∇ operates on $\mathbf{x}$ while treating $\mathbf{x}'$ as constant. This gives us

$$\mathbf{u}(\mathbf{x}) = \nabla \times \mathbf{A}(\mathbf{x}) = \frac{1}{4\pi} \int \nabla \times \left(\frac{\boldsymbol{\omega}'}{|\mathbf{r}|} \right) d\mathbf{x}' = \frac{1}{4\pi} \int \frac{\boldsymbol{\omega}' \times \mathbf{r}}{|\mathbf{r}|^3} d\mathbf{x}' \qquad (2.42)$$

as required.

2.7.2 The Vorticity Evolution Equation

We now turn to dynamics, restricting ourselves to incompressible fluids. As noted in §2.4.3, we may regard the Navier–Stokes equation as an evolution equation for $\mathbf{u}$, of the form $\partial \mathbf{u}/\partial t = \mathbf{f}(\mathbf{u}, p(\mathbf{u}))$. However, each time we calculate a new velocity field from the old, we must update the pressure field using the non-local integral equation, (2.24), which can be awkward. An alternative strategy, advocated by Helmholtz (who largely developed the subject of vortex dynamics), is to take the curl of the Navier–Stokes equation. This eliminates the pressure gradient at the same time as providing an evolution equation for the vorticity field. As we shall see, this has several advantages. Of course, given $\boldsymbol{\omega}(\mathbf{x}, t)$, we may use the Biot–Savart law to determine $\mathbf{u}(\mathbf{x}, t)$.

The simplest way to get an evolution equation for $\boldsymbol{\omega}(\mathbf{x}, t)$ is to rewrite (2.20) in the same form as (1.24),

$$\partial \mathbf{u}/\partial t = \mathbf{u} \times \boldsymbol{\omega} - \nabla H + \nu \nabla^2 \mathbf{u}, \qquad (2.43)$$

where H is Bernoulli's function. Taking the curl of (2.43), and noticing that the Laplacian commutes with the curl operator, gives us the evolution equation,

$$\boxed{\frac{\partial \boldsymbol{\omega}}{\partial t} = \nabla \times (\mathbf{u} \times \boldsymbol{\omega}) + \nu \nabla^2 \boldsymbol{\omega}} \text{ (form I).} \qquad (2.44)$$

However, both $\mathbf{u}$ and $\boldsymbol{\omega}$ are solenoidal and so the identity $\nabla \times (\mathbf{u} \times \boldsymbol{\omega}) = (\boldsymbol{\omega} \cdot \nabla)\mathbf{u} - (\mathbf{u} \cdot \nabla)\boldsymbol{\omega}$ yields the alternative expression,

$$\boxed{\frac{D\boldsymbol{\omega}}{Dt} = (\boldsymbol{\omega} \cdot \nabla)\mathbf{u} + \nu \nabla^2 \boldsymbol{\omega}} \text{ (form II).} \qquad (2.45)$$

We may regard either (2.44) or (2.45) as our evolution equation for vorticity, in the sense that we can advance $\boldsymbol{\omega}$ in time by writing either equation in the form $\partial \boldsymbol{\omega}/\partial t = F(\mathbf{u}, \boldsymbol{\omega})$. Of course, we must use (2.39) to update the velocity field in parallel with the vorticity.

Since $\boldsymbol{\omega}(\mathbf{x}, t)$ is twice the angular velocity of a fluid blob passing through point $\mathbf{x}$ at time t, we might anticipate that (2.45) is related to angular momentum conservation, and this is indeed the case. Consider a blob of fluid that is *instantaneously* spherical, as shown in Figure 2.4. Since $\boldsymbol{\omega}$ is twice the angular velocity of the blob, the angular momentum of the sphere is $\mathbf{H} = I\boldsymbol{\omega}/2$, where I is its moment of inertia. Moreover, $\mathbf{H}$ can change only as a result of tangential surface stresses acting on the blob, the pressure playing no role whilst the blob is spherical. We conclude that, *for as long as the fluid blob remains spherical*, $D\mathbf{H}/Dt$ is determined entirely by a viscous torque. Now the convective derivative operating on a product acts just like a conventional derivative, and so differentiating out the product $I\boldsymbol{\omega}/2$, and dividing through by $I/2$, gives us

$$\frac{D\boldsymbol{\omega}}{Dt} = -\boldsymbol{\omega}\frac{D}{Dt}(\ln I) + \frac{\text{viscous torque on sphere}}{I/2}. \tag{2.46}$$

This is the physical interpretation of (2.45). It tells us that the vorticity of a fluid element can change either because its moment of inertia changes or because the viscous stresses spin up or spin down that element.

Actually, this argument is not rigorous, as the spherical blob will not *stay* spherical. However, it does at least suggest that the first term on the right of (2.45), $\boldsymbol{\omega} \cdot \nabla \mathbf{u}$, represents the product of $|\boldsymbol{\omega}|$ with the rate of change of the moment of inertia of a fluid blob. To show that this is indeed the case, consider a thin *vortex tube* whose boundary is composed of an aggregate of *vortex lines* (lines drawn parallel to $\boldsymbol{\omega}$), just like a stream-tube is composed of streamlines. Let $u_\parallel$ be the component of velocity parallel to the tube (*i.e.* parallel to $\boldsymbol{\omega}$) and s a coordinate measured along the tube (Figure 2.13). Then,

$$(\boldsymbol{\omega} \cdot \nabla)u_\parallel = |\boldsymbol{\omega}|\frac{du_\parallel}{ds}. \tag{2.47}$$

Now, the vortex tube is stretched whenever $du_\parallel/ds > 0$, i.e. $u_B > u_A$ in Figure 2.13, and it is compressed when $du_\parallel/ds < 0$. It follows that, when $(\boldsymbol{\omega} \cdot \nabla)u_\parallel > 0$, the vortex tube is stretched and thinned, causing a reduction in its moment of inertia and a rise in angular velocity. This is consistent with $\boldsymbol{\omega} \cdot \nabla \mathbf{u}$ producing changes in $|\boldsymbol{\omega}|$ through changes in the moment of inertia of a fluid element.

In summary, then, the stretching of a fluid element in the direction of its vorticity causes it to spin up, as illustrated in Figure 2.14. This is called *vortex-line stretching* and it is one of the most important processes in turbulence. As we shall see in Chapter 13, the chaotic velocity field in a turbulent flow tends to stretch the vortex tubes more than it compresses them, and so there is a continual intensification of the vorticity through vortex-line stretching. There is then a competition between vorticity intensification by stretching and the viscous destruction of *enstrophy*, defined as $\boldsymbol{\omega}^2/2$ (see §2.7.4).

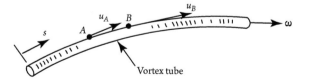

Figure 2.13 The stretching of a tube of vorticity.

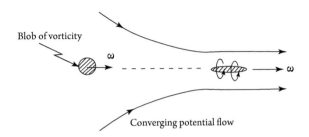

Figure 2.14 Stretching a fluid element can intensify its vorticity.

Perhaps it is worth making another point about turbulence. In the absence of vorticity, the non-linear inertial term in (2.43) vanishes, yet it is precisely this non-linearity which drives the chaos we call turbulence. In short, we cannot have turbulence without a vorticity field. Indeed, it is common to define turbulence as a *chaotic vorticity field* evolving in accordance with (2.45). The velocity appearing in this equation is then partly that *imposed* on the turbulence, say by a simple shear flow, and partly the turbulent velocity field associated directly with the chaotic vorticity through the Biot–Savart law.

Note that there is one degenerate case in which there is no vortex-line stretching. This is two-dimensional motion of the form $\mathbf{u}(x,y) = (u_x, u_y, 0)$. Here, $\boldsymbol{\omega}(x,y) = (0,0,\omega)$ and so (2.45) reduces to the scalar equation

$$\frac{D\omega}{Dt} = \nu\nabla^2\omega. \tag{2.48}$$

This might be compared with the governing equation for the temperature, T, in a thermally conducting fluid,

$$\frac{DT}{Dt} = \alpha\nabla^2 T, \tag{2.49}$$

where α is the thermal diffusivity and the term on the right of (2.49) represents the diffusion of heat in or out of fluid elements as they move around. Equations of this type are referred to as *advection–diffusion equations*. Evidently, in a two-dimensional flow, vorticity is materially transported by the flow (we say it is *advected* by $\mathbf{u}$) while simultaneously diffusing between fluid elements. In an ideal fluid the two-dimensional vorticity equation is simply $D\omega/Dt = 0$, which says that each fluid element holds on to its vorticity as it moves around. Equation (2.48) then shows how, in the presence of viscous diffusion, the vorticity in a given fluid element may rise or fall as it diffuses in or out of that element.

Note, however, that neither the advection of, nor the diffusion within, an *isolated patch* of vorticity can change the net amount of vorticity within that patch, just as the advection and diffusion of heat cannot change the net thermal energy in an isolated region of heated fluid. Rather, these two processes merely act to redistribute a given amount of the vorticity (or heat) in space. This may be proved by rewriting (2.48) as

$$\frac{\partial \omega}{\partial t} = -\nabla \cdot (\omega \mathbf{u}) + \nabla \cdot (\nu \nabla \omega), \tag{2.50}$$

and then integrating (2.50) over an isolated vortex patch while applying Gauss' theorem to the two divergences. The resulting surface integrals vanish, leaving the net vorticity in the patch conserved.

2.7.3 Where does the Vorticity come from?

To summarize, vorticity may be redistributed throughout space by advection and diffusion. Moreover, in three-dimensional motion, it may be intensified or diminished through the stretching or compression of the vortex tubes. However, none of these processes can create vorticity in a region of fluid which had none to start with. One might ask, therefore, where the vorticity in the Kármán vortex street shown in Figure 2.15 has come from.

Here, once again, the analogy between heat transport and vorticity in a two-dimensional flow is useful. Heat cannot be created or destroyed within the interior of a fluid by advection or diffusion. However, if the cylinder in Figure 2.15 were heated, then heat would get into the fluid by diffusing in from the surface of the cylinder. The same is true of vorticity. In the absence of rotational body forces, vorticity can get into a fluid only by diffusing in from the boundaries. So, just as there is a *thermal* boundary layer on a heated cylinder, through which heat diffuses into the fluid, so there is a *viscous* boundary layer, though which vorticity diffuses into the flow. Indeed, all of the vorticity downstream of the cylinder shown in Figure 2.15 started off in a thin viscous boundary layer on the upstream surface of the cylinder.

This gives us an alternative way of thinking about viscous boundary layers: they are diffusion layers for vorticity. Consider the boundary layer shown in Figure 2.16. The no-slip condition at the wall means that high velocity gradients form at that surface, and so we may think of the wall as a potential source of vorticity. Vorticity can then diffuse from the wall

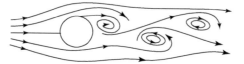

Figure 2.15 Vortex shedding behind a cylinder.

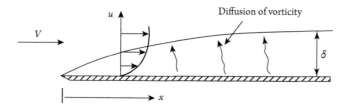

Figure 2.16 A boundary layer is a diffusion layer for vorticity (see Exercise 2.5).

into the fluid while being swept downstream by the flow. When Re is large, this diffusion is slow by comparison with advection, and so the boundary layer is thin.

Let us return to Figure 2.15 and reinterpret it with this in mind. As the fluid passes over the front of the cylinder, vorticity is generated at the cylinder's surface. This then diffuses into the fluid, so the boundary layer thickens as the fluid passes from the forward stagnation point to the top and bottom of the cylinder. By the time we get to the edges of the cylinder, very high levels of vorticity have diffused out from the surface and into the boundary layer. Separation then occurs and all of this vorticity is ejected into the flow.

Similar processes occur for other bluff bodies, causing either vortex shedding or a fully developed turbulent wake, depending on the Reynolds number. Either way, the intense vorticity created by flow over a surface finds its way into the bulk motion through boundary-layer separation. This is why, on a windy day, any street is full of vorticity: vorticity that has diffused into boundary layers on the sides and roofs of the buildings, and then spilled out into the street.

2.7.4 Enstrophy and its Governing Equation

The intensification of vorticity by vortex-line stretching is often quantified using enstrophy, $\omega^2/2$. A budget equation for enstrophy comes from the dot product of ω with (2.45),

$$\frac{D}{Dt}\left(\frac{\omega^2}{2}\right) = \omega_i\omega_j\frac{\partial u_i}{\partial x_j} + \nu\omega\cdot\left(\nabla^2\omega\right), \qquad (2.51)$$

which is usually rewritten as

$$\frac{D}{Dt}\left(\frac{\omega^2}{2}\right) = \omega_i\omega_j S_{ij} - \nu\left(\nabla\times\omega\right)^2 + \nu\nabla\cdot\left[\omega\times\left(\nabla\times\omega\right)\right]. \qquad (2.52)$$

The first term on the right of (2.52) corresponds to the generation of enstrophy through vortex-line stretching, or else a reduction in enstrophy via compression of the vortex lines. The second term, by contrast, represents the destruction of enstrophy by viscous forces, while the third term is often unimportant as the divergence integrates to zero for a localized distribution of ω. We conclude that enstrophy, just like mechanical energy, is destroyed by friction. However, unlike mechanical energy, there is a natural generation mechanism in the absence of body forces. Enstrophy plays an important role in turbulence where, as noted in §2.7.2, there is a delicate balance between vorticity intensification by chaotic vortex-line stretching and the viscous destruction of enstrophy.

2.7.5 A Glimpse at Potential (Vorticity Free) Flow and its Limitations

We close our discussion of vortex dynamics with a gentle warning about the dangers of potential flow theory, i.e. the traditional study of inviscid flows devoid of vorticity. Consider Figure 2.7 in which there is a boundary layer filled with vorticity, and an external flow. In classical aerodynamics the flow upstream of the aerofoil is usually assumed to be free

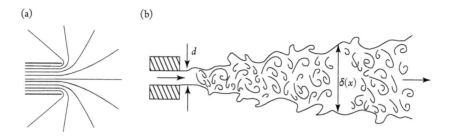

Figure 2.17 (a) The potential flow for a two-dimensional flow entering or leaving a duct. (b) Schematic of the actual flow for a submerged jet at large Re. (From Davidson, 2017.)

of vorticity, and since the vorticity generated on the surface of the wing is confined to a boundary layer, the entire external flow is then a potential flow. Computing the external velocity field is now reduced to the simple problem of solving the two *kinematic* equations, $\nabla \cdot \mathbf{u} = 0$ and $\nabla \times \mathbf{u} = 0$, subject to $\mathbf{u} \cdot d\mathbf{S} = 0$. Crucially, Newton's second law plays no role in mapping out the velocity field, as discussed below.

Such potential flows have the advantage of being easy to solve, but they are rare in nature. Arguably, a streamlined body moving through still fluid, plus certain types of free surface flows, represent the only common examples of incompressible potential flow. Most real flows are heavily laden with vorticity; often vorticity that has been generated in a boundary layer and then released into the bulk flow by boundary-layer separation. So potential flow theory can be useful in aerodynamics, but much less so in other fields.

The inherent dangers of potential flow theory are illustrated by Figure 2.17. Figure 2.17(a) appears in many fluids texts as the two-dimensional potential flow at the entrance to a duct. The same figure also appears in books on magnetism, representing a static magnetic field **B** entering or leaving a pole, governed by $\nabla \cdot \mathbf{B} = 0$ and $\nabla \times \mathbf{B} = 0$. Of course, there is no concept of inertia associated with **B**, and this reminds us that inertia plays almost no role in the calculation of a potential flow, other than to ensure that a state of zero initial rotation is maintained by the fluid particles. Indeed, Newton's second law enters the problem only *retrospectively*, when we use Bernoulli's equation to infer a pressure distribution from a given velocity field. Moreover, if we change the sign of **u** in Figure 2.17(a), the flow still satisfies $\nabla \cdot \mathbf{u} = 0$ and $\nabla \times \mathbf{u} = 0$, and this now represents fluid *leaving* a two-dimensional duct. Yet it is never presented as such because, in reality, such a flow looks nothing like Figure 2.17(a). Rather, it takes the form of a submerged jet laden with vorticity, vorticity which has been stripped off the walls of the duct (Figure 2.17(b)). So extreme caution must be exercised if you plan to use potential flow theory.

2.8 Summing up: Real versus Ideal Fluid Mechanics

We have covered a lot of ground in Chapters 1 and 2, so perhaps we should conclude by briefly recalling the different categories of motion that one encounters in incompressible flow theory. Perhaps the most fundamental division is that between the idealization of an

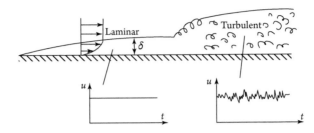

Figure 2.18 Schematic of the development of a boundary layer on a flat plate. (From Davidson, 2017.)

inviscid fluid and the dynamics of real fluids. The large value of Re associated with most natural flows makes ideal theory an appealing option, and indeed it can be useful in, for example, free-surface gravity waves and vortex dynamics. However, we have seen that the assumption of an ideal fluid is fraught with difficulties. In particular, the need to satisfy the no-slip condition for a real fluid leads directly to the existence of boundary layers, and to the related phenomena of drag, boundary-layer separation, and turbulence.

Then there is the more subtle distinction between ideal flow and potential flow. In traditional aerodynamics texts, the ideal flow outside a boundary layer is usually equated to a potential flow devoid of vorticity, but this is simply a matter of convenience. As Helmholtz (1858) emphasised, often there is no justification for this assumption. Indeed, the translation of a streamlined body through still air, admittedly an important special case, plus certain types of free-surface flows are the only commonly encountered incompressible potential flows. By way of contrast, we shall see that there are many natural flows which are reasonably well approximated by ideal theory, yet are not potential flows, in that they owe their very existence to vorticity. So potential flow is a useful concept for the aerodynamicist, but it can be very misleading if applied elsewhere.

Finally, from a practical point of view, perhaps the most important subdivision is between laminar and turbulent motion. As we have emphasised, turbulence is to be expected whenever Re rises above a few thousand, and this means that most natural flows include regions of turbulence. For example, a flat-plate boundary layer will become turbulent if subject to large enough perturbations, or if the plate is sufficiently long (Figure 2.18). So turbulence is of immense practical importance. Some texts have little to say on the subject, which is understandable given the difficulties inherent in describing such chaotic flows. Nevertheless, somewhat surprisingly, it turns out that predictions of a rather general nature can be made about certain turbulent flows, and so perhaps it is important that these are discussed. We shall do just that in Chapters 13 and 14.

○ ○ ○

This concludes our introduction to real fluids. Readers looking for more background will find a particularly accessible overview in Chapter 41 of Feynman et al. (1964).

....

EXERCISES

2.1 *The symmetric nature of the stress tensor.* Consider the angular equation of motion about the y-axis for the small cube shown in Figure 2.6. Show that, even if the cube possesses an angular acceleration about that axis, the leading-order torque balance still requires $\tau_{xz} = \tau_{zx}$. Now show that, more generally, the stress tensor τ_{ij} is symmetric, even in the presence of angular acceleration of the fluid, or of a volumetric body force.

2.2 *The laminar boundary layer on a flat plate.* Consider the boundary layer on a thin, flat plate of semi-infinite extent aligned with a uniform external flow, V. The only geometric length scale in this problem is the distance from the leading edge of the plate, x, so the boundary-layer thickness, δ, must be some function of ν, x, and V. Use (2.22), plus dimensional analysis, to show that this functional relationship *must* be of the form $\delta \sim \sqrt{\nu x / V}$. Since the external flow is uniform, there is no external pressure gradient. Look for a self-similar solution of (2.37) of the form

$$u_x / V = f(y/\delta_0) = f(\eta), \quad \delta_0 = \sqrt{\nu x / V},$$

where y is measured from the plate and the boundary conditions are $f(0) = 0$ and $f(\infty) = 1$. Show that, if $f = g'(\eta)$, then g satisfies $gg'' + 2g''' = 0$, which is readily solved numerically. This yields $f(4) = 0.956$ and $f'(0) = 0.332$. Evidently the boundary-layer thickness and wall shear stress are $\delta \approx 4\sqrt{\nu x / V}$ and $\tau_w = 0.332\rho\sqrt{\nu V^3 / x}$.

2.3 *Comparing the drag force on a flat plate with the momentum deficit in its wake.* Use the results of Exercise 2.2 to show that the drag force on both sides of a thin, flat plate of length L, aligned with a uniform flow V, is $F_D = 1.33\rho\sqrt{\nu V^3 L}$ (per unit width). Given that the shape of the wake just behind the plate is given by $f(\eta)$ at $x = L$, use $gg'' + 2g''' = 0$ to show that this drag force is exactly that predicted by the wake deficit equation, (2.36).

2.4 *Laminar boundary-layer separation caused by an external pressure rise.* It is an experimental observation that the point of separation of a laminar boundary layer is characterized by the onset of a region of reverse flow near the wall, as shown in Figure 2.19. The *zone of separation* downstream of that point has a negative wall shear stress, and hence $(\partial u_x / \partial y)_{y=0} < 0$. Sketch the boundary-layer velocity profile at the point of

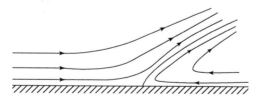

Figure 2.19 Boundary-layer separation caused by an external pressure rise.

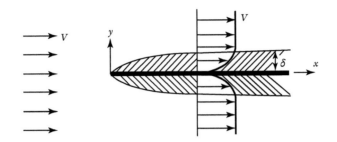

Figure 2.20 A flat-plate boundary layer.

separation, and also slightly upstream and downstream of that point. Show that the separated velocity profiles must contain an inflection point at which $\partial^2 u_x/\partial y^2 = 0$, with $\partial^2 u_x/\partial y^2 > 0$ close to the wall and $\partial^2 u_x/\partial y^2 < 0$ near the top of the boundary layer. Now use (2.37) to show that all of the velocity profiles must satisfy $\rho\nu(\partial^2 u_x/\partial y^2)_{y=0} = dp/dx$. Hence confirm that laminar boundary-layer separation occurs in a region of rising external pressure, $dp/dx > 0$. We shall see in Chapter 11 that velocity profiles containing an inflection point tend to be unstable, and hence act as a source of turbulence.

2.5 *The flat-plate boundary layer as a diffusion layer for vorticity.* Heat diffuses a distance of order $\sqrt{\alpha t}$ in a time t. This is known as the diffusion length. By implication, vorticity diffuses a distance $\sim \sqrt{\nu t}$ in a time t. Consider the boundary layer on a thin, flat plate aligned with a uniform external flow, V, as shown in Figure 2.20. A fluid particle located at a height y above the leading edge of the plate at $t = 0$ will first acquire vorticity by diffusion from the plate at a time $t \sim y^2/\nu \sim x/V$, where x is the stream-wise distance travelled by the particle in the time t. Use this estimate to explain the boundary-layer scaling $\delta \sim \sqrt{\nu x/V}$ of Exercise 2.2.

2.6 *The Biot–Savart law.* The Biot–Savart law (2.39) not only ensures that $\nabla \times \mathbf{u} = \boldsymbol{\omega}$, it also enforces $\nabla \cdot \mathbf{u} = 0$, as we now show. Rewrite (2.39) in the form

$$\mathbf{u}(\mathbf{x}) = \frac{1}{4\pi} \int \nabla\left(1/|\mathbf{r}|\right) \times \boldsymbol{\omega}' \, d\mathbf{x}', \quad \mathbf{r} = \mathbf{x} - \mathbf{x}',$$

where ∇ operates on $\mathbf{x}$ while treating $\mathbf{x}'$ as constant. Hence show that

$$\nabla \cdot \mathbf{u} = \frac{1}{4\pi} \int \nabla \cdot \left(\nabla\left(1/|\mathbf{r}|\right) \times \boldsymbol{\omega}'\right) d\mathbf{x}' = 0.$$

....

REFERENCES

Batchelor, G.K., 1967, *An introduction to fluid dynamics*, Cambridge University Press.
Davidson, P.A., 2015, *Turbulence: an introduction for scientists and engineers*, 2nd Ed., Oxford University Press.

Davidson, P.A., 2017, *An introduction to magnetohydrodynamics*, 2nd Ed., Cambridge University Press.

Feynman, R.P., Leighton, R.B., & Sands, M., 1964, *The Feynman lectures on physics*, Vol. II, Addison-Wesley.

Helmholtz, H., 1858, On integrals of the hydrodynamical equations which express vortex motion. Translated into English in: *Phil Mag.* (series 4), 1867, **33**, 485–512.

Stokes, G.G., 1845, On the theories of the internal friction of fluids in motion, and of the equilibrium and motion of elastic solids. *Trans. Camb. Phil. Soc.*, **8**, 287–305.

3

Some Elementary Solutions of the Navier–Stokes Equation

(a)　　　　　　　　　　　　　　(b)

A new chapter in fluid mechanics followed the widespread acceptance of the Navier–Stokes equation. Two important players were (a) the applied mathematician Sir Horace Lamb (1849–1934) and (b) the mathematical physicist Joseph Valentin Boussinesq (1842–1929). While Lamb made important contributions in many areas of physics, he is now best known for his book on hydrodynamics. First published in 1879 under a different title, the sixth edition is still in print and in common usage. Boussinesq, along with his mentor Saint-Venant, was one of the early pioneers of turbulence research. Publishing ahead of Osborne Reynolds, he introduced the idea of an eddy viscosity. Moreover, following John Russell's discovery of solitary waves, a precursor to the soliton, Boussinesq provided the first detailed mathematical model of such waves, including a derivation of the now famous KdeV equation, several years before the much-quoted paper by Korteweg and de Vries.

3.1 Some Simple Laminar Flows

3.1.1 Planar Viscous Flow

Let us start with a particularly simple flow—that of steady, fully developed flow between two parallel plates, in which $\mathbf{u}(y) = (u_x, 0, 0)$. We shall allow the top plate to move at the speed V while the bottom plate is held stationary, as shown in Figure 3.1. We shall also allow for a *negative* pressure gradient in the x direction, so that the fluid moves partly because it is dragged along by the top plate and partly because it is pushed through the gap by a pressure difference.

Since $u_y = 0$ and u_x is not a function of x, the inertial term in the Navier–Stokes equation is zero, $\mathbf{u} \cdot \nabla \mathbf{u} = 0$. It follows that there is no pressure gradient in the transverse direction and the Navier–Stokes equation reduces to the simple linear equation

$$\frac{dp}{dx} = \rho \nu \frac{d^2 u_x}{dy^2} = \frac{d\tau_{xy}}{dy}. \tag{3.1}$$

This represents the force balance for a small rectangular element of fluid, $dxdy$, in which the pressure difference between the front and back of the element is balanced by a difference in shear stress on the top and bottom, as indicated in Figure 3.1. Integrating (3.1) twice and demanding that $u_x = 0$ at the bottom plate, $y = 0$, now yields

$$u_x = \frac{1}{2\rho\nu} \left| \frac{dp}{dx} \right| (cy - y^2), \tag{3.2}$$

where c is a constant of integration. It remains to satisfy $u_x = V$ at the top plate, $y = h$, and after a little algebra this gives us

$$u_x = \frac{1}{2\rho\nu} \left| \frac{dp}{dx} \right| (hy - y^2) + \frac{Vy}{h}. \tag{3.3}$$

This is often called *plane Couette flow*. Evidently, the two driving forces for motion, the drag of the top plate and the negative pressure gradient, have produced two distinct contributions to u_x.

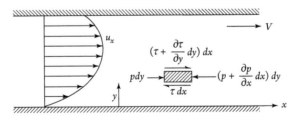

Figure 3.1 Laminar flow between parallel plates.

The special case where the top plate is stationary, $V = 0$, is sometimes called *plane Poiseuille flow*. This is best expressed using a vertical coordinate measured from the mid-plane of the gap, say $y^* = y - h/2$. In terms of y^*, (3.3) becomes

$$u_x = \frac{1}{2\rho\nu} \left| \frac{dp}{dx} \right| ((h/2)^2 - (y^*)^2),$$

(3.4)

yielding the familiar parabolic profile which is symmetric about the centreline. The total volumetric flow rate, $Q = \dot{m}/\rho$, for plane Poiseuille flow is evidently

$$Q = \frac{h^3}{12\rho\nu} \left| \frac{dp}{dx} \right|,$$

(3.5)

per unit depth into the page. Actually, the scaling $Q \sim (h^3/\rho\nu) |dp/dx|$ follows directly from dimensional analysis and the observation that Q is proportional to the pressure gradient, as discussed in Exercise 3.6.

Let us now consider a steady, laminar flow driven by gravity. Consider a thin film of fluid running down an inclined plate, as shown in Figure 3.2. We take x and y to be measured parallel and perpendicular to the inclined surface, with $y = 0$ representing the wall. Observations suggests that, under the right conditions, a plane flow develops in which, once again, $\mathbf{u}(y) = (u_x, 0, 0)$, and we seek $u_x(y)$. We shall work with gauge pressure and ignore any frictional drag exerted by the air on the moving film, so that $p = \tau_{xy} = 0$ at the free surface, $y = h$.

Once again, the inertial forces are zero, $\mathbf{u} \cdot \nabla \mathbf{u} = 0$, and the Navier–Stokes equation reduces to

$$\frac{1}{\rho}\frac{\partial p}{\partial x} = g\sin\alpha + \nu\frac{\partial^2 u_x}{\partial y^2}, \qquad \frac{1}{\rho}\frac{\partial p}{\partial y} = -g\cos\alpha,$$

(3.6)

where α is the angle of inclination. Given that $p = 0$ at $y = h$, the second of these integrates to give $p = \rho g (h - y) \cos\alpha$, and so p is independent of x. The first equation now simplifies to

$$\nu\frac{d^2 u_x}{dy^2} = -g\sin\alpha.$$

(3.7)

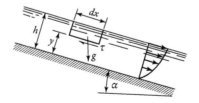

Figure 3.2 Laminar flow of a thin film down an inclined surface.

This integrates once to give

$$\nu \frac{du_x}{dy} = \frac{\tau_{xy}}{\rho} = g \sin \alpha (h - y), \tag{3.8}$$

where we have demanded that the shear stress is zero at the free surface. A glance at Figure 3.2 confirms that (3.8) represents a force balance between gravity and the shear stress acting on the rectangular element $(h - y)dx$. Integrating a second time, while setting $u_x = 0$ at $y = 0$, gives the velocity profile

$$u_x = \frac{g \sin \alpha}{2\nu} (2h - y)y, \tag{3.9}$$

whose volumetric flow (per unit width) is

$$Q = \frac{gh^3 \sin \alpha}{3\nu}. \tag{3.10}$$

3.1.2 The Boundary Layer near a Two-dimensional Stagnation Point

Let us now consider the flow in the vicinity of a two-dimensional stagnation point. The inviscid irrotational flow near such a point is $\mathbf{u} = (\alpha x, -\alpha y, 0)$, where α is a positive constant (see Figure 3.3). This clearly satisfies $\nabla \cdot \mathbf{u} = 0$, $\nabla \times \mathbf{u} = 0$, and $\mathbf{u} \cdot d\mathbf{S} = 0$ on the plane $y = 0$. The question now arises: can we find a steady solution for the viscous flow near the wall which satisfies the no-slip condition and merges into the inviscid solution at large distances from the wall? Physically, this is a well-posed problem as vorticity will tend to diffuse away from the wall, while being swept back down towards the wall by u_y. So, to find a steady boundary-layer solution, we simply need to balance these two processes.

Let us see if we can estimate the boundary-layer thickness, δ. The downward advection of vorticity, ω, is

$$u_y \partial \omega / \partial y \sim (\alpha \delta)(\omega/\delta) \sim \alpha \omega,$$

while the upward diffusion is of the order of $\nu \partial^2 \omega / \partial y^2 \sim \nu \omega / \delta^2$. Equating these two estimates yields $\delta \sim \sqrt{\nu/\alpha}$. Note that this suggests that the boundary layer thickens if

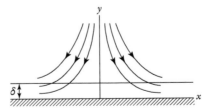

Figure 3.3 Stagnation point flow.

ν is increased, but thins if the flow speed, as measured by α, rises. Both of these trends are plausible, so let us look for a solution of the form

$$u_x = \alpha x f'(\eta), \quad \eta = y/\delta_0, \quad \delta_0 = \sqrt{\nu/\alpha}, \tag{3.11}$$

where $f'(\eta)$ runs smoothly from 0 at the wall to 1 at large η. Continuity then requires

$$u_y = -\alpha \delta_0 f(\eta), \quad f(0) = f'(0) = 0, \quad f'(\infty) = 1, \tag{3.12}$$

while the vorticity is

$$\omega = -\frac{\partial u_x}{\partial y} = -\frac{\alpha x}{\delta_0} f''(\eta). \tag{3.13}$$

A little algebra now shows that the advection and diffusion of vorticity are given by the expressions

$$\mathbf{u} \cdot \nabla \omega = \frac{\alpha^2 x}{\delta_0} \frac{d}{d\eta}\left(f f'' - f'^2\right), \quad \nu \nabla^2 \omega = -\nu \frac{\alpha x}{\delta_0^3} \frac{d}{d\eta} f'''.$$

Equating these two terms and integrating then yields

$$f''' + f f'' - f'^2 = c = -1, \tag{3.14}$$

where the constant of integration, c, is fixed by the requirement that $f'(\infty) = 1$. (This may also be obtained directly from Prandtl's boundary-layer equation, (2.37), as discussed in Exercise 3.1.) Equation (3.14) is readily integrated numerically, subject to the boundary conditions in (3.12). This reveals that $f'(\eta)$ grows monotonically from 0 at the wall to 1 at large η, with $f'(2) = 0.973$. We conclude that the boundary-layer thickness is independent of x and of the order of $\delta \approx 2\delta_0$.

3.2 The Diffusion of Vorticity from a Moving Surface

3.2.1 The Impulsively Started Plate: Stokes' First Problem

We now discuss a problem, first analysed by Stokes (1851), which illustrates the generation of vorticity by a moving solid surface. Consider a plate of infinite area which is immersed in a still fluid and sits on the plane $y = 0$. At time $t = 0$ it (somehow) acquires, and then maintains, a constant velocity, V, in its own plane and we wish to find the subsequent motion of the fluid.

Before seeking an exact solution for the velocity $\mathbf{u}(y, t)$, let us build up a physical picture of what happens for $t > 0$. The no-slip condition requires that the fluid adjacent to the plate adheres to it, while fluid remote from the plate remains stationary. Evidently, moving the

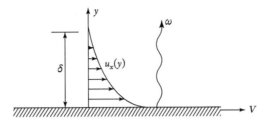

Figure 3.4 The diffusion of vorticity from a moving plate.

plate creates a gradient in velocity, and hence a layer of vorticity, adjacent to the surface $y = 0$. The plate has become a source of vorticity, which must then diffuse into the fluid, as shown in Figure 3.4. Now u_x is not a function of x and so continuity requires $\partial u_y / \partial y = 0$. Since $u_y = 0$ at infinity, we conclude that $u_y = 0$ everywhere and the vorticity is simply $\omega(y,t) = -\partial u_x / \partial y$. Moreover, since ω is not a function of x and $u_y = 0$, we have $\mathbf{u} \cdot \nabla \omega = 0$. The two-dimensional vorticity equation, (2.48), now simplifies to the diffusion equation

$$\frac{\partial \omega}{\partial t} = \nu \frac{\partial^2 \omega}{\partial y^2}, \tag{3.15}$$

which describes the diffusion of vorticity away from the plate and into the fluid.

This situation is reminiscent of the diffusion of heat into a stagnant fluid from a stationary flat plate whose surface temperature is suddenly raised from $T = T_0$, the initial temperature of the fluid, to $T = T_0 + \Delta T$, where ΔT is a constant. In this thermal analogue we must solve exactly the same diffusion equation,

$$\frac{\partial T}{\partial t} = \alpha \frac{\partial^2 T}{\partial y^2}, \tag{3.16}$$

but subject to $T = T_0 + \Delta T$ at $y = 0$ and an initial condition of $T = T_0$ in the fluid. The only differences between the two problems are that we have replaced a kinematic viscosity by a thermal diffusivity, α, and the nature of the boundary condition at $y = 0$ is a little different. Since the spreading of heat by thermal conduction is more intuitive than the diffusion of vorticity, let us spend a few moments thinking about this thermal analogue. Our brief detour into heat transfer will also serve to illustrate the power of dimensional analysis.

This sort of thermal diffusion problem can be solved by looking for a *self-similar solution* of the form

$$(T - T_0)/\Delta T = f\left(y/\sqrt{\alpha t}\right). \tag{3.17}$$

In fact, dimensional analysis tells us that we are not just *free* to look for solutions of the form (3.17), we are *obliged* to look for solutions of the form (3.17). The argument goes as follows. Let $\ell(t)$ be the characteristic distance the heat has diffused away from the plate in a time t, which is known as the *diffusion length*. Since there is no geometric length scale in this problem, the only parameters on which ℓ can depend are α and t: $\ell = \ell(\alpha, t)$. Moreover, α has the dimensions of $\text{m}^2\,\text{s}^{-1}$, and so the three parameters ℓ, α, and t contain between them only two dimensions: length and time. The Buckingham Pi theorem then tells us that the number of independent dimensionless groups we can form from ℓ, α, and t is

number of groups (G) = number of parameters (P) − number of dimensions (D) = 1.

By inspection, this dimensionless group can be written as $\Pi = \ell/\sqrt{\alpha t}$. Now, dimensional analysis also tells us that one dimensionless group can depend only on another dimensionless group, but in this problem there is no other dimensionless group on which $\ell/\sqrt{\alpha t}$ can depend. (The linearity of (3.16) rules out the possibility of $\ell/\sqrt{\alpha t}$ depending on $\Delta T/T_0$, say.) So we conclude that $\ell/\sqrt{\alpha t}$ is simply a constant in this problem, and multiplying out we obtain a scaling law for ℓ, i.e. $\ell \sim \sqrt{\alpha t}$. That is to say, the distance the heat diffuses from the plate grows with time according to $\ell \sim \sqrt{\alpha t}$.

Next, we ask what the dimensionless temperature field, $(T - T_0)/\Delta T$, can depend on. It is clearly a function of y, but because $(T - T_0)/\Delta T$ is dimensionless, it can only depend on a dimensionless form of y. So we are obliged to look for a length scale with which to normalize y, and the only length available to us is ℓ, there being no geometric length scale in the problem. We conclude, therefore, that the solution for $(T - T_0)/\Delta T$ *must* take the self-similar form

$$(T - T_0)/\Delta T = f(y/\ell), \quad \ell = \sqrt{\alpha t}.$$

Note that time does not appear explicitly in this solution, but rather implicitly through ℓ, which is the essence of a self-similar solution. (By definition, a self-similar solution looks the same at all times when position is appropriately normalized.) Substituting (3.17) into the thermal diffusion equation (3.16) then yields the precise form of the self-similar solution, which happens to be an error function.

Now, what is true for the diffusion of heat must also be true for the diffusion of vorticity from our moving plate. That is, we can deploy precisely the same dimensional arguments to the flow in Figure 3.4. It is necessary only to swap ν for α, and dimensionless vorticity, $\omega\ell/V$, for $(T - T_0)/\Delta T$. It follows that the solution of our vorticity equation *must* take the form

$$\omega = \frac{V}{\ell} f(y/\ell), \; \ell = \sqrt{2\nu t}, \tag{3.18}$$

where the factor of 2 in the definition of ℓ is included to make the algebra cleaner. Substituting (3.18) into the diffusion equation, (3.15), and integrating once, then yields

$$f'(\eta) + \eta f(\eta) = 0, \; \eta = y/\ell, \tag{3.19}$$

which may be integrated a second time to give

$$\omega = c\frac{V}{\ell} \exp\left[-\eta^2/2\right]. \tag{3.20}$$

To find the constant of integration, c, we need to integrate (3.20) to give the velocity distribution. Setting $u_x(y=0)$ to V then yields $c^2 = 2/\pi$, and so the vorticity distribution is

$$\omega = \frac{V}{\sqrt{\pi \nu t}} \exp\left[-y^2/(4\nu t)\right]. \qquad (3.21)$$

Evidently, the layer of vorticity adjacent to the plate, which is created by the no-slip condition, diffuses into the interior of the fluid in exactly the same way as heat diffuses in from a heated surface. This problem not only illustrates how a solid surface can act as a source of vorticity, it also highlights the power of dimensional analysis.

3.2.2 The Oscillating Plate: Stokes' Second Problem

The case in which the plate oscillates, rather than moves impulsively, was also analysed by Stokes in 1851. This kind of oscillatory diffusion is common in nature, an obvious thermal example being the diurnal fluctuation in air temperature above the ground. If ϖ is the frequency of the daily temperature cycle, and α the thermal diffusivity of the ground, then the estimate $\ell \sim \sqrt{\alpha t}$ for the diffusion length suggests that the fluctuations in air temperature will diffuse only a distance $\delta \sim \sqrt{\alpha/\varpi}$ into the ground before the fluctuation in surface air temperature reverses sign. A wave of reverse temperature fluctuation then diffuses downward, partially eradicating the previous half cycle. The net result is that the fluctuations in air temperature only ever penetrate a distance of order $\delta \sim \sqrt{\alpha/\varpi}$ below the surface. Another common example of oscillatory diffusion arises in electrical engineering, where an oscillating magnetic field will penetrate only a distance $\delta \sim \sqrt{\alpha_m/\varpi}$ into an electrically conducting solid, α_m being the so-called magnetic diffusivity of the conductor. Electrical engineers call this the skin effect, and δ the skin depth.

All of this suggests that a plate immersed in a viscous fluid, and oscillating at a frequency ϖ, will produce an oscillatory vorticity field adjacent to the plate which is restricted to a layer of thickness $\delta \sim \sqrt{\nu/\varpi}$. We shall now confirm that this is indeed the case. Suppose that, as before, the plate sits on the plane $y = 0$ and is adjacent to a fluid of semi-infinite extent. It oscillates in its own plane with a velocity of $V\cos(\varpi t)$, as shown in Figure 3.5 and we consider the long-term response, ignoring any initial transient

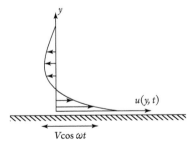

Figure 3.5 A plate oscillates with velocity $V\cos(\varpi t)$ adjacent to a fluid.

As in the case of an impulsively started plate, we have $u_y = 0$ while u_x and ω_z are functions of y and t only. This time, however, because of the boundary condition, it is easier to solve for u_x, rather than ω_z, whose governing equation is

$$\frac{\partial u_x}{\partial t} = \nu \nabla^2 u_x, \quad u_x(y = 0) = V \cos \varpi t. \tag{3.22}$$

Given the oscillatory boundary condition, (3.22) is most readily solved by looking for solutions of the form $u_x(y, t) = \Re\{\hat{u}_x(y) \exp(j \varpi t)\}$, where $j = \sqrt{-1}$, $\hat{u}_x(y)$ is complex, and $\Re$ indicates the real part of what follows. Clearly, $\hat{u}_x''(y) = (j \varpi / \nu) \hat{u}_x$, which we rewrite as

$$\hat{u}_x''(y) = \frac{(1 + j)^2}{\delta^2} \hat{u}_x, \quad \delta = \sqrt{2\nu / \varpi}. \tag{3.23}$$

This then integrates to give

$$u_x(y, t) = V \exp\left[-y/\delta\right] \cos\left[\varpi t - y/\delta\right]. \tag{3.24}$$

Note that, as anticipated, the oscillatory motion is more or less restricted to a layer of thickness $\delta \sim \sqrt{\nu / \varpi}$.

3.3 The Navier–Stokes Equation in Cylindrical Polar Coordinates

3.3.1 Moving from Cartesian to Cylindrical Polar Coordinates

So far we have restricted ourselves to problems in Cartesian coordinates. Of course, there are many flows which have axial symmetry and for which cylindrical polar coordinates, (r, θ, z), are more appropriate. Let us now consider such flows, starting with a derivation of the Navier–Stokes equation in cylindrical polar coordinates.

The key to switching to polar coordinates is to recall that, unlike in Cartesians, the unit vectors $(\hat{\mathbf{e}}_r, \hat{\mathbf{e}}_\theta, \hat{\mathbf{e}}_z)$ are not all constant. In particular,

$$\frac{\partial \hat{\mathbf{e}}_r}{\partial \theta} = \hat{\mathbf{e}}_\theta, \qquad \frac{\partial \hat{\mathbf{e}}_\theta}{\partial \theta} = -\hat{\mathbf{e}}_r, \tag{3.25}$$

from which

$$\frac{D\hat{\mathbf{e}}_r}{Dt} = \frac{u_\theta}{r} \hat{\mathbf{e}}_\theta, \qquad \frac{D\hat{\mathbf{e}}_\theta}{Dt} = -\frac{u_\theta}{r} \hat{\mathbf{e}}_r. \tag{3.26}$$

Moreover, the convective derivative operates on products in the same way as a conventional derivative, and so we have

$$\frac{D\mathbf{u}}{Dt} = \frac{Du_r}{Dt}\hat{\mathbf{e}}_r + u_r\frac{D\hat{\mathbf{e}}_r}{Dt} + \frac{Du_\theta}{Dt}\hat{\mathbf{e}}_\theta + u_\theta\frac{D\hat{\mathbf{e}}_\theta}{Dt} + \frac{Du_z}{Dt}\hat{\mathbf{e}}_z,$$

which combines with (3.26) to give

$$\frac{D\mathbf{u}}{Dt} = \frac{Du_r}{Dt}\hat{\mathbf{e}}_r + \frac{u_r u_\theta}{r}\hat{\mathbf{e}}_\theta + \frac{Du_\theta}{Dt}\hat{\mathbf{e}}_\theta - \frac{u_\theta^2}{r}\hat{\mathbf{e}}_r + \frac{Du_z}{Dt}\hat{\mathbf{e}}_z. \tag{3.27}$$

Equation (3.27) provides us with the left-hand side of the Navier–Stokes equation. On the right we have the Laplacian of $\mathbf{u}$, which is most simply evaluated in polar coordinates by noting that, since $\mathbf{u}$ is solenoidal, $\nabla^2\mathbf{u} = -\nabla \times \nabla \times \mathbf{u}$. We now recall that

$$\nabla \times \mathbf{u} = \left(\frac{1}{r}\frac{\partial u_z}{\partial \theta} - \frac{\partial u_\theta}{\partial z}\right)\hat{\mathbf{e}}_r + \left(\frac{\partial u_r}{\partial z} - \frac{\partial u_z}{\partial r}\right)\hat{\mathbf{e}}_\theta + \left(\frac{1}{r}\frac{\partial}{\partial r}(ru_\theta) - \frac{1}{r}\frac{\partial u_r}{\partial \theta}\right)\hat{\mathbf{e}}_z,$$

and take the curl again to give

$$\nabla^2\mathbf{u} = \left(\nabla^2 u_r - \frac{u_r}{r^2} - \frac{2}{r^2}\frac{\partial u_\theta}{\partial \theta}\right)\hat{\mathbf{e}}_r + \left(\nabla^2 u_\theta - \frac{u_\theta}{r^2} + \frac{2}{r^2}\frac{\partial u_r}{\partial \theta}\right)\hat{\mathbf{e}}_\theta + (\nabla^2 u_z)\hat{\mathbf{e}}_z.$$

The Navier–Stokes equation therefore has the following components in cylindrical polar coordinates:

$$\frac{\partial u_r}{\partial t} + \left[(\mathbf{u}\cdot\nabla)u_r - \frac{u_\theta^2}{r}\right] = -\frac{1}{\rho}\frac{\partial p}{\partial r} + \nu\left[\nabla^2 u_r - \frac{u_r}{r^2} - \frac{2}{r^2}\frac{\partial u_\theta}{\partial \theta}\right]$$

$$\frac{\partial u_\theta}{\partial t} + \left[(\mathbf{u}\cdot\nabla)u_\theta + \frac{u_r u_\theta}{r}\right] = -\frac{1}{\rho r}\frac{\partial p}{\partial \theta} + \nu\left[\nabla^2 u_\theta - \frac{u_\theta}{r^2} + \frac{2}{r^2}\frac{\partial u_r}{\partial \theta}\right]$$

$$\frac{\partial u_z}{\partial t} + (\mathbf{u}\cdot\nabla)u_z = -\frac{1}{\rho}\frac{\partial p}{\partial z} + \nu\nabla^2 u_z$$

When the motion is axisymmetric, these simplify considerably. Perhaps the most convenient form is then

$$\frac{Du_r}{Dt} - \frac{u_\theta^2}{r} = -\frac{1}{\rho}\frac{\partial p}{\partial r} + \nu\frac{1}{r}\nabla_*^2(ru_r), \tag{3.28}$$

$$\frac{Dru_\theta}{Dt} = \nu\nabla_*^2(ru_\theta), \tag{3.29}$$

$$\frac{Du_z}{Dt} = -\frac{1}{\rho}\frac{\partial p}{\partial z} + \nu\nabla^2 u_z, \tag{3.30}$$

where

$$\frac{D}{Dt} = \frac{\partial}{\partial t} + u_r\frac{\partial}{\partial r} + u_z\frac{\partial}{\partial z},$$

and the operator,

$$\nabla_*^2 f = r \frac{\partial}{\partial r} \frac{1}{r} \frac{\partial f}{\partial r} + \frac{\partial^2 f}{\partial z^2} = r^2 \nabla \cdot \left(\frac{1}{r^2} \nabla f \right) = \nabla \cdot \left(r^2 \nabla \left(\frac{f}{r^2} \right) \right), \qquad (3.31)$$

is sometimes called the Stokes operator. In terms of the rate-of-strain tensor, S_{ij}, these axisymmetric equations are

$$\frac{D u_r}{Dt} - \frac{u_\theta^2}{r} = -\frac{1}{\rho} \frac{\partial p}{\partial r} + 2\nu \left[\frac{1}{r} \frac{\partial}{\partial r} (r S_{rr}) + \frac{\partial S_{rz}}{\partial z} - \frac{S_{\theta\theta}}{r} \right], \qquad (3.32)$$

$$\frac{D r u_\theta}{Dt} = 2\nu \left[\frac{1}{r} \frac{\partial}{\partial r} (r^2 S_{\theta r}) + \frac{\partial}{\partial z} (r S_{\theta z}) \right], \qquad (3.33)$$

$$\frac{D u_z}{Dt} = -\frac{1}{\rho} \frac{\partial p}{\partial z} + 2\nu \left[\frac{1}{r} \frac{\partial}{\partial r} (r S_{zr}) + \frac{\partial S_{zz}}{\partial z} \right], \qquad (3.34)$$

where

$$S_{rr} = \frac{\partial u_r}{\partial r}, \qquad S_{\theta\theta} = \frac{u_r}{r}, \qquad S_{zz} = \frac{\partial u_z}{\partial z},$$

$$S_{r\theta} = \frac{r}{2} \frac{\partial}{\partial r} \left(\frac{u_\theta}{r} \right), \qquad S_{\theta z} = \frac{1}{2} \frac{\partial u_\theta}{\partial z}, \qquad S_{zr} = \frac{1}{2} \left[\frac{\partial u_r}{\partial z} + \frac{\partial u_z}{\partial r} \right].$$

Finally, we note that, in an axisymmetric flow, **u** may be decomposed into its *azimuthal*, u_θ, and *poloidal*, $\mathbf{u}_p = (u_r, 0, u_z)$, components, which are individually solenoidal. Moreover, recall from §1.2.3 that, because $\nabla \cdot \mathbf{u}_p = 0$ in an axisymmetric flow, the poloidal velocity may be expressed in terms of the *Stokes streamfunction*, $\Psi(r, z)$, defined by

$$\mathbf{u}_p = \nabla \times \mathbf{A}_\theta = \nabla \times \left[(\Psi(r, z)/r)\, \hat{\mathbf{e}}_\theta \right], \qquad (3.35)$$

or in component form,

$$\mathbf{u}_p = \left(-\frac{1}{r} \frac{\partial \Psi}{\partial z}, \, 0, \, \frac{1}{r} \frac{\partial \Psi}{\partial r} \right). \qquad (3.36)$$

Evidently, $\mathbf{u}_p \cdot \nabla \Psi = 0$, and so the projection of the streamlines in the $r - z$ plane are lines of constant Ψ.

3.3.2 Hagen–Poiseuille Flow in a Pipe

Perhaps the simplest axisymmetric motion is steady, laminar flow in a circular pipe of radius R, as shown in Figure 3.6. Here, $\mathbf{u}(r) = (0, 0, u_z)$ and so the inertial term in the Navier–Stokes equation is zero: $\mathbf{u} \cdot \nabla \mathbf{u} = 0$. It then follows from (3.28) that there is no pressure gradient in the radial direction.

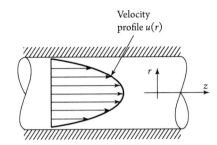

Figure 3.6 Steady laminar flow in a pipe.

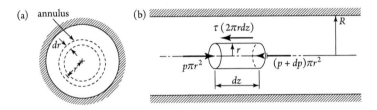

Figure 3.7 (a) A small annulus of fluid, of volume $2\pi r dr dz$. (b) Force balance for a cylindrical element of fluid in a laminar pipe flow.

In the absence of inertial forces, the Navier–Stokes equation reduces to the simple linear equation

$$\frac{dp}{dz} = \rho \nu \frac{1}{r}\frac{d}{dr}r\frac{du_z}{dr} = \frac{1}{r}\frac{d}{dr}(r\tau_{zr}).\qquad(3.37)$$

It is readily confirmed that this relationship between p and τ_{zr} represents a force balance for a small annulus of fluid of volume $2\pi r dr dz$ (Figure 3.7(a)), in which the difference in the pressure forces between the front and back of the element, $(dp/dz)\,2\pi r dr dz$, is balanced by a difference in the shear force, $\tau_{zr}(2\pi r dz)$, at r and $r + dr$. Integrating (3.37) once gives

$$\frac{dp}{dz}\frac{r^2}{2} = \rho \nu r \frac{du_z}{dr} = r\tau_{zr},$$

which again can be thought of as a force balance, this time of a cylindrical element of fluid of radius r and volume $\pi r^2 dz$, as shown in Figure 3.7(b). Integrating a second time, noting that the axial pressure gradient is negative, and demanding that the velocity is zero at $r = R$, yields the well-known parabolic profile

$$u_z = \frac{1}{4\rho\nu}\left|\frac{dp}{dz}\right|(R^2 - r^2).\qquad(3.38)$$

This is known as *Hagen–Poiseuille flow* after the German engineer Hagen and the French physicist Poiseuille who first investigated it experimentally, the latter being motivated by an interest in blood flow. The mean speed in the pipe, defined as $V = \dot{m}/\rho\pi R^2$, is one half of the centreline velocity,

$$V = \frac{R^2}{8\rho\nu}\left|\frac{dp}{dz}\right|.$$

From a practical point of view, the primary interest here lies in the relationship between the mean speed in the pipe, V, and the resulting pressure drop and mechanical energy loss. To that end, we note that the global force balance for the fluid in a section of the pipe of length L requires

$$\Delta p(\pi R^2) = (2\pi RL)\tau_w,$$

where Δp is the drop in pressure over the length, L, and τ_w is the magnitude of the viscous shear stress at the pipe wall. It is conventional to rewrite this using the dimensionless *skin-friction coefficient*, c_f, defined through $\tau_w = c_f\left(\frac{1}{2}\rho V^2\right)$. Noting that our expression for the mean velocity may be rearranged to give $\Delta p = 8\rho\nu VL/R^2$, this tells us that, for fully developed laminar flow,

$$\Delta p = \frac{c_f L}{R}\rho V^2, \quad c_f = \frac{16}{2RV/\nu} = \frac{16}{\mathrm{Re}}. \tag{3.39}$$

Equation (3.39) turns out to be accurate up to Reynolds numbers of $\mathrm{Re} \sim 2000\rightarrow2500$, above which the flow is typically turbulent. For turbulent flow, c_f is defined through the expression $\bar{\tau}_w = c_f\left(\frac{1}{2}\rho V^2\right)$, where $\bar{\tau}_w$ is the *time-averaged* wall stress. The first equation in (3.39) then still applies. It is an empirical observation that the skin-friction coefficient for fully developed turbulent flow in a *smooth* pipe, say for $4000 < \mathrm{Re} < 10^5$ in commercial piping, is given by the so-called Blasius law, $c_f \approx 0.079\,\mathrm{Re}^{-1/4}$. For pipes whose walls are *hydraulically rough*, with an average roughness height of $\hat{k}$, the turbulent friction coefficient, c_f, is observed to be a function of both Re and $\hat{k}/R$, $c_f = c_f(\mathrm{Re}, \hat{k}/R)$, as indicted in Figure 3.8. However, for sufficiently small viscosity, say $\hat{k}V/\nu > 1000$, the skin-friction coefficient in a turbulent pipe flow is found to be more or less independent of viscosity and a function of $\hat{k}/R$ only, $c_f = c_f(\hat{k}/R)$, with c_f being a monotonically increasing function of $\hat{k}/R$. This regime lies above the dashed curve in Figure 3.8.

To obtain an expression for the rate of dissipation of mechanical energy we return to (2.28) which, for steady flow in a pipe, simplifies to

$$-\oint_S p\mathbf{u}\cdot d\mathbf{S} = \int_V \rho\varepsilon dV = \text{rate of dissipation of energy in volume } V.$$

For the particular case of laminar flow in a pipe of length L, this yields

$$\text{rate of dissipation of energy} = \dot{m}\Delta p/\rho = 8\pi\rho\nu LV^2.$$

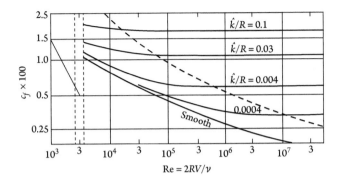

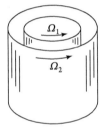

Figure 3.9 Rotating Couette flow

More generally, when written in terms of c_f, this becomes

$$\text{rate of dissipation of energy} = \dot{m}\Delta p/\rho = \pi c_f LR\rho V^3,$$

which may be applied also to turbulent pipe flow, provided, of course, that an appropriate estimate for c_f is used.

3.3.3 Rotating Couette Flow

Let us now consider steady laminar flow between two concentric cylinders of radius R_1 and R_2, which rotate with angular velocities Ω_1 and Ω_2, as shown in Figure 3.9. The cylinders are taken to be infinitely long and the velocity field is simply $\mathbf{u}(r) = (0, u_\theta, 0)$. This is often called *rotating Couette flow*.

Once again, the inertial contribution to the Navier–Stokes equation is zero, except for the centrifugal term, u_θ^2/r, which is balanced by a radial pressure gradient. The azimuthal component of the Navier–Stokes equation now reduces to

$$\rho \nu r \frac{d}{dr}\frac{1}{r}\frac{d}{dr}(r u_\theta) = \frac{1}{r}\frac{d}{dr}(r^2 \tau_{\theta r}) = 0, \tag{3.40}$$

which represents the azimuthal torque balance for a small annular element of fluid, of volume $2\pi r dr dz$. Equation (3.40) integrates to give $u_\theta = ar + b/r$, where a and b are constants of integration. Application of the no-slip condition now yields

$$u_\theta = \frac{\Omega_2 R_2^2 - \Omega_1 R_1^2}{R_2^2 - R_1^2}r - \frac{(\Omega_2 - \Omega_1)R_1^2 R_2^2}{R_2^2 - R_1^2}\frac{1}{r}. \tag{3.41}$$

Note that the viscous torque (per unit axial length) exerted between successive annuli of different radii is

$$T_z = 2\pi r (r\tau_{\theta r}) = 2\pi r \left[\rho\nu r^2 \frac{d}{dr}\left(\frac{u_\theta}{r}\right)\right] = 4\pi\rho\nu\frac{(\Omega_2 - \Omega_1)R_1^2 R_2^2}{R_2^2 - R_1^2}, \tag{3.42}$$

which is independent of radius, as expected.

If $R_1 = 0$, so we remove the inner cylinder, then (3.41) and (3.42) yield $u_\theta = \Omega_2 r$ and $T_z = 0$. Conversely, if we take $\Omega_2 = 0$ and let $R_2 \to \infty$, which removes the outer cylinder, we obtain the irrotational velocity distribution $u_\theta = R_1^2\Omega_1/r$ and a finite torque of $T_z = -4\pi\rho\nu\Omega_1 R_1^2$. The reason why there is a finite torque acting on the inner cylinder in the steady state is related to the fact that $u_\theta = R_1^2\Omega_1/r$ integrates to give an infinite angular momentum, and so this torque is required to continually propagate angular momentum out to infinity.

It turns out that rotating Couette flow is readily destabilized when $\Omega_1 > \Omega_2$, resulting in so-called *Taylor–Couette flow*. We shall study this instability in Chapter 11.

3.3.4 The Diffusion of a Long, Thin, Cylindrical Vortex

We now return to the theme of the diffusion of vorticity in a two-dimensional flow. Consider the evolution of a long, thin, axisymmetric tube of vorticity centred on the z-axis. It has the vorticity distribution $\boldsymbol{\omega}(r,t) = (0, 0, \omega_z)$ and the corresponding velocity field is $\mathbf{u}(r,t) = (0, u_\theta, 0)$. Thus, the vortex lines are parallel to the z-axis while the streamlines are circles centred on the same axis, as shown in Figure 3.10.

As in rotating Couette flow, the symmetry of the motion means that the inertial contribution to the Navier–Stokes equation is zero, except for the centrifugal term, u_θ^2/r, which is balanced by a radial pressure gradient. Equation (3.29) now yields

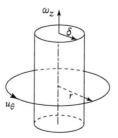

Figure 3.10 A vortex tube centred on the z-axis.

$$\frac{\partial \Gamma}{\partial t} = \nu \nabla_*^2 \Gamma = \nu r \frac{\partial}{\partial r} \frac{1}{r} \frac{\partial \Gamma}{\partial r} \tag{3.43}$$

for the angular momentum density $\Gamma = r u_\theta$. Note that Γ is related to the axial vorticity by

$$\omega_z = \frac{1}{r} \frac{\partial \Gamma}{\partial r}. \tag{3.44}$$

Equation (3.43) could be solved directly for Γ, but it is more instructive to take its radial derivative and substitute for Γ using (3.44). This yields the familiar diffusion equation for vorticity,

$$\frac{\partial \omega_z}{\partial t} = \nu \nabla^2 \omega_z = \nu \frac{1}{r} \frac{\partial}{\partial r} r \frac{\partial \omega_z}{\partial r}, \tag{3.45}$$

which we could have written down using (2.48) without any knowledge of (3.43), on the grounds that the advection of vorticity vanishes along with the inertial forces in the Navier–Stokes equation. It is natural to look for a self-similar solution of (3.45) in which r is normalized by the diffusion length, say

$$\omega_z = \frac{\Phi_0}{\pi \delta^2} f(r/\delta), \quad \delta = \sqrt{\nu t}, \tag{3.46}$$

where Φ_0 is the total flux of vorticity along the vortex tube,

$$\Phi_0 = \int \omega_z 2\pi r dr. \tag{3.47}$$

Notice that Φ_0 is a constant, as diffusion merely redistributes vorticity, without creating or destroying it. Substituting (3.46) into the diffusion equation, (3.45), yields

$$2 \frac{d}{d\eta} \left(\eta f'(\eta) \right) + \eta^2 f'(\eta) + 2\eta f = 0, \quad \eta = r/\delta, \tag{3.48}$$

which may be integrated once to give $2f'(\eta) + \eta f(\eta) = 0$. Integrating a second time gives a Gaussian distribution for the vorticity,

$$\omega_z = c \frac{\Phi_0}{\pi \delta^2} \exp\left[-\eta^2/4\right], \tag{3.49}$$

where the constant of integration, c, is determined by the normalization condition (3.47), which requires $c = \frac{1}{4}$. We conclude that the vorticity diffuses outward from the z-axis governed by

$$\omega_z = \frac{\Phi_0}{4\pi \nu t} \exp\left[-r^2/(4\nu t)\right], \tag{3.50}$$

which might be compared with (3.21) for the diffusion of vorticity from an impulsively started plate. Note that the initial distribution of ω_z, which can be found by taking $t \to 0$ in (3.50), corresponds to a two-dimensional delta function.

Finally, combining (3.44) with (3.50) gives us the angular momentum density Γ. Since $\Gamma(r = 0) = 0$, we find

$$\Gamma = \frac{\Phi_0}{2\pi} \left[1 - \exp\left(-\eta^2/4\right)\right], \tag{3.51}$$

and the azimuthal velocity is evidently

$$u_\theta = \frac{\Phi_0}{2\pi r} \left[1 - \exp\left(-r^2/4\nu t\right)\right]. \tag{3.52}$$

It is instructive to consider the limiting cases of $r \gg \delta$ and $r \ll \delta$. For large radii we have

$$u_\theta = \frac{\Phi_0}{2\pi r}, \tag{3.53}$$

which is, of course, the irrotational flow associated with, and external to, a thin line vortex sitting on the z-axis and of strength Φ_0. The $1/r$ fall-off in velocity for such a line vortex follows directly from Stokes' theorem applied to the surface enclosed by a circular streamline of radius r,

$$\oint \mathbf{u} \cdot d\mathbf{r} = 2\pi r u_\theta = \int \omega_z dA = \Phi_0. \tag{3.54}$$

For a small radius, on the other hand, (3.52) simplifies to

$$u_\theta = \frac{\Phi_0 r}{8\pi \nu t},$$

which represents rigid-body rotation whose angular velocity decays with time as $1/t$.

3.3.5 A Thin Film on a Spinning Disc

Let us now turn to a problem somewhat reminiscent of the viscous, free-surface flow shown in Figure 3.2. A thin film of fluid covers the surface of a disc which rotates with angular velocity Ω and we adopt cylindrical polar coordinates, (r, θ, z), centred on the disc. There is negligible air resistance and so the fluid acquires the rotational velocity of the disc, Ωr, and since there is no radial pressure gradient to balance the centrifugal force, $\rho \Omega^2 r$, the fluid is centrifuged radially outward, as shown in Figure 3.11. In effect, the centrifugal force plays the role of gravity in the inclined-plane flow of Figure 3.2. The magnitude of the radial velocity is now set by the requirement that the force $\rho \Omega^2 r$, integrated through the depth of the fluid, is balanced by a radial shear stress, τ_{rz}, exerted on the fluid by the surface of the disc.

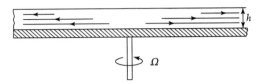

Figure 3.11 A film of fluid is centrifuged radially outward on the surface of a rotating disc.

Of course, the outward radial flow requires that the film thickness decreases with time, but we shall focus on the highly viscous, quasi-steady regime in which the fluid drains only slowly and so time derivatives may be neglected, $u_z \approx 0$, and $u_r \ll \Omega r$. Moreover, in order to simplify the analysis, we shall ignore surface tension and gravity. In short, we seek a quasi-steady, axisymmetric flow in which $\mathbf{u} = (u_r(r, z), \Omega r, 0)$ and $p = 0$. Finally, we shall assume that there exists a regime in which the film thickness, h, is independent of radius, an assumption which we shall check retrospectively.

Since the centrifugal force, $\rho \Omega^2 r h$, is balanced by the surface stress, τ_{rz}, and h is assumed to be independent of r, we might expect both τ_{rz} and u_r to increase linearly with radius, and we shall see shortly that this is indeed the case. The radial contribution to the Stokes operator in (3.28) is then exactly zero, and given that $u_r \ll \Omega r$ and $p = 0$, the equation of motion reduces to

$$\rho \nu \frac{\partial^2 u_r}{\partial z^2} = \frac{\partial \tau_{rz}}{\partial z} = -\rho \Omega^2 r. \tag{3.55}$$

Now, $u_r = 0$ at the surface of the disc, $z = 0$, while $\tau_{rz} = 0$ at the surface of the film, $z = h$. It follows that (3.55) integrates to give

$$u_r = \frac{\Omega^2 r}{2\nu} \left(2hz - z^2\right), \tag{3.56}$$

and so the mass flow rate through a cylindrical surface of radius r is

$$\dot{m} = 2\pi \frac{\rho \Omega^2 r^2 h^3}{3\nu}. \tag{3.57}$$

Conservation of mass applied to an annulus of volume $2\pi r h dr$ now requires that

$$\frac{\partial}{\partial t}(2\pi r h) = -\frac{\partial}{\partial r}\left(2\pi \frac{\Omega^2 r^2 h^3}{3\nu}\right). \tag{3.58}$$

We conclude that, if at some initial instant h is independent of r, it remains so thereafter, which justifies our assumption of a uniform h. In such cases the film drains away at the rate

$$\frac{dh}{dt} = -\frac{2\Omega^2 h^3}{3\nu}, \tag{3.59}$$

from which

$$h = \left[\frac{1}{h_0^2} + \frac{4\Omega^2 t}{3\nu} \right]^{-1/2}. \tag{3.60}$$

This flow is used in certain commercial operations to deposit a thin layer of fluid of uniform thickness on a substrate. The process may be considered as quasi-steady, with $u_r \ll \Omega r$, provided that the Reynolds number, $\Omega h^2/\nu$, is small, as discussed in Exercise 3.5.

3.3.6 The Azimuthal–poloidal Decomposition of Axisymmetric Flows

We conclude our discussion of axisymmetric flow with a brief description of an important decomposition, first introduced in §3.3.1. An axisymmetric flow written in terms of cylindrical polar coordinates is most naturally decomposed into its azimuthal, $\mathbf{u}_\theta$, and poloidal, $\mathbf{u}_p = (u_r, 0, u_z)$, components, each of which are solenoidal. The vorticity may be likewise decomposed into two fields, $\boldsymbol{\omega}_\theta$ and $\boldsymbol{\omega}_p$, which are also individually solenoidal. When this is done a simple, but important, observation is that the curl of a poloidal field is azimuthal while the curl of an azimuthal field is poloidal. It follows that $\boldsymbol{\omega}_\theta = \nabla \times \mathbf{u}_p$ and $\boldsymbol{\omega}_p = \nabla \times \mathbf{u}_\theta$, which is easily confirmed by direct evaluation. For example, introducing the angular momentum density $\Gamma = r u_\theta$, we have

$$\boldsymbol{\omega}_p = \left(-\frac{1}{r}\frac{\partial \Gamma}{\partial z}, 0, \frac{1}{r}\frac{\partial \Gamma}{\partial r} \right), \tag{3.61}$$

and a comparison with (3.36) tells us that Γ is the Stokes streamfunction for the poloidal vorticity field, $\boldsymbol{\omega}_p$. Moreover, combining the azimuthal component of $\boldsymbol{\omega} = \nabla \times \mathbf{u}$ with (3.36) gives us

$$\omega_\theta = \frac{\partial u_r}{\partial z} - \frac{\partial u_z}{\partial r} = -\frac{1}{r}\nabla_*^2 \Psi, \tag{3.62}$$

where ∇_*^2 is the Stokes operator defined by (3.31). Thus, the Stokes streamfunction determines both the poloidal velocity field, through (3.36), and the azimuthal vorticity.

Now, the instantaneous velocity distribution is uniquely determined by the two scalar functions $\Gamma = r u_\theta$ and ω_θ/r, since $\boldsymbol{\omega}_\theta = \nabla \times \mathbf{u}_p$ can always be inverted to give $\mathbf{u}_p$ by solving the Poisson-like equation (3.62). So, when considering axisymmetric flows, it is natural to focus on the two scalar fields, Γ and ω_θ/r. Evolution equations for these fields can be found from the azimuthal components of the Navier–Stokes and vorticity equations. For example, the azimuthal component of the momentum equation, (3.29), is the simple advection–diffusion-like equation

$$\boxed{\frac{D\Gamma}{Dt} = \nu \nabla_*^2 \Gamma}. \tag{3.63}$$

On the other hand, remembering that the curl of a poloidal field is azimuthal, it becomes evident that the azimuthal component of the vorticity equation, (2.44), is

$$\partial \boldsymbol{\omega}_\theta / \partial t = \nabla \times [\mathbf{u}_p \times \boldsymbol{\omega}_\theta] + \nabla \times [\mathbf{u}_\theta \times \boldsymbol{\omega}_p] + \nu \nabla^2 \boldsymbol{\omega}_\theta. \tag{3.64}$$

Moreover, noting that $\mathbf{u}_p$ and $\boldsymbol{\omega}_\theta$ are solenoidal, and using (3.25) to evaluate $\partial \hat{\mathbf{e}}_r / \partial \theta$, gives

$$\nabla \times [\mathbf{u}_p \times \boldsymbol{\omega}_\theta] = \boldsymbol{\omega}_\theta \cdot \nabla \mathbf{u}_p - \mathbf{u}_p \cdot \nabla \boldsymbol{\omega}_\theta = -r \mathbf{u}_p \cdot \nabla (\omega_\theta / r) \, \hat{\mathbf{e}}_\theta. \tag{3.65}$$

Similarly,

$$\nabla \times [\mathbf{u}_\theta \times \boldsymbol{\omega}_p] = \boldsymbol{\omega}_p \cdot \nabla \mathbf{u}_\theta - \mathbf{u}_\theta \cdot \nabla \boldsymbol{\omega}_p = r \boldsymbol{\omega}_p \cdot \nabla (u_\theta / r) \, \hat{\mathbf{e}}_\theta. \tag{3.66}$$

The Laplacian in (3.64), on the other hand, transforms in the same way as the Laplacian in the Navier–Stokes equation, and so (3.29) tells us that

$$\nabla^2 \boldsymbol{\omega}_\theta = r^{-1} \nabla_*^2 (r \omega_\theta) \hat{\mathbf{e}}_\theta. \tag{3.67}$$

Putting all of this together gives us the scalar evolution equation

$$\frac{D}{Dt} \left(\frac{\omega_\theta}{r} \right) = \boldsymbol{\omega}_p \cdot \nabla \left(\frac{u_\theta}{r} \right) + \frac{\nu}{r^2} \nabla_*^2 (r \omega_\theta). \tag{3.68}$$

Finally, substituting for $\boldsymbol{\omega}_p$ using (3.61) yields the most convenient form of the evolution equation for azimuthal vorticity:

$$\boxed{ \frac{D}{Dt} \left(\frac{\omega_\theta}{r} \right) = \frac{\partial}{\partial z} \left(\frac{\Gamma^2}{r^4} \right) + \frac{\nu}{r^2} \nabla_*^2 (r \omega_\theta) }. \tag{3.69}$$

Note that, unlike the evolution equation for Γ, this has a source term in the form of axial gradients in angular momentum. Physically, this represents the spiralling-up of the poloidal vortex lines by an axial gradient in swirl, which sweeps out a component of ω_θ from $\boldsymbol{\omega}_p$. Equations (3.63) and (3.69) are central to many axisymmetric flows, including the dynamics of vortex rings and tropical cyclones, as discussed in §8.7.5 and §10.6, respectively.

For the special case of inviscid flow, our two evolution equations reduce to

$$\boxed{ \frac{D\Gamma}{Dt} = 0, \qquad \frac{D}{Dt} \left(\frac{\omega_\theta}{r} \right) = \frac{\partial}{\partial z} \left(\frac{\Gamma^2}{r^4} \right) }. \tag{3.70}$$

If the flow is also steady then $\mathbf{u} \cdot \nabla \Gamma = 0$ and so Γ is constant on a streamline. In such cases we may write $\Gamma = \Gamma(\Psi)$, since $\mathbf{u}_p \cdot \nabla \Psi = 0$. The azimuthal vorticity equation is then

$$\mathbf{u} \cdot \nabla \left(\frac{\omega_\theta}{r} \right) = \frac{2 \Gamma \Gamma'(\Psi)}{r^4} \frac{\partial \Psi}{\partial z} = -\Gamma \Gamma'(\Psi) \frac{2 u_r}{r^3},$$

which may be rewritten as

$$\mathbf{u} \cdot \nabla \left(\frac{\omega_\theta}{r} - \frac{\Gamma\Gamma'(\Psi)}{r^2} \right) = 0. \tag{3.71}$$

It follows that the azimuthal vorticity in a steady, axisymmetric, inviscid flow satisfies

$$r\omega_\theta = -\nabla_*^2 \Psi = \Gamma\Gamma'(\Psi) + r^2 F(\Psi), \tag{3.72}$$

for some function $F(\Psi)$.

We shall pick up this story again in Chapter 8, where we shall see that F is related to Bernoulli's function, H, which is constant along a streamline in an inviscid flow, $H = H(\Psi)$. In particular, we shall show that $F = -H'(\Psi)$, and hence

$$\nabla_*^2 \Psi = r^2 H'(\Psi) - \Gamma\Gamma'(\Psi). \tag{3.73}$$

It would seem that, if we know the variation of angular momentum and energy across the streamlines, *i.e.* $\Gamma(\Psi)$ and $H(\Psi)$ are prescribed at some upstream location, then (3.73) may be solved for Ψ, and hence for the poloidal velocity, $\mathbf{u}_p$. This provides a systematic way of determining steady, axisymmetric, inviscid flows with swirl, and several examples of its use are given in Batchelor (1967).

○ ○ ○

This concludes our discussion of elementary solutions of the Navier–Stokes equation. Most of the examples we have covered are fairly standard and are also discussed in, say, Chapter 5 of Schlichting (1979), Chapter 2 of Acheson (1990), and Chapter 3 of White (1991).

··

EXERCISES

3.1 *Prandtl's boundary-layer equation applied to stagnation-point flow.* Deduce the stagnation-point equation, (3.14), directly from Prandtl's boundary-layer equation, (2.37), assuming scalings (3.11) and (3.12). (Hint: first find the inviscid pressure gradient $\partial p/\partial x$.)

3.2 *The diffusion of vorticity from a plate which exerts a constant shear stress on the fluid.* The model problem of §3.2.1, in which vorticity diffuses away from an impulsively started plate, is not in practice realizable as the plate has an infinite acceleration at $t = 0$. A more realistic problem to consider is a large flat plate sitting at $y = 0$ which is initially stationary and then subject to a constant force for $t > 0$. The boundary condition on the fluid is then $\omega(y = 0) = |\tau_0|/\rho\nu$, where τ_0 is the constant shear stress exerted by the plate on the fluid for $t > 0$. Suppose the fluid fills the semi-infinite space $y > 0$ and is initially stationary. Use dimensional analysis to show that the solution for the vorticity field must take the form $\rho\nu\omega/|\tau_0| = f(\eta)$, where $\eta = y/2\sqrt{\nu t}$ and f is a function to be determined. Now show that, if erfc is the complimentary error function, then the vorticity is in fact given by

$$\omega = \frac{|\tau_0|}{\rho \nu} \mathrm{erfc} \left(y \big/ 2\sqrt{\nu t} \right).$$

3.3 *The spin-down of a viscous fluid in a long cylinder.* A fluid sits in a long cylindrical container of radius a and is subject to the no-slip boundary condition $\mathbf{u} = 0$ at $r = a$. Its initial velocity field is purely azimuthal and a function of radius only. For $t > 0$ the fluid starts to spin down due to the frictional drag of the outer boundary. If the motion remains axisymmetric and purely azimuthal then the spin-down process is governed by (3.43). Rewrite this equation in terms of u_θ and use separation of the variables to show that one possible solution for u_θ is

$$u_\theta(r,t) = A_n J_1 \left(\delta_n r/a \right) \exp \left(-\delta_n^2 \nu t/a^2 \right),$$

where A_n is a constant, J_1 is the usual Bessel function, and δ_n is the nth zero of J_1. Now use the orthogonality property of Bessel functions to show that the general solution is

$$u_\theta(r,t) = \sum_{n=1}^{\infty} A_n J_1 \left(\delta_n r/a \right) \exp \left(-\delta_n^2 \nu t/a^2 \right),$$

where A_n are the coefficients in a Bessel-series representation of the initial condition.

3.4 *The decay of kinetic energy during spin-down in a long cylinder.* Show that, for the spin-down problem above, the kinetic energy per unit length of the cylinder decays as

$$\frac{dE}{dt} = \frac{d}{dt} \int \frac{u_\theta^2}{2} 2\pi r dr = -\nu \int \left[r \frac{d}{dr} \left(\frac{u_\theta}{r} \right) \right]^2 2\pi r dr$$

$$= -\nu \int \left[\left(\frac{du_\theta}{dr} \right)^2 + \left(\frac{u_\theta}{r} \right)^2 \right] 2\pi r dr.$$

Hence show that

$$\frac{dE}{dt} + \frac{2\nu}{a^2} E \leq 0$$

and find an exponentially decaying upper bound on $E(t)$.

3.5 *The quasi-steady centrifugal spinning of a thin film.* Use (3.56) and (3.59) to estimate $\partial u_r/\partial t$ in the spinning disc problem of §3.3.5. Hence show that the assumption of quasi-steady dynamics requires that the Reynolds number, $\Omega h^2/\nu$, is small. Also show that, in this regime, we have $u_r \ll \Omega r$ and the rate of dissipation of kinetic energy on a disc of radius R is equal to $(\pi/6)\rho \Omega^4 R^4 h^3/\nu$.

3.6 *The scaling law for pressure-driven flow between parallel plates determined by dimensional analysis.* Consider the pressure-driven flow shown in Figure 3.1, in which the top plate is stationary. The volumetric flow rate, Q, is clearly a function of h, ν, and the pressure gradient $\rho^{-1}|dp/dx|$. Show that there are only two dimensionless groups that can be constructed from these parameters, and that two candidates are Q/ν and $\left(h^3/\rho\nu^2\right)|dp/dx|$. Noting that Q is proportional to $|dp/dx|$, deduce the (3.5) to within an unknown pre-factor.

REFERENCES

Acheson, D.J., 1990, *Elementary fluid dynamics*, Oxford University Press.
Batchelor, G.K., 1967, *An introduction to fluid dynamics*, Cambridge University Press.
Schlichting, H., 1979, *Boundary-layer theory*, Seventh Ed., McGraw-Hill.
Stokes, G.G., 1851, On the effect of internal friction of fluids on the motion of pendulums. *Trans. Camb. Phil. Soc.*, **IX**, 8.
White, F.M., 1991, *Viscous fluid flow*. 2nd Ed., McGraw-Hill.

4
· · • · ·

Flows with Negligible Inertia
Stokes Flow, Lubrication Theory, and Thin Films

(a)   (b)

(a) The engineer Osborne Reynolds (1842–1912). (b) The theoretical physicist Arnold Sommerfeld (1868–1951). Both made important contributions to the theory of lubrication.

Flows in which inertia may be neglected are rather rare, but they do occur and so warrant our attention. Roughly speaking, these fall into one of two categories. The most obvious class is where the Reynolds number based on some geometrical length scale is small. One may then neglect inertia in those parts of the flow where the velocity gradients scale on the geometric length scale. This includes cases where the scale of the motion is small, for example swimming sperm or the sedimentation of tiny particles, and where the viscosity is large, say mantle convection or the flow of syrup or tar. The second class of flows are those where the streamlines are almost parallel. Here we can often neglect inertia even when the Reynolds number is large. Consider, for example, fully developed laminar flow in a circular pipe, as shown in Figure 3.6. Here the inertia is exactly zero, even though the Reynolds number may be as large as 10^3. It follows that, if the pipe were *slightly* tapered, the inertial forces may still be smaller than their viscous counterparts, even when Re $\gg$ 1, provided, of course, that the taper is not too excessive. Lubrication theory falls into this second category. Let us start, though, with cases where the Reynolds number is small.

4.1 Motion at Low Reynolds Number: Stokes Flow

4.1.1 The Governing Equations at Low Reynolds Number

If inertia, $D\mathbf{u}/Dt$, is dropped from the Navier–Stokes equation we are left with the linear equation

$$\boxed{\nabla\left(p/\rho\nu\right) = \nabla^2\mathbf{u} = -\nabla\times\boldsymbol{\omega}}\,, \qquad (4.1)$$

along with continuity, $\nabla\cdot\mathbf{u} = 0$. Since $\nabla\cdot(\nabla\times\boldsymbol{\omega}) = 0$ and $\nabla\times(\nabla p) = 0$, the divergence and curl of (4.1) tell us that p and $\boldsymbol{\omega}$ are both harmonic:

$$\boxed{\nabla^2 p = 0, \quad \nabla^2\boldsymbol{\omega} = 0}\,. \qquad (4.2)$$

A flow governed by these equations is called a *Stokes flow*, or a *creeping flow*.

Note that, in this approximation, the advection of vorticity, $D\boldsymbol{\omega}/Dt$, as well as $\boldsymbol{\omega}\cdot\nabla\mathbf{u}$, are dropped from the transport equation for $\boldsymbol{\omega}$ on the grounds that Re $\ll$ 1, i.e.

$$D\boldsymbol{\omega}/Dt = \boldsymbol{\omega}\cdot\nabla\mathbf{u} + \nu\nabla^2\boldsymbol{\omega}$$

is replaced by $\nu\nabla^2\boldsymbol{\omega} = 0$. Crucially, this not only ignores $\mathbf{u}\cdot\nabla\boldsymbol{\omega}$, but also $\partial\boldsymbol{\omega}/\partial t$, which implicitly assumes that the redistribution of vorticity by diffusion has had ample time to complete. In short, the evolution of the vorticity field is assumed to proceed in a *quasi-static manner*, instantaneously adjusting to any changes in the boundary conditions.

The linearity of (4.1) and (4.2) means that low-Re flows are much better behaved than their non-linear counterparts. For example, as we shall see, it is possible to establish a

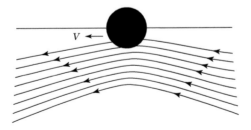

Figure 4.1 Stokes' flow arising from the slow translation of a sphere through a viscous fluid. In this frame of reference, the fluid is stationary at infinity.

uniqueness theorem which says that, if you have found one solution which satisfies the boundary conditions, then this is *the* solution to the problem. Moreover, such flows are reversible in the sense that, if we drive motion through movement of the boundaries, then if we stop that motion and reverse the boundary conditions, the fluid particles will return to their initial positions. So low-Re flows have much in common with the theory of elasticity.

Perhaps the first and most famous application of (4.1) is Stokes' solution for the slow translation of a solid sphere through a viscous fluid (Figure 4.1), so let us start with this problem. In order to make the flow steady, we shall change to a frame of reference moving with the sphere.

4.1.2 Flow past a Sphere at Low Reynolds Number

Consider a stationary sphere of radius R which is immersed in a flow that approaches with a uniform velocity $\mathbf{V}$, as shown in Figure 4.2. The motion is steady and axisymmetric and we adopt *cylindrical polar coordinates* (r, θ, z) centred on the sphere and aligned with the oncoming flow.

Evidently, the velocity field is poloidal, $\mathbf{u} = \mathbf{u}_p = (u_r, 0, u_z)$, while the vorticity is azimuthal, $\boldsymbol{\omega} = \omega_\theta = \nabla \times \mathbf{u}_p$. The boundary conditions on $\mathbf{u}$ are

$$\mathbf{u} = V\hat{\mathbf{e}}_z, \ |\mathbf{x}| \to \infty, \quad \mathbf{u} = 0, \ |\mathbf{x}| = R, \tag{4.3}$$

and we choose a datum for pressure which sets the far-field pressure to zero. Finally, we take $\mathrm{Re} = 2RV/\nu \ll 1$ and then *assume* that this allows us to drop the inertial forces at *all* points in the flow, an assumption we need to check retrospectively.

Let us start with the pressure field and see how far purely dimensional arguments will take us. A moment's thought confirms that the normalized pressure, $p/\rho\nu$, is a function of $\mathbf{x}$, R, and $\mathbf{V}$ only. Moreover, the linearity of the problem tells us that it is a linear function of $\mathbf{V}$. It follows that $p/\rho\nu$ must be of the general form

$$p/\rho\nu = (\mathbf{V} \cdot \mathbf{x})F(\mathbf{x}, R), \tag{4.4}$$

for some function F. Since $p/(\rho\nu\mathbf{V} \cdot \mathbf{x})$ had the dimensions of m^{-2}, we can create only two dimensionless groups from the parameters $\mathbf{x}$, R, and $p/(\rho\nu\mathbf{V} \cdot \mathbf{x})$. These may be

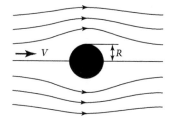

Figure 4.2 Stokes flow over a sphere.

taken as $\mathbf{x}^2/R^2$ and $pR^2/(\rho\nu\mathbf{V}\cdot\mathbf{x})$. As one dimensionless group can depend only on another dimensionless group, dimensional analysis now demands that (4.4) adopts the specific form

$$\frac{p}{\rho\nu} = \frac{\mathbf{V}\cdot\mathbf{x}}{R^2}f\left(\eta\right), \; \eta = \frac{\mathbf{x}^2}{R^2}, \tag{4.5}$$

where f is subject to the far-field boundary condition $f_\infty = 0$. We are almost there. Substituting (4.5) into (4.2) now yields, after a little algebra,

$$\frac{d}{d\eta}\left(\eta^{5/2}f'(\eta)\right) = 0,$$

and integrating twice we conclude that $f = c\eta^{-3/2}$ for some dimensionless constant c. (The second constant of integration is set to zero by virtue of $f_\infty = 0$.) It follows that

$$\frac{p}{\rho\nu} = c\frac{(\mathbf{V}\cdot\mathbf{x})R}{|\mathbf{x}|^3}. \tag{4.6}$$

A similar line of reasoning tells us that the vorticity is a function of $\mathbf{x}$, R, and $\mathbf{V}$ only, and is linear in $\mathbf{V}$. It is also a pseudo-vector, i.e. a vector that changes sign on moving from the right-hand convention to a left-hand rule for cross products. A little thought then confirms that $\boldsymbol\omega$ must take the form

$$\boldsymbol\omega = (\mathbf{V}\times\mathbf{x})G(\mathbf{x}, R), \tag{4.7}$$

for some function G. Once again we deploy dimensional analysis, and this tells us that (4.7) must adopt the specific form

$$\boldsymbol\omega = \frac{\mathbf{V}\times\mathbf{x}}{R^2}g\left(\eta\right), \tag{4.8}$$

where the boundary condition at infinity is $g_\infty = 0$. On substituting (4.5) and (4.8) into (4.1) we find, after some algebra, that $g = f$ and so

$$\boldsymbol\omega = c\frac{(\mathbf{V}\times\mathbf{x})R}{|\mathbf{x}|^3} = c\frac{VRr}{|\mathbf{x}|^3}\hat{\mathbf{e}}_\theta. \tag{4.9}$$

It remains to determine the constant, c, which requires that we find the velocity field and implement the no-slip condition at the surface of the sphere. To that end, we introduce the Stokes streamfunction, Ψ, defined by (3.36):

$$\mathbf{u}_p = \left(-\frac{1}{r}\frac{\partial\Psi}{\partial z}, 0, \frac{1}{r}\frac{\partial\Psi}{\partial r}\right). \tag{4.10}$$

Equation (3.62) then gives the governing equation for Ψ as

$$\nabla_*^2 \Psi = -r\omega_\theta = -\frac{cVRr^2}{|\mathbf{x}|^3}, \tag{4.11}$$

with $\Psi = \frac{1}{2}Vr^2$ in the far field. The r^2 dependence of the source term in (4.11), as well as the appearance of r^2 in the far-field boundary condition, suggests looking for a solution of the form

$$\Psi(r,z) = \frac{1}{2}Vr^2 h(\eta), \quad h_\infty = 1. \tag{4.12}$$

We shall also take $h(1) = 0$ in order to ensure that the normal component of velocity on the surface of the sphere is zero. Substituting (4.12) into (4.11), and imposing the boundary conditions $h(1) = 0$ and $h_\infty = 1$, yields

$$\Psi = \frac{1}{2}Vr^2 \left[1 + c\frac{R}{|\mathbf{x}|} - (c+1)\frac{R^3}{|\mathbf{x}|^3} \right], \tag{4.13}$$

from which,

$$u_r = \frac{1}{2}Vrz \left[c\frac{R}{|\mathbf{x}|^3} - 3(c+1)\frac{R^3}{|\mathbf{x}|^5} \right]. \tag{4.14}$$

Since we have already ensured that there is no normal component of velocity at the surface of the sphere, the no-slip condition is guaranteed provided that we demand $u_r = 0$ on the surface. Evidently, this requires $c = -3/2$, and so we conclude that

$$\frac{p}{\rho\nu} = -\frac{3VRz}{2|\mathbf{x}|^3}, \quad \boldsymbol{\omega} = -\frac{3VRr}{2|\mathbf{x}|^3}\hat{\mathbf{e}}_\theta, \tag{4.15}$$

$$\Psi = \frac{1}{2}Vr^2 \left[1 - \frac{3R}{2|\mathbf{x}|} + \frac{R^3}{2|\mathbf{x}|^3} \right]. \tag{4.16}$$

This flow is shown in Figure 4.2. Note the relatively slow decline in the radial velocity, as $u_r \sim |\mathbf{x}|^{-1}$. This means that well-separated spheres will tend to interact, and that seemingly remote boundaries can have an unexpected influence on the motion. Note also that the flow over a gas bubble is similar to that over a rigid sphere. Indeed, the entire analysis above is unchanged up to and including (4.13) and (4.14). The only difference between the two flows is that, in order to satisfy a zero-stress condition at the bubble surface, we must take $c = -1$ in (4.13), rather than $c = -3/2$. This is discussed in Exercise 4.5.

Given the velocity field, we should check retrospectively that the inertial forces are indeed negligible by comparison with the viscous forces when $\text{Re} = 2RV/\nu \ll 1$. Here we get a surprise. It is readily confirmed that

$$\nu\nabla^2\mathbf{u} \sim \nu V R/|\mathbf{x}|^3, \quad \mathbf{u}\cdot\nabla\mathbf{u} \sim V^2 R \Big/ |\mathbf{x}|^2, \tag{4.17}$$

and so the ratio of inertial to viscous forces is

$$\frac{|\mathbf{u} \cdot \nabla \mathbf{u}|}{|\nu \nabla^2 \mathbf{u}|} \sim \frac{|\mathbf{x}|}{R} \cdot \left(\frac{VR}{\nu}\right) = \frac{V|\mathbf{x}|}{\nu}. \tag{4.18}$$

We conclude that taking Re $\ll$ 1 only allows us to neglect inertia in the near and intermediate fields. In the far field, inertia is significant and our solution is inconsistent with its starting assumptions. (The physical origins of this inconsistency are discussed in Exercise 4.4.) Fortunately, this embarrassing inconsistency can be patched up, as we shall see in the next section.

It remains to determine the drag force on the sphere. Given the pressure and velocity fields, this is a routine calculation and we refer the reader to Batchelor (1967) for the details. We merely note here that one-third of the drag comes from the anti-symmetric pressure distribution over the sphere (see Exercise 4.1) and two-thirds from the shear stresses. The total drag force is

$$\boxed{F_D = 6\pi \rho \nu RV}, \tag{4.19}$$

which is often called Stokes' drag law. Actually, given that the vorticity field is linear in V and independent of viscosity, this law follows directly from dimensional analysis and symmetry, except, of course, for the pre-factor of 6π, which requires detailed calculation.

The Stokes flow over a cylinder is discussed in Exercise 4.2. As in the case above, the growth of inertia in the far field limits the solution to the near field. In fact, as Stokes discovered, flow over a cylinder is rather delicate as the low-inertia solution diverges in the far field and so cannot be directly matched to the far-field condition $\mathbf{u} = \mathbf{V}$. As Stokes (1851) put it:

> I first tried a long cylinder, because the solution of the problem appeared likely to be simpler than in the case of a sphere. But after having proceeded a good way towards the result, I was stopped by a difficulty relating to the determination of the arbitrary constants... Having failed in the case of a cylinder, I tried a sphere, and presently found that the corresponding differential equation admitted of integration in finite terms, so that the solution of the problem could be completely effected. The result, I found, agreed very well with Baily's experiments.

The problems raised by the far-field divergence of Stokes flow over a cylinder can be resolved using the so-called Oseen correction. The same correction also resolves the inconsistencies inherent in Stokes' solution for flow over a sphere, as we now discuss.

4.1.3 The Oseen Correction for Flow over a Sphere at Low Re

The difficulty encountered by Stokes, of non-negligible inertia in the far field of a sphere, was partially resolved by Oseen in 1910. He continued to assume that Re $= 2RV/\nu \ll 1$, but proposed retaining inertia in a linearized form, i.e. $\mathbf{u} \cdot \nabla \mathbf{u}$ is replaced by the linear approximation $\mathbf{V} \cdot \nabla \mathbf{u}$. This linearization may be a poor approximation close to the sphere, but then inertia in this region is negligible anyway, and it is a good approximation in the far field, where $\mathbf{u} \approx \mathbf{V}$ and inertia is important. Equation (4.1) is then replaced by

$$\mathbf{V}\cdot\nabla\mathbf{u} + \nabla(p/\rho) = \nu\nabla^2\mathbf{u} = -\nu\nabla\times\boldsymbol{\omega}, \qquad (4.20)$$

from which

$$\nabla^2 p = 0, \qquad (4.21)$$

and

$$\mathbf{V}\cdot\nabla\boldsymbol{\omega} = \nu\nabla^2\boldsymbol{\omega}. \qquad (4.22)$$

Physically, the main difference to Stokes' flow is that vorticity in now advected by $\mathbf{V}$ as well as diffusing away from the sphere. This advection plays a leading role only in the far field, but it means that the flow no longer possesses upstream–downstream symmetry, with the far-field vorticity being swept into a wake at the rear of the sphere (see Figure 4.3).

Although equations (4.20)→(4.22) are linear, they do not appear to have a solution in closed form. However, there are simple closed-form approximations to Oseen's solution which are easy to work with (again, see Batchelor, 1967). It turns out that, unlike Stokes' solution, Oseen's analysis is self-consistent in the sense that, for small Re, the neglected non-linear term $(\mathbf{u} - \mathbf{V})\cdot\nabla\mathbf{u}$ is *everywhere* small by comparison with the terms retained in (4.20). In the vicinity of the sphere, where $|\mathbf{x}| \sim R$, Oseen's solution takes the form

$$\Psi = \frac{1}{2}Vr^2\left[1 - \frac{3R}{2|\mathbf{x}|} + \frac{R^3}{2|\mathbf{x}|^3} + O(\mathrm{Re})\right], \qquad \mathrm{Re} \ll 1, \qquad (4.23)$$

which reverts to Stokes' solution (4.16) for Re→0. Although the Oseen approximation cannot be justified when $|\mathbf{x}| \sim R$, it turns out that, somewhat fortuitously, it provides an accurate $O(\mathrm{Re})$ correction to Stokes' drag law.

The Oseen modification to Stokes' drag law takes the form

$$F_D = 6\pi\rho\nu RV + \frac{9}{4}\pi R^2\rho V^2 = 6\pi\rho\nu RV\left[1 + \frac{3}{8}\frac{VR}{\nu}\right]. \qquad (4.24)$$

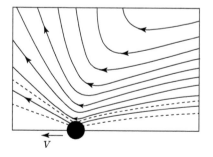

Figure 4.3 Oseen's solution for the slow translation of a sphere through a viscous fluid. In this frame of reference, the fluid is stationary at infinity. (Adapted from Rosenhead, 1963.)

This modified drag law is normally rewritten in terms of a dimensionless *drag coefficient*. By convention, such drag coefficients are defined as the drag force exerted on an object normalized by the frontal area of the object and by $\frac{1}{2}\rho V^2$. Thus, by definition,

$$C_D = \frac{F_D}{\frac{1}{2}\rho V^2 \pi R^2},$$
(4.25)

and our modified drag force now becomes

$$C_D = \frac{24}{\mathrm{Re}}\left[1 + \frac{3}{16}\mathrm{Re}\right], \quad \mathrm{Re} = 2RV/\nu.$$
(4.26)

This compares favourably with experimental data up to a Reynolds number of $\mathrm{Re} \approx 0.5$.

The Oseen correction for flow over a cylinder, which is discussed in some detail in Exercise 4.3, is rather more involved. In any event, the end result is that, as with flow over a sphere, the correction smooths out the worst inconsistencies in the Stokes solution.

4.1.4 The Uniqueness and Minimum Dissipation Theorems for Low-Re Flows

We have already suggested that there exists a *uniqueness theorem* for low-Re flows. Such uniqueness theorems are common for linear problems in continuum mechanics and electrodynamics, and they are usually established using a proof by contradiction. In particular, one starts by assuming that there exist two distinct solutions that satisfy both the governing equation and boundary conditions. One then focusses on the difference between these two solutions, which satisfies the same homogeneous governing equation, but is subject to null boundary conditions. A consideration of the integral properties of this difference solution is then usually sufficient to show that it is identically zero at all points.

For the case of low-Re flows we proceed as follows. Suppose that the fluid occupies the domain, V, and is bounded by a surface, S, on which the prescribed boundary condition is $\mathbf{u} = \mathbf{u}_B$. We now assume that there exist two distinct solutions of (4.1) which both satisfy $\mathbf{u} = \mathbf{u}_B$ on S, say, $\mathbf{u}_1$ and $\mathbf{u}_2$, with corresponding pressure distributions, p_1 and p_2. If $\mathbf{v} = \mathbf{u}_2 - \mathbf{u}_1$, then this difference solution satisfies

$$\nabla(p_2 - p_1) = \rho\nu\nabla^2\mathbf{v}, \quad \mathbf{v} = 0 \text{ on } S.$$
(4.27)

Taking the dot product of (4.27) with $\mathbf{v}$ now yields

$$\nabla \cdot [(p_2 - p_1)\mathbf{v}] = \rho\nu\left[\mathbf{v}\cdot\nabla^2\mathbf{v}\right] = 2\rho\nu\left[\frac{\partial}{\partial x_j}\left(v_i S_{ij}^{(v)}\right) - S_{ij}^{(v)}S_{ij}^{(v)}\right],$$
(4.28)

where $S_{ij}^{(v)}$ is the rate-of-strain tensor of the difference solution $\mathbf{v}$. On integrating this over the fluid domain V, applying Gauss' theorem, and noting that $\mathbf{v} = 0$ on S, we obtain

$$\int S_{ij}^{(v)} S_{ij}^{(v)} dV = 0.$$
(4.29)

This demands that all six components of $S_{ij}^{(v)}$ are everywhere zero (see (2.29). However, the only motion in which there is no deformation of any fluid element is one consisting of rigid-body rotation plus uniform translation, which is excluded by virtue of the boundary condition $\mathbf{v} = 0$ on S. We conclude, therefore, that $\mathbf{v} = 0$ and $\mathbf{u}_2 = \mathbf{u}_1$ at all points in the domain V, which establishes the uniqueness of the original flow $\mathbf{u}$.

The uniqueness theorem tells us that low-Re flows are reversible, in the sense that reversing the boundary condition, $\mathbf{u}_B \rightarrow -\mathbf{u}_B$, will reverse the motion, $\mathbf{u} \rightarrow -\mathbf{u}$. That is to say, if $\mathbf{u}$ and p represent the unique solution in V corresponding to $\mathbf{u} = \mathbf{u}_B$ on S, then the linearity of (4.1) tells us that $-\mathbf{u}$ and $-p$ represent one possible solution in V corresponding to $\mathbf{u} = -\mathbf{u}_B$ on S. The uniqueness theorem now says that, to within an unimportant constant in the pressure, $-\mathbf{u}$ and $-p$ constitute *the* solution corresponding to $\mathbf{u} = -\mathbf{u}_B$ on S. This reversibility has some interesting consequences. For example, suppose we fill the gap between two concentric cylinders with syrup, and rotate the outer cylinder clockwise by an angle α. The syrup is then sheared as it is dragged around by the outer cylinder. If we now reverse the motion of the outer cylinder and rotate it by an angle of $-\alpha$, then the flow is perfectly reversed and so all of the fluid particles will return to their original positions.

There is a second theorem for low-Re flow which is closely related to the uniqueness theorem, called the *minimum dissipation theorem*. Suppose that $\mathbf{u}$ is the unique solution of (4.1) which satisfies $\mathbf{u} = \mathbf{u}_B$ on S, while $\mathbf{u}^*$ is *any* solenoidal velocity field in V which satisfies the prescribed boundary condition on S. In particular, $\mathbf{u}^*$ need not satisfy (4.1). As before, we let $\mathbf{v}$ be the difference solution, $\mathbf{v} = \mathbf{u} - \mathbf{u}^*$, which is solenoidal and satisfies $\mathbf{v} = 0$ on S. Then the dot product of $\mathbf{v}$ with (4.1) yields

$$\nabla \cdot [p\mathbf{v}] = \rho\nu \left[\mathbf{v} \cdot \nabla^2 \mathbf{u} \right] = 2\rho\nu \left[\frac{\partial}{\partial x_j}(v_i S_{ij}^{(u)}) - S_{ij}^{(v)} S_{ij}^{(u)} \right], \tag{4.30}$$

where the superscripts (v) and (u) indicate that $S_{ij}^{(v)}$ and $S_{ij}^{(u)}$ are the rate-of-strain tensors for $\mathbf{v}$ and $\mathbf{u}$, respectively. As in the proof of the uniqueness theorem, we now integrate this over the fluid domain V, apply Gauss' theorem, and note that $\mathbf{v} = 0$ on S. This yields

$$\int S_{ij}^{(v)} S_{ij}^{(u)} dV = 0. \tag{4.31}$$

Moreover, if $S_{ij}^{(*)}$ is the rate-of-strain tensor for $\mathbf{u}^*$, then

$$S_{ij}^{(*)} S_{ij}^{(*)} = S_{ij}^{(u)} S_{ij}^{(u)} - 2S_{ij}^{(u)} S_{ij}^{(v)} + S_{ij}^{(v)} S_{ij}^{(v)},$$

which, in the light of (4.31), integrates to give

$$\int S_{ij}^{(*)} S_{ij}^{(*)} dV = \int S_{ij}^{(u)} S_{ij}^{(u)} dV + \int S_{ij}^{(v)} S_{ij}^{(v)} dV. \tag{4.32}$$

Finally, we note that (2.29) tells us that all three integrals in (4.32) are non-negative, and so have the inequality

$$2\rho\nu \int S_{ij}^{(*)} S_{ij}^{(*)} dV \geq 2\rho\nu \int S_{ij}^{(u)} S_{ij}^{(u)} dV, \tag{4.33}$$

where both integrals represent the net dissipation rate of the flow to which they refer. Evidently, the rate of dissipation of mechanical energy in the motion governed by the low-Re equation, (4.1), is less than any other solenoidal flow which satisfies the same boundary conditions. Both the uniqueness and minimum dissipation theorems were established by Helmholtz.

4.1.5 Two-dimensional Flow in a Wedge at Low Reynolds Number

We now consider two-dimensional flow in a wedge of half angle α, driven by an applied shear stress at the outer radius. It is observed experimentally that such corner flows exhibit eddies (regions of closed streamlines) provided that α is not too large, as shown schematically in Figure 4.4.

Moffatt (1964) considered this problem in some detail and showed that the critical angle for the formation of corner vortices is $\alpha \leq 73.2°$. The analysis proceeds as follows. Since the motion is two-dimensional, it is natural to use the streamfunction, (1.15), to describe the flow, which is related to the vorticity by $\omega = -\nabla^2 \psi$. Our low-Re equation of motion, (4.2), now requires that ψ satisfies the biharmonic equation $\nabla^4 \psi = 0$, which in polar coordinates (r, θ) reads

$$\left(\frac{\partial^2}{\partial r^2} + \frac{1}{r} \frac{\partial}{\partial r} + \frac{1}{r^2} \frac{\partial^2}{\partial \theta^2} \right)^2 \psi = 0.$$

This biharmonic equation is homogeneous in r and so there exist separable solutions of the form $\psi = r^{\mu+1} f(\theta)$, and these solutions are

$$\psi = r^{\mu+1} \left[A\cos(\mu+1)\theta + B\sin(\mu+1)\theta + C\cos(\mu-1)\theta + D\sin(\mu-1)\theta \right]. \tag{4.34}$$

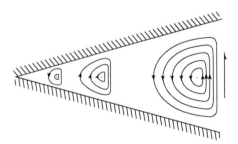

Figure 4.4 Moffatt's corner eddies.

(We shall exclude the degenerate cases of $\mu = -1, 0, 1$, where the solution takes a different form.) Now

$$\mathbf{u} = (u_r, u_\theta) = \left(\frac{1}{r} \frac{\partial \psi}{\partial \theta}, -\frac{\partial \psi}{\partial r} \right), \tag{4.35}$$

and we are interested in solutions in the domain $-\alpha < \theta < \alpha$ for which u_θ is symmetric about $\theta = 0$. We may therefore take $B = D = 0$ in (4.34), giving

$$\psi = r^{\mu+1} \left[A \cos(\mu+1)\theta + C \cos(\mu-1)\theta \right]. \tag{4.36}$$

The no-slip condition on $\theta = \pm \alpha$ now requires

$$u_r(\theta = \pm\alpha) = 0 : (\mu+1)A\sin(\mu+1)\alpha + (\mu-1)C\sin(\mu-1)\alpha = 0,$$
$$u_\theta(\theta = \pm\alpha) = 0 : A\cos(\mu+1)\alpha + C\cos(\mu-1)\alpha = 0,$$

from which,

$$(\mu+1)\alpha \tan(\mu+1)\alpha = (\mu-1)\alpha \tan(\mu-1)\alpha. \tag{4.37}$$

Given α, this may be solved for μ. If μ is real then u_θ evaluated at $\theta = 0$ does not change sign as r increases, and so corner vortices of the type shown in Figure 4.4 cannot form. On the other hand, if μ is complex, say $\mu = p + jq$, then u_θ evaluated at $\theta = 0$ is composed of terms like

$$\Re\left(r^{p+jq} \right) = r^p \Re \left[\left(e^{\ln r} \right)^{jq} \right] = r^p \Re \left(e^{jq \ln r} \right) = r^p \cos(q \ln r), \tag{4.38}$$

where $j = \sqrt{-1}$ and $\Re$ indicates the real part of what follows. Clearly, complex solutions of (4.37) are indicative of an infinite hierarchy of nested vortices of alternating sign. It turns out that μ is complex when $\alpha < 73.2°$, which therefore constitutes the criterion for the formation of corner vortices.

Perhaps some comments are in order. First, we have not considered the precise nature of the driving force for the flow, which dictates the outer boundary condition at large r. Nevertheless, we might place our faith in the intuitive hypothesis that sufficiently far into the wedge, the flow will be more or less independent of the nature of the external driving. Second, laboratory experiments do indeed confirm the existence of alternating vortices for $\alpha < 73.2°$. Third, in practice the experiments rarely show more than two alternating vortices, rather than the infinite hierarchy predicted by (4.38), which is consistent with the fact that the predicted intensity of the vortices decays extremely rapidly as $r \to 0$.

4.1.6 Suspensions

Consider a suspension of small spherical particles in a Newtonian fluid (a gas or a liquid) which are swept around by the fluid. We take the suspension to be dilute in the sense that the particles have a size much less than their mean separation, yet locally homogeneous in the sense that their mean separation is much less than the global scale of the flow (Figure 4.5).

We also take the particles to be sufficiently small that they have negligible inertia and so move with the local fluid velocity. However, even though the particles move with the fluid, they still perturb the background flow as the no-slip condition must be satisfied on the surface of each particle. It is of interest to calculate this perturbation in velocity since, as we shall see, it changes the bulk characteristics of the motion. We take the motion *relative to a particle* to be Stokesian, *i.e.* $uR/\nu \ll 1$, where u is a typical velocity and R is the particle radius.

Since the separation between the particles is much larger than R, we may ignore interactions between particles and focus on a single sphere, which we place at the origin at some instant. Moreover, since the global scale of the flow is also much larger than R, we may consider the *background* fluid motion to be a linear function of the coordinates throughout the domain of influence of the sphere. Thus we write the background motion as $u_i = u_i\left(|\mathbf{x}| = 0\right) + \left(\partial u_i/\partial x_j\right)_0 x_j$, where $\left(\partial u_i/\partial x_j\right)_0$ is a constant. Now, a sphere subject to uniform translation or rotation will not perturb the background velocity field as the particle will simply translate or rotate with the fluid. So, if our aim is to calculate the perturbation to the background velocity created by the presence of a sphere, the obvious thing to do is to move into a frame of reference that moves and rotates with the background flow. The sphere is then stationary in such a frame. Given that the local rotation of the fluid is captured by the vorticity, this choice of frame of reference leaves the background motion devoid of vorticity and composed of a pure *straining motion*, $\mathbf{u}^{(s)} = S_{ij} x_j$, in accordance with (2.7). So the question becomes: how is the imposed straining motion $\mathbf{u}^{(s)} = S_{ij} x_j$, S_{ij} being a constant, perturbed by the presence of a sphere centred on the origin?

Let us write the perturbed velocity field as $\mathbf{u} = \mathbf{u}^{(s)} + \mathbf{u}' = S_{ij} x_j + \mathbf{u}'$, where $\mathbf{u}'$ is the perturbation to the imposed flow created by the presence of the sphere. The perturbed and unperturbed flows satisfy

$$\frac{\partial \mathbf{u}}{\partial t} + \mathbf{u} \cdot \nabla \mathbf{u} = -\nabla\left(p/\rho\right) + \nu \nabla^2 \mathbf{u},$$

and

$$\frac{\partial \mathbf{u}^{(s)}}{\partial t} + \mathbf{u}^{(s)} \cdot \nabla \mathbf{u}^{(s)} = -\nabla\left(p^{(s)}\big/\rho\right) + \nu \nabla^2 \mathbf{u}^{(s)},$$

and subtracting the two gives us

$$\frac{\partial \mathbf{u}'}{\partial t} + \mathbf{u} \cdot \nabla \mathbf{u}' + u'_j S_{ij} = -\nabla\left(p'/\rho\right) + \nu \nabla^2 \mathbf{u}'.$$

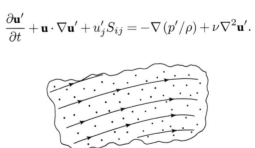

Figure 4.5 A dilute suspension of spherical particles of radius R in a slowly varying flow.

We now limit ourselves to the Stokes regime in which $S_{ij}R^2/\nu \ll 1$, and so we recover the usual governing equations,

$$\nabla p' = \rho\nu\nabla^2\mathbf{u}' = -\rho\nu\nabla \times \boldsymbol{\omega}', \tag{4.39}$$

$$\nabla^2 p' = 0, \quad \nabla^2\boldsymbol{\omega}' = 0, \tag{4.40}$$

which must be solved subject to $\mathbf{u}' = -\mathbf{u}^{(s)} = -S_{ij}x_j$ on $|\mathbf{x}| = R$ and $\mathbf{u}' = p' = 0$ for $|\mathbf{x}| \to \infty$.

Arguments based on linearity and dimensional considerations, similar to those deployed in §4.1.2, suggest that we look for solutions of the form

$$\frac{p'}{\rho\nu} = \frac{\mathbf{u}^{(s)}\cdot\mathbf{x}}{R^2}f(\eta), \quad \eta = \frac{\mathbf{x}^2}{R^2}, \tag{4.41}$$

and

$$\boldsymbol{\omega}' = \frac{\mathbf{u}^{(s)}\times\mathbf{x}}{R^2}g(\eta), \tag{4.42}$$

where f and g are dimensionless and subject to the far-field boundary conditions $f_\infty = 0$ and $g_\infty = 0$. On substituting (4.41) and (4.42) into (4.39) and (4.40), and demanding that $f_\infty = 0$ and $g_\infty = 0$, we find (after some algebra) that $g = f = c\eta^{-5/2}$. It follows that

$$\frac{p'}{\rho\nu} = c\frac{R^3}{|\mathbf{x}|^5}\mathbf{u}^{(s)}\cdot\mathbf{x}, \quad \boldsymbol{\omega}' = c\frac{R^3}{|\mathbf{x}|^5}\mathbf{u}^{(s)}\times\mathbf{x}, \tag{4.43}$$

for some constant c. To pin down c we need to apply the no-slip boundary condition on the surface of the sphere, which requires that we uncurl $\boldsymbol{\omega}' = \nabla \times \mathbf{u}'$ to give $\mathbf{u}'$. Again, arguments based on linearity and dimensional considerations suggest that $\mathbf{u}'$ may take the form

$$\mathbf{u}' = h_1(\eta)\mathbf{u}^{(s)} + h_2(\eta)\frac{(\mathbf{u}^{(s)}\cdot\mathbf{x})\mathbf{x}}{R^2}. \tag{4.44}$$

Note that the vector contributions to (4.44), which are linear in $\mathbf{u}^{(s)}$, are taken to be proportional to $\mathbf{u}^{(s)}$ and $(\mathbf{u}^{(s)}\cdot\mathbf{x})\mathbf{x}$, whereas the equivalent term in the expression for $\boldsymbol{\omega}'$ is proportional to $\mathbf{u}^{(s)}\times\mathbf{x}$. This is because $\mathbf{u}'$ is a true vector whereas $\boldsymbol{\omega}'$ is a pseudo-vector. In any event, (4.44) turns out to be a good guess. It is readily confirmed that

$$\mathbf{u}' = -\frac{R^5}{|\mathbf{x}|^5}\mathbf{u}^{(s)} - \frac{5R^3}{2|\mathbf{x}|^5}\left[1 - \frac{R^2}{|\mathbf{x}|^2}\right](\mathbf{u}^{(s)}\cdot\mathbf{x})\mathbf{x} \tag{4.45}$$

satisfies the no-slip and far-field conditions $\mathbf{u}' = -\mathbf{u}^{(s)}$ on $|\mathbf{x}| = R$ and $\mathbf{u}' = 0$ for $|\mathbf{x}| \to \infty$, as well as $\boldsymbol{\omega}' = \nabla \times \mathbf{u}'$, provided that we take $c = -5$. We conclude that the perturbations in pressure and vorticity created by the presence of the sphere are

$$\frac{p'}{\rho \nu} = -5\frac{R^3}{|\mathbf{x}|^5}\mathbf{u}^{(s)} \cdot \mathbf{x}, \quad \boldsymbol{\omega}' = -5\frac{R^3}{|\mathbf{x}|^5}\mathbf{u}^{(s)} \times \mathbf{x}. \tag{4.46}$$

Perhaps it is worth making two comments at this point. First, as with uniform Stokes flow over a sphere, this solution is not self-consistent in the far field, because inertia becomes important there. However, for $S_{ij}R^2/\nu \ll 1$, this does not change the velocity in the near and intermediate fields. Second, essentially the same analysis can be performed for gas bubbles in a liquid, water droplets in a gas (a mist), and liquid droplets in another liquid (an emulsion). One can move from one situation to another by changing the boundary conditions on the surface of the sphere.

These perturbations in the vorticity and velocity fields are useful in a number of respects, but perhaps the most important is that it allows us to estimate the additional viscous dissipation that occurs in the fluid as a result of the presence of a random, but sparse, distribution of tiny spherical particles. This, in turn, allows us to calculate the effective viscosity of a suspension, say $\bar{\mu}$, defined such that the total dissipation rate per unit volume of the dispersion is equal to $2\bar{\mu}S_{ij}^{(s)}S_{ij}^{(s)}$. That is to say, we define $\bar{\mu}$ so that we recover the true dissipation rate of the suspension when we use the rate-of-strain tensor for the unperturbed velocity distribution (squared) and multiply this by the effective bulk viscosity, $\bar{\mu}$.

We shall not pause to calculate the additional dissipation caused by the presence of tiny spherical particles, but rather refer the reader to Batchelor (1967) for a detailed discussion. We merely note here that the end result of the calculation is that the effective bulk viscosity is given by

$$\bar{\mu} = \mu\left(1 + 5\alpha/2\right), \tag{4.47}$$

where α is the volume fraction occupied by the spheres. This expression was first derived by Einstein (1906) and so (4.47) is known as *Einstein's equation*.

Perhaps some comments are in order. First, experiments suggest that (4.47) is accurate up to volume fractions of $\alpha \sim 0.03$. Second, the equivalent result for gas bubbles in a liquid is $\bar{\mu} = \mu(1+\alpha)$. Finally, when small but finite particle–particle interactions are allowed for, (4.47) generalizes to $\bar{\mu} = \mu\left(1 + 5\alpha/2 + k\alpha^2\right)$, where the value of k depends on the nature of the background flow.

4.1.7 The Subtleties of Self-propulsion at Low Reynolds Number

The reversibility of Stokes flow, and the associated absence of inertial effects, means that micro-organisms are forced to adopt unusual strategies for swimming. For example, at low Re it turns out that the simple fish-like flapping of a tail produces no net propulsion since whatever is achieved by one flap is countered by the reverse stroke. So bacteria, sperm, and ciliated protozoa, all of which swim at low Reynolds number (Table 4.1), adopt strategies

Table 4.1 Typical lengths, speeds, and Reynolds numbers for some micro-organisms.

	Length, ℓ (mm)	Swimming speed, V (mm/s)	$Re = V\ell/\nu$
Bacteria	10^{-4}	$0.01 \rightarrow 0.1$	10^{-5}
Sperm	$0.01 \rightarrow 0.1$	0.1	$10^{-3} \rightarrow 10^{-2}$
Ciliated protozoa	0.1	1	0.1

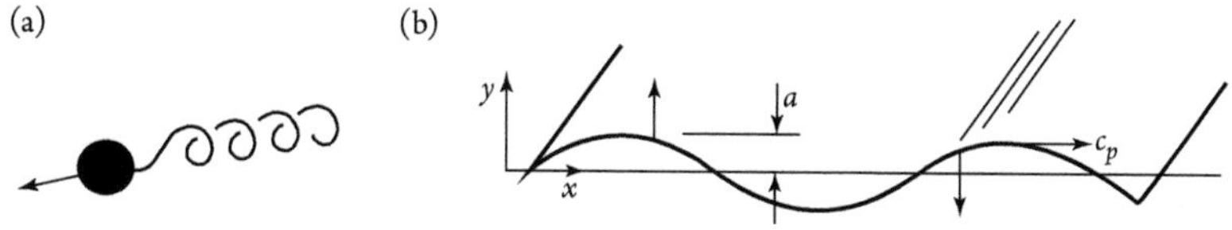

Figure 4.6 (a) Sperm swim by sending helical waves along their tails. (b) G. I. Taylor's swimming sheet provides a simple model of propulsion at low Re.

quite different to, say, a fish. Often this strategy takes the form of the transverse ripples of filaments (cilia or flagella), or else helical waves propagating along a tail (Figure 4.6 (a). We shall focus here on the flapping of a thin membrane, as shown in Figure 4.6 (b).

An illustrative example of self-propulsion at low Re, motivated by flagellar propulsion (the beating of a filament), is the so-called *swimming sheet*. This was first introduced as a model problem for low-Re swimming by G.I. Taylor (1951), but our description follows that of Childress (1981) and Acheson (1990). Consider a thin, flexible sheet of infinite extent which is immersed in a viscous fluid. It is centred on the x–z plane and undergoes strictly transverse oscillations, as shown in Figure 4.6 (b).

The transverse displacement of the sheet takes the form $y_s = a\sin(kx - \varpi t)$, $ka \ll 1$, which is a travelling wave of small amplitude moving along the x-axis from left to right. The resulting motion of the fluid is two-dimensional and the velocity of the sheet, and hence of the fluid in contact with it, is

$$\mathbf{u}_s = (u_x, u_y) = (0, -\varpi a \cos(kx - \varpi t)).$$

Although the motion of the sheet is strictly transverse, it is plausible that the propagating nature of this displacement pushes some of the ambient fluid ahead of it, and we shall see that this is indeed the case. In fact, such an infinite sheet, if left to its own devices for long enough, induces a uniform horizontal velocity in the far field, of magnitude

$$u_\infty = (ka)^2 \frac{\varpi}{2k} = \frac{1}{2}(ka)^2 c_p, \tag{4.48}$$

where $c_p = \varpi/k$ is the *phase velocity* of the traveling wave. (We note in passing that dimensional analysis demands that $u_\infty = F(ka)c_p$ for some dimensionless function F, while (4.48) tells us that u_∞ is *second order* in the small parameter ka.) At first sight, (4.48) seems unremarkable, but if we now change the frame of reference so that the fluid in the far field has no mean motion, then the sheet propagates to the left at a speed of $(ka)^2 c_p/2$ through a fluid which is motionless at large distances from the sheet. In short, the sheet can swim.

Let us now see where (4.48) comes from. We shall restrict attention to the region above the sheet, $y > 0$. Since the motion is two-dimensional, it is natural to use the streamfunction ψ, to describe the flow, which is defined by $\mathbf{u}(x, y) = \nabla \times (\psi \hat{\mathbf{e}}_z)$. Our low-Re equation of motion, $\nabla^2 \omega = 0$, now requires that ψ satisfies the biharmonic equation, $\nabla^4 \psi = 0$, and our task is to solve this equation subject to the boundary condition

$$\mathbf{u}(x, y = y_s) = (\partial\psi/\partial y, -\partial\psi/\partial x) = (0, -\varpi a \cos(kx - \varpi t)). \qquad (4.49)$$

We may find the flow to leading order in the small parameter $\varepsilon = ka$ in more or less the same way as one obtains solutions for small-amplitude surface gravity waves. The trick is to linearize the boundary condition (4.49) by applying it at $y = 0$, rather than at $y = y_s$. If we then look for solutions of the form

$$\psi = f(y) \sin(kx - \varpi t),$$

the biharmonic equation requires

$$f = (A + By) \exp(-ky) + (C + Dy) \exp(ky).$$

For $y > 0$ we must take $C = D = 0$, and applying (4.49) at $y = 0$ fixes A and B. A little algebra then gives us

$$\psi = \varepsilon \frac{\varpi}{k^2}(1 + ky) \exp(-ky) \sin(kx - \varpi t). \qquad (4.50)$$

Evidently, there is no mean flow associated with this solution, and so we must push the analysis to second order in ε. Let us write

$$\psi = \varepsilon \psi^{(1)} + \varepsilon^2 \psi^{(2)} + O(\varepsilon^3), \qquad (4.51)$$

where $\varepsilon \psi^{(1)}$ is given by (4.50). We now perform a truncated Taylor expansion of the velocity components in (4.49) about $y = 0$:

$$\left(\frac{\partial\psi}{\partial x}\right)_{y=y_s} = \left(\frac{\partial\psi}{\partial x}\right)_{y=0} + y_s \left(\frac{\partial^2\psi}{\partial x \partial y}\right)_{y=0} = \varepsilon \frac{\varpi}{k} \cos(kx - \varpi t),$$

$$\left(\frac{\partial\psi}{\partial y}\right)_{y=y_s} = \left(\frac{\partial\psi}{\partial y}\right)_{y=0} + y_s \left(\frac{\partial^2\psi}{\partial y^2}\right)_{y=0} = 0.$$

To leading order in ε, these boundary conditions are those used to obtain (4.50), while gathering second-order terms yields

$$\left(\frac{\partial \psi^{(2)}}{\partial x}\right)_{y=0} = -\frac{y_s}{\varepsilon}\left(\frac{\partial^2 \psi^{(1)}}{\partial x \partial y}\right)_{y=0} = -\frac{1}{k}\left(\frac{\partial^2 \psi^{(1)}}{\partial x \partial y}\right)_{y=0} \sin(kx - \varpi t),$$

$$\left(\frac{\partial \psi^{(2)}}{\partial y}\right)_{y=0} = -\frac{y_s}{\varepsilon}\left(\frac{\partial^2 \psi^{(1)}}{\partial y^2}\right)_{y=0} = -\frac{1}{k}\left(\frac{\partial^2 \psi^{(1)}}{\partial y^2}\right)_{y=0} \sin(kx - \varpi t).$$

Substituting for $\psi^{(1)}$ using (4.50) now gives us $\left(\partial \psi^{(2)}/\partial x\right)_{y=0} = 0$ and

$$\left(\frac{\partial \psi^{(2)}}{\partial y}\right)_{y=0} = \frac{\varpi}{k}\sin^2(kx - \varpi t) = \frac{\varpi}{2k}\left[1 - \cos 2(kx - \varpi t)\right]. \tag{4.52}$$

Now $\psi^{(1)}$ and $\psi^{(2)}$ individually satisfy the biharmonic equation. The solution of this equation for $\psi^{(2)}$, which satisfies the boundary conditions at $y = 0$, is readily shown to be

$$\psi^{(2)} = \frac{\varpi}{2k^2}ky\left[1 - \exp(-2ky)\cos 2(kx - \varpi t)\right]. \tag{4.53}$$

This gives a mean horizontal motion in the far field of $(ka)^2 c_p/2$, exactly as anticipated by (4.48). In a frame of reference in which the far-field fluid is stationary, the sheet moves to the left at a speed of $(ka)^2 c_p/2$, which is much less than the phase velocity, c_p.

In 1952 G.I. Taylor extended this analysis from a sheet to a thin filament, thus showing how beating cilia or flagella enable micro-organisms to swim. Sperm, on the other hand, propel themselves forward by sending helical waves along their tails, as shown in Figure 4.6 (a). Both of these intriguing mechanisms are discussed in some detail in Childress (1981).

4.2 Lubrication Theory

We now turn to flows in which inertia is negligible, not because the Reynolds number is small, but rather because the streamlines are almost parallel. The most important application of such flows is lubrication theory, which was first developed by Reynolds in the 1880s.

4.2.1 The Approximations and Governing Equations of Lubrication Theory

Consider the flow shown in Figure 4.7, in which oil fills the gap between two plane surfaces in a so-called slipper bearing. The upper surface, that is to say the base of the slipper block, is

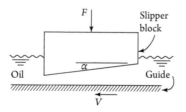

Figure 4.7 Oil fills the gap in a slipper bearing.

inclined at an angle α to the horizontal, with $\alpha \ll 1$, and it moves relative to the guide with a speed of V. For simplicity, we assume that the motion is steady when viewed in a frame of reference moving with the slipper, and also two-dimensional, with the surface of the guide at $y = 0$.

Vertical gradients in velocity are much larger than horizontal gradients, by a factor of α, and so the two components of the Navier–Stokes equation simplify to

$$\mathbf{u} \cdot \nabla u_x = -\frac{1}{\rho}\frac{\partial p}{\partial x} + \nu \frac{\partial^2 u_x}{\partial y^2}, \tag{4.54}$$

$$\mathbf{u} \cdot \nabla u_y = -\frac{1}{\rho}\frac{\partial p}{\partial y} + \nu \frac{\partial^2 u_y}{\partial y^2}, \tag{4.55}$$

in a frame of reference moving with the slipper. Moreover, u_y is much smaller than u_x, also by a factor of α, and so comparing (4.54) with (4.55) tells us that vertical pressure gradients are negligible by comparison with horizontal gradients. The horizontal momentum equation now simplifies to

$$\mathbf{u} \cdot \nabla u_x = -\frac{1}{\rho}\frac{dp}{dx} + \nu \frac{\partial^2 u_x}{\partial y^2}, \tag{4.56}$$

with $\partial p/\partial y \approx 0$. Finally, we compare the ratio of inertial to viscous forces in (4.56). If $h(x)$ is the local thickness of the oil film, and $L \sim h/\alpha$ a characteristic horizontal dimension, say the length of the bearing, then the inertial forces are of order $u_x^2/L \sim \alpha u_x^2/h$. The ratio of these to the viscous forces is then

$$\frac{\alpha u_x^2/h}{\nu u_x/h^2} \sim \alpha \frac{Vh}{\nu} = \alpha \, \mathrm{Re}, \tag{4.57}$$

where Re is based on h, rather than L. We conclude that inertia is negligible provided that $\mathrm{Re} \ll 1/\alpha$, which in practice tends to limit $\mathrm{Re} = Vh/\nu$ to values of a few hundred, or less.

In the so-called *lubrication approximation* it is assumed that we satisfy the combined limits

$$h/L \ll 1, \quad \alpha \, \mathrm{Re} \ll 1, \tag{4.58}$$

and so in this regime the streamlines are almost parallel and we may neglect inertia. Our governing equation is then the quasi-parallel approximation

$$\frac{1}{\rho}\frac{dp}{dx} = \nu \frac{\partial^2 u_x}{\partial y^2} = \frac{\partial}{\partial y}\left(\frac{\tau_{xy}}{\rho}\right), \tag{4.59}$$

which must be complemented by continuity of mass.

Although we have chosen to present these scaling arguments in the context of a slipper bearing, they clearly generalize to many similar quasi-parallel flows. Let us now consider some applications of (4.59), including Reynolds' pioneering analysis of slipper bearings, Sommerfeld's masterful description of journal (or shaft) bearings, and Rayleigh's stepped bearing.

4.2.2 Reynolds' Analysis of the Slipper Bearing

Let us start by returning to the slipper (or slider) bearing. We shall take a frame of reference in which the slipper block is stationary, with an origin of coordinates located below the rear of the slipper and at the surface of the guide. The guide has a speed of V in the positive x direction, as shown in Figure 4.8, and the oil gap $h(x)$ runs from $h(0) = h_1$ to $h(L) = h_2$, where L is the length of the bearing and $h_1 > h_2$. In practice, the oil gap is typically around $h \sim 0.01 \rightarrow 0.1$ mm.

Perhaps the first point to note is that (4.59) tells us that the normal force supported by the bearing, F, is much greater than the drag force associated with the shear stress acting on the guide. In particular,

$$F = \int_0^L p\,dx \sim -\frac{1}{\alpha}\int_0^L \tau_{xy}\,dx = \frac{D}{\alpha}, \tag{4.60}$$

where D is the drag force and $\alpha = (h_1 - h_2)/L$ is the angle the lower face of the slipper makes to the horizontal. Typically $\alpha \sim 0.001$, and so the coefficient of friction for the bearing, which is of order α, is very low.

Next we note that (4.59) in the form

$$\frac{\partial^2 u_x}{\partial y^2} = \frac{1}{\rho\nu}\frac{dp}{dx}$$

integrates to give

$$u_x = V\left(1 - y/h\right) - \frac{1}{2\rho\nu}\frac{dp}{dx}\left(hy - y^2\right). \tag{4.61}$$

Integrating a second time yields the volumetric flow rate,

$$Q = \frac{Vh}{2} - \frac{h^3}{12\rho\nu}\frac{dp}{dx}, \tag{4.62}$$

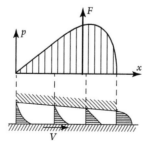

Figure 4.8 A slipper bearing and the associated pressure distribution viewed in a frame of reference moving with the slipper block. Adapted from Prandtl (1952).

which may be rearranged to give

$$h^3 \frac{dp}{dx} = 6\rho\nu \left(Vh - 2Q \right). \tag{4.63}$$

This is valid for any gap profile, $h(x)$, and is usually rewritten as

$$\frac{d}{dx}\left(h^3 \frac{dp}{dx} \right) = 6\rho\nu V \frac{dh}{dx}, \tag{4.64}$$

which is the one-dimensional version of the so-called *Reynolds equation*. Alternatively, (4.63) may be integrated directly to give

$$\frac{p(x) - p(0)}{6\rho\nu} = V \int_0^x \frac{dx}{h^2} - 2Q \int_0^x \frac{dx}{h^3}, \tag{4.65}$$

where the requirement that $p(0) = p(L) = 0$, say, fixes Q through

$$2Q \int_0^L \frac{dx}{h^3} = V \int_0^L \frac{dx}{h^2}. \tag{4.66}$$

Turning to the particular case where the bottom of the slipper is flat, $dh/dx = -\alpha$, (4.66) and (4.65) yield

$$Q = \frac{Vh_1 h_2}{h_1 + h_2}, \tag{4.67}$$

and a pressure distribution of

$$p(x) = 6\rho\nu VL \frac{(h_1 - h)(h - h_2)}{h^2(h_1^2 - h_2^2)}. \tag{4.68}$$

The force supported by the bearing is then

$$F = \int_0^L p\,dx = \frac{6\rho\nu V}{\alpha^2}\left[\ln\left(\frac{h_1}{h_2} \right) - 2\frac{h_1 - h_2}{h_1 + h_2} \right]. \tag{4.69}$$

Note that the pressure rises steadily up to a maximum of

$$\frac{p}{\rho\nu VL/h_1^2} = \frac{3h_1(h_1 - h_2)}{2h_2(h_1 + h_2)} \quad \text{at} \quad h = \frac{2h_1 h_2}{h_1 + h_2}, \tag{4.70}$$

and then drops back down to zero. In those regions where the pressure is rising, the pressure gradient opposes the flow and the oil is pulled through the gap by the frictional drag of the guide. Equation (4.61) then tells us that the velocity profile is concave. Conversely, the pressure gradient is negative for larger x, and acts to push the fluid through the gap. Here the velocity distribution is convex; that is to say, it has a fuller profile, as shown in Figure 4.8.

4.2.3 Sommerfeld's Analysis of the Journal Bearing

The case of a shaft rotating in a bearing is rather more complicated and its theory was first developed by the theoretical physicist Arnold Sommerfeld. Here we have a narrow oil film between a fixed outer cylinder (the bearing) and a rotating inner cylinder (the shaft or journal) whose axis is offset from that of the outer cylinder. The geometry is shown in Figure 4.9, where the difference in the radii of the two cylinders, which in reality is very small, is greatly exaggerated for clarity. Let R be the radius of the outer cylinder, a the radius of the inner cylinder, b the offset of the two axes, and $h(\theta)$ the oil gap between the two cylinders. Clearly these variables are related by $R = a + b + h(0)$.

It is convenient to introduce some additional notation. We write $R = a(1 + \varepsilon)$, where $\varepsilon \ll 1$, and introduce the *shaft eccentricity*, β, defined by $b = \beta \varepsilon a = \beta(R - a)$. The gap, $h(0)$, is then given by $h(0) = \varepsilon a(1 - \beta)$ and we see that $0 \le \beta \le 1$, with $\beta = 0$ corresponding to concentric cylinders and $\beta = 1$ to zero clearance. We now apply the cosine rule to a triangle whose vertices are the centres of the two cylinders and a point on the housing:

$$(a + h)^2 = R^2 + b^2 - 2Rb\cos\theta.$$

This yields, to leading order in ε,

$$h(\theta) = \varepsilon a(1 - \beta\cos\theta). \tag{4.71}$$

With this specification of $h(\theta)$, we are ready to consider the dynamics of the bearing. If we neglect the effects of curvature on the grounds that the oil gap is very thin, we may reuse much of our analysis of a slider bearing, but with x replaced by $a\theta$. In particular,

$$\frac{\partial^2 u_\theta}{\partial y^2} = \frac{1}{\rho \nu a} \frac{dp}{d\theta}$$

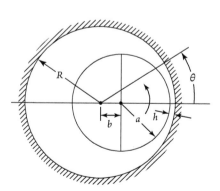

Figure 4.9 A journal bearing.

integrates to give

$$u_\theta = V\left(1 - y/h\right) - \frac{1}{2\rho v a}\frac{dp}{d\theta}\left(hy - y^2\right), \tag{4.72}$$

where y is the radial distance measured from the surface of the shaft and V is the peripheral speed of the shaft. Integrating a second time yields the volumetric flow rate in the gap,

$$Q = \frac{Vh}{2} - \frac{h^3}{12\rho v a}\frac{dp}{d\theta}. \tag{4.73}$$

Evidently, the pressure distribution is given by

$$\frac{p(\theta) - p(0)}{6\rho v a} = V\int_0^\theta \frac{d\theta}{h^2} - 2Q\int_0^\theta \frac{d\theta}{h^3}. \tag{4.74}$$

We now eliminate Q using the fact that $p(0) = p(2\pi)$, which gives us

$$2Q\int_0^{2\pi} \frac{d\theta}{h^3} = V\int_0^{2\pi} \frac{d\theta}{h^2}. \tag{4.75}$$

The integrals in (4.75) may be evaluated using the particular integral

$$\int_0^{2\pi} \frac{d\theta}{(1 - \beta\cos\theta)^n} = \frac{2\pi}{(1 - \beta^2)^{n/2}} P_{n-1}\left(1\Big/\sqrt{1 - \beta^2}\right),$$

where P_n are the Legendre polynomials, i.e. $P_1(x) = x$ and $P_2 = \frac{1}{2}(3x^2 - 1)$. This yields

$$Q = \varepsilon a V \frac{1 - \beta^2}{2 + \beta^2}. \tag{4.76}$$

The pressure and velocity distributions can now be found. In particular, substituting for Q in (4.74) gives us

$$p(\theta) - p(0) = \frac{6\rho v V}{\varepsilon^2 a}\left\{\int_0^\theta \frac{d\theta}{(1 - \beta\cos\theta)^2} - 2\frac{1 - \beta^2}{2 + \beta^2}\int_0^\theta \frac{d\theta}{(1 - \beta\cos\theta)^3}\right\}. \tag{4.77}$$

Evaluating these integrals is straight forward and gives $p(\theta)$ in closed form (Exercise 4.6).

A typical pressure distribution is shown in Figure 4.10. Note that, since $h(\theta)$ is an even function of θ, (4.73) requires that $p(\theta)$ is an odd function, as indicated in the figure. There is a sharp rise in pressure as one approaches the minimum gap at $\theta = 0$, which is similar to the behaviour of a slipper bearing, and then a rapid fall in pressure either side of the minimum gap. The resulting low pressure region can readily lead to cavitation.

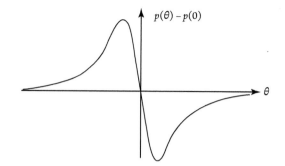

Figure 4.10 The pressure distribution in a journal bearing.

Since $p(\theta)$ is an odd function of θ, there is no net pressure force acting on the shaft in the direction of $\theta = 0$. There is, however, a net force acting on the shaft in the vertical direction, *i.e.* $\theta = \pi/2$, and integrating (4.77) gives this force as

$$F = \frac{12\pi\rho\nu V}{\varepsilon^2} \cdot \frac{\beta}{(2+\beta^2)\sqrt{1-\beta^2}}$$

(Sommerfeld, 1964). Note that F is a monotonically increasing function of the eccentricity, β, and the closer β gets to unity, the smaller the gap $h(0)$. Thus the bearing will automatically respond to changes in the external load by adjusting the eccentricity. For example, if the external load increases, then the gap $h(0)$ reduces and F rises. Moreover, the appearance of the factor $\sqrt{1-\beta^2}$ in the denominator tells us that extremely large bearing forces can be accommodated by letting the oil film become particularly thin.

4.2.4 Rayleigh's Analysis of the Stepped Bearing

We conclude our discussion of lubrication theory by considering the stepped bearing shown in Figure 4.11. As in §4.2.2, we adopt a frame of reference in which the pad is stationary and we take an origin of coordinates located below the rear of the pad. The speed of the guide is V and the oil gap $h(x)$ takes values from $h(0) = h_1$ to $h(L) = h_2$, where $L = L_1 + L_2$ is the length of the bearing. This time, however, the change from h_1 to h_2 occurs abruptly at location $x = L_1$.

If we assume steady conditions, we may integrate Reynolds equation, (4.63), over the ranges $0 < x < L_1$ and $L_1 < x < L$ to give two equations relating the pressure at $x = L_1$ to the flow rate Q. Eliminating Q from these equations shows that the pressure varies linearly from zero (the reference pressure at $x = 0$) to a maximum of

$$\frac{p_{max}}{6\rho\nu V} = \frac{h_1 - h_2}{h_1^3/L_1 + h_2^3/L_2} \tag{4.78}$$

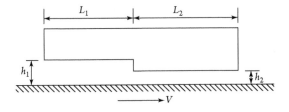

Figure 4.11 Rayleigh's stepped bearing.

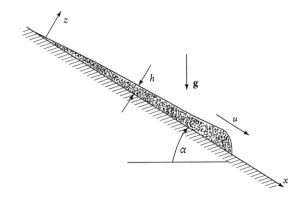

Figure 4.12 A thin film sliding down an incline.

at $x = L_1$, and then declines linearly back down to zero at $x = L$. The net force supported by the bearing, per unit width of the bearing, is evidently $p_{\max} L/2$.

It is instructive to consider the particular case of $L_1 = L_2$ and $h_1 = 2h_2$. The maximum pressure generated in a stepped bearing is then

$$p_{\max} = \frac{4\rho\nu VL}{3h_1^2},$$

which is higher than that given by (4.70) for the equivalent inclined bearing, *i.e.*

$$p_{\max} = \rho\nu VL/h_1^2.$$

Lord Rayleigh was probably the first to point out that a stepped bearing is capable of carrying a higher load than an inclined plane bearing.

4.3 Thin Films with a Free Surface

Let us now turn to thin films, by which we mean thin layers of a viscous fluid with a free surface. We restrict ourselves to plane layers and neglect surface tension.

4.3.1 Approximations and Governing Equations

The conditions under which inertia can be ignored in a thin fluid film are similar to those in which inertia can be neglected in lubrication theory. However, there is no imposed taper angle in a thin film, and so we cannot express those conditions in the form of (4.58).

Let the depth of the fluid film be h, a typical length scale parallel to the film be L, and a characteristic velocity in the plane of the film be u. By definition, a film has $h \ll L$, and so we have the estimates $\left| \nu \nabla^2 \mathbf{u} \right| \sim \nu u / h^2$ and $\left| \mathbf{u} \cdot \nabla \mathbf{u} \right| \sim u^2 / L$, from which

$$\frac{\left| \mathbf{u} \cdot \nabla \mathbf{u} \right|}{\left| \nu \nabla^2 \mathbf{u} \right|} \sim \frac{uh}{\nu} \cdot \frac{h}{L} = \mathrm{Re} \frac{h}{L}. \tag{4.79}$$

Evidently, we may ignore the non-linear inertial term, provided

$$\mathrm{Re} \ll L/h. \tag{4.80}$$

This may not, however, be enough to guarantee that $\partial \mathbf{u} / \partial t$ is appropriately small. So we shall adopt a strategy in which we start out by neglecting $\partial \mathbf{u} / \partial t$ in our analysis and then check retrospectively that this assumption is indeed self-consistent.

Let the base of the fluid film lie on the x–y plane, with $z = h(x, y, t)$ locating the surface of the film, as shown in Figure 4.12. When both the temporal and spatial inertial terms are ignored, our equation of motion simplifies to

$$\rho \nu \frac{\partial^2 \mathbf{u}}{\partial^2 z} + \rho \mathbf{g} = \nabla p. \tag{4.81}$$

Continuity of mass tells us that the z-component of velocity is small, of order hu/L, and so we dispense with the z-component of the viscous term and rewrite (4.81) as

$$\nu \frac{\partial^2 u_x}{\partial^2 z} + g_x = \frac{1}{\rho} \frac{\partial p}{\partial x}, \tag{4.82}$$

$$\nu \frac{\partial^2 u_y}{\partial^2 z} + g_y = \frac{1}{\rho} \frac{\partial p}{\partial y}, \tag{4.83}$$

$$p = \rho g_z (z - h), \tag{4.84}$$

where we have taken the pressure at the surface of the film, $z = h$, to be zero. Finally, we evaluate the pressure gradients in the x and y directions using (4.84). This gives the viscous thin-film equations:

$$\nu \frac{\partial^2 u_x}{\partial^2 z} + g_x = -g_z \frac{\partial h}{\partial x}, \tag{4.85}$$

$$\nu \frac{\partial^2 u_y}{\partial^2 z} + g_y = -g_z \frac{\partial h}{\partial y}. \tag{4.86}$$

There are two sources of motion in these equations: the components of **g** parallel to the film, which will drive flow down an incline, for example, and the normal component of **g**. It is tempting to drop the terms on the right involving the normal component, on the grounds that $\partial h/\partial x$ and $\partial h/\partial y$ are of order h/L. However, this is not legitimate when the film is almost horizontal, and hence $|g_z| \gg |g_x|, |g_y|$. In such cases, the dominant process is a slumping of the fluid film under the influence of gravity and the terms on the right of (4.85) and (4.86) are crucial.

When appropriate boundary conditions are given at the base and surface of the fluid film, which are typically no-slip at the base and zero shear stress at the surface, these equations of motion may be integrated to give u_x and u_y in terms of $h(x, y, t)$. To close the problem, one must then evaluate the mass fluxes associated with u_x and u_y and use continuity of mass to relate the divergence of those mass fluxes to $\partial h/\partial t$. The way in which this is done is, perhaps, best illustrated by example.

4.3.2 The Gravity-driven Spreading of a Circular Pool

Let us start with the case of an axisymmetric film spreading on a horizontal surface under the action of gravity. In this case $g_x = g_y = 0$ while $g_z = -g$. Ignoring any air resistance on the free surface of the film, and adopting cylindrical polar coordinates, our boundary conditions become $u_r = 0$ at $z = 0$ and $\partial u_r/\partial z = 0$ at $z = h$. Our thin-film equation,

$$\nu \frac{\partial^2 u_r}{\partial^2 z} = g \frac{\partial h}{\partial r},$$

now integrates to yield

$$u_r = -\frac{g}{2\nu} \frac{\partial h}{\partial r} (2hz - z^2), \tag{4.87}$$

and so the mass flow rate through a cylindrical surface of radius r is

$$\dot{m} = 2\pi r \rho \int_0^h u_r dz = -2\pi r \rho \frac{gh^3}{3\nu} \frac{\partial h}{\partial r}. \tag{4.88}$$

We conclude that the rate of depletion of mass in an annulus of width dr is

$$\frac{\partial \dot{m}}{\partial r} dr = -2\pi \rho \frac{\partial}{\partial r} \left[r \frac{gh^3}{3\nu} \frac{\partial h}{\partial r} \right] dr,$$

and so conservation of mass demands

$$\frac{\partial h}{\partial t} = \frac{1}{r} \frac{\partial}{\partial r} \left[r \frac{gh^3}{3\nu} \frac{\partial h}{\partial r} \right]. \tag{4.89}$$

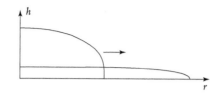

Figure 4.13 The radial spreading of a film on a horizontal substrate according to (4.90).

This is our governing equation for $h(r,t)$, which must be solved subject to the constraint of conservation of total film volume. Huppert (1982) has shown that, if $R(t)$ is the outer radius of the film, one solution of (4.89) is

$$h = \frac{4m}{3\pi\rho R^2}\left[1 - \left(\frac{r}{R}\right)^2\right]^{1/3}, \quad R = R_0\left[1 + \frac{16}{9}\frac{h_0^3}{R_0^2}\frac{gt}{\nu}\right]^{1/8}, \tag{4.90}$$

where $R_0 = R(t=0)$, h_0 is the initial depth at the centre of the film, and m is the net mass of the film. Although $\partial h/\partial r$ in (4.90) is infinite at $r = R$, as illustrated in Figure 4.13, it is assumed that surface tension regularizes the solution in this region. Indeed, it turns out that solution (4.90) is a good fit to the experimental data, provided sufficient time has elapsed.

More generally, rewriting (4.89) in the form

$$\frac{\partial h}{\partial t} = \nabla \cdot (\kappa \nabla h), \quad \kappa = \frac{gh^3}{3\nu}, \tag{4.91}$$

shows that h is governed by a non-linear diffusion equation, with a diffusivity κ which is proportional to h^3. From our intuitive understanding of solutions of the diffusion equation, we might expect that gradients in h will be progressively eradicated, and that is indeed the case. Moreover, this process will occur more rapidly in regions where h, and hence κ, is relatively large. Of course, physically, this corresponds to the film slumping under gravity. Note that we have ignored the role of surface tension in the discussion above, though this must play a decisive role at the edges of the film. The effects of surface tension are discussed in, for example, Guyon et al. (2015).

4.3.3 A Film on an Incline

Let us now return to a problem first introduced in §3.1.1. Consider two-dimensional flow down an incline of angle α, as shown in Figure 4.14. Here $h = h(x,t)$, $u_x = u_x(x,z,t)$, and we may neglect the term on the right of (4.85) on the grounds that $\partial h/\partial x$ is of order h/L. (We shall exclude very small values of α.) Our thin-film equation now simplifies to

$$\nu\frac{\partial^2 u_x}{\partial z^2} = -g\sin\alpha. \tag{4.92}$$

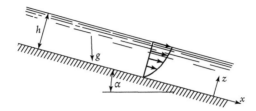

Figure 4.14 The flow of a viscous film down an inclined surface.

If we demand that the shear stress is zero at the free surface, (4.92) integrates to give

$$u_x = \frac{g \sin \alpha}{2\nu}(2h - z)z, \tag{4.93}$$

whose corresponding mass flow rate is

$$\dot{m} = \frac{\rho g h^3 \sin \alpha}{3\nu}. \tag{4.94}$$

We conclude that the rate of depletion of mass in a control volume of area hdx is

$$\frac{\partial \dot{m}}{\partial x} dx = \frac{\rho g h^2 \sin \alpha}{\nu} \frac{\partial h}{\partial x} dx, \tag{4.95}$$

and so conservation of mass requires

$$\frac{\partial h}{\partial t} = -\frac{g h^2 \sin \alpha}{\nu} \frac{\partial h}{\partial x}. \tag{4.96}$$

This is our governing equation for $h(x,t)$ and it is readily solved to give the evolution of the film surface.

Perhaps the simplest way to find the solutions of (4.96) is to rewrite it in the form

$$\frac{\partial h}{\partial t} + V(h) \frac{\partial h}{\partial x} = 0, \quad V = \frac{g h^2 \sin \alpha}{\nu}. \tag{4.97}$$

It is readily confirmed that the general solution of this equation is $h = F(x - Vt)$, where F is an arbitrary function of its argument. This rather general expression is somewhat reminiscent of d'Alembert's solution of the wave equation, in the sense that, if V were a constant, a given profile of h would travel without change of shape in the positive x direction at the speed V. The main difference here is that V is not a constant, but rather proportional to h^2, and so features of $h(x,t)$ will propagate faster in those regions where the film is relatively deep, and more slowly where h is shallow. We conclude that the overall profile of h will, in general, distort as the film spreads.

Suppose, for example, that the film occupies the region $0 \le x \le L(t)$, and that the film slowly thickens with increasing x before thinning near the front, as shown in Figure 4.12. Then the deeper parts of the film will propagate faster than those at the front or at the trailing edge. It follows that the front of the film will tend to steepen, potentially leading to a singularity, while the rear of the film will progressively develop a rather gentle slope. Of course, the tendency for the front to become singular is offset by surface tension, and so (4.97) soon breaks down near the front. In those regions where (4.97) does hold, the decline in $\partial h/\partial x$ where $\partial h/\partial x > 0$, and growth of $|\partial h/\partial x|$ when $\partial h/\partial x < 0$, follows from

$$\left(\frac{\partial}{\partial t} + V \frac{\partial}{\partial x} \right) \frac{\partial h}{\partial x} = -\frac{2V}{h} \left(\frac{\partial h}{\partial x} \right)^2, \tag{4.98}$$

which in turn follows directly from (4.97).

Now, suppose that the film occupies $0 \le x \le L(t)$, with $h(x = 0, t) = 0$. Then we have $h(0, t) = F(0) = 0$. Moreover, the film becomes relatively thin for large t and so, in those regions where (4.97) holds, $F \to 0$ for $t \to \infty$. This suggests that $x \to Vt$, or

$$h^2 \to \nu x/(gt \sin \alpha), \tag{4.99}$$

for $t \to \infty$. In practice, films often approach this solution at large times (Huppert, 1986), except near the front of the film where surface tension is important. An estimate for $L(t)$ can then be obtained from (4.99) by noting that the mass of the film is conserved,

$$m/\rho = \int_0^L h(x)dx = \left(\frac{4\nu L^3}{9gt \sin \alpha} \right)^{1/2} = \text{constant}, \tag{4.100}$$

which yields $L \sim \left(m^2 gt \sin \alpha/\rho^2 \nu \right)^{1/3}$.

4.3.4 A Thin Film on a Rotating Disc (Reprise)

In §3.3.5 we showed that the radial flow of a thin film on the surface of a spinning disc satisfies

$$\frac{\partial h}{\partial t} = -\frac{1}{r} \frac{\partial}{\partial r} \left(\frac{\Omega^2 r^2 h^3}{3\nu} \right), \tag{4.101}$$

where h is the film thickness, Ω is the rate of rotation of the disc, and we use cylindrical polar coordinates. We noted that, when h is independent of radius, it decreases with time as

$$\frac{dh}{dt} = -\frac{2\Omega^2 h^3}{3\nu}, \tag{4.102}$$

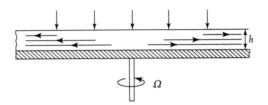

Figure 4.15 A film of fluid is centrifuged radially outward on the surface of a rotating disc.

as shown in Figure 4.15. It is readily confirmed that a more general solution for $h(r,t)$ is

$$\frac{h^2}{h_0^2} = (1-\tau)F\left((r/r_0)^{4/3}(1-\tau)\right), \quad \tau = \frac{4\Omega^2 h^2 t}{3\nu}, \quad (4.103)$$

where h_0 and r_0 are vertical and radial length scales associated with $h(r)$ at $t = 0$ and F is an arbitrary function of its argument. If, at large times, r_0 ceases to be a relevant variable, then this solution reduces to the case considered in §3.3.5, in which (4.102) integrates to give $h^2/h_0^2 = 1 - \tau$.

It is instructive to consider the particular case of $F(x) = 1/(1+x)$, corresponding to an initial condition of $h^2/h_0^2 = \left(1 + (r/r_0)^{4/3}\right)^{-1}$. We leave it as an exercise for the reader to confirm that, on rearranging (4.103) for $h(r,t)$, we recover the uniform depth solution of §3.3.5 for $\Omega t \to \infty$ (with r fixed).

○ ○ ○

This concludes our brief introduction to flows without inertia. It is a rich topic which repays cultivation. Excellent overviews may be found in Guyon et al. (2015) and Acheson (1990), while Batchelor (1967) provides many of the detailed arguments sometimes missing in textbooks. A particularly lucid discussion of lubrication theory may be found in the theoretical physics text by Sommerfeld (1964), reflecting his interest in the topic.

···

EXERCISES

4.1 *Pressure drag on a sphere in Stokes flow.* It is tempting to assume that the Stokes drag on a sphere is due entirely to the shear stresses. However, this is not the case. Use (4.15) to show that the pressure drag on a sphere at low Re is $F_D = 2\pi\rho\nu RV$.

4.2 *The breakdown of two-dimensional Stokes flow over a cylinder.* Following the dimensional arguments of §4.1.2, the pressure and vorticity fields for two-dimensional Stokes flow over a stationary cylinder of radius R must take the form

$$\frac{p}{\rho\nu} = \frac{\mathbf{V}\cdot\mathbf{x}}{R^2}f(\eta), \quad \boldsymbol{\omega} = \frac{\mathbf{V}\times\mathbf{x}}{R^2}g(\eta), \quad \eta = \frac{r^2}{R^2},$$

where $r = |\mathbf{x}|$, $\mathbf{V}$ is the far-field velocity, $VR/\nu \ll 1$, and f and g satisfy the far-field boundary conditions $f_\infty = 0$ and $g_\infty = 0$. If we adopt (x, y) coordinates centred on the cylinder, with $\mathbf{V}$ aligned with x, show that (4.1) and (4.2) require

$$\frac{p}{\rho\nu} = c\frac{Vx}{r^2}, \quad \omega = c\frac{Vy}{r^2}.$$

We expect the constant of integration, c, to be negative, since the pressure must drop in the direction of the flow. Now solve $\omega = -\nabla^2\psi$ for the streamfunction, subject to the no-slip condition on the surface of the cylinder. Hence show that, in the Stokes flow approximation

$$\psi = \frac{cVy}{4}\left[1 - \frac{1}{\eta} - \ln\eta\right].$$

We now encounter two problems. First, in line with Stokes flow over a sphere, the inertial forces are predicted to be of the order of the viscous forces for $Vr/\nu > 1$, which violates the Stokes flow approximation. Evidently, our solution is valid only in the vicinity of the cylinder. Second, this solution diverges for large r, and so we cannot match it to the outer boundary condition of $\mathbf{u} = \mathbf{V}$, which leaves the constant c undetermined. To match this flow to the far field, and hence determine c, we need to adopt the Oseen approximation.

4.3 *The Oseen approximation for flow over a cylinder at low Re.* In the Oseen approximation the vorticity is governed by $\mathbf{V} \cdot \nabla\omega = \nu\nabla^2\omega$. Since ω is odd in y for flow over a cylinder, let us write $\omega = \partial\chi/\partial y$ and look for a symmetric solution of the form

$$\chi = Vh(s)\exp(Vx/2\nu), \quad s = Vr/\nu.$$

Show that

$$\chi = -aV\mathrm{K}_0(Vr/2\nu)\exp(Vx/2\nu),$$

(see Figure 4.16) and hence

$$\omega = aV\left[\frac{Vr}{2\nu}\mathrm{K}_1(Vr/2\nu)\exp(Vx/2\nu)\right]\frac{y}{r^2},$$

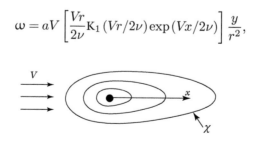

Figure 4.16 Lines of constant χ for Oseen's approximation to flow over a cylinder.

where a is a dimensionless constant of integration and K_0 and K_1 are the usual modified Bessel functions. Confirm that, for $Vr/\nu \ll 1$, the vorticity distribution given above may be matched to the Stokes flow vorticity of Exercise 4.2 by taking $a = c$. In fact, Lamb (1932) goes further and shows how to determine c by matching the Oseen and Stokes *velocity fields* in the vicinity of the cylinder and in the limit of $\mathrm{Re} = 2RV/\nu \ll 1$. Lamb's solution yields

$$c = -\frac{2}{\ln(8/\mathrm{Re}) - \gamma + 1/2} = -\frac{2}{\ln(7.41/\mathrm{Re})},$$

where γ is Euler's constant, $\gamma \approx 0.5772$.

4.4 *A physical interpretation of the breakdown of Stokes flow over a sphere at* $V|\boldsymbol{x}|/\nu \sim 1$. In the Stokes approximation the advection of vorticity, $\mathbf{u} \cdot \nabla \boldsymbol{\omega}$, is dropped from the transport equation for $\boldsymbol{\omega}$, (2.45), on the grounds that it is too weak. When the time derivative of $\boldsymbol{\omega}$ is also dropped, to yield the governing equation $\nu\nabla^2\boldsymbol{\omega} = 0$, it is implicitly assumed that the redistribution of vorticity by diffusion has had adequate time to complete. Since the diffusion length is $\ell \sim \sqrt{\nu t}$, this amounts to an assumption that the time to diffuse across a domain of size L, *i.e.* L^2/ν, is much less than any other relevant timescale.

For flow in an infinite domain, these assumptions inevitably break down in the far-field. Consider steady flow over a sphere. Show, through a consideration of the advection and diffusion of vorticity, that the vorticity generated at the surface of the sphere cannot propagate a distance upstream greater than $|\mathbf{x}| \sim \nu/V$. Hence show that the Stokes approximation must break down upstream of the sphere at points where $V|\mathbf{x}|/\nu \sim 1$.

4.5 *Stokes flow over a spherical gas bubble.* Suppose we replace the rigid sphere of §4.1.2 with a gas bubble of radius R. Assume that surface tension keeps the bubble spherical and that the dynamic viscosity of the gas is so small that there is negligible tangential stress exerted by the gas on the interface. As in §4.1.2, we adopt a frame of reference in which the sphere is stationary. In such a situation, the only change as far as the external flow is concerned is that the no-slip condition at $|\mathbf{x}| = R$ must be replaced by one of zero tangential stress on the bubble surface. The analysis therefore remains unchanged up to and including equations (4.13) and (4.14), which ensure zero normal velocity at the interface $|\mathbf{x}| = R$.

Note that equations (4.13) and (4.14) are expressed in cylindrical polar coordinates (r, θ, z). Rewrite these equations in terms of spherical polar coordinates (r, θ, ϕ) and deduce an expression for the poloidal velocity, u_θ, tangential to spherical surfaces centred on the bubble. Hence find the shear stress tangential to the surface of the bubble,

$$\tau_{r\theta} = \rho\nu r \frac{\partial}{\partial r}\left(\frac{u_\theta}{r}\right), \quad r = R,$$

as a function of the constant c in (4.13). Show that the choice of $c = -1$ in these equations ensures that the tangential stress vanishes at $|\mathbf{x}| = R$, and hence corresponds to flow over a gas bubble. More generally, if the dynamic viscosity of the fluid in the bubble (or droplet) is finite—let us call it μ_{sphere}—then the condition of zero tangential stress at $|\mathbf{x}| = R$ must be replaced by one of continuity of tangential stress across the interface. On calculating the flow in the bubble or droplet, and matching tangential stresses at the interface, the constant c now turns out to be

$$c = -\frac{2\mu + 3\mu_{\text{sphere}}}{2\mu + 2\mu_{\text{sphere}}}, \quad \mu = \rho\nu.$$

(See, for example, the discussion in Batchelor, 1967.) This coincides with the special cases of a rigid sphere ($c = -3/2$) and a gas bubble of low density ($c = -1$).

4.6 *The pressure distribution in a journal bearing.* Show that (4.77) gives the pressure distribution in a journal bearing as

$$p(\theta) - p(0) = -\frac{6\rho\nu V}{\varepsilon^2 a} \left\{ \frac{1}{2 + \beta^2} \frac{\beta \sin\theta(2 - \beta\cos\theta)}{(1 - \beta\cos\theta)^2} \right\}.$$

4.7 *Stokes flow induced by a spinning sphere.* Consider a rigid sphere of radius R which is centred on the origin, sits in an infinite fluid, and rotates with angular velocity $\boldsymbol{\Omega} = \Omega\mathbf{k}$. In the Stokes limit, in which inertia is neglected, only the azimuthal component of velocity is generated in the fluid by the rotation of the sphere, as shown in Figure 4.17.

In cylindrical polar coordinates (r, θ, z) we have $\mathbf{u} = (0, \Gamma/r, 0)$, where $\Gamma = ru_\theta$ is the angular momentum density. Equation (3.63) then demands that $\nabla_*^2\Gamma = 0$, where ∇_*^2 is the Stokes operator defined by (3.31). Show that the distribution of Γ which is consistent with the no-slip condition on the sphere is $\Gamma = \Omega R^3 r^2/|\mathbf{x}|^3$. Now use symmetry and dimensional analysis to show that the viscous torque exerted on the sphere by the fluid must be of the form $\mathbf{T} = -C\rho\nu R^3\boldsymbol{\Omega}$ for some dimensionless constant C. Find the constant C.

4.8 *The coefficient of friction for a plane slipper bearing.* Show that the shear stress exerted on the guide in a plane, inclined slipper bearing is given by

$$\frac{\tau_{xy}}{2\rho\nu V} = \frac{3h_1 h_2}{h_1 + h_2} \cdot \frac{1}{h^2} - \frac{2}{h},$$

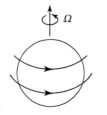

Figure 4.17 Azimuthal flow induced by a rotating sphere.

and that the drag force acting on the guide is consequently

$$D = -\int_0^L \tau_{xy}dx = \frac{2\rho\nu V}{\alpha}\left[2\ln\left(\frac{h_1}{h_2}\right) - 3\frac{h_1 - h_2}{h_1 + h_2}\right].$$

Compare this with the bearing load given by (4.69) and hence derive an expression for the coefficient of friction for the bearing.

....

REFERENCES

Acheson, D.J., 1990, *Elementary fluid dynamics*, Oxford University Press.

Batchelor, G.K., 1967, *An introduction to fluid dynamics*, Cambridge University Press.

Childress, S., 1981, *Mechanics of swimming and flying*, Cambridge University Press.

Einstein, A, 1906, A new determination of the molecular dimensions. *Ann. Physik*, **19**(2), 289–306.

Guyon, E., Hulin, J.-P., Petit, L., & Mitescu, C.D., 2015, *Physical hydrodynamics*, Oxford University Press.

Huppert, H.E., 1982, The propagation of two-dimensional and axisymmetric viscous gravity currents over a rigid horizontal surface. *J. Fluid Mech.*, **121**, 43–58.

Huppert, H.E., 1986, The intrusion of fluid mechanics into geology. *J. Fluid Mech.*, **173**, 557–94.

Lamb, H., 1932, *Hydrodynamics*, 6th Ed., Cambridge University Press.

Moffatt, H.K., 1964, Viscous and resistive eddies near a sharp corner. *J. Fluid Mech.*, **18**, 1–18.

Prandtl, L., 1952, *Essentials of Fluid Dynamics*, Blackie & Son.

Sommerfeld, A., 1964, *Mechanics of deformable bodies*, Lectures on Theoretical Physics Vol. II, Academic Press.

Rosenhead, L., 1963, *Laminar Boundary Layers*, Oxford University Press.

Stokes, G.G., 1951, On the effect of the internal friction of fluids on the motion of pendulums. *Trans. Cambridge Phil. Soc.*, IX, 8–93.

Taylor, G.I., 1951, Analysis of the swimming of microscopic organisms. *Proc. Roy. Soc. A*, **209**, 447–61.

5

Laminar Flow at High Reynolds Number
Boundary Layers

(a)   1601   (b)

(a) The German engineer Ludwig Prandtl (1875–1953) contributed greatly to the early development of aerodynamics. His introduction of the boundary layer in 1904 revolutionized fluid mechanics. (Image rights held by DLR-Archiv.) (b) The Hungarian engineer Theodore von Kármán (1881–1963) championed aerodynamics in the USA.

5.1 Prandtl's Boundary Layer and a Revolution in Fluid Dynamics

Navier and Stokes established the viscous equations of motion in the first half of the 19th century. Yet, as late as 1904 there remained a vast chasm between the engineers who were studying the flow of real fluids, such as in the fledgling field of aerodynamics, and the applied mathematicians working on theoretical hydrodynamics (inviscid theory). As Rayleigh put it in 1916:

> During the last few years much work has been done in connection with artificial flight. We may hope that before long this may be coordinated and brought into closer relationship with theoretical hydrodynamics. In the meantime, one can hardly deny that much of the latter science is out of touch with reality.

Note that Rayleigh's comment is dated 1916, some 12 years after Prandtl first introduced the ideas of the boundary layer and boundary-layer separation. This reflects the surprisingly long time that it took the fluid dynamics community to fully appreciate the significance of Prandtl's ideas.

We introduced the basic idea of a boundary layer in §2.6.1, and we now build on that discussion. Consider a high Reynolds number flow over a streamlined body, such as the wing shown in Figure 5.1. Let L be the geometric scale of that body, say the chord of the wing, and V a typical free-stream velocity, so that the Reynolds number is $\mathrm{Re} = VL/\nu$. Because Re is large, viscous stresses are negligible in regions where the velocity gradients scale on L. Prandtl's proposal was to divide the flow into an external region, which may be treated as effectively inviscid, and thin boundary layers adjacent to the surfaces of the wing, where the transverse velocity gradients are large and so the shear stresses play a decisive role despite the small viscosity. The role of the boundary layer is then to pull the free-stream velocity down to zero at the surface of the wing, as demanded by the no-slip condition. In such a scheme one first solves the external inviscid problem subject to the free-slip boundary condition $\mathbf{u} \cdot d\mathbf{S} = 0$ on the surface. This then provides the outer boundary conditions for the boundary layers top and bottom of the wing, and in particular it provides the tangential velocity at the outer edges of the boundary layers.

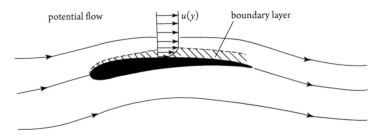

potential flow $u(y)$ boundary layer

Figure 5.1 The boundary layer on an aerofoil.

Of course, the viscous forces must compete with the inertial forces within the boundary layers in order to reduce the tangential velocity to zero at the surface of the wing. This force balance allows us to estimate the boundary-layer thickness, δ. In particular, the inertial forces scale as $\mathbf{u} \cdot \nabla \mathbf{u} \sim V^2/L$ while the viscous forces are of the order of $\nu \nabla^2 \mathbf{u} \sim \nu V/\delta^2$, and equating the two yields

$$\delta \sim L\sqrt{\nu/VL} \sim L\mathrm{Re}^{-1/2}. \tag{5.1}$$

We conclude that, when $\mathrm{Re} \gg 1$, the boundary layers are indeed very thin.

Prandtl took advantage of the fact that $\delta \ll L$ to develop a simplified set of equations for boundary layers, as we now discuss. In view of the smallness of δ we might treat the surface as if it were locally flat, and so we employ Cartesian coordinates where x is aligned with the surface and y points in the normal direction. Moreover, for simplicity, we shall restrict ourselves to steady, two-dimensional flow. Continuity then requires

$$u_y \sim \frac{\partial u_y}{\partial y}\delta \sim \frac{\partial u_x}{\partial x}\delta \sim \frac{\delta}{L}V,$$

and so, as expected, the normal component of velocity in the boundary layer is small. Comparing the magnitude of the acceleration terms in the two components of the Navier–Stokes equation,

$$u_x \frac{\partial u_x}{\partial x} + u_y \frac{\partial u_x}{\partial y} = -\frac{1}{\rho}\frac{\partial p}{\partial x} + \nu\left(\frac{\partial^2 u_x}{\partial x^2} + \frac{\partial^2 u_x}{\partial y^2}\right),$$

$$u_x \frac{\partial u_y}{\partial x} + u_y \frac{\partial u_y}{\partial y} = -\frac{1}{\rho}\frac{\partial p}{\partial y} + \nu\left(\frac{\partial^2 u_y}{\partial x^2} + \frac{\partial^2 u_y}{\partial y^2}\right),$$

now tells us that

$$\frac{\partial p}{\partial y} \sim \frac{\delta}{L}\frac{\partial p}{\partial x},$$

and so we may ignore transverse gradients in pressure. In short, to leading order in δ/L, the pressure in the boundary layer at any one streamwise location is independent of y and *imposed* on the fluid by the pressure just outside the boundary layer. Moreover, the viscous forces are clearly dominated by the large transverse gradient in velocity. So, for a steady, laminar, two-dimensional boundary layer, Prandtl gave the governing equations as

$$\boxed{\begin{aligned} u_x \frac{\partial u_x}{\partial x} + u_y \frac{\partial u_x}{\partial y} &= -\frac{1}{\rho}\frac{dp}{dx} + \nu\frac{\partial^2 u_x}{\partial y^2}, \\ \frac{\partial u_x}{\partial x} + \frac{\partial u_y}{\partial y} &= 0, \end{aligned}} \tag{5.2}$$

where an ordinary derivative is used for the pressure gradient. These equations must be solved subject to a known, imposed external pressure gradient, dp/dx, and imposed external velocity, $V(x)$, both of which are obtained from the external inviscid flow, and which are related by

$$V\frac{dV}{dx} = -\frac{1}{\rho}\frac{dp}{dx}.$$

We note, in passing, that these boundary-layer equations are often rewritten in *integral form*, which then lends itself to approximate methods of solution. The resulting integral equation, and its utility in approximate analysis, is discussed in Exercise 5.2.

Of course, there is not a sharp edge to a boundary layer, but rather a gradual transition from the viscous flow near the wall to the inviscid external motion. So, in practice one asks that u_x within the boundary layer blends smoothly into $V(x)$ at a distance which is large by comparison with a notional estimate of the boundary-layer thickness, but small by comparison with L, i.e. in the region $\delta \ll y \ll L$.

This entire procedure works well when there is no boundary-layer separation, because the external flow can then be found by solving the inviscid equations of motion subject to the free-slip boundary condition $\mathbf{u} \cdot d\mathbf{S} = 0$ on the surface of the body. However, when separation occurs, as in flow over a bluff body, the inviscid flow is confined to the region outside the wake, whose spatial extent may not be known in advance. As noted in §2.6.1, boundary-layer separation occurs at points where there is a positive pressure gradient in the free stream, causing the external flow to slow down. The same pressure gradient acts to decelerate the fluid adjacent to the boundary, but because that has very little momentum, it soon reverses direction, causing the entire boundary layer to detach.

5.2 The Archetypal Boundary Layer: a Flat Plate Aligned with a Uniform Flow

A thin, flat plate of semi-infinite extent aligned with a uniform external flow, V, provides a simple illustration of Prandtl's boundary-layer equations (Figure 5.2). Here, the only possible choice for L is the distance from the leading edge of the plate, x, as there are no other length scales available. The force balance that leads to (5.1) then suggests that we take $\delta \sim \sqrt{\nu x/V}$. Alternatively, we might treat the boundary layer as a diffusion layer for

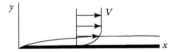

Figure 5.2 A flat-plate boundary layer.

vorticity, as discussed in §2.7.3. Vorticity created at the leading edge of the plate travels downstream a distance $x \sim Vt$ in a time t. In the same time it diffuses a distance $y \sim \sqrt{\nu t}$ from the surface of the plate (see §3.2.1), and eliminating t from these estimates yields, once again, $\delta \sim \sqrt{\nu x/V}$. Either way, the scaling $\delta \sim \sqrt{\nu x/V}$ provides estimates for the wall shear stress, τ_w, and the net drag exerted on the plate in the range $0 \to x$:

$$\tau_w(x) \sim \rho \nu \left(V/\delta\right) \sim \rho V^2 \sqrt{\nu/Vx}\,, \quad F_D(x) \sim \rho V^2 x \sqrt{\nu/Vx}\,. \tag{5.3}$$

We shall now show that the estimates in (5.3) are correct. Since the external flow is uniform, it is natural to look for a self-similar solution of equations (5.2) of the form

$$u_x/V = f\left(y/\delta_0\right) = f(\eta), \quad \delta_0 = \sqrt{\nu x/V},$$

where $f(0) = 0$ and $f(\infty) = 1$. We now introduce a function g defined by $f = g'(\eta)$ and $g(0) = 0$. Continuity then requires

$$u_y = \frac{V \delta_0}{2\,x}\left(\eta\, g'(\eta) - g\right),$$

while the equation of motion,

$$u_x \frac{\partial u_x}{\partial x} + u_y \frac{\partial u_x}{\partial y} = \nu \frac{\partial^2 u_x}{\partial y^2}\,,$$

yields $2g''' + gg'' = 0$. This must be solved subject to the boundary conditions

$$g(0) = 0,\ g'(0) = 0,\ g'(\infty) = 1.$$

Numerical integration shows that $f(\eta)$ grows monotonically from the surface of the plate to the edge of the boundary layer, with $f(4) = 0.956$ and $f'(0) = 0.332$ (see Exercise 5.3). We conclude that the boundary-layer thickness and wall shear stress are

$$\delta \approx 4\sqrt{\nu x/V}, \quad \tau_w = 0.332\rho V^2 \sqrt{\nu/Vx}\,.$$

The drag on a plate of length ℓ is then

$$F_D = 2\int_0^\ell \tau_w\,dx = 1.33\rho V^2 \ell \sqrt{\nu/V\ell}\,,$$

which is consistent with scaling (5.3). This solution is due to Blasius and we shall see that it is indicative of the behaviour of more complex boundary layers.

5.3 A Generalization of Prandtl's Boundary Layer to Other Physical Systems

5.3.1 A Popular Model Problem and the Concept of Matched Asymptotic Expansions

As Prandtl was well aware, the concept of a boundary layer may be generalized to other physical systems in which a small parameter, say ε, multiplies the highest derivative in the governing differential equation. The general idea is the following. Putting $\varepsilon = 0$ in the governing equation reduces the order of that equation, and so the solution corresponding to $\varepsilon = 0$ cannot satisfy all of the boundary conditions. For example, in the case of fluid flow, the inviscid solution cannot meet the no-slip condition at a solid surface, merely the boundary condition of zero normal velocity. The exact solution for a small but finite ε must then contain a thin transitional layer adjacent to the surface or point where the $\varepsilon = 0$ solution fails to meet the additional boundary condition. The role of that thin layer is to adapt the $\varepsilon = 0$ solution to the additional boundary condition demanded by a finite ε, and of course this must all occur within a narrow region so that the product of a small ε with a high derivative leaves the term involving ε of order unity in the governing equation.

There is a simple model problem, popularized by Van Dyke (1975), which helps get across the general idea. Consider the equation

$$\varepsilon f''(x) + f'(x) = a\,, \quad f(0) = 0, \quad f(1) = 1, \tag{5.4}$$

where $0 < a < 1$ and ε is small and positive. The solution corresponding to $\varepsilon = 0$ and satisfying $f(1) = 1$, but not $f(0) = 0$, is

$$f_{\text{out}}(x) = (1 - a) + ax. \tag{5.5}$$

We refer to this as the *outer solution*, and to the corresponding boundary condition at $x = 1$ as the *outer boundary condition*. This is analogous to the solution for inviscid flow over a streamlined body. The exact solution for a finite ε which satisfies *both* boundary conditions is readily shown to be

$$f(x) = (1 - a)\frac{1 - e^{-x/\varepsilon}}{1 - e^{-1/\varepsilon}} + ax. \tag{5.6}$$

Away from the origin, $\varepsilon \ll x < 1$, this reverts to the outer solution, $f_{\text{out}}(x) = (1 - a) + ax$, whereas close to the origin, within a thin layer of thickness $x = O(\varepsilon)$, we have

$$f_{\text{in}}(x) = (1 - a)\left(1 - e^{-x/\varepsilon}\right), \tag{5.7}$$

plus a small correction of order ε. This satisfies the boundary condition $f(0) = 0$, but not the outer boundary condition $f(1) = 1$. We refer to this as the *inner solution*, and to the corresponding boundary condition at $x = 0$ as the *inner boundary condition*.

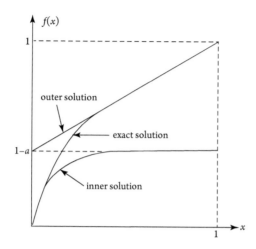

Figure 5.3 The inner, outer, and exact solutions of (5.4). (Adapted from Van Dyke, 1975.)

Note that the outer solution satisfies the outer boundary condition but not the inner boundary condition, while the inner solution satisfies the inner boundary condition but not the outer boundary condition. Note also that the inner solution holds only in a thin layer of thickness ε adjacent to $x = 0$, i.e. in a 'generalized boundary layer', whereas the outer solution is a good approximation everywhere outside this boundary layer. This is illustrated in Figure 5.3.

Perhaps the most direct route to the inner solution is to rescale x as $X = x/\varepsilon$, so that our governing equation, (5.4), becomes

$$f''(X) + f'(X) = \varepsilon a, \quad f(0) = 0, \quad f(X = 1/\varepsilon) = 1. \tag{5.8}$$

We now let $\varepsilon \to 0$ while restricting X to be of order unity. We must then abandon the outer boundary condition and our governing equation becomes

$$f''(X) + f'(X) = 0, \quad f(0) = 0, \tag{5.9}$$

whose solution is

$$f_{\text{in}}(X) = d\left(1 - e^{-X}\right), \tag{5.10}$$

d being a constant of integration. In order to recover the inner solution (5.7) we must now choose $d = 1 - a$. However, we know this only because we obtained the inner solution from the exact solution (5.6). Could we have predicted that $d = 1 - a$ without knowing the exact solution, but merely the outer solution? The answer turns out to be yes, because there is a general matching principle that ensures that the inner solution blends smoothly into the outer solution. This principle demands that the $X \to \infty$ limit of the inner solution is equal to the $x \to 0$ limit of the outer solution. In this particular example, (5.5) and (5.10) yield $f_{\text{out}}(x \to 0) = 1 - a$ and $f_{\text{in}}(X \to \infty) = d$, so that the matching principle requires $d = 1 - a$. The process of separately finding the inner and outer solutions, and then fixing the constants of integration through this matching procedure, is known as the method of *matched asymptotic expansions*.

5.3.2 Prandtl's Generalization of the Boundary Layer: Another Model Problem

Much of the formal mathematical apparatus of matched asymptotic expansions was developed in the 1950s. However, some twenty years earlier, Prandtl already had a good understanding of the role of 'generalized boundary-layers' in systems in which a small parameter multiplies the highest derivative in the governing equation. Indeed, according to Schlichting (1979), Prandtl's preferred model system which illustrates this behaviour was the simple damped mass–spring oscillator shown in Figure 5.4 (a). This is governed by

$$m\ddot{x}(t) + c\dot{x}(t) + kx = 0, \quad x(0) = 0, \tag{5.11}$$

where m is the mass, c a viscous drag coefficient, and k the spring stiffness. Prandtl considered the limit of a small mass, $m \to 0$, so that, once again, we have a small parameter multiplying the highest derivative.

Introducing the *short* timescale $\tau = m/c$, (5.11) may be rewritten as

$$\tau\ddot{x}(t) + \dot{x}(t) + (k/c)x = 0, \quad x(0) = 0, \tag{5.12}$$

which might be compared with (5.4). The exact solution of this equation, valid for any m or τ, is evidently

$$x(t) = Ae^{-t/2\tau}\left[e^{\lambda t/2\tau} - e^{-\lambda t/2\tau}\right], \quad \lambda = \sqrt{1 - 4\tau k/c},$$

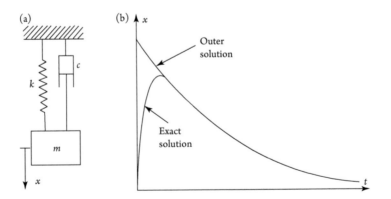

Figure 5.4 (a) A damped mass–spring oscillator. (b) The behaviour of the mass–spring oscillator showing the exact solution (5.13) and outer solution (5.15).

where the constant of integration A is fixed by the initial velocity, $u_0 = \dot{x}(0)$. However, when $m \ll c^2/k$, or equivalently $\tau \ll c/k$, we have $\lambda = 1 - 2k\tau/c$, and so our exact solution simplifies to

$$x(t) = \tau u_0 \left[\exp(-kt/c) - \exp(-t/\tau)\right], \quad \tau \ll c/k. \tag{5.13}$$

This solution takes different forms depending on whether we are close to $t = 0$, i.e. $t \sim \tau$, or else $t \gg \tau$. The inner and outer solutions corresponding to (5.13) are evidently

$$\text{inner solution: } x_{\text{in}}(t) = \tau u_0 \left[1 - \exp(-t/\tau)\right], \quad t \sim \tau \ll c/k, \tag{5.14}$$

$$\text{outer solution: } x_{\text{out}}(t) = \tau u_0 \exp(-kt/c), \quad t \sim c/k \gg \tau. \tag{5.15}$$

The outer solution (5.15) is compared with the exact solution in Figure 5.4 (b) and the boundary-layer behaviour near $x = 0$ is evident. At short times the mass moves downward because of its initial momentum, but eventually that momentum is lost, primarily through viscous drag, and the spring then pulls the mass back up towards $x = 0$. During the initial downward motion the primary force balance is between inertia and viscous drag, while the return motion is characterized by a balance between the spring force and drag. Since the mass has very little inertia, it does not take long before its initial velocity reverses, which is the origin of the boundary-layer behaviour in this case.

In order to illustrate the matching principle of §5.3.1, we now construct the inner and outer solutions without reference to the exact solution (5.13), fixing any unknown constants of integration using the method of matched asymptotic expansions. Our first step is to set $m = 0$ in (5.11), which then becomes $c\dot{x}(t) + kx = 0$. This has the solution

$$x_{\text{out}}(t) = B\exp(-kt/c), \quad m = 0, \tag{5.16}$$

for some B, which is as yet undetermined. Note that this represents the return motion, and that the balance is between the spring force and viscous drag. Note also that this does not satisfy the 'inner boundary condition' of $x(0) = 0$. On the other hand, if we rescale time according to $T = t/\tau$, our governing equation becomes

$$\ddot{x}(T) + \dot{x}(T) + \left(km/c^2\right)x = 0, \quad x(0) = 0.$$

For small m this simplifies to $\ddot{x}(T) + \dot{x}(T) = 0$, which represents the initial outward motion and constitutes a balance between inertia and drag. It may be solved subject to the inner boundary condition $x(0) = 0$ to yield

$$x_{\text{in}}(T) = \tau u_0 \left[1 - \exp(-T)\right], \quad t \sim \tau \ll c/k, \tag{5.17}$$

which is the same as (5.14). We now demand that our inner and outer solutions merge smoothly in the sense that $x_{\text{in}}(T \to \infty) = x_{\text{out}}(t \to 0)$, from which we obtain $B = \tau u_0$ and (5.16) becomes

$$x_{\text{out}}(t) = \tau u_0 \exp(-kt/c), \quad t \gg \tau. \tag{5.18}$$

We have recovered the inner and outer solutions, (5.14) and (5.15), but without the need to first find the exact solution, (5.13).

5.4 The Effects of an Accelerating External Flow on Boundary-layer Development

We now ask what happens if the external flow accelerates or decelerates. As we shall see, an accelerated external flow slows the growth of the boundary layer, while deceleration leads to separation.

5.4.1 The Falkner–Skan Solutions for Flow over a Two-dimensional Wedge

Consider the inviscid, irrotational flow over a wedge of half angle α, as shown in Figure 5.5. If the wedge occupies the region $-\alpha < \theta < \alpha$ then, in polar coordinates centred on the apex of the wedge, the streamfunction in the region $\alpha < \theta < \pi$ is readily shown to be

$$\psi \sim r^{n+1} \sin\left(\frac{\pi(\theta - \alpha)}{\pi - \alpha}\right), \quad n = \frac{\alpha}{\pi - \alpha}. \tag{5.19}$$

The radial velocity on the surface of the wedge is then proportional to r^n. If we now rotate our coordinate system and use Cartesian coordinates in which x is aligned with the upper surface of the wedge, then the inviscid surface velocity can be written in the form $V(x) = Cx^n$, $0 \leq n \leq 1$, for some constant C. Evidently, the inviscid flow accelerates over the surface of the wedge and so this provides a useful model problem with which to study the effects of acceleration on the development of a boundary layer.

It turns out that there is an exact solution for the boundary layer in this case. The force balance

$$V\frac{dV}{dx} \sim \nu\frac{V}{\delta^2}, \tag{5.20}$$

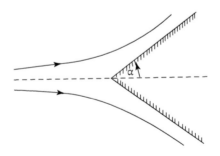

Figure 5.5 Schematic of inviscid low over a wedge.

gives a nominal boundary-layer thickness of $\delta_0 = \sqrt{\nu x / V}$ and suggests the trial solution

$$u_x = V(x)f\left(y/\delta_0\right) = V(x)f(\eta), \tag{5.21}$$

where $f(0) = 0$ and $f(\infty) = 1$. In terms of the streamfunction we have $\psi = V\delta_0 g(\eta)$, where $f = g'(\eta)$ and $g(0) = 0$. For the case where V is constant (*i.e.* $n = \alpha = 0$), this is identical to the self-similar solution for the boundary-layer on a flat plate. The difference here is that V varies as $V(x) = Cx^n$. On substituting this into the boundary-layer equations (5.2) we find, after a little algebra,

$$2g''' + 2n\left(1 - g'^2\right) + (n+1)gg'' = 0, \tag{5.22}$$

which must be solved subject to the boundary conditions

$$g(0) = 0,\ g'(0) = 0,\ g'(\infty) = 1. \tag{5.23}$$

Once again, this reverts to the Blasius solution for a flat-plate boundary layer when $n = 0$. At the other extreme, for $n = 1$ (*i.e.* $\alpha = \pi/2$) we recover the stagnation-point flow of §3.1.2. The solutions of (5.22) are called the Falkner–Skan solutions.

The velocity profiles for $n = 0, 1$ are shown in Figure 5.6 and it is clear that these profiles become fuller as n increases from 0 to 1. This is an inevitable consequence of (5.2) applied at the boundary:

$$\nu\left[\frac{\partial^2 u_x}{\partial y^2}\right]_0 = \frac{1}{\rho}\frac{dp}{dx} = -V\frac{dV}{dx}. \tag{5.24}$$

The corresponding wall shear stress, τ_w, scales as

$$\tau_w = \rho\nu\left[\frac{\partial u_x}{\partial y}\right]_0 \sim \rho\nu\frac{V}{\delta_0} \sim x^{(3n-1)/2}, \tag{5.25}$$

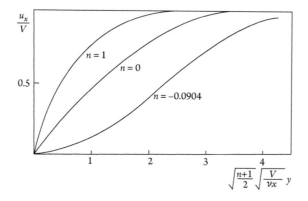

Figure 5.6 Boundary-layer profiles for the external velocity $V(x) = Cx^n$. For $0 < n < 1$ this corresponds to flow over a wedge. (Adapted from Schlichting, 1979.)

with uniform wall shear when $n = 1/3$ (*i.e.* $\alpha = \pi/4$). However, the main effect of an accelerated external flow is to *suppress the growth of the boundary layer*, with

$$\delta_0 = \sqrt{\nu x / V} \sim x^{(1-n)/2}. \tag{5.26}$$

The fastest growth rate corresponds to zero pressure gradient (*i.e.* to a thin, flat plate aligned with the flow), in which case $\delta_0 \sim x^{1/2}$, and the slowest to stagnation-point flow, in which case the boundary layer has a constant thickness. This partial suppression of the growth of δ is a direct result of the external flow sweeping the vorticity generated at the boundary back towards that boundary. That is to say, as the vorticity tries to diffuses out from the surface, the external flow sweeps it back inward, thus inhibiting the growth in δ.

In principle one can also look for decelerating solutions of (5.22), in which $n < 0$, although some thought is required to picture the corresponding external flow. It turns out that such boundary-layer solutions exist, but only for $n > -0.0904$ (see Figure 5.6). These decelerating flows all exhibit a point of inflection, which is a direct consequence of (5.24) requiring a positive value of $\partial^2 u_x / \partial y^2$ at the boundary. As n decreases below zero the wall shear stress is observed to fall along with n, and at $n = -0.0904$ that stress reaches zero. There are no boundary-layer solutions of (5.22) for lower values of n, although there are solutions that exhibit regions of reverse flow. Thus $n = -0.0904$ corresponds to the largest adverse (*i.e.* positive) pressure gradient that can be accommodated without boundary-layer separation. This suggests that *even a gentle deceleration of the external flow is sufficient to trigger separation*. We shall return to this point shortly.

5.4.2 The Boundary Layer near the Forward Stagnation Point of a Circular Cylinder

So far we have considered only very simple geometries, and it is natural to ask if analytical solutions of Prandtl's boundary-layer equations exist for more complex flows. As early as 1908, Blasius suggested an approximate procedure for calculating the boundary-layer development on two-dimensional bodies which are symmetric with respect to a uniform crossflow, such as a circular cylinder in a uniform stream.

Let x represent the distance along the surface of the body, measured from the forward stagnation point. In the vicinity of that stagnation point, the solution of §3.1.2 applies and so we have $V = \alpha x$, and hence

$$\psi = V \delta_0 g(\eta), \quad u_x / V = g'(\eta), \tag{5.27}$$

$$g''' - g'^2 + g g'' = -1, \tag{5.28}$$

where α is the strain-rate at the forward stagnation point, $\delta_0 = \sqrt{\nu/\alpha}$, and $\eta = y/\delta_0$. Of course, this is simply the case of $n = 1$ in the analysis of §5.4.1. Blasius assumed that the external inviscid flow adjacent to the surface of such a body could be represented by the power series

$$V(x) = \alpha x + \beta x^3 + \gamma x^5 + \dots \tag{5.29}$$

and he looked for a solution of the boundary-layer equations of the form

$$\psi = \delta_0 \left[\alpha x g_1(\eta) + 4\beta x^3 g_3(\eta) + 6\gamma x^5 g_5(\eta) + ... \right], \quad \eta = y \big/ \delta_0 = y\sqrt{\alpha/\nu}. \quad (5.30)$$

On substituting this into (5.2) we obtain a series of recursive equations for the functions g_n, of which the first two in the sequence turn out to be

$$g_1''' - g_1'^2 + g_1 g_1'' = -1, \tag{5.31}$$

$$g_3''' - 4g_1' g_3' + g_1 g_3'' + 3g_3 g_1'' = -1 \tag{5.32}$$

(Schlichting, 1979). The first of these is simply the governing equation for stagnation-point flow, as it must be. All subsequent equations in this series are linear in g_n, given that for a prescribed value of n the forms of g_i are all known for $i < n$.

It turns out that such an approach is of little value for slender bodies, as the series is then painfully slow to converge, but it can be informative for smooth bodies with an aspect ratio of order unity. Consider the case of a circular cylinder of radius R, sitting in a uniform crossflow, V_∞. Using the irrotational solution for flow over a circular cylinder to determine the coefficients in (5.29), we find $\alpha = 2V_\infty/R$, $\beta = -V_\infty/3R^3$ etc. The resulting solution of the Blasius recursive equations, taken up to order x^{11}, is shown in Figure 5.7.

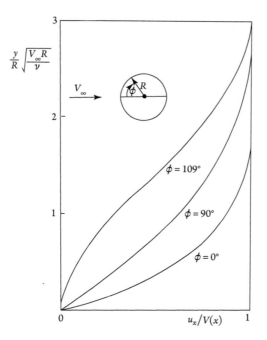

Figure 5.7 Boundary-layer profiles for flow over a circular cylinder estimated using the Blasius power-series approximation, terminated at x^{11}. (Adapted from Schlichting, 1979.)

As we might have expected from the discussion in §5.4.1, there is an inflection point in the velocity profiles in the region of adverse pressure gradient, which in this case corresponds to a polar angle in excess of $90°$. The surface shear stress is then predicted to fall to zero at $109°$, heralding the onset of reverse flow and of boundary-layer separation.

While giving a qualitatively correct picture for the development of a laminar boundary layer on a circular cylinder, some of the quantitative predictions are at odds with reality; for example, in practice, boundary-layer separation occurs earlier, at a polar angle of $70° \rightarrow 80°$ for laminar boundary layers. This discrepancy is largely due to the fact that, because of the formation of a wake, the pressure distribution outside the boundary layer differs significantly from that predicted by the inviscid flow over a circular cylinder. When the *measured* boundary-layer pressure distribution is used instead of that associated with irrotational flow, Blasius' expansion, terminated at order x^{11}, yields accurate predictions.

5.5 Jeffery–Hamel Flow in a Convergent or Divergent Channel

The main message of §5.4 is that an accelerated external flow slows the growth of a boundary layer, while deceleration leads to separation. The same behaviour is seen here. Suppose we have a steady, radial flow in a convergent channel bounded by plane walls at $\theta = \pm\alpha$, as shown in Figure 5.8. Then, in polar coordinates, continuity requires that

$$u_r = \frac{F(\theta)}{r}, \quad \omega_z = -\frac{F'(\theta)}{r^2}, \tag{5.33}$$

and the two-dimensional vorticity equation,

$$\mathbf{u} \cdot \nabla\omega_z = \nu\nabla^2\omega_z,$$

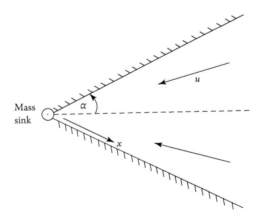

Figure 5.8 Two-dimensional radial flow in a convergent channel.

is readily shown to yield

$$2FF' + \nu (F''' + 4F') = 0, \quad F(\pm\alpha) = 0. \tag{5.34}$$

If the flow is radially inward, drained at the origin, then any vorticity that diffuses in from the walls is swept back towards the boundary. We are therefore at liberty to seek a boundary-layer solution in which the boundary layers are bounded by an inviscid core flow which is devoid of vorticity. From (5.33), that inviscid flow must take the form $F'(\theta) = 0$, or $F = -C$, where C is a positive constant.

If we now adopt a Cartesian coordinate system in which x is aligned with the lower boundary, pointing radially outward, we may seek a solution of Prandtl's boundary-layer equations in which the inviscid outer velocity is $V(x) = -C/x$. The boundary-layer solutions of §5.4.1, in which $V(x) = Cx^n$, suggest looking for a solution of the form

$$u_x/V = -xu_x/C = f(\eta), \quad \eta = y/\delta_0, \quad \delta_0 = x\sqrt{\nu/C}, \tag{5.35}$$

subject to the boundary conditions $f(0) = 0$, $f(\infty) = 1$. Continuity then yields

$$u_y = -\frac{\sqrt{\nu C}}{x}\eta f(\eta), \tag{5.36}$$

which is *towards* the boundary. On substituting (5.35) and (5.36) into the boundary-layer equation, (5.2), we obtain $f'' + 1 = f^2$, which integrates once to give

$$\tfrac{1}{2}f'^2 + f - \tfrac{1}{3}f^3 = \tfrac{2}{3}. \tag{5.37}$$

A second integration then yields

$$f = 3\tanh^2\left(c + \eta/\sqrt{2}\right) - 2, \quad \tanh^2 c = 2/3, \tag{5.38}$$

as discussed in Exercise 5.7. This boundary layer has the notable feature that $\delta \sim x$, which tells us that δ *thins* in the direction of flow. This is due to the core flow sweeping the vorticity back towards the boundaries, and it is consistent with $u_y < 0$ in (5.36).

Interestingly, no such boundary-layer solution exists for the case of a radial *outflow*. This is because the vorticity generated at the walls is now swept away from the boundaries and eventually spreads to permeate the entire flow. Consequently, to obtain solutions involving outflow, we must return to (5.34), which integrates twice to yield

$$\tfrac{2}{3}F^3 + \nu\left(F'^2 + 4F^2\right) = 2cF + d, \quad F(\pm\alpha) = 0, \tag{5.39}$$

where c and d are constants of integration. The solutions of (5.39) are discussed in, for example, Batchelor (1967). In the case of inflow we recover the boundary-layer solution above. For outflow, one might have expected a symmetric flow consisting of outward radial motion only. However, such solutions exist only for $\alpha^2 F(0)/\nu < 10.3$. At higher Reynolds numbers the solutions are complex and non-unique, typically involving thin, alternating layers of inflow and outflow. These layers have a characteristic width of order $\nu^{1/2}$, and so viscous forces remain of order unity despite the nominally large value of Re.

5.6 Boundary-layer Separation and Pressure Drag

We have already emphasized that almost any deceleration of the external flow quickly leads to boundary-layer separation. The point is that, in order to decelerate the external flow, there must be a positive gradient in the free-stream pressure (a so-called *adverse pressure gradient*), and that same pressure gradient is impressed on the fluid within the boundary layer. However, the fluid near the wall has much less momentum than that in the free stream and so it soon comes to a halt and reverses direction. This then pushes the vorticity of the upstream boundary layer away from the wall to form a detached shear layer, as shown in Figure 5.9. In the case of a bluff body, such as a sphere or circular cylinder, this occurs close to the outer edges of the body and the detached boundary layer rapidly degenerates into a turbulent wake. The pressure in the wake is roughly that of the free stream and thus considerably lower than that at the forward stagnation point. This then leads to a *pressure drag* on the body, which usually dominates the viscous drag associated with the upstream boundary layer, giving rise to a drag force of $F_D \sim \frac{1}{2}\rho V^2 A$, where V is the upstream speed and A the frontal area.

Schematic velocity profiles in the vicinity of a separation point are shown in Figure 5.10. At the separation point the surface shear stress falls to zero, subsequently turning negative in the region of reverse flow. Moreover, throughout the region of adverse pressure gradient, the velocity profiles contain an inflection point. This is an inevitable consequence of the boundary-layer equation,

$$\nu \left[\frac{\partial^2 u_x}{\partial y^2} \right]_{y=0} = \frac{1}{\rho}\frac{dp}{dx} = -V\frac{dV}{dx}, \tag{5.40}$$

which requires a positive value of $\partial^2 u_x / \partial y^2$ at the surface. These inflection points are a source of instability and hence turbulence, as discussed in Chapter 11.

The flow in the vicinity of the separation point cannot be readily analysed using conventional boundary-layer theory. For example, there is the difficulty that, after separation, the

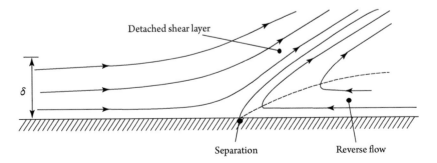

Figure 5.9 Boundary layer separation caused by an external pressure rise.

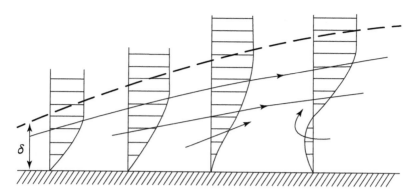

Figure 5.10 Velocity profiles near a separation point.

vorticity is no longer confined to a thin surface layer, and so it is no longer clear that we satisfy the various assumptions that underpin the boundary-layer equations (5.2). Worse still, in the case of a bluff body, the detached shear layer and subsequent wake reshapes the region of inviscid flow, so that the pressure field calculated assuming irrotational flow around the body is often nothing like the true pressure field. Crucially, this is true even upstream of the point of separation, and so the separation point itself cannot be accurately determined by irrotational theory. This is why the laminar boundary layer on a sphere or circular cylinder tends to separate at a polar angle of $70° \rightarrow 80°$, rather than $\sim 100°$, as suggested by an exclusively irrotational external flow (see §5.4.2). High-Re theories of separation suggest that the detached shear layer is of the same thickness as the upstream boundary layer, while the region of reverse flow is somewhat thinner. However, precise predictions are difficult as these high-Re flows tend to be unsteady and often develop turbulence within the boundary layer ahead of the separation point.

One unexpected phenomenon arising from boundary-layer separation is the so-called *drag crisis*. This is associated with the fact that a turbulent boundary layer is less prone to separation than a laminar one, because the turbulent mixing of momentum reduces the volume of low momentum fluid near the boundary. In the case of a sphere, turbulence starts to develop in the boundary layer at a Reynolds number of $\mathrm{Re_{crit}} \approx 3 \times 10^5$, where $\mathrm{Re} = 2RV/\nu$, V is the upstream speed, and R the radius of the sphere. At a Reynolds number somewhat below $\mathrm{Re_{crit}}$, the laminar boundary layer separates at an angle of around $80°$. This results in a wake which is slightly wider than the sphere and a drag coefficient, based on the frontal area $A = \pi R^2$, of

$$C_D = \frac{F_D}{\frac{1}{2}\rho V^2 A} \approx 0.45 - 0.48.$$

Above the critical Reynolds number, however, separation occurs on the downstream side of the sphere, with the turbulent boundary layer remaining attached up to a polar angle of $\sim 110°$. The resulting wake is then substantially thinner and the drag coefficient correspondingly smaller; around $C_D \approx 0.1$. This transition is illustrated in Figure 5.11.

That this reduction in drag is caused by boundary-layer turbulence can be demonstrated using a simple experiment, as described in Prandtl (1952). Here $\mathrm{Re} < \mathrm{Re_{crit}}$ and a thin trip-wire is attached to the surface of the sphere just ahead of the point where the laminar

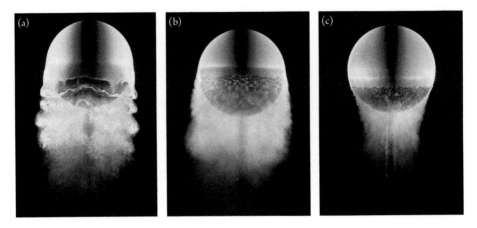

Figure 5.11 Flow past a sphere for (a) Re $= 2 \times 10^4$, (b) Re $= 2 \times 10^5$ and (c) a turbulent boundary layer. (Photographs by H. Werlé of ONERA, courtesy of J. Delery.)

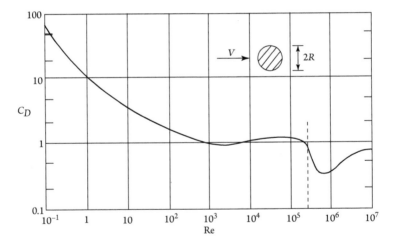

Figure 5.12 Drag coefficient $C_D = F_D / \frac{1}{2} \rho V^2 A$ for a circular cylinder as a function of Re.

boundary layer normally separates. The trip-wire triggers turbulence in the boundary layer and so causes it to remain attached up to polar angles of $\sim 110°$. This, in turn, causes a sharp drop in drag, despite the Reynolds number being subcritical.

A circular cylinder undergoes a similar transition at around the same Reynolds number, $\text{Re}_{\text{crit}} \approx 3 \times 10^5$, with the separation point moving from an upstream to a downstream location. In the process, the drag coefficient, based on a frontal area of $A = 2R$, drops by a factor of roughly four, from $C_D \approx 1.23$ to $C_D \approx 0.35$. This is illustrated in Figure 5.12.

5.7 Thermal Boundary Layers

Let us now turn to thermal boundary layers. We shall adopt the *Boussinesq approximation* in which variations in ρ are assumed small and so are important only to the extent that they introduce a buoyancy force, $-\beta(T - T_0)\mathbf{g} = -\beta\Theta\mathbf{g}$, where β is the expansion coefficient, and $\Theta = T - T_0$ is the departure from a reference temperature T_0. Our governing equations are then

$$\frac{D\mathbf{u}}{Dt} = -\nabla(p/\rho) + \nu\nabla^2\mathbf{u} - \beta(T - T_0)\mathbf{g}, \tag{5.41}$$

and the energy conservation equation

$$\frac{D}{Dt}(\rho c_p T) = -\nabla \cdot \dot{\mathbf{q}}, \quad \dot{\mathbf{q}} = -k_c\nabla T, \tag{5.42}$$

where $\nabla \cdot \mathbf{u} = 0$, k_c is the thermal conductivity, c_p the specific heat, and $\dot{\mathbf{q}}$ the conducted heat per unit area. Equation (5.42) tells us that the rate of loss of internal energy from a small material element must equal the integral of the heat conducted out through the surface of that element. (We ignore here the generation of internal energy by viscous forces.) For uniform material properties, (5.42) simplifies to

$$\frac{DT}{Dt} = \alpha\nabla^2 T, \quad \alpha = k_c/\rho c_p, \tag{5.43}$$

where α is the thermal diffusivity. The ratio of the two diffusivities, ν and α, is known as the *Prandtl number*, Pr, where $\mathrm{Pr} = \nu/\alpha$, and we expect a thermal boundary layer to be thicker than the corresponding mechanical one when $\mathrm{Pr} \ll 1$ (liquid metals), but thinner than the mechanical boundary layer when $\mathrm{Pr} \gg 1$ (lubricating oil). Note that $\mathrm{Pr} \approx 0.71$ for air.

5.7.1 Forced Convection

Convective heat transfer naturally divides into forced and free convection. We start with the former and consider the simple case of uniform steady flow over a flat plate aligned with the flow, as shown in Figure 5.13. There is no free-stream pressure gradient and we shall neglect the buoyancy force. The plate sits on the x-axis, starting at $x = 0$, and is held at the uniform temperature $T_0 + \Delta T$, while the free stream has speed V and temperature T_0. Our boundary-layer equations are

$$u_x\frac{\partial u_x}{\partial x} + u_y\frac{\partial u_x}{\partial y} = \nu\frac{\partial^2 u_x}{\partial y^2}, \tag{5.44}$$

$$u_x\frac{\partial\Theta}{\partial x} + u_y\frac{\partial\Theta}{\partial y} = \alpha\frac{\partial^2\Theta}{\partial y^2}, \tag{5.45}$$

where $\Theta(y = 0) = \Delta T$, $\Theta(y \to \infty) = 0$.

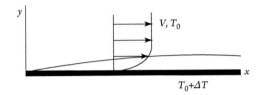

Figure 5.13 Boundary layer on a flat plate aligned with a uniform flow.

Since the velocity field is decoupled from the temperature distribution we may use the self-similar boundary-layer solution of §5.2, in which $\delta_0 = \sqrt{\nu x/V}$ and

$$u_x/V = f(y/\delta_0) = f(\eta). \tag{5.46}$$

In terms of the streamfunction this becomes $\psi = V\delta_0 g(\eta)$, where $f = g'(\eta)$ and a prime indicates differentiation with respect to η. The governing equation for g was shown in §5.2 to be $2g''' + gg'' = 0$ and the shape of the resulting velocity profile is discussed in Exercise 5.3. The similarity of (5.44) and (5.45) tells us that, when Pr $= 1$, the temperature and velocity boundary layers have the same shape, with $\Theta/\Delta T = 1 - u_x/V$. More generally, if we seek a self-similar solution for the thermal boundary layer of the form $\Theta/\Delta T = h(\eta)$, then (5.45) yields

$$2h'' + \Pr g(\eta)h' = 0. \tag{5.47}$$

This has the simple solution $h = 1 - f$ for Pr $= 1$, as it must, while the solution for arbitrary Pr is readily shown to be

$$h(\eta) = \int_\eta^\infty [f']^{\Pr}\,d\eta \bigg/ \int_0^\infty [f']^{\Pr}\,d\eta. \tag{5.48}$$

The temperature profiles for various values of Pr are given in, for example, Schlichting (1979). As expected, the thermal boundary-layer thickness δ_T is larger than δ_0 for Pr $\ll 1$, but thinner when Pr $\gg 1$. Indeed, it turns out that $\delta_T \sim \Pr^{-1/3}\delta_0$ for Pr $\gg 1$, as shown in Exercise 5.4. Of particular interest is the fact that

$$h'(0) = -[f'(0)]^{\Pr} \bigg/ \int_0^\infty [f']^{\Pr}\,d\eta = -[0.332]^{\Pr} \bigg/ \int_0^\infty [f']^{\Pr}\,d\eta = -C(\Pr),$$
$$\tag{5.49}$$

for some function C of Pr. It transpires that a good approximation to C for $0.6 < \Pr < \infty$ is $C(\Pr) \approx 0.332\Pr^{1/3}$ (see Schlichting, 1979), which is accurate to within 2% and consistent with $\delta_T \sim \Pr^{-1/3}\delta_0$. Hence the local rate of heat transfer from the plate, per unit area of the plate, is given by

$$\dot{q} = C(\mathrm{Pr})\frac{k_c\Delta T}{\delta_0} \approx 0.332\,\mathrm{Pr}^{1/3}\frac{k_c\Delta T}{x}\sqrt{\frac{Vx}{\nu}}, \qquad \mathrm{Pr} > 0.6. \qquad (5.50)$$

This is usually expressed in terms of the dimensionless *Nusselt number*, defined locally by the expression $\mathrm{Nu} = \dot{q}x/k_c\Delta T$. Equation (5.50) then yields $\mathrm{Nu} = C(\mathrm{Pr})\sqrt{Vx/\nu}$.

5.7.2 Free Convection

We now turn to free convection and consider the simple case of a flat, vertical plate which is heated. The plate is held at the uniform temperate $T_0 + \Delta T$, which is higher than the ambient temperature T_0. If x is anti-parallel to $\mathbf{g}$ and y normal to the plate, our steady boundary-layer equations are

$$u_x\frac{\partial u_x}{\partial x} + u_y\frac{\partial u_x}{\partial y} = \nu\frac{\partial^2 u_x}{\partial y^2} + g\beta\Theta, \qquad (5.51)$$

$$u_x\frac{\partial\Theta}{\partial x} + u_y\frac{\partial\Theta}{\partial y} = \alpha\frac{\partial^2\Theta}{\partial y^2}, \qquad (5.52)$$

where $\Theta(y=0) = \Delta T, \Theta(y\to\infty) = 0$. Note that the velocity profile now takes the form of a *thermal wall jet*, in which u_x vanishes at both $y=0$ and $y\to\infty$, and that the width of the wall jet is set by the *thermal* boundary-layer thickness, δ_T, as shown in Figure 5.14. The width of the *mechanical* boundary layer, by contrast, is set by the location of the peak in u_x.

Let us start our analysis by introducing a characteristic velocity and width associated with the thermal wall jet, say $V(x)$ and δ_0, based on the triple force balance

$$V\frac{dV}{dx} \sim g\beta\Delta T \sim \nu\frac{V}{\delta_0^2}, \qquad (5.53)$$

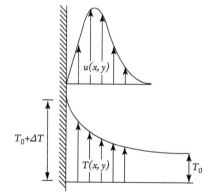

Figure 5.14 The thermal and mechanical boundary layers on a heated vertical plate.

which holds whenever Pr ~ 1. We may then define $V(x)$ and δ_0 as

$$V = \sqrt{2g\beta x \Delta T}, \quad \delta_0 = \sqrt{\nu x / V} = \left(\frac{\nu^2 x}{2g\beta \Delta T}\right)^{1/4}, \tag{5.54}$$

where $x = 0$ marks the lower edge of the plate. Consider now the self-similar solution

$$u_x = V(x)f(y/\delta_0) = V(x)f(\eta), \quad \Theta = \Delta T \, h(\eta), \tag{5.55}$$

where $f(0) = 0$, $f(\infty) = 0$, $h(0) = 1$ and $h(\infty) = 0$. Equivalently, in terms of a stream-function, ψ, we write $\psi = V\delta_0 g(\eta)$ where $f = g'(\eta)$ and $g(0) = 0$. Substituting our guesses into (5.51) and (5.52) yields, after a little algebra, the coupled equations

$$4g''' + 3gg'' - 2g'^2 + 2h = 0,$$
$$4h'' + 3\Pr g(\eta)h' = 0,$$

which are readily integrated numerically. This self-similar solution holds for *any* Pr, although V and δ_0 need not be a representative speed and width of the wall jet when Pr differs greatly from unity. Indeed, for Pr $\gg 1$ the wall-jet thickness is $\delta_T \sim \Pr^{-1/4} \delta_0$, rather than $\delta_T \sim \delta_0$, as noted in Exercise 5.5. Since the surface heat flux is of order $\dot{q} \sim k_c \Delta T / \delta_T$, the local Nusselt number then scales as

$$\mathrm{Nu} = \frac{\dot{q}x}{k_c \Delta T} \sim \frac{x}{\delta_T} \sim \Pr^{1/4}\frac{x}{\delta_0} \sim \Pr^{1/4}\left(\frac{g\beta\Delta T x^3}{\nu^2}\right)^{1/4} = \Pr^{1/4}\mathrm{Gr}^{1/4}, \tag{5.56}$$

for Pr $\gg 1$. The dimensionless parameter Gr $= g\beta\Delta T x^3/\nu^2$ is known as the local *Grashof number*. The equivalent scaling law for Pr $\ll 1$ is Nu $\sim \Pr^{1/2} \mathrm{Gr}^{1/4}$ (see Exercise 5.5).

5.8 Submerged Laminar Jets

We close this chapter with a brief discussion of submerged, laminar jets. It might seem a little odd to include such a discussion in a chapter on boundary layers, as there are no boundaries involved. However, the thinness of a submerged jet at large Re means that it may be analysed using the boundary-layer equations (5.2), in which transverse gradients in pressure are neglected, along with longitudinal gradients in the viscous stresses. So it is common to group jets along with boundary layers. One further simplification, that is particular to jets, arises from the fact that the pressure is uniform in the far field, which allows pressure gradients to be ignored everywhere within a thin jet.

As with boundary-layer theory, we have in mind the case where a suitably defined Reynolds number is large. It should be emphasized, however, that jets at even modest values of Re tend to be unstable, and so this is somewhat of an academic exercise. Nevertheless, such an analysis is of some value as it provides a starting point for stability calculations and because turbulent jets have much in common with their laminar counterparts. Let us start with the simple case of a two-dimensional jet.

5.8.1 The Two-dimensional Jet

Consider a two-dimensional, submerged jet created by injecting fluid through a slot, as shown in Figure 5.15. We take the x-axis to lie on the centreline of the jet, with the jet initiated at $x = 0$.

Our equation of motion is (5.2) in which the pressure gradient is set to zero, which we now rewrite as

$$\frac{\partial}{\partial x}\left(u_x^2\right) + \frac{\partial}{\partial y}\left(u_y u_x\right) = \nu\frac{\partial^2 u_x}{\partial y^2}. \tag{5.57}$$

Noting that both u_x and the viscous stresses vanish at large $|y|$, this may be integrated across the jet to give

$$\frac{d}{dx}\int_{-\infty}^{\infty} u_x^2 dy = 0, \quad \text{or} \quad M = \int_{-\infty}^{\infty} \rho u_x^2 dy = \text{constant}, \tag{5.58}$$

where M is the momentum flux of the jet. Next, we introduce $\delta(x)$ as a local measure of the jet thickness and let $u_0(x)$ be the centreline velocity. It is then natural to look for a self-similar solution of the form

$$u_x = u_0(x)f(y/\delta) = u_0(x)f(\eta), \tag{5.59}$$

or equivalently,

$$\psi = u_0(x)\delta(x)g(\eta), \quad f(\eta) = g'(\eta), \tag{5.60}$$

where (5.58) requires

$$u_0^2(x)\delta(x) = \text{constant}. \tag{5.61}$$

We now substitute our assumed self-similar solution into (5.57). Noting that the boundary conditions on g are

$$g(0) = g'(\infty) = g''(0) = 0, \quad g'(0) = 1, \tag{5.62}$$

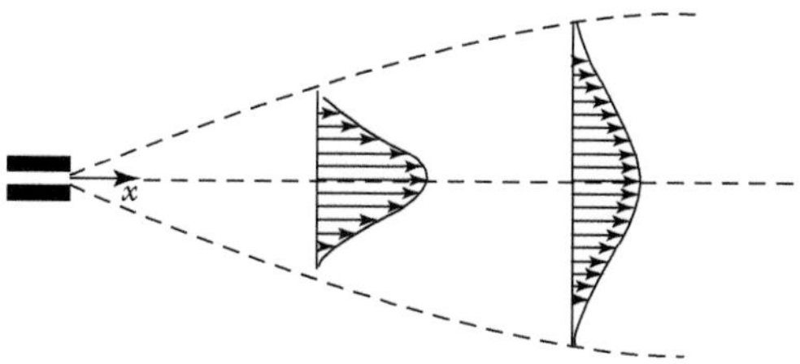

Figure 5.15 A two-dimensional jet. The spreading angle is exaggerated for clarity.

(5.57) and (5.61) give us

$$g''(\eta) = \frac{\delta^2}{\nu} \frac{du_0}{dx} gg'(\eta), \qquad (5.63)$$

where we have used $g(0) = g''(0) = 0$. This, in turn, yields two equations,

$$\frac{\delta^2}{2\nu} \frac{du_0}{dx} = -c^2, \quad g'' = -2c^2 gg', \qquad (5.64)$$

where c is a constant. The second of these integrates to $g' = c^2 \left(g_\infty^2 - g^2\right)$, where we have used $g'(\infty) = 0$, and the remaining boundary conditions, $g(0) = 0$ and $g'(0) = 1$, tell us that $c = 1/g_\infty$. Our equation for g now integrates a second time to give $g = g_\infty \tanh(\eta/g_\infty)$, which yields

$$u_x(x, y) = u_0(x) \operatorname{sech}^2 (\eta/g_\infty). \qquad (5.65)$$

It remains to pin down g_∞ and this we do by substituting the velocity profile (5.65) into (5.58). This gives

$$M = \tfrac{4}{3} g_\infty \rho u_0^2 \delta, \qquad (5.66)$$

which allows us to substitute for both g_∞ and c in terms of M. We now eliminate c from (5.64) to give

$$\left(\frac{3M}{4\rho}\right)^2 \frac{du_0}{dx} = -2\nu u_0^4, \qquad (5.67)$$

from which we obtain

$$u_0 = \left[\frac{3\left(M/\rho\right)^2}{32\nu x}\right]^{1/3}. \qquad (5.68)$$

Equation (5.66) now tells us that the jet width and dimensionless transverse coordinate η scale as

$$\frac{\delta}{x} \sim \left[\frac{Mx}{\rho\nu^2}\right]^{-1/3}, \quad \frac{\eta}{g_\infty} = \frac{y}{2} \left[\frac{M/\rho}{6\nu^2 x^2}\right]^{1/3}. \qquad (5.69)$$

Evidently the jet may be treated as thin, and (5.57) is justified, provided $Mx/\rho\nu^2 \gg 1$.

Finally, it is of interest to determine the mass flux in the jet, which we can do by combining (5.65) with (5.69) and integrating across the jet. After a little algebra we find

$$\dot{m} = \int_{-\infty}^{\infty} \rho u_x dy = \left(36\rho^2 M\nu x\right)^{1/3}. \qquad (5.70)$$

This growth in mass flux with x seems, at first sight, somewhat paradoxical. However, this is to be expected as u_y does not vanish at large $|y|$, but rather is given by

$$u_y(y \to \pm\infty) = \mp \left(M\nu/6\rho x^2\right)^{1/3}. \tag{5.71}$$

Evidently, the jet entrains fluid from the far field, causing its mass flux to increase with x.

5.8.2 The Axisymmetric Jet

An axisymmetric jet can be treated in a similar manner using cylindrical polar coordinates in which z is aligned with the axis of the jet, as shown in Figure 5.16. As before, we take u_0 to be the centreline velocity and δ the characteristic width of the jet.

The governing equation of motion is now

$$\frac{1}{r}\frac{\partial}{\partial r}(r u_r u_z) + \frac{\partial}{\partial z}(u_z^2) = \nu \frac{1}{r}\frac{\partial}{\partial r}\left(r\frac{\partial u_z}{\partial r}\right), \tag{5.72}$$

which integrates across the jet to give

$$M = 2\pi\rho \int_0^\infty u_z^2 r \, dr = \text{constant}. \tag{5.73}$$

A self-similar solution, if it exists, now requires

$$u_0^2(z)\delta^2(z) = \text{constant}, \tag{5.74}$$

which is different to the planar case. Nevertheless, following the same procedure as for the plane jet yields yet another self-similar solution. We shall not pause to spell out the details, but rather refer the reader to Exercise 5.6, or Goldstein (1965). We merely note here that taking $\Psi = u_0(z)\delta^2(z)g(r/\delta)$, where Ψ is the Stokes streamfunction, yields

$$\eta g''(\eta) - g'(\eta) = \frac{\delta^2}{\nu}\frac{du_0}{dz}gg'(\eta), \quad \eta = r/\delta(z),$$

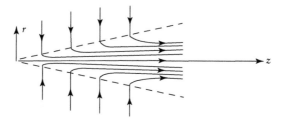

Figure 5.16 A round jet. Note that the jet entrains fluid from the far field.

which might be compared with (5.63). This, in turn, leads to

$$u_0(z) \sim \frac{M/\rho}{\nu z}, \quad \delta(z) \sim \frac{\nu z}{\sqrt{M/\rho}}, \quad \frac{u_z}{u_0(z)} = \frac{1}{\left[1 + \eta^2/2g_\infty\right]^2}. \quad (5.75)$$

Note that δ grows linearly with z. Clearly the jet is thin, *i.e.* $\delta \ll z$, and the use of (5.72) is justified, provided $M/\rho\nu^2 \gg 1$.

○ ○ ○

This concludes our introduction to boundary layers and jets. Readers seeking a more complete treatment might consult Schlichting (1979) or Rosenhead (1963). As always, both Prandtl (1952) and Batchelor (1967) are also well worth a look.

....

EXERCISES

5.1 *Measures of boundary-layer thickness and vorticity flux.* Two common measures of the thickness of a boundary layer are:

$$\text{displacement thickness, } \delta_1(x) = \int_0^\infty \left(1 - \frac{u_x}{V}\right) dy,$$

$$\text{momentum thickness, } \delta_2(x) = \int_0^\infty \frac{u_x}{V}\left(1 - \frac{u_x}{V}\right) dy,$$

where $V(x)$ is the free-stream velocity. Show that $\rho\delta_1 V$ is the reduction in mass flow rate introduced by the boundary layer, and $\rho\delta_2 V^2$ is the flux of the deficit of momentum within the boundary layer as compared with that of the inviscid flow. Confirm that, for the linear velocity profile $u_x/V = y/\delta$, we have $\delta_1 = \delta/2$ and $\delta_2 = \delta/6$.

Also show that the net vorticity and flux of vorticity at any given location are, to leading order in δ/L,

$$\int_0^\infty \omega\, dy = -V(x), \quad \int_0^\infty u_x \omega\, dy = -\tfrac{1}{2}V^2(x).$$

5.2 *The momentum integral equation for boundary layers and Pohlhausen's approximate method of solution.* Show that the boundary-layer equation, (5.2), rewritten as

$$\frac{\partial}{\partial x}\left(u_x^2\right) + \frac{\partial}{\partial y}\left(u_y u_x\right) = V\frac{dV}{dx} + \frac{\partial}{\partial y}\left(\frac{\tau_{xy}}{\rho}\right),$$

can be manipulated into the form

$$\frac{\partial}{\partial x}\left(u_x\left(V - u_x\right)\right) + \left(V - u_x\right)\frac{dV}{dx} + \frac{\partial}{\partial y}\left[\left(V - u_x\right)u_y\right] = -\frac{\partial}{\partial y}\left(\frac{\tau_{xy}}{\rho}\right).$$

Hence confirm that

$$\frac{d}{dx}\left(\delta_2 V^2\right) + \delta_1 V \frac{dV}{dx} = \frac{\tau_w}{\rho},$$

where τ_w is the shear stress at the wall. This is known as the *momentum-integral equation* and it is the starting point for several approximate methods of calculating boundary-layer behaviour. For example, consider the velocity profile

$$\frac{u_x(x,y)}{V(x)} = \left(2\eta - 2\eta^3 + \eta^4\right) + \frac{1}{6}\Lambda\eta\left(1-\eta\right)^3, \quad \Lambda = \frac{\delta^2}{\nu}\frac{dV}{dx},$$

where $\eta = y/\delta(x)$. Show that this velocity profile satisfies the boundary conditions

$$y = \delta: \quad u_x = V, \; \partial u_x/\partial y = 0, \; \partial^2 u_x/\partial y^2 = 0,$$

$$y = 0: \quad u_x = 0, \; \nu \frac{\partial^2 u_x}{\partial y^2} = -V\frac{dV}{dx},$$

where the second condition at $y = 0$ is (5.24). If we adopt this velocity profile as a reasonable approximation for a laminar boundary layer, then we can relate $\tau_w, \delta_1,$ and δ_2 to $V, \Lambda,$ and δ. Show that

$$\frac{\delta}{V}\frac{\tau_w}{\rho\nu} = 2 + \frac{\Lambda}{6}, \quad \frac{\delta_1}{\delta} = \frac{3}{10} - \frac{\Lambda}{120}, \quad \frac{\delta_2}{\delta} = \frac{37}{315} - \frac{\Lambda}{945} - \frac{\Lambda^2}{9072}.$$

Substituting for $\tau_w, \delta_1,$ and δ_2 in the momentum integral equation, and assuming that $V(x)$ is prescribed, yields an ordinary differential equation for $\delta(x)$. This can be integrated to give $\delta(x)$ and $\Lambda(x)$, and hence the streamwise development of τ_w and of the velocity profile. This approximate method of analysis is usually attributed to Pohlhausen and over the years many different versions of the scheme have been proposed.

5.3 *The Blasius velocity profile for a flat-plate boundary layer.* In §5.2 we showed that the velocity in a flat-plate boundary layer takes the form

$$u_x/V = f\left(y/\delta_0\right) = f(\eta), \quad \delta_0 = \sqrt{\nu x/V},$$

and if $f = g'(\eta)$ then g satisfies $2g''' + gg'' = 0$, subject to the boundary conditions $g(0) = 0, g'(0) = 0, g'(\infty) = 1$. Show that a power series solution for g which satisfies the boundary conditions at $\eta = 0$ is

$$g(\eta) = \gamma\frac{\eta^2}{2!} - \frac{\gamma^2}{2}\frac{\eta^5}{5!} + \frac{11\gamma^3}{4}\frac{\eta^8}{8!} - \frac{375\gamma^4}{8}\frac{\eta^{11}}{11!} + \frac{27897\gamma^5}{16}\frac{\eta^{14}}{14!} + \cdots,$$

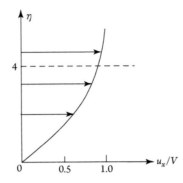

Figure 5.17 The Blasius velocity profile for a flat plate boundary layer.

and hence

$$f(\eta) = \gamma\eta - \frac{\gamma^2}{2}\frac{\eta^4}{4!} + \frac{11\gamma^3}{4}\frac{\eta^7}{7!} - \frac{375\gamma^4}{8}\frac{\eta^{10}}{10!} + \frac{27897\gamma^5}{16}\frac{\eta^{13}}{13!} + \cdots.$$

This expansion for $f(\eta)$, terminated at η^{13}, is accurate to 0.03% at $\eta = 3$ and 2.4% at $\eta = 4$. The constant γ is determined by the boundary condition $f(\infty) = 1$ and numerical integration shows that $\gamma = 0.33206$, giving $\tau_w = 0.332\rho V^2\sqrt{\nu/Vx}$. The shape of the velocity profile is shown in Figure 5.17. The boundary-layer thickness is $\delta \sim 4\delta_0$, at which the truncated expansion above remains accurate, while the displacement thickness is $\delta_1 = 1.72\,\delta_0$.

5.4 *Forced convection in a flat-plate boundary layer.* Consider the thermal boundary layer on a heated flat plate, as discussed in §5.7.1. For $\Pr \gg 1$ the thermal boundary layer is much thinner than δ_0 and so sits entirely within the uniform shear region of the Blasius velocity profile. Use the results of Exercise 5.3 to show that (5.47) becomes $4h'' + \gamma\Pr\eta^2 h' = 0$, where $\gamma = 0.332$, and hence confirm that $\delta_T \sim \Pr^{-1/3}\delta_0$ for $\Pr \gg 1$. The local Nusselt number, Nu, Stanton number, St, and skin friction coefficient, c_f, for such a boundary layer are defined through the expressions $\mathrm{Nu} = \dot{q}x/k_c\Delta T$, $\mathrm{St} = \alpha\dot{q}/k_c V\Delta T$, and $\tau_w = c_f\frac{1}{2}\rho V^2$. Use (5.50) to show that, for $\Pr > 0.6$, we have $\mathrm{Nu} \approx \frac{1}{2}\Pr^{1/3}c_f\mathrm{Re}$ and $\mathrm{St} \approx \frac{1}{2}\Pr^{-2/3}c_f$, where $\mathrm{Re} = Vx/\nu$. More generally, in forced convection *Reynolds' analogy* states that $\mathrm{St} \sim c_f$ when $\Pr \sim 1$. Why should this be so?

5.5 *Free convection from a heated vertical plate.* Consider the thermal boundary layer on a heated vertical plate, as discussed in §5.7.2. For $\Pr \gg 1$, inertia is weak and the primary force balance is between viscosity and buoyancy. Show that the heat equation, combined with this force balance, gives the wall-jet scaling laws as $\tilde{V} \sim \Pr^{-1/2}\sqrt{g\beta x\Delta T}$ and $\delta_T \sim \Pr^{-1/2}\sqrt{\nu x/\tilde{V}} \sim \Pr^{-1/4}\delta_0$, rather than (5.54). These turn out to be a good approximation in the range $0.7 < \Pr < \infty$. For $\Pr \ll 1$, on the other hand, the primary force balance is between inertia and buoyancy, with the viscous forces confined to a thin surface layer. Show that the heat equation, combined with this force balance, gives the wall-jet scaling laws as $\delta_T \sim \mathrm{Pr}^{-1/2}\,\delta_0$ and $V \sim \sqrt{g\beta x \Delta T}$. Hence show that, for $\mathrm{Pr} \ll 1$, the local Nusselt number, $\mathrm{Nu} = \dot{q}x/k_c\Delta T$, scales as

$$\mathrm{Nu} \sim \left(g\beta \Delta T x^3/\alpha^2\right)^{1/4} = \mathrm{Pr}^{1/2}\,\mathrm{Gr}^{1/4}.$$

5.6 *The axisymmetric jet.* Consider a self-similar solution of equation (5.72) of the form $\Psi = u_0(z)\delta^2(z)g(\eta)$, where $g(0) = 0$, $\eta = r/\delta$, and Ψ is the Stokes streamfunction (3.36). Show, with the help of (5.74), that

$$u_z = u_0(z)\frac{g'(\eta)}{\eta}, \quad u_r = -\delta\frac{du_0}{dz}\left[g'(\eta) - g/\eta\right],$$

and that (5.72) then yields

$$\eta g''(\eta) - g'(\eta) = \frac{\delta^2}{\nu}\frac{du_0}{dz}gg'(\eta).$$

Compare this with (5.63) and hence, following the logic of §5.8.1, show that

$$\frac{\delta^2}{2\nu}\frac{du_0}{dz} = -c^2, \quad \eta g' - 2g + c^2 g^2 = 0,$$

where c is a constant. Finally, show that $c^2 = 2/g_\infty$ and hence deduce that

$$u_0(z) \sim \frac{M/\rho}{\nu z}, \quad \frac{u_z}{u_0(z)} = \frac{1}{\left[1 + \eta^2/2g_\infty\right]^2},$$

where M is the conserved momentum flux in the jet. This solution of (5.72) is valid provided that the jet is thin, which in turn requires that $M/\rho\nu^2 \gg 1$. It is, in fact, the limiting case of an exact jet-like solution of the Navier–Stokes equation, called the Landau–Squire solution, which is valid for all values of $M/\rho\nu^2$. (See, for example, the discussion in Rosenhead, 1963.)

5.7 *The shape of the boundary layer in a converging channel flow.* In §5.5 we showed that the boundary layer in a converging channel flow is governed by

$$u_x = V(x)f(\eta), \quad \tfrac{1}{2}f'^2 + f - \tfrac{1}{3}f^3 = \tfrac{2}{3},$$

where $V = -C/x$ is the external velocity, $\eta = y/\delta_0$, $\delta_0 = x\sqrt{\nu/C}$, and the boundary conditions on f are $f(0) = 0$ and $f(\infty) = 1$. Show that the governing equation for f can be rewritten as

$$f'(\eta) = \sqrt{2/3}\,(1 - f)\,\sqrt{2 + f},$$

and that this integrates to give

$$f = 3\tanh^2\left(c + \eta/\sqrt{2}\right) - 2.$$

The constant of integration, c, is fixed by the boundary condition $f(0) = 0$. Find expressions for the displacement thickness and wall shear stress for this boundary layer.

....

REFERENCES

Batchelor, G.K., 1967, *An introduction to fluid dynamics*, Cambridge University Press.
Goldstein, S., 1965, *Modern developments in fluid dynamics*, Vol. 1, Dover.
Prandtl, L., 1952, *Essentials of fluid dynamics*, Blackie & Son.
Rosenhead, L., 1963, *Laminar boundary layers*, Oxford University Press.
Schlichting, H., 1979, *Boundary-layer theory*, 7th Ed., McGraw-Hill.
Van Dyke, M., 1975, *Perturbation methods in fluid mechanics*, Parabolic Press.

6

· · • · ·

Potential Flow Theory with Applications to Aerodynamics

(a)   (b)

(a) The Italian mathematical physicist Joseph-Louis Lagrange (1736–1813) was, in many ways, Euler's successor. (b) Around the same time, Pierre-Simon Laplace (1749–1827) dominated French mathematical physics. Both played a central role in potential theory.

6.1 Some Elementary Ideas in Potential Flow Theory

6.1.1 The Physical Basis for, and Dangers of, Potential Flow Theory

As we have seen, the vorticity $\boldsymbol{\omega}(\mathbf{x}, t)$ at location $\mathbf{x}$ and time t is twice the angular velocity of a fluid particle passing through that point at that instant (see Figure 2.4). By definition, an *irrotational flow* is one that has no vorticity, so that fluid elements have no intrinsic spin as they slide along streamlines. The classic example of irrotational flow is the two-dimensional motion of an aerofoil through still air, viewed in a frame of reference moving with the foil (Figure 6.1). The Reynolds number is assumed to be large, and so outside the boundary layers the fluid may be treated as inviscid, for which the two-dimensional vorticity equation is $D\omega/Dt = 0$. Since, by construction, there is no vorticity upstream of the foil, and vorticity is materially conserved in a two-dimensional inviscid flow, the entire motion outside the boundary layers (and associated wake) is devoid of vorticity. Physically, since we demand that the fluid elements approaching the wing have no intrinsic angular momentum, and there are no viscous torques in the free stream that can spin up those fluid elements, the fluid particles retain their state of zero rotation as they sweep over the wing. A similar situation arises when a uniform flow approaches a three-dimensional wing. Outside the boundary layers the vorticity is governed by

$$\frac{D\boldsymbol{\omega}}{Dt} = (\boldsymbol{\omega} \cdot \nabla)\mathbf{u}, \tag{6.1}$$

and if the upstream flow is devoid of vorticity then the entire free stream is irrotational.

Notice, however, that very special conditions are required in order to realize irrotational motion. For example, the flow in the boundary layer in Figure 6.1 is packed full of vorticity, vorticity which has diffused out from the surface of the wing, and that vorticity is thrown into the free stream at the rear of the body to form a wake. For bluff bodies, such as a wing in stall, the wake occupies a substantial region of space and its size and shape is often unknown in advance. In such cases, much of the flow is rotational. Worse still, if the flow upstream of the wing in Figure 6.1 contains vorticity, perhaps because the wing is passing through atmospheric turbulence, then the entire free stream contains vorticity. Thus, for example, as a plane climbs up through the atmospheric boundary layer, the flow over its wings is very far from irrotational, and indeed that vorticity is so intense it can cause a fully laden airliner to shudder violently.

Outside inviscid aerodynamics and the classical theory of surface gravity waves, flows are rarely irrotational. For example, steady, two-dimensional, internal flows are never irrotational, as can be seen by integrating **u** along a closed streamline and applying Stokes' theorem. Moreover, flows which are subject to rotational body forces, such as buoyancy, are rarely devoid of vorticity, as the body force generates vorticity within the interior of the fluid. Finally, we should note that turbulence is usually defined as a *self-advecting chaotic vorticity field*, not a chaotic velocity field, since, as we shall see, there can be no turbulence without vorticity. So irrotational flow, which is the subject of this chapter, is the exception

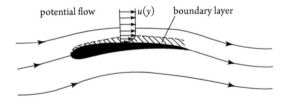

Figure 6.1 Irrotational flow over an aerofoil.

rather than the rule, and erroneous results are readily obtained if a flow is assumed to be irrotational when it is not. Indeed, d'Alembert's paradox reminds us that the apparently rigorous application of mathematics without due attention to the physics can so readily render paradoxical results. Nevertheless, irrotational flow theory can be very useful if used with care, especially in aerodynamics, and so we shall devote this entire chapter to its study.

Let us, for the moment, focus on the issue of computing the velocity field in an incompressible, irrotational flow, temporarily setting aside the pressure field. Irrotational flow theory then reduces to the simple problem of solving the two *kinematic* equations

$$\nabla \cdot \mathbf{u} = 0, \quad \nabla \times \mathbf{u} = 0,$$

subject to the boundary condition $\mathbf{u} \cdot d\mathbf{S} = 0$ on a stationary solid surface. Since $\nabla \cdot (\nabla \times \mathbf{A}) = 0$ for any $\mathbf{A}$, we can use $\nabla \cdot \mathbf{u} = 0$ to introduce a *vector potential* for $\mathbf{u}$, defined by the two equations $\nabla \times \mathbf{A} = \mathbf{u}$ and $\nabla \cdot \mathbf{A} = 0$. (Note that a vector field is uniquely determined if its curl and divergence are specified and suitable boundary conditions given.) Similarly, one can use $\nabla \times \mathbf{u} = 0$ to introduce a *scalar potential* for $\mathbf{u}$, defined by $\mathbf{u} = \nabla \varphi$. Since $\mathbf{A}$ is solenoidal we have the identity $\nabla \times \nabla \times \mathbf{A} = -\nabla^2 \mathbf{A}$, and thus $\mathbf{A}$ and φ are both harmonic:

$$\boxed{\mathbf{u} = \nabla \times \mathbf{A} = \nabla \varphi, \qquad \nabla^2 \mathbf{A} = \nabla^2 \varphi = 0}. \tag{6.2}$$

(Note that, in other branches of mechanics, scalar potentials are normally defined through expressions like $\mathbf{F} = -\nabla \varphi$, the minus sign being included so that φ can be interpreted as the potential energy of $\mathbf{F}$ in cases where $\mathbf{F}$ is a force. In fluid mechanics, however, it has become common to omit the minus sign in the interests of simplicity.) Evidently $\mathbf{u}$ may be determined by solving Laplace's equation subject to the appropriate boundary conditions. Because of the appearance of a scalar potential in (6.2), the study of inviscid, irrotational motion is called *potential flow theory*.

It is a curious feature of potential flow theory that Newton's second law appears to play almost no role in mapping out the velocity field. This oddity is illustrated by Figure 6.2 (a) which shows the two-dimensional potential flow at the entrance to a submerged duct. The same figure also appears in books on electromagnetism, representing a static magnetic field $\mathbf{B}$ entering or leaving a magnetic pole face, governed by $\nabla \cdot \mathbf{B} = 0$ and $\nabla \times \mathbf{B} = 0$. Just as there is no concept of inertia associated with a magnetostatic field, inertia plays almost no role in the calculation of a potential flow, other than to ensure that a state of zero initial rotation is maintained. Indeed, as we shall see, Newton's second law enters the problem only retrospectively, when Bernoulli's equation is used to infer a pressure distribution from a given irrotational velocity field.

Figure 6.2 illustrates a second point. If we change the sign of $\mathbf{u}$ in Figure 6.2 (a), the flow still satisfies $\nabla \cdot \mathbf{u} = 0$ and $\nabla \times \mathbf{u} = 0$ as well as the inviscid boundary conditions, and this now represents fluid leaving a duct. While potential flow theory yields a reasonable approximation to the inflow problem, it provides a terrible estimate of the outflow, which looks more like Figure 6.2 (b). In reality, the outflow takes the form of a submerged jet which is laden with vorticity, vorticity that has been stripped off the walls of the duct. This reminds us that great care must be exercised when seeking to apply potential flow theory.

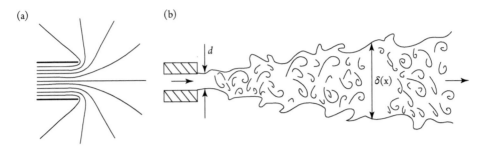

Figure 6.2 (a) The potential flow for a two-dimensional motion entering or leaving a duct. (b) Schematic of the actual flow for a submerged jet at large Re. (From Davidson, 2017.)

6.1.2 The Retrospective Application of Newton's Second Law: Bernoulli Revisited

Euler's equation for an inviscid fluid, in the form of (1.24), simplifies in the absence of vorticity to

$$\frac{\partial \mathbf{u}}{\partial t} = \frac{\partial}{\partial t}\nabla\varphi = -\nabla H, \quad H = p/\rho + \mathbf{u}^2/2, \tag{6.3}$$

where H is Bernoulli's function and we have neglected gravity. We conclude that, in a potential flow, the gradient of $\partial\varphi/\partial t + p/\rho + \mathbf{u}^2/2$ is everywhere zero, and hence

$$\boxed{\frac{\partial\varphi}{\partial t} + \frac{p}{\rho} + \frac{1}{2}\mathbf{u}^2 = f(t)}. \tag{6.4}$$

Thus, if we (somehow) solve (6.2) for $\mathbf{u}$, we can *retrospectively* ensure that Newton's second law is everywhere satisfied by choosing the pressure field in accordance with (6.4). Note that (6.4) generalizes the Bernoulli equation of Chapter 1 in two respects. First, it extends the result to unsteady flows. Second, in a steady, inviscid flow with vorticity, H is constant *along* a streamline, but varies *across* the streamlines. By contrast, in a steady potential flow, H is uniform throughout the entire flow field.

It is sometimes helpful to consider Euler's equation in intrinsic coordinates attached to a particular streamline. For a *steady* flow (1.27) tells us that

$$\boxed{\frac{\partial p}{\partial s} = -\rho V \frac{\partial V}{\partial s}, \quad \frac{\partial p}{\partial n} = \rho\frac{V^2}{R}}, \tag{6.5}$$

where $V = |\mathbf{u}|$, R is the local radius of curvature of the streamline, s is measured along the streamline, and n is in the direction of the principal normal to the streamline, directed *away* from the centre of curvature. Since, in an irrotational flow, $p + \rho V^2/2$ is constant *across* the streamlines, this yields

$$\frac{\partial p}{\partial n} = -\rho V \frac{\partial V}{\partial n} = \rho \frac{V^2}{R}, \qquad (6.6)$$

from which,

$$\boxed{\frac{\partial V}{\partial n} + \frac{V}{R} = 0}. \qquad (6.7)$$

This reflects the fact that, in a rotational flow, the component of vorticity normal to the plane that locally contains the streamlines (*i.e.* the component of vorticity in the direction of the binormal) turns out to be

$$\omega_{\text{binormal}} = \frac{\partial V}{\partial n} + \frac{V}{R}. \qquad (6.8)$$

Equation (6.7) is therefore a *kinematic* consequence of setting the vorticity to zero. (Readers sceptical about the authenticity of (6.8) might want to consider the simple case of concentric circular streamlines, or else consult Exercise 1.10.) The main consequence of (6.7) is that, for a curved streamline, *the speed is higher on the inside of the bend than the outside.*

6.1.3 Some Simple Examples of Two-dimensional Potential Flow

Consider a two-dimensional flow of the form $\mathbf{u}(x, y) = (u_x, u_y, 0)$. If we take the vector potential to be $\mathbf{A} = \psi(x, y)\hat{\mathbf{e}}_z$, where ψ is the streamfunction introduced in §1.2.3, then (6.2) simplifies to

$$\mathbf{u} = \left(\frac{\partial \psi}{\partial y}, -\frac{\partial \psi}{\partial x} \right) = \left(\frac{\partial \varphi}{\partial x}, \frac{\partial \varphi}{\partial y} \right), \quad \nabla^2 \psi = \nabla^2 \varphi = 0. \qquad (6.9)$$

Note that $\mathbf{u} \cdot \nabla \psi = 0$, so ψ is constant along a streamline, and since $\nabla \varphi \cdot \nabla \psi = 0$, the streamlines and equipotential lines must be orthogonal (except when $|\mathbf{u}| = 0$). The boundary conditions on a stationary solid surface are then

$$\psi = \text{constant}, \quad \nabla \varphi \cdot d\mathbf{S} = 0. \qquad (6.10)$$

It is often convenient to adopt 2D polar coordinates (r, θ), in which case (6.9) becomes

$$\mathbf{u} = (u_r, u_\theta) = \left(\frac{1}{r} \frac{\partial \psi}{\partial \theta}, -\frac{\partial \psi}{\partial r} \right) = \left(\frac{\partial \varphi}{\partial r}, \frac{1}{r} \frac{\partial \varphi}{\partial \theta} \right). \qquad (6.11)$$

Perhaps the simplest examples of two-dimensional potential flow correspond to various types of singularities placed at the origin, with the velocity field expressed in polar coordinates. For example, a mass source placed at the origin produces the radial outflow of $u_r = Q/2\pi r$, where $Q = \dot{m}/\rho$ is the two-dimensional volumetric flowrate. To within an arbitrary constant (which does not matter), the corresponding scalar potential and streamfunction are

$$\varphi = \frac{Q}{2\pi}\ln r, \quad \psi = \frac{Q}{2\pi}\theta. \tag{6.12}$$

The equipotential surfaces are therefore concentric circles and the streamlines radial.

Similarly, a point vortex (or line vortex if viewed in three dimensions) of strength $\Gamma = \int \omega dA$ will induce a flow with circular streamlines centred on the origin. Symmetry plus Stokes' theorem then requires $\Gamma = \oint \mathbf{u} \cdot d\mathbf{r} = u_\theta 2\pi r$, or equivalently $u_\theta = \Gamma/2\pi r$. The corresponding distributions of φ and ψ are evidently

$$\varphi = \frac{\Gamma}{2\pi}\theta, \quad \psi = -\frac{\Gamma}{2\pi}\ln r. \tag{6.13}$$

Superposition may now be used to find more complicated flows. For example, the two-dimensional analogue of a dipole in fluid mechanics is called a *doublet*, and it consists of a source, Q, on the x-axis at $x = d/2$, plus an equal and opposite sink at $x = -d/2$. If we interpret r in (6.12) to be the distance from the sink or source, the doublet potential is

$$\varphi = \frac{Q}{4\pi}\ln\frac{y^2 + (x - d/2)^2}{y^2 + (x + d/2)^2} \approx \frac{Q}{4\pi}\ln\frac{x^2 + y^2 - xd}{x^2 + y^2 + xd},$$

for $d \ll r$. If we take the limit of $d \to 0$, with $\mu = Qd$ held fixed, we obtain

$$\varphi = -\frac{\mu}{2\pi}\frac{x}{r^2} = -\frac{\mu}{2\pi}\frac{\cos\theta}{r}, \tag{6.14}$$

where μ is the strength of the doublet. The corresponding flow is shown in Figure 6.3.

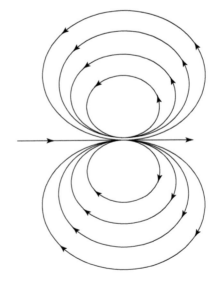

Figure 6.3 The streamlines for a doublet.

Another simple example of superposition involves a uniform flow in the x direction, of magnitude V, combined with a mass source Q at the origin. On the x-axis we have $u_x = V + Q/2\pi x$ and $u_y = 0$, which give a stagnation point, S, at $x_s = -Q/2\pi V$. The streamfunction is then

$$\psi = Vy + \frac{Q}{2\pi}\theta = (y + |x_s|\theta)\,V, \tag{6.15}$$

and the resulting flow pattern is shown in Figure 6.4.

Note that the streamline which passes through the stagnation point $y = 0$, $\theta = \pi$ is $\psi_s = \pi|x_s|V$, and this streamline can be taken to represent a solid surface located at $y = |x_s|(\pi - \theta)$. This is known as the Rankine half-body.

6.1.4 D'Alembert's Paradox

We have already mentioned d'Alembert's paradox, and so perhaps we should now explain how this paradox arises. Consider a three-dimensional body which translates with velocity $\mathbf{V}$ through an inviscid fluid which is stationary at infinity. In a frame of reference moving with the body, we have a steady potential flow over the body which is uniform at infinity. Such a potential flow, which is governed by (6.2) and the boundary condition $\mathbf{u} \cdot d\mathbf{S} = 0$ on the body, is uniquely determined by $\mathbf{V}$ and by the size and shape of the body. Since $\mathbf{V}$ is constant, this potential flow is independent of time. Moreover, there is no dissipation of energy in an inviscid fluid. Moving back into a frame of reference in which the fluid at infinity is *stationary*, the fluid has constant kinetic energy and there is zero dissipation of energy. Conservation of energy then demands that no work is done by the body on the fluid, and it follows that there can be no net drag force on the body in the direction of $\mathbf{V}$. As d'Alembert put it: 'I do not see how one can explain the resistance of fluids by the theory . . . a singular paradox which I leave to the geometricians to explain.'

In the case of a two-dimensional body this argument needs to be refined somewhat, in part because the kinetic energy integral may not be convergent, and in part because there can now be a finite circulation around the body, $\Gamma = \oint \mathbf{u} \cdot d\mathbf{r}$. The potential flow over the body is then determined by $\mathbf{V}$, Γ, and the shape of the body. We shall pick up this story again in §6.3.2, where we shall see that the same result is obtained in two dimensions: *i.e.*

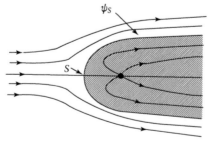

Figure 6.4 Flow around a Rankine half-body.

there is no drag force on a two-dimensional body in the direction of **V**. There is, however, a lift force perpendicular to **V** when Γ is finite.

6.2 The Kinematics of Two-dimensional Potential Flow

6.2.1 The Complex Potential

We shall now focus on the kinematics of two-dimensional potential flows, for which there exists a cornucopia of known solutions. This abundance of solutions arises, in part, because there is a particularly powerful technique for planar flows, known as the *complex potential*. We start by returning to (6.9) in the form

$$\frac{\partial \varphi}{\partial x} = \frac{\partial \psi}{\partial y}, \quad \frac{\partial \varphi}{\partial y} = -\frac{\partial \psi}{\partial x}, \qquad (6.16)$$

which ensures that φ and ψ are individually harmonic. Such relationships are well known in the theory of complex variables, where they are called the *Cauchy–Riemann equations*.

The significance of (6.16) is the following. Suppose that the two functions φ and ψ have continuous derivatives. Then it may be shown that the Cauchy–Riemann equations provide both necessary and sufficient conditions for the complex function

$$F(z) = \varphi + j\psi$$

to be an analytic function of $z = x + jy$, i.e. dF/dz has a unique value at all points, independent of the direction of dz in the complex plane. To show that (6.16) is indeed a necessary condition, let us assume that F is an analytic function of z. Then

$$\frac{\partial F}{\partial x} = \frac{\partial \varphi}{\partial x} + j\frac{\partial \psi}{\partial x} = \frac{dF}{dz}\frac{\partial z}{\partial x} = \frac{dF}{dz},$$

and

$$\frac{\partial F}{\partial y} = \frac{\partial \varphi}{\partial y} + j\frac{\partial \psi}{\partial y} = \frac{dF}{dz}\frac{\partial z}{\partial y} = j\frac{dF}{dz},$$

from which,

$$\frac{\partial \varphi}{\partial x} + j\frac{\partial \psi}{\partial x} = \frac{\partial \psi}{\partial y} - j\frac{\partial \varphi}{\partial y}.$$

Equating real and imaginary parts brings us back to the Cauchy–Riemann equations.

We conclude that *any* analytic function of the form $F(z) = \varphi + j\psi$ represents some kind of two-dimensional potential flow. Such a function is known as a *complex potential*. Moreover, since

$$\frac{dF}{dz} = \frac{\partial \varphi}{\partial x} + j\frac{\partial \psi}{\partial x} = u_x - ju_y, \tag{6.17}$$

we have $|dF/dz| = |\mathbf{u}|$, which provides a simple method for evaluating $|\mathbf{u}|$ at all points.

Of course, this is an indirect approach to potential flow theory, in that we do not know in advance what flow a given function $F(z)$ will yield. Rather, we start out with a given F and see where it leads. Nevertheless, over the last 150 years a multitude of interesting potential flows have been discovered in this way.

6.2.2 Some Elementary Examples of the Complex Potential

A moment's thought confirms that all of the elementary flows introduced in §6.1.3 have particularly simple representations in terms of the complex potential. These are listed in Table 6.1.

Note that, to shift the centre of the flow from the origin to (x_0, y_0), it is necessary only to replace z by $z - z_0$ in the complex potential, where $z_0 = x_0 + jy_0$. Thus, for example, the complex potential for a mass source Q located at (x_0, y_0) is $(Q/2\pi)\ln(z - z_0)$. Similarly, we can rotate a flow by an angle θ_0 by replacing $z = re^{j\theta}$ by $re^{j(\theta-\theta_0)} = ze^{-j\theta_0}$. For example, a uniform flow V at an angle θ_0 to the x-axis has the complex potential $Vze^{-j\theta_0}$.

The final entry in Table 6.1 is the stagnation-point flow shown in Figure 6.5. This has a velocity field, scalar potential, and streamfunction given by

$$\mathbf{u} = (\alpha x, -\alpha y), \quad \varphi = \tfrac{1}{2}\alpha\left(x^2 - y^2\right), \quad \psi = \alpha xy, \tag{6.18}$$

Table 6.1 The complex potentials for some simple flows.

Description of flow	Complex potential
Uniform flow along the x-axis with speed V	Vz
Mass source of strength Q located at the origin	$\frac{Q}{2\pi}\ln z$
Point vortex of strength Γ located at the origin	$-j\frac{\Gamma}{2\pi}\ln z$
Doublet of strength μ located at the origin and aligned with the $x-$axis	$-\frac{\mu}{2\pi z}$
Stagnation-point flow in the half-plane $y > 0$	$\tfrac{1}{2}\alpha z^2$

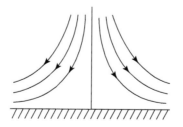

Figure 6.5 Stagnation point flow.

and hence a complex potential of $F = \frac{1}{2}\alpha z^2$. Stagnation-point flow is just one example of a broader class of complex potentials that take the form of a power-law,

$$F(z) = Cz^n, \tag{6.19}$$

where C is a positive constant. In terms of two-dimensional polar coordinates, this yields

$$\varphi = Cr^n \cos n\theta, \quad \psi = Cr^n \sin n\theta. \tag{6.20}$$

Let us rewrite n as $n = \pi/\alpha$, thus defining an angle α. Then the streamfunction ψ is equal to zero on the radial surfaces $\theta = 0$ and $\theta = \alpha$, while symmetric about the bisector $\alpha/2$. We conclude that, for $\alpha < \pi$, or equivalently $n > 1$, (6.19) represents flow in the corner region defined by two intersecting surfaces located at $\theta = 0$ and $\theta = \alpha$. This is illustrated in Figure 6.6 (a) for $\alpha = 2\pi/3$. The case of $n = 1$ ($\alpha = \pi$) can then be interpreted as a uniform flow in the half-plane $y > 0$, in accordance with Table 6.1. Conversely, for $\alpha > \pi$, or $n < 1$, we have the solution for flow around a two-dimensional wedge composed of two surfaces separated by the half-angle $\beta = \pi - \alpha/2$. This is shown in Figure 6.6 (b) for the case where $\beta = \pi/4$. Since $\beta = \pi(1 - 1/2n)$, the limiting case of $\beta \to 0$, that is to say flow around the edge of a semi-infinite flat plate, corresponds to $n = 1/2$.

Since $|dF/dz| = |\mathbf{u}|$, all solutions of (6.19) have a speed that is a function of radius only, given by $|\mathbf{u}| = nCr^{n-1}$. Thus, for the corner-like flows ($n > 1$) the velocity goes to zero along with r, whereas the wedge-like flows ($n < 1$) exhibit a singularity at the tip of the wedge. The most severe case is $\beta \to 0$, which represents flow around the edge of a semi-infinite plate, as shown in Figure 6.7. Here the singularity grows as $|\mathbf{u}| \sim r^{-1/2}$ and this singular growth in velocity gives rise to large negative pressures near the edge of the plate, of the form $p \sim -r^{-1}$. Of course, in reality, the flow separates at the edge to form a wake.

As our final elementary example of a complex potential we consider the superposition of a uniform flow V along the x-axis with a doublet of strength $-\mu$ located at the origin. According to Table 6.1, this has a complex potential and streamfunction given by

(a)   (b)

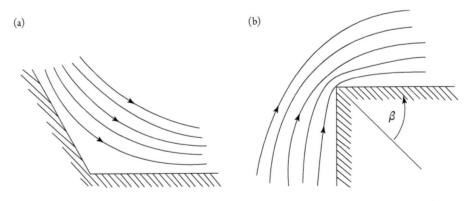

Figure 6.6 (a) A corner flow ($n = 3/2$, $\alpha = 2\pi/3$). (b) A wedge flow ($n = 2/3$, $\beta = \pi/4$).

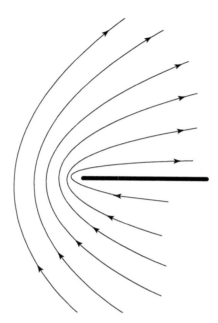

Figure 6.7 The case of $n = 1/2$ in (6.19) corresponds to flow around the edge of a thin plate.

$$F(z) = Vz + \frac{\mu}{2\pi z}, \quad \psi = Vy - \frac{\mu y}{2\pi r^2}. \tag{6.21}$$

If we now make the substitution $a^2 = \mu/2\pi V$, these expressions become

$$F(z) = Vz + V\frac{a^2}{z}, \quad \psi = Vy\left(1 - \frac{a^2}{r^2}\right), \tag{6.22}$$

and we see that $\psi = 0$ on $r = a$. Evidently, this represents flow over a circular cylinder centred on the origin and with the uniform velocity $\mathbf{u} = V\hat{\mathbf{e}}_x$ for $r \gg a$ (Figure 6.8). The upstream–downstream symmetry of the flow means that there is no net pressure drag on the cylinder, in accordance with d'Alembert's paradox. As noted in Chapter 5, potential flow gives a passable approximation to the real flow on the upstream side of the cylinder (outside the boundary layer), but a terrible approximation on the downstream side, as boundary-layer separation leads to a wide wake which dominates the downstream flow.

The expression for $F(z)$ in (6.22) is a particular example of a more general result known as the *circle theorem*. This theorem says the following. Suppose that $F(z)$ is the complex potential for a flow in an infinite domain, and that all of the singularities of $F(z)$ lie outside the circle $|z| = a$. Then the complex potential for the same flow, but bounded internally by a cylinder of radius a placed at the origin, is

$$F(z) + \overline{F(a^2/\bar{z})}, \tag{6.23}$$

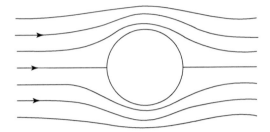

Figure 6.8 Potential flow over a circular cylinder.

where the overbar indicates a complex conjugate. The proof of this statement rests on the fact that $z = a^2/\bar{z}$ on the cylinder, and so our complex potential on $|z| = a$ is $F(z) + \overline{F(z)}$, which is real, thus ensuring $\psi = 0$. In the case of *uniform flow* over a cylinder we have $F(z) = Vz$, which has no singularities within the circle $|z| = a$. Also $F(a^2/\bar{z}) = Va^2/\bar{z}$, and hence $\overline{F(a^2/\bar{z})} = Va^2/z$, so (6.22) is indeed a special case of the circle theorem.

6.2.3 Flow Normal to a Flat Plate of Finite Width

Let us return to Figure 6.7, only now we consider the case of flow around a plate of *finite width*. Consider the complex potential

$$F(z) = -jV\sqrt{z^2 - a^2}. \tag{6.24}$$

For $|z| \gg a$ we have $F(z) = -jVz$, and given that $F'(z) = u_x - ju_y$, we conclude that $\mathbf{u} = V\hat{\mathbf{e}}_y$ in the far field. Moreover, if we shift the origin to $(\pm a, 0)$, replacing z by $z \pm a$, and let $a \to \mp\infty$, then (6.24) becomes $F(z) = V\sqrt{2|a|}z^{1/2}$. We have seen this potential before: it represents flow past a semi-infinite flat plate, as shown in Figure 6.7. We might anticipate, therefore, that complex potential (6.24) corresponds to flow over a plate located on the x-axis at $-a < x < a$, as shown in Figure 6.9 (a). It turns out that this is indeed the case.

The streamfunction and equipotential surfaces are most readily obtained from the expression $F^2 = V^2 (a^2 - z^2)$, which yields

$$\varphi^2 - \psi^2 = V^2 (a^2 + y^2 - x^2), \qquad \psi\varphi = -V^2 xy. \tag{6.25}$$

In particular, on the x-axis we find

$$\psi = 0, \quad \varphi^2 = V^2 (a^2 - x^2), \quad \text{for } |x| < a,$$
$$\varphi = 0, \quad \psi^2 = V^2 (x^2 - a^2), \quad \text{for } |x| > a,$$

and so $\psi = 0$ for $|x| < a$, consistent with flow over a plate of width $2a$. Moreover, the components of $\mathbf{u}$ at the surface of the plate are

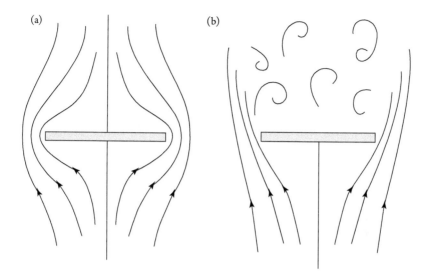

Figure 6.9 (a) Potential flow over a plate. (b) Actual flow over a plate.

$$u_x = \frac{\partial \varphi}{\partial x} = \mp V \frac{x}{\sqrt{a^2 - x^2}}, \quad u_y = -\frac{\partial \psi}{\partial x} = 0, \quad |x| < a, \tag{6.26}$$

where the upper sign in (6.26) applies to the top surface of the plate and the lower sign to the bottom surface. On the other hand, on the x-axis either side of the plate we find

$$u_x = \frac{\partial \varphi}{\partial x} = 0, \quad u_y = -\frac{\partial \psi}{\partial x} = \pm V \frac{x}{\sqrt{x^2 - a^2}}, \quad |x| > a, \tag{6.27}$$

with $\mathbf{u} = V \hat{\mathbf{e}}_y$ for $|x| \gg a$. Note that, as with the flow in Figure 6.7, we have a $|\mathbf{u}| \sim r^{-1/2}$ singularity at the edges of the plate, where in this case $r = |x| - a$.

This potential flow exhibits upstream–downstream symmetry and hence exerts zero drag, in accordance with d'Alembert's paradox. In reality, the flow separates at the edges of the plate to form a downstream wake, as shown in Figure 6.9 (b). The pressure in the wake is below ambient and so the plate experiences a particularly large drag force due to the high pressure at the forward stagnation point and the low wake pressure. This force is around $F_D \approx \rho V^2 (2a)$, corresponding to a drag coefficient (based on the frontal area) of

$$C_D = \frac{F_D}{\frac{1}{2}\rho V^2 (2a)} \approx 2.0.$$

The large drag associated with separation of the flow is one reason why the stalling of a wing has such a catastrophic effect. Stall typically occurs at angles of attack in excess of $\sim 12°$ and corresponds to the separation of the boundary layer on the upper surface of the wing, close to the leading edge. It is associated with a dramatic loss of lift and rise in drag (see Figure 6.10).

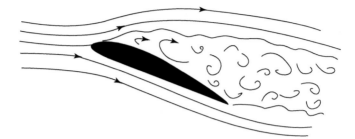

Figure 6.10 A wing in stall.

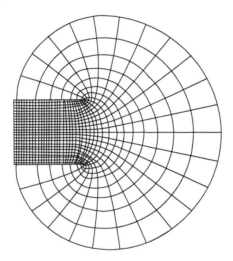

Figure 6.11 The streamlines and equipotential surfaces for potential flow at the entrance to a submerged, two-dimensional duct. The same solution gives the magnetic field lines entering or leaving a magnetic pole face. (Reproduced from Maxwell, 1891.)

6.2.4 A Not so Simple Example: the Intake to a Submerged Duct

So far we have restricted the discussion to rather simple flows. We shall now consider an altogether more complex case: that of irrotational flow from a large reservoir into a submerged, two-dimensional duct (Figure 6.11), as first formulated by Helmholtz in 1868. The same complex potential also finds applications in electrostatics and magnetostatics. Indeed, Figure 6.11 is taken from Maxwell (1891), where it represents the electric field at the edge of a parallel-plate capacitor. The same potential also represents a static magnetic field $\mathbf{B}$ entering or leaving a magnetic pole face, governed by $\nabla \cdot \mathbf{B} = 0$ and $\nabla \times \mathbf{B} = 0$. It turns out that the complex potential for all of these problems takes the form of the inverse function $z = -F + \exp(-F)$, whose real and imaginary components are

$$x = -\varphi + e^{-\varphi}\cos\psi, \quad y = -\psi - e^{-\varphi}\sin\psi. \tag{6.28}$$

To determine the flow associated with (6.28) we first note that the streamline $\psi = 0$ sits on the x-axis, with φ running from $\varphi \to \infty$ for large negative x to $\varphi \to -\infty$ for large positive x, in accordance with $x = -\varphi + e^{-\varphi}$. Next, we observe that the streamline $\psi = -\pi$ lies on the line $y = \pi$, with φ running from $\varphi \to -\infty$ for large negative x to $\varphi = 0$ at $x = -1$, and then back to large negative x for $\varphi \to \infty$, in accordance with $x = -\varphi - e^{-\varphi}$. So this is a streamline that comes in from $x = -\infty$, bends back on itself at $x = -1$, and then returns to large negative x. Similarly, the streamline $\psi = \pi$ lies on the line $y = -\pi$, with φ running from $\varphi \to -\infty$ for large negative x to $\varphi = 0$ at $x = -1$, and finally back to large negative x for $\varphi \to \infty$. These two streamlines are the top and bottom surfaces of the duct, with $\varphi > 0$ on the inside of the two surfaces and $\varphi < 0$ outside the duct (see Figure 6.12).

Next we note that, for $\varphi \to \infty$, $F(z)$ takes the form $F = -z$, which represents uniform flow in the negative x direction. Since $\varphi = -x$ when $F = -z$, this occurs when x is large and negative while φ is large and positive. In short, this represents the uniform flow within the duct. Also, for large negative φ we have $z = \exp(-F)$, i.e. $F = -\ln z$ (or $\varphi = -\ln r$). This represents a sink located near the origin, as seen from large r. So we conclude that the complex potential (6.28) does indeed represent the intake to a duct, with the edges of the duct located at $y = \pm\pi$, and the entrance to the duct at $x = -1$.

6.2.5 The Method of Images for Plane and Cylindrical Boundaries

The *method of images* was introduced by Kelvin and subsequently developed by Helmholtz and Maxwell, motivated largely by problems in electromagnetism, rather than fluid dynamics. In its simplest embodiment it is a means of obtaining the velocity field for a mass source or line vortex in the half-space $y > 0$ when $y = 0$ is a solid boundary. For example, in the case of a mass source located a distance d above a boundary at $y = 0$, one replaces the semi-infinite problem by an infinite domain in which there are two sources of equal magnitude located one above the other at $y = \pm d$. Symmetry then ensures that the surface $y = 0$ is a streamline, and so the solution for $y > 0$ is that of the single source in a half-space. Similarly, for a line vortex a distance d above a solid boundary at $y = 0$, one introduces a second vortex

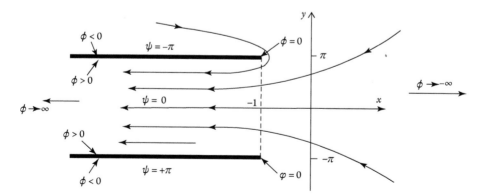

Figure 6.12 Salient values of ψ and φ corresponding to (6.28). (Adapted from Bird et al., 1960.)

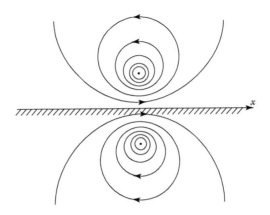

Figure 6.13 A line vortex near a boundary and its image vortex.

of equal magnitude but *opposite* sign at location $y = -d$, as shown in Figure 6.13. Again, symmetry ensures that the two vortices in an infinite domain yield the same velocity (for $y > 0$) as the single vortex in a half-space. The vortex at $y = -d$ is called the *image vortex*.

The problem of a mass source or line vortex adjacent to a cylindrical surface is less trivial. The key breakthrough was an elegant paper by a young Lord Kelvin (then William Thomson) in 1848. Kelvin was interested in the electric field created by a point charge q adjacent to a conducting sphere at constant potential (Figure 6.14). Suppose that the sphere is centred on the origin and has a radius of a, while the charge sits on the x-axis at $x = d$, with $d > a$. Inspired by a theorem of Apollonius of Perga (*circa* 200 BCE), which says that a sphere is the locus of all points for which the ratio of the distances from two fixed points is a constant, Kelvin showed that, as far as the electric field outside the sphere is concerned, the conducting sphere can be replaced by two 'image charges' located on the x-axis at $x = 0$ and $x = a^2/d$. The same idea can be applied to potential flow.

In the case of planar flow, consisting of a mass source or line vortex adjacent to a cylinder, one could simply adapt Kelvin's analysis to two dimensions. However, this is not necessary as the circle theorem of §6.2.2 (which was published in 1940) delivers the same result. Suppose we have a cylinder of radius a, centred on the origin, and a source of strength Q located on the x-axis at $x = d$, with $d > a$. The complex potential for the mass source in the absence of the cylinder is $(Q/2\pi)\ln(z - d)$, and so the circle theorem tells us that the complex potential for the source in the presence of the cylinder must be

$$F(z) = \frac{Q}{2\pi}\ln(z - d) + \frac{Q}{2\pi}\overline{\ln\left(\frac{a^2}{\bar{z}} - d\right)} = \frac{Q}{2\pi}\ln(z - d) + \frac{Q}{2\pi}\ln\left(\frac{a^2}{z} - d\right),$$

where we have used Schwarz's reflection principle to handle the complex conjugates. Subtracting $(Q/2\pi)\ln(-d)$ then yields

$$F(z) = \frac{Q}{2\pi}\ln(z - d) + \frac{Q}{2\pi}\ln\left(z - \frac{a^2}{d}\right) - \frac{Q}{2\pi}\ln(z), \tag{6.29}$$

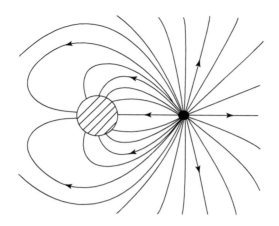

Figure 6.14 The electric field of a point charge adjacent to a conducting sphere. (From Davidson, 2019.)

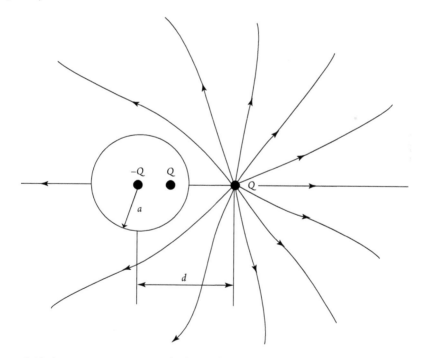

Figure 6.15 A mass source next to a cylinder can be represented by two sources and a sink.

which represents the original source at $x = d$ plus two image sources within the circle, one at the origin, of strength $-Q$, and one at $x = a^2/d$, of strength Q (see Figure 6.15). It is readily confirmed retrospectively that the imaginary part of (6.29), *i.e.* ψ, is constant on the surface of the cylinder, and hence $|\mathbf{x}| = a$ is indeed a streamline. The flow is shown in Figure 6.15.

Similarly, a line vortex of strength Γ, instantaneously located on the x-axis at $x = d$ and adjacent to the cylinder $|\mathbf{x}| = a$, produces a velocity field whose complex potential is

$$F(z) = -\frac{j\Gamma}{2\pi}\ln(z-d) - \frac{\overline{j\Gamma}}{2\pi}\overline{\ln\left(\frac{a^2}{\bar{z}} - d\right)} = -\frac{j\Gamma}{2\pi}\ln(z-d) + \frac{j\Gamma}{2\pi}\ln\left(\frac{a^2}{z} - d\right),$$

from which

$$F(z) = -j\frac{\Gamma}{2\pi}\ln(z-d) + j\frac{\Gamma}{2\pi}\ln\left(z - \frac{a^2}{d}\right) - j\frac{\Gamma}{2\pi}\ln(z) + \text{const}.. \qquad (6.30)$$

The image system consists of a vortex of strength Γ at the origin, plus a second vortex at $x = a^2/d$, of strength $-\Gamma$.

6.3 The Lift Force Exerted on a Body by a Uniform Incident Flow

We now turn from kinematics to dynamics and consider the important topic of *lift*.

6.3.1 Two-dimensional Flow over a Cylinder with Circulation: an Illustrative Example

Let us start by revisiting the discussion in §6.2.2 of irrotational flow over a circular cylinder. It is clear that the boundary conditions are still satisfied if we superimpose on the left-to-right motion the velocity field associated with a line vortex located at the origin. Equation (6.22) then generalizes to

$$F(z) = V\left(z + \frac{a^2}{z}\right) - j\frac{\Gamma}{2\pi}\ln z, \quad \psi = V\sin\theta\left(r - \frac{a^2}{r}\right) - \frac{\Gamma}{2\pi}\ln r, \qquad (6.31)$$

with velocity components

$$u_r = V\left(1 - \frac{a^2}{r^2}\right)\cos\theta, \quad u_\theta = -V\left(1 + \frac{a^2}{r^2}\right)\sin\theta + \frac{\Gamma}{2\pi r}. \qquad (6.32)$$

With one eye to flow over an aerofoil, it is normal to consider the case in which $\Gamma < 0$, and we shall conform to that convention. The resulting flow pattern clearly depends on the ratio of Γ to V, but in all cases the velocity associated with the line vortex supplements the left-to-right flow above the cylinder, but opposes that flow below. Bernoulli's equation then tells us that the pressure below the cylinder is higher than that above. Three typical examples of this flow are shown in Figure 6.16. It is readily confirmed that, for $|\Gamma| < 4\pi Va$, there are two stagnation points on the lower side of the cylinder, while at $|\Gamma| = 4\pi Va$ those stagnation points merge at the base of the cylinder. For $|\Gamma| > 4\pi Va$, on the other hand, the circulation around the cylinder is so strong that the central stagnation point moves off the surface of the cylinder and sits in the free stream.

Of course, it is far from clear which, if any, of these flow patterns could exist, even approximately, in a real viscous fluid. Nor is it obvious how one might try to recreate any of these flows in practice. (One way to induce circulation in a *real* fluid is to rotate the cylinder, although the resulting flow looks very different to any of those in Figure 6.16.) Nevertheless, the interesting thing about this hypothetical flow is that it is simple to calculate the net pressure force acting on the cylinder, and here we obtain an intriguing result.

From Figure 6.16 it is clear that the upstream–downstream symmetry of the flow excludes any stream-wise pressure force on the cylinder (d'Alembert's paradox, yet again). It is equally clear from the emergence of stagnation points on or near the lower surface of the cylinder that there is an up–down pressure asymmetry, and that the mean pressure on the lower side of the cylinder is higher than that on the top. Thus, we expect a net upward, or *lift*, force on the cylinder. Let us calculate that lift force. We can find the pressure on the surface of the cylinder using Bernoulli's equation, and this yields

$$p = p_0 - \frac{1}{2}\rho \mathbf{u}^2 = p_0 - \frac{1}{2}\rho\left[4V^2\sin^2\theta + \frac{2V|\Gamma|}{\pi a}\sin\theta + \left(\frac{\Gamma}{2\pi a}\right)^2\right], \qquad (6.33)$$

where p_0 is the pressure at the stagnation point. The upward pressure force on a small part of the surface, $ad\theta$, is evidently $-pa\sin\theta d\theta$, and on integrating this over the surface of the cylinder we obtain a lift force per unit length of the cylinder of

$$F_L = \int_0^{2\pi}\left[2\rho V^2\sin^2\theta + \frac{\rho V|\Gamma|}{\pi a}\sin\theta\right]a\sin\theta d\theta = \rho V|\Gamma|. \qquad (6.34)$$

As we shall now show, the expression $F_L = \rho V|\Gamma|$ is a specific example of a famous and rather general result in aerodynamics, known as the *Kutta–Joukowski lift theorem*.

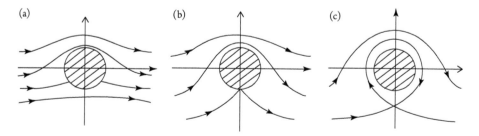

Figure 6.16 Flow over a circular cylinder with negative circulation. (a) $|\Gamma| < 4\pi Va$, (b) $|\Gamma| = 4\pi Va$, and (c) $|\Gamma| > 4\pi Va$.

6.3.2 Flow over a Planar Body of Arbitrary Shape: the Kutta–Joukowski Lift Theorem

Consider the steady, two-dimensional, inviscid, irrotational flow over a body of arbitrary shape, as shown in Figure 6.17. Let the velocity at infinity be $V_\infty \hat{\mathbf{e}}_x$ and the velocity elsewhere be $\mathbf{u} = V_\infty \hat{\mathbf{e}}_x + \mathbf{u}'$.

We shall apply the momentum theorem to a circular control volume of radius R centred on the body, taking R to be very much larger than the characteristic size of the body. We may then take advantage of the fact that the components of $\mathbf{u}'$ evaluated on the circular control surface S are of order $1/R$, or smaller, and neglect second-order terms in $1/R$ when evaluating the pressure acting on, or momentum flux through, the surface S. For example, from Bernoulli's theorem the pressure on S is

$$p = \tfrac{1}{2}\rho V_\infty^2 - \tfrac{1}{2}\rho \mathbf{u}^2 = -\rho V_\infty u_x' + O(R^{-2}), \tag{6.35}$$

where we have taken the pressure at infinity to be zero. Using polar coordinates centred on the body, the net pressure force exerted on the control surface by the external fluid is then

$$\mathbf{F}_p = -\int_0^{2\pi} p\,(\hat{\mathbf{e}}_x \cos\theta + \hat{\mathbf{e}}_y \sin\theta)\,R d\theta = \rho V_\infty \int_0^{2\pi} u_x'\,(\hat{\mathbf{e}}_x \cos\theta + \hat{\mathbf{e}}_y \sin\theta)\,R d\theta.$$

Similarly, the flux of momentum out through the surface S is

$$\oint_S (\rho\mathbf{u})\mathbf{u}\cdot d\mathbf{S} = \rho\oint_S (\mathbf{u} - V_\infty \hat{\mathbf{e}}_x)\mathbf{u}\cdot d\mathbf{S} = \rho\oint_S \mathbf{u}'(V_\infty \hat{\mathbf{e}}_x)\cdot d\mathbf{S},$$

where we have used the fact that $\oint \mathbf{u}\cdot d\mathbf{S} = 0$ and neglected second-order terms in $1/R$. In terms of polar coordinates this gives

$$\oint_S (\rho\mathbf{u})\mathbf{u}\cdot d\mathbf{S} = \rho V_\infty \int_0^{2\pi} \mathbf{u}'\cos\theta\,R d\theta. \tag{6.36}$$

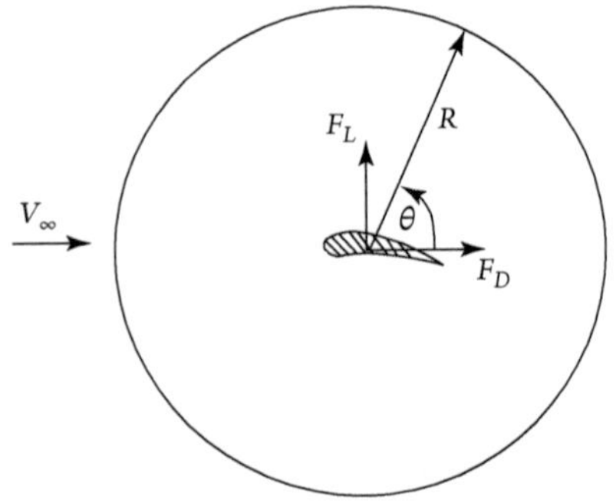

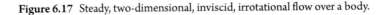

Figure 6.17 Steady, two-dimensional, inviscid, irrotational flow over a body.

Now suppose that $\mathbf{R}$ is the net *external force* applied to the body to keep it in equilibrium. Then the momentum theorem yields

$$\mathbf{R} + \mathbf{F}_p = \oint_S (\rho\mathbf{u})\mathbf{u} \cdot d\mathbf{S}, \tag{6.37}$$

and substituting for the pressure force and momentum flux we obtain

$$\mathbf{R} + \rho V_\infty \int_0^{2\pi} u'_x (\hat{\mathbf{e}}_x \cos\theta + \hat{\mathbf{e}}_y \sin\theta) R d\theta = \rho V_\infty \int_0^{2\pi} (u'_x \hat{\mathbf{e}}_x + u'_y \hat{\mathbf{e}}_y) \cos\theta R d\theta. \tag{6.38}$$

Of course, the net force exerted on the body by the flow, say $\mathbf{F}$, is equal and opposite to $\mathbf{R}$, and so (6.38) yields

$$\mathbf{F} = \rho V_\infty \hat{\mathbf{e}}_y \int_0^{2\pi} (u'_x \sin\theta - u'_y \cos\theta) R d\theta. \tag{6.39}$$

Finally, we introduce some notation. First, we write $(F_x, F_y) = (F_D, F_L)$, thus defining the lift and drag forces acting on the body. Second, we note that the circulation around the control surface, S, evaluated in the *anti-clockwise* direction, is

$$\Gamma = \oint \mathbf{u} \cdot d\mathbf{r} = \int_0^{2\pi} (u'_y \cos\theta - u'_x \sin\theta) R d\theta.$$

Our expression for $\mathbf{F}$ now reduces to the remarkably simple result

$$\boxed{F_D = 0, \quad F_L = -\rho V_\infty \Gamma}. \tag{6.40}$$

The absence of a drag force is, of course, d'Alembert's paradox, while the expression for the lift force is known as the Kutta–Joukowski lift theorem, of which (6.34) is a special case.

We first met the expression $F_L = -\rho V_\infty \Gamma$ in §1.3.6, where we derived it for a single blade embedded within a cascade of blades. It is clear from that derivation that the lift force arises physically because a blade or wing which has camber, or else an angle of attack relative to the oncoming flow, deflects the momentum of the flow downwards, and the lift force is a direct reaction to this deflection. The Kutta–Joukowski theorem tells us that there must be a relationship between the downward deflection of the momentum in the fluid and the circulation around the blade or aerofoil. We shall explore this relationship shortly. First, however, we need to introduce Kelvin's circulation theorem.

6.3.3 Kelvin's Circulation Theorem

In a three-dimensional inviscid flow, *Kelvin's circulation theorem* asserts that the circulation around any closed material curve is an invariant of the motion. The proof of the theorem proceeds as follows. Let C_m be a closed material curve, always composed of the same fluid particles. If the short displacement vector $\delta\mathbf{l}$ is part of C_m, then the fact that $\delta\mathbf{l}$ moves with the fluid, like a dye line, means that it obeys the equation

$$\frac{D}{Dt}(\delta\mathbf{l}) = \delta\mathbf{l}\cdot\nabla\mathbf{u}, \tag{6.41}$$

as discussed in Exercise 6.5. It therefore follows that

$$\frac{D}{Dt}(\mathbf{u}\cdot\delta\mathbf{l}) = \frac{D\mathbf{u}}{Dt}\cdot\delta\mathbf{l} + \mathbf{u}\cdot(\delta\mathbf{l}\cdot\nabla\mathbf{u}) = -\delta\mathbf{l}\cdot\nabla(p/\rho) + \delta\mathbf{l}\cdot\nabla(u^2/2),$$

where we have used Euler's equation to substitute for $D\mathbf{u}/Dt$. On integrating this around the material curve C_m, and noting that D/Dt commutes with the integral sign, we obtain

$$\frac{d}{dt}\oint_{C_m}\mathbf{u}\cdot d\mathbf{l} = \oint_{C_m}\nabla(u^2/2 - p/\rho)\cdot d\mathbf{l}. \tag{6.42}$$

However, the line integral on the right is zero since $u^2/2 - p/\rho$ is single valued, and so we conclude that

$$\boxed{\frac{d}{dt}\oint_{C_m}\mathbf{u}\cdot d\mathbf{r} = 0}. \tag{6.43}$$

Is short, the circulation around any closed material curve is conserved.

Note that this proof does not require that C_m is spanned by a surface lying wholly within the fluid, and so (6.43) may be applied to a material curve that encircles an aerofoil, for example. Note also that we require Euler's equation to hold only on the curve C_m, and so we may use (6.43) even in cases where viscous effects are important, such as in a boundary layer, provided that those viscous effects do not occur anywhere on the curve C_m.

6.3.4 The Role of Boundary-layer Vorticity in Establishing Circulation round an Aerofoil

Armed with Kelvin's theorem, we now consider how a circulation, Γ, is created around an aerofoil. In the process, we shall see why there is a connection between circulation and the downward deflection of momentum in the fluid passing over a wing, and hence a connection between circulation and lift.

An aerofoil is a slender, streamlined body with a rounded leading edge and a sharp trailing edge. The smooth leading edge helps prevent boundary-layer separation on the upper surface of the wing. Lift is generated either because the aerofoil is cambered (*i.e.* its centreline is curved) or because the wing is slightly inclined to the oncoming flow, typically with an *angle of attack* somewhat less than $10°$. Either way, the momentum in the fluid passing over the wing is deflected downward, generating lift. For higher angles of attack the boundary layer tends to separate on the upper surface of the wing near the leading edge, a phenomenon known as *stall* (see Figure 6.10). This leads to a dramatic loss of lift and an increase in drag.

For a wing with camber, and/or a small angle of attack, the potential flow in which the circulation is zero places the rear stagnation point on the top surface of the wing, as shown in Figure 6.18 (a). While such a flow pattern can be observed immediately after motion begins, it is clearly unsustainable, with a singularity at the trailing edge and a rapid deceleration of the fluid as it passes from the trailing edge to the rear stagnation point. It is observed that, very quickly, the stagnation point is swept back to the trailing edge and the flow settles down to the pattern shown in Figure 6.18 (b), in which fluid passes smoothly over the trailing edge. While both flows in Figure 6.18 may be regarded as a potential flow, at least outside the boundary layers, that on the right possesses a finite circulation around the foil, $\Gamma = \oint \mathbf{u} \cdot d\mathbf{r} \neq 0$. As we shall see, for a given flow speed, wing shape, and angle of attack, there is one and only one value of Γ that ensures a smooth trailing-edge flow. (For a thin, symmetrical wing of chord L, and angle of attack α, this turns out to be $\Gamma = -\pi V_{\infty} L \sin \alpha$.) The *Kutta condition* asserts that it is this value of Γ which is found in practice, and it is clear from Figure 6.18 that this circulation is negative (clockwise). Notice that the momentum of the fluid in Figure 6.18 (b) is deflected downward, consistent with an upward lift force on the wing. Notice also that a negative circulation augments the flow above the wing but diminishes the motion below the wing, and so the pressure below the foil is increased relative to the potential flow with zero circulation, while that above the wing is decreased. Again, this is consistent with an upward lift force.

The appearance of circulation around an aerofoil immediately following the initiation of motion seems, at first sight, to contravene Kelvin's theorem. However, this matter was resolved by Prandtl who observed that, as soon as the wing starts to move, the boundary layer separates near the rear stagnation point and vorticity is ejected into the free stream (Figure 6.19 (b)). This process shifts the rear stagnation point to the trailing edge while creating a downstream vortex with positive vorticity, as shown in Figure 6.19 (c). The shed vortex is known as the *starting vortex*. Now consider a material contour that encircles both the wing and the starting vortex, as shown in Figure 6.19 (d). Clearly, there is zero circulation around this contour when the fluid is stationary, and Kelvin's theorem tells us that this circulation must remain zero once there is motion. However, if this contour is split into one loop that encircles the shed vortex and another that encircles the wing, then there is positive circulation around the starting vortex, of magnitude $\Gamma_{SV} = \int \omega dA$. Since the net

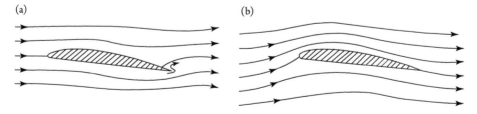

Figure 6.18 Potential flow over an aerofoil. (a) Flow without circulation. (b) Flow with a circulation Γ whose magnitude is chosen to ensure smooth motion at the trailing edge.

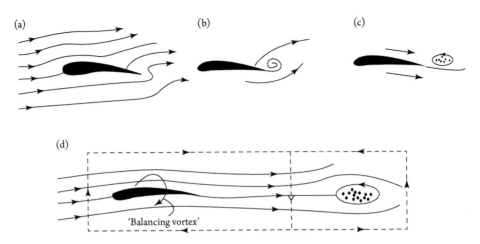

Figure 6.19 The formation of a starting vortex and the associated bound vortex.

circulation is zero, there must be an equal and opposite circulation around the aerofoil, and this is known as the *bound vorticity* of the wing.

Notice that, while the notion of lift arising from a combination of mean flow and circulation is an inviscid concept, it is viscous dynamics which establishes the Kutta condition and hence determines the magnitude of the lift. Moreover, the bound vorticity is nothing more than the net vorticity within the boundary layers that form on the top and bottom surfaces of the wing. This can be seen from the following argument. Consider the upper boundary layer and let δ be the boundary layer thickness, L the chord of the aerofoil, and s and n curvilinear coordinates measured along, and normal to, the upper surface of the wing. Then the net vorticity in the upper boundary layer is

$$\int_{\text{top}} \omega dA = \int_0^L \int_0^\delta \left(-\frac{\partial u_s}{\partial n} \right) dn ds = -\int_{\text{top}} \mathbf{u} \cdot d\mathbf{r}, \qquad (6.44)$$

where the line integral on the right of (6.44) is taken along the inviscid streamline that sits just above the boundary layer, running from the leading edge of the wing to the trailing edge. Similarly, the net vorticity in the lower boundary layer is

$$\int_{\text{bottom}} \omega dA = +\int_{\text{bottom}} \mathbf{u} \cdot d\mathbf{r},$$

where the line integral is again from the leading edge to the trailing edge, taken along an inviscid streamline that lies just outside the boundary layer. Because of the smooth trailing-edge flow, the anticlockwise circulation around a contour that lies just outside the two boundary layers is

$$\oint \mathbf{u} \cdot d\mathbf{r} = \int_{\text{bottom}} \mathbf{u} \cdot d\mathbf{r} - \int_{\text{top}} \mathbf{u} \cdot d\mathbf{r} = \int_{\text{bottom}} \omega dA + \int_{\text{top}} \omega dA. \qquad (6.45)$$

So the bound vorticity is the sum of the negative vorticity in the top boundary layer and positive vorticity in the lower boundary layer. (See, also, Exercise 6.6 for an alternative derivation of this result.)

Finally, we might note that, because of the Kutta condition, there is zero net vorticity in the wake behind the trailing edge. The point is that the smooth flow leaving the trailing edge cannot support any transverse pressure gradient, and so Bernoulli's equation applied along a streamline tells us that the free-stream velocity just outside the top and bottom boundary layers must be equal at the trailing edge. This is sufficient to ensure that the net negative vorticity in the top boundary layer at the trailing edge is equal and opposite to the net positive vorticity in the bottom boundary layer, as discussed in Exercise 6.7.

6.3.5 *The Lift Generated by a Slender Aerofoil*

Thin, symmetrical aerofoils are particularly amenable to analysis using complex potential theory, and in particular to the technique of *conformal mapping*, as discussed in Exercise 6.9. However, we shall not pause to detail that analysis here, but rather state the outcome and provide a more direct explanation for the results. Suppose that L is the chord (*i.e.* width) of the aerofoil, and α the angle of attack to a uniform incident flow, V_∞, as shown in Figure 6.20. Then we might reasonably assert that the lift force per unit span of the aerofoil is proportional to a representative difference in pressure between the upper and lower surfaces of the wing, say the forward stagnation pressure, and to the frontal area presented to the flow, *i.e.*

$$F_L \sim \rho V_\infty^2 L \sin\alpha. \tag{6.46}$$

It turns out that, for a thin, symmetrical aerofoil, the Kutta condition combined with irrotational flow theory yields

$$\boxed{F_L = \pi \rho V_\infty^2 L \sin\alpha}. \tag{6.47}$$

Perhaps the simplest way to see where the pre-factor of π comes from is to invoke the following argument. Let s be a stream-wise coordinate measured along the centreline of the wing from the leading edge to the trailing edge at $s = L$. If the aerofoil is thin and

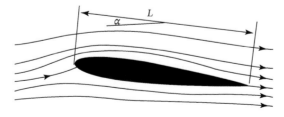

Figure 6.20 Flow around a symmetrical aerofoil at an angle of attack of α.

symmetrical then, to a first approximation, the bound vorticity per unit s, say $\gamma(s)$, which satisfies the Kutta condition, as well as zero normal velocity at the surface of the foil, is

$$\gamma(s) = -2V_\infty \sin\alpha\sqrt{L/s - 1}, \quad \text{where} \quad \Gamma = \int_0^L \gamma(s)ds = -\pi V_\infty L \sin\alpha. \quad (6.48)$$

To see why, we first note that the velocity induced by a point vortex is, in polar coordinates, $u_\theta = \Gamma/2\pi r$. So the vorticity distribution (6.48) generates a velocity normal to the chord of the wing of

$$u_n(s) = \int_0^L \frac{\gamma(s')ds'}{2\pi(s - s')} = -\frac{V_\infty \sin\alpha}{\pi} \int_0^L \frac{\sqrt{L - s'}ds'}{\sqrt{s'}(s - s')},$$

positive u_n being taken as upward. Changing variables to $\cos\theta = 1 - 2s/L$, and invoking the definite integral

$$\int_0^\pi \frac{\cos n\theta' d\theta'}{\cos\theta' - \cos\theta} = \pi\frac{\sin n\theta}{\sin\theta},$$

yields, after a little algebra,

$$u_n(s) = -\frac{V_\infty \sin\alpha}{\pi} \int_0^\pi \frac{(1 + \cos\theta')d\theta'}{\cos\theta' - \cos\theta} = -V_\infty \sin\alpha. \quad (6.49)$$

This exactly counters the normal component of velocity associated with the incident flow. Moreover, since $\gamma = 0$ at the trailing edge, this vorticity distribution ensures that the

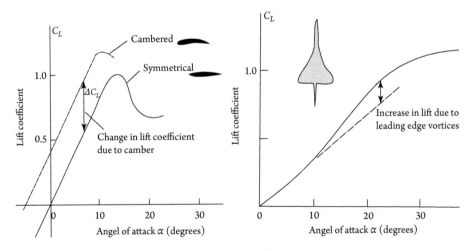

Figure 6.21 Lift coefficient versus angle of attack for: (a) two-dimensional aerofoils with and without camber, and (b) a delta wing. (From Barnard and Philpott, 1995.)

tangential velocity is the same above and below the wing at the trailing edge. Thus (6.48) yields a smooth trailing-edge flow, as demanded by the Kutta condition (see Exercise 6.7).

Finally, the Kutta–Joukowski theorem shows that the lift per unit span of the wing is

$$\boxed{F_L = -\rho V_\infty \Gamma = \pi \rho V_\infty^2 L \sin \alpha}, \tag{6.50}$$

in accordance with (6.47). This force is normally expressed in terms of a dimensionless lift coefficient, defined as $C_L = F_L / \frac{1}{2}\rho V_\infty^2 L$, from which $C_L = 2\pi \sin \alpha$. When the small but finite thickness of the wing, T, is allowed for, this is modified to read (see Exercise 6.9)

$$C_L = 2\pi \sin \alpha \, (1 + 0.77 T/L). \tag{6.51}$$

Predictions (6.50) and (6.51) are found to be reasonable approximations to the actual lift coefficient, to within 10%, provided that the angle of attack is not too large. Angles of attack higher than $\alpha \sim 10°$ are prone to stall, which results in a sharp reduction in lift.

More generally, for thin, two-dimensional aerofoils, with or without camber, potential flow theory predicts

$$C_L = 2\pi \sin(\alpha + \beta)\,(1 + 0.77 T/L),$$

where β is a measure of the angle between the centreline of the foil and the straight line that runs from the leading edge to the trailing edge, with $\beta = 0$ for a symmetrical wing. Thus, cambered wings have finite lift at zero angle of attack, as indicated in Figure 6.21 (a). Such wings tend to stall at angles of $10° \to 16°$. By comparison, three-dimensional delta wings can sustain much higher angles of attack without stall, as shown in Figure 6.21 (b).

So far, we have limited the discussion to two-dimensional flows. Of course, any real wing has a finite span and so three-dimensional effects can be important. Perhaps the most

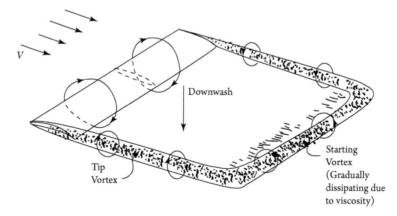

Figure 6.22 The vortex system for a wing of finite length, including the tip vortices, the starting vortex and the bound vortex. (Figure by G.D. Matthew.)

striking three-dimensional phenomenon is the vorticity that gets stripped off the wing tips to form a pair of *tip vortices*, as shown in Figure 6.22. The tip vortices take the form of vortex tubes of opposite sign that eventually link up with the starting vortex, at least for those times before the starting vortex is dissipated by viscosity. The entire vortex system in Figure 6.22 forms a closed loop, with the vorticity directed downstream in the right-hand trailing vortex, towards the reader in the starting vortex, and upstream in the left-hand trailing vortex. This vortex system creates a *downwash* in the wake of the wing, causing the tip vortices to propagate towards the ground (see Exercise 6.3). The kinetic energy of the tip vortices is generated by the rate of working of the aerofoil, which therefore experiences an *induced drag*, supplementing the conventional frictional drag.

○ ○ ○

This concludes our brief excursion into the realms of potential flow theory. There is a truly vast literature on the subject, partly reflecting the historical interest in the topic, and partly because it continues to be a crucial part of aerodynamics to this day. As always, Batchelor (1967) provides a thorough discussion of the subject, while Lamb (1932) and Milne-Thomson (1968) offer encyclopaedic coverage.

....

EXERCISES

6.1 *The Rankine oval.* Consider the superposition of a mass source Q at $x = -a$ and mass sink $-Q$ at $x = a$, both on the x-axis, with a uniform flow V in the x direction. Starting from Figure 6.4 for flow around a Rankine half-body, argue that this represents flow over an oval body which is symmetric about the x and y axes. Show that the length and width of the body are

$$\frac{\ell}{2a} = \left(1 + \frac{Q}{\pi a V}\right)^{1/2}, \quad \frac{w}{2a} = \cot\left(\frac{\pi a V}{Q}\frac{w}{2a}\right).$$

Consider the limit of small a and show that the Rankine oval tends to a circular cylinder of radius $\sqrt{\mu/2\pi V}$, $\mu = 2aQ$, consistent with (6.22).

6.2 *A vortex pair.* Two line vortices of opposite sign are a distance $2d$ apart in an infinite domain. Show that they propagate as a pair and find the speed of propagation. Consider a frame of reference in which the vortices are at rest and located on the x-axis at $x = \pm d$. In this frame there is an oval streamline that encloses the two vortices. Show that this streamline has the equation

$$\frac{x}{d} = \ln\frac{(x+d)^2 + y^2}{(x-d)^2 + y^2}.$$

6.3 *A vortex pair propagating towards a wall.* The two line vortices of Exercise 2 are symmetrically placed either side of the y-axis and propagate directly towards a wall located at the x-axis. Their instantaneous location is $(\pm x_{vor}(t), y_{vor}(t))$. Use the method of images to determine the instantaneous velocity of the right-hand vortex and hence show that the vortices follow the trajectory

$$\frac{1}{x_{vor}^2} + \frac{1}{y_{vor}^2} = \frac{1}{d^2},$$

where $2d$ is their initial separation at large distances from the wall.

6.4 *Blasius' theorem for the two-dimensional force on a body.* Consider the steady, inviscid, two-dimensional, irrotational flow over a body of arbitrary shape. If the perimeter of the body is marked by the contour C, and the components of the pressure force acting on the body in the x and y directions are denoted F_x and F_y, then we have

$$F_x - jF_y = -\oint_C p(dy + jdx) = -j\oint_C p\,d\bar{z},$$

or, using Bernoulli's theorem,

$$F_x - jF_y = \tfrac{1}{2}j\rho\oint_C \mathbf{u}^2 d\bar{z}.$$

Let the complex potential for the flow be $F(z)$, so that (6.17) gives us

$$\mathbf{u}^2 = \frac{dF}{dz}\frac{\overline{dF}}{dz}.$$

Show that the net pressure force acting on the body can be written as

$$F_x - jF_y = \tfrac{1}{2}j\rho\oint_C \left(\frac{dF}{dz}\right)^2 dz.$$

(You may find it helpful to note that, on the contour C, $\overline{dF/dz} = u_x + ju_y$ and $dz = dx + jdy$ are complex numbers of equal argument.) This expression for the force is known as Blasius' theorem.

6.5 *The rate of change of part of a dye line.* Suppose the short displacement vector $\delta\mathbf{l}$ is part of a material curve, so that $\delta\mathbf{l}$ moves with the fluid, like a dye line. The change in $\delta\mathbf{l}$ in a time δt is $\delta\mathbf{u}\delta t$, where $\delta\mathbf{u}$ is the difference in velocity across $\delta\mathbf{l}$. Show that $\delta\mathbf{u} = \delta\mathbf{l}\cdot\nabla\mathbf{u}$ and hence

$$\frac{D}{Dt}(\delta\mathbf{l}) = \delta\mathbf{l}\cdot\nabla\mathbf{u}.$$

6.6 *Relating the circulation around a wing to the net vorticity in the boundary layers.* Suppose there exist two unconnected closed curves, C_1 and C_2, with C_1 lying 'inside' C_2. Let S be a simple two-sided surface that spans the gap between C_1 and C_2. Show that

$$\oint_{C_2} \mathbf{u} \cdot d\mathbf{r} - \oint_{C_1} \mathbf{u} \cdot d\mathbf{r} = \int_S \boldsymbol{\omega} \cdot d\mathbf{S},$$

where the two line integrals and the surface flux are evaluated in accordance with the right-hand rule. Consider the aerofoil shown in Figure 6.1. Use the result above to show that the circulation is the same for all closed curves which lie outside the boundary layers. Next, taking C_1 to be the surface of the wing and C_2 to be any closed curve lying outside the boundary layers, show that the circulation around the aerofoil is the net vorticity in the top and bottom boundary layers.

6.7 *The Kutta condition requires that the bound vorticity $\gamma(s)$ is zero at the trailing edge of a wing.* Show that the smooth flow at the trailing edge of a wing, demanded by the Kutta condition, ensures that (locally) there is no cross-stream gradient in pressure. Now use Bernoulli's equation, applied *along* a streamline, to show that the trailing-edge velocities just outside the top and bottom boundary layers are equal. Hence show that the net negative vorticity in the top boundary layer is equal and opposite to the net positive vorticity in the lower layer, and so the bound vorticity $\gamma(s)$ must be zero at the trailing edge.

6.8 *A paradox?* Consider a thin, flat plate sitting in a uniform, two-dimensional flow at a finite angle of incidence. Outside the boundary layers the flow is irrotational and satisfies the Kutta condition. The fact that pressure acts normal to a surface suggests that the net pressure force acting on the plate is perpendicular to the plate, yet d'Alembert's paradox tells us that the net pressure force is directed normal to the flow. Discuss, qualitatively, how this apparent paradox may be resolved. (Hint: first consider flow over a plate with a *small but finite* thickness, noting that the forward stagnation point sits on the front surface of the plate, as shown in Figure 6.23. Next consider the limiting process of the plate thickness tending to zero, allowing a singularity in velocity to develop at the leading edge, as in Figure 6.7.)

6.9 *The lift on a flat plate at low incidence by the method of conformal mapping.* We can map points in one complex plane, $z = x + jy$, to another complex plane, $Z = X + jY$, through expressions of the form $Z = h(z)$. If h is analytic, then this is called a *conformal transformation* and it has three properties: (i) except at certain critical points, the angles between intersecting curves are preserved as the curves map from the z-plane to the Z-plane; (ii) if a function $g(x,y)$ is harmonic in the z-plane, then the corresponding function $G(X,Y)$ is also harmonic; and (iii) the mapping conserves circulation. Property (i) tells us that the orthogonality of the φ and ψ lines in the

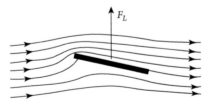

Figure 6.23 Potential flow over a flat plate which satisfies the Kutta condition.

z-plane ensures that the corresponding potential lines and streamlines are orthogonal in the Z-plane, while property (ii) ensures that φ and ψ remain harmonic after the mapping. Thus, if we have a complex potential $F(z)$ representing a known flow in the z-plane, we can use such a mapping to generate a new potential flow in the Z-plane.

Show that the *Joukowski transformation*, $Z = z + a^2/z$, maps the circle $|z| = a$ to the straight-line segment $-2a < X < 2a, Y = 0$, with $(\pm a, 0)$ mapping onto $(\pm 2a, 0)$. Since z maps to Z for large $|z|$, the Joukowski transformation maps flow over a circular cylinder to flow around a thin, flat plate of length $4a$ sitting on the X-axis (Figure 6.24). Confirm that the complex potential for flow over a circular cylinder with far-field velocity V_∞, circulation Γ, and angle of inclination α, is

$$
F_c(z) = V_\infty \left(z e^{-j\alpha} + \frac{a^2}{z} e^{j\alpha} \right) - j \frac{\Gamma}{2\pi} \ln z.
$$

Let $F_p(Z) = F_c(z(Z))$ be the corresponding potential for flow over a plate in the Z-plane, with velocity components

$$
u_x - j u_y = \frac{dF_p}{dZ} = \frac{d}{dZ}(F_c(z)) = \frac{dF_c}{dz} \left(\frac{dZ}{dz} \right)^{-1} = \frac{dF_c/dz}{1 - a^2/z^2}.
$$

The Kutta condition for flow over the plate forbids an infinite velocity at the trailing edge, $Z = 2a$, which evidently translates to the requirement that $dF_c/dz = 0$ at $z = a$ for flow over a cylinder. Show that this demands $\Gamma = -4\pi V_\infty a \sin \alpha$, and by implication,

$$
F_L = -\rho V_\infty \Gamma = \pi \rho V_\infty^2 (4a) \sin \alpha.
$$

We have arrived back at (6.47) for the lift force on a thin, symmetric aerofoil.

In practice, boundary-layer separation occurs just behind the leading edge of a thin plate for all but the smallest of angles of inclination, and so this theoretical lift force is of limited practical interest for a plate. However, for a *symmetrical aerofoil with a rounded leading edge*, separation is supressed up to angles of $10° \to 16°$, and this estimate of F_L is then of more practical importance. A slightly more intricate conformal mapping can be used to analyse such rounded, symmetrical aerofoils of small but finite thickness, T. It leads to essentially the same lift force,

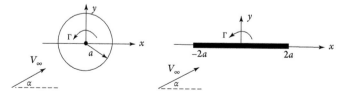

Figure 6.24 The Joukowski mapping of flow over a cylinder to flow around a flat plate.

$$F_L = \pi \rho V_\infty^2 L \sin \alpha \left(1 + 4T \big/ 3\sqrt{3}L \right),$$

where L is the chord and the correction involving T is modest (see Batchelor, 1967).

6.10 *Irrotational flow over a sphere.* Consider irrotational flow over a sphere of radius a, where the incident velocity is uniform and equal to V. Using *cylindrical* polar coordinates (r, θ, z) centred on the sphere, where z is aligned with the oncoming flow, show that

$$\frac{u_r}{V} = -\frac{3a^3 r z}{2\,|\mathbf{x}|^5}, \quad \frac{u_z}{V} = 1 + \frac{a^3}{2\,|\mathbf{x}|^3} - \frac{3a^3 z^2}{2\,|\mathbf{x}|^5}$$

satisfies $\nabla \cdot \mathbf{u} = 0$, $\nabla \times \mathbf{u} = 0$, and the boundary condition on the surface of the sphere. Find the corresponding velocity potential and Stokes streamfunction for the flow.

REFERENCES

Barnard, R.H. & Philpott, D.R., 1995, *Aircraft flight*, 2nd Ed., Longman.
Batchelor, G.K., 1967, *An introduction to fluid dynamics*. Cambridge University Press.
Bird, R.B., Stewart, W.E., & Lightfoot, E.N., 1960, *Transport phenomena*. Wiley.
Davidson, P.A., 2017, *An introduction to magnetohydrodynamics*. 2nd Ed., Cambridge University Press.
Davidson, P.A., 2019, *An introduction to electrodynamics*. Oxford University Press.
Lamb, H., 1932, *Hydrodynamics*. 6th Ed., Cambridge University Press.
Maxwell, J.C., 1891, *A treatise on electricity and magnetism*. 3rd Ed., Clarendon Press.
Milne-Thomson, L.M., 1968, *Theoretical hydrodynamics*. 5th Ed, MacMillan Press.

7

Surface Gravity Waves in Deep and Shallow Water

(a)   (b)

Many names are associated with water waves, but two of the key pioneers in the UK were (a) John Scott Russell and (b) George Biddell Airy (print by Lock & Whitfield). Russell (1808–1882) was a Scottish civil engineer who discovered the solitary wave, a precursor to the soliton. Airy (1801–1892) was an English mathematician and astronomer whose 1845 memoire on tides and waves was, for some time, a prominent authority on the subject. It is probably fair to say that the two did not see eye to eye.

7.1 The Wave Equation and Dispersive versus Non-dispersive Waves

7.1.1 The Wave Equation and d'Alembert's Solution

Prior to discussing surface gravity waves, perhaps it is worth briefly discussing waves in general, in order to establish a few basic ideas. Consider the particular case of waves on a stretched string (Figure 7.1). These are governed by the *one-dimensional wave equation*

$$\frac{\partial^2 \eta}{\partial t^2} = c^2 \frac{\partial^2 \eta}{\partial x^2}, \tag{7.1}$$

where $\eta(x, t)$ is the lateral displacement of the string and c the wave speed. For waves on a string $c = \sqrt{T_s/\rho_s}$, where T_s is the tension in the string and ρ_s its mass per unit length.

An important feature of the wave equation is that it supports the general solution

$$\eta(x, t) = F_f(x - ct) + F_b(x + ct), \tag{7.2}$$

where F_f and F_b are arbitrary functions of their arguments. This is known as *d'Alembert's solution* and its validity is readily established. If we write $\chi^\pm = x \pm ct$, then (7.2) yields

$$\frac{\partial^2 \eta}{\partial t^2} = c^2 \left[F_f''(\chi^-) + F_b''(\chi^+) \right] = c^2 \frac{\partial^2 \eta}{\partial x^2}, \tag{7.3}$$

which does indeed satisfy (7.1). From Figure 7.2 we see that $F_f(x - ct)$ represents a *forward-travelling* disturbance that propagates at the speed c without change of shape, while $F_b(x + ct)$ is a *backward-travelling* pulse which also propagates at the speed c without change of shape.

The generality of solution (7.2) is rather remarkable. Waves that can realize such a solution, which is any wave governed by the one-dimensional wave equation, (7.1), are called *non-dispersive waves* because wave packets travel without change of shape. Another example of a non-dispersive wave is an inviscid, long-wavelength wave of small amplitude propagating on shallow water. As we shall see, such a wave is governed by

$$\boxed{\frac{\partial^2 \eta}{\partial t^2} = c^2 \frac{\partial^2 \eta}{\partial x^2}, \quad c = \sqrt{gh_0}}, \tag{7.4}$$

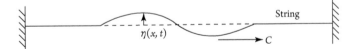

Figure 7.1 Wave on a stretched string.

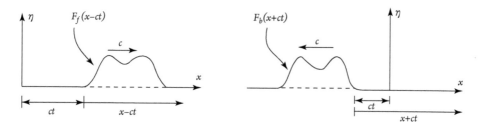

Figure 7.2 Forward- and backward-travelling pulses which travel at the wave speed c.

where η is now the vertical displacement of the water surface, g is gravity, and h_0 is the undisturbed water depth. It is important to note, however, that waves governed by a wave-like partial differential equation (PDE), but not by *the* wave equation, do not usually admit d'Alembert's solution.

7.1.2 Two Classes of Waves: Dispersive versus Non-dispersive Waves

One of the most fundamental distinctions in wave theory is between those waves that are governed by *the* wave equation, (7.1), and those which are governed by a wave-like PDE which is similar to, but distinct from, the wave equation. The latter are called *dispersive waves*, because wave packets spread out (*i.e.* disperse) as they propagate. This is a consequence of the fact that d'Alembert's solution works only for *the* wave equation, but not for other wave-like PDEs. In order to explore this distinction, let us introduce some simple examples of dispersive waves, starting with the flexural vibrations of a beam.

Consider a thin, straight beam of mass per unit length ρ_b, Young's modulus E, and a cross-section whose second moment of area is I (Figure 7.3). Transverse oscillations of the beam are governed by

$$\rho_b \frac{\partial^2 \eta}{\partial t^2} + IE \frac{\partial^4 \eta}{\partial x^4} = 0, \tag{7.5}$$

where $\eta(x, t)$ is the lateral displacement of the centreline of the beam and the coordinate x runs along the centreline. Let us look for a travelling-wave disturbance of wavelength λ, of the form $\eta(x, t) \sim \cos(kx - \varpi t)$, where ϖ is the angular frequency, taken to be *positive*, and $|k| = 2\pi/\lambda$. We take the wavenumber, k, to be positive for forward-travelling waves and negative for backward-travelling waves. Substituting our trial solution into (7.5) yields the *dispersion relationship*,

$$\varpi = \sqrt{IE/\rho_b}\, k^2. \tag{7.6}$$

More generally, the dispersion relationship for any wave-bearing system, usually written in the form $\varpi = \varpi(k)$, is the relationship between ϖ and k corresponding to a sinusoidal travelling wave. The velocity of the wave crests, $c_p = \varpi/k$, is called the *phase velocity*, and

Figure 7.3 Transverse vibrations of a beam.

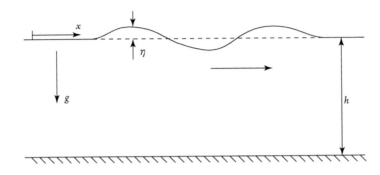

Figure 7.4 A wave on the surface of deep water.

in the case of flexural vibrations of a beam this is given by $c_p = \sqrt{IE/\rho_b}k$. By way of contrast, the dispersion relationship for either (7.1) or (7.4) is simply $\varpi = c|k|$, with a phase velocity of $c_p = \pm c$.

As a second example of dispersive waves, consider inviscid, small-amplitude, surface gravity waves propagating across *deep* water of undisturbed depth h_0, with $h_0 \gg \lambda$. Let $\eta(x, t)$ be the surface displacement (see Figure 7.4) and consider travelling waves of the form $\eta \sim \cos(kx - \varpi t)$. Since $\lambda \ll h_0$, h_0 cannot be important and dimensional analysis requires $\varpi \sim \sqrt{g/\lambda}$. In fact, the dispersion relationship turns out to be $\varpi = \sqrt{g|k|}$, giving a phase velocity of $c_p = \varpi/k = \pm\sqrt{g/|k|}$. Like waves on a beam, this is a function of k.

The key difference between waves on a string and dispersive waves on deep water is that the phase velocity for the non-dispersive waves is independent of k and equal to $\pm c$, i.e. equal to *the* wave speed for the system. For dispersive waves, by contrast, the phase velocity depends on k. In short, for any wave governed by *the* wave equation, (7.1), the wave crests corresponding to different wavenumbers all travel at the same speed. On the other hand, in a wave-bearing system governed by a wave-like PDE, but not *the* wave equation, the wave crests corresponding to different wavenumbers travel at different speeds. This is the basis of the distinction between dispersive and non-dispersive waves.

We now come to a key point: we shall see in §7.3 that when c_p is a function of k, different Fourier modes in a disturbance travel at different speeds, whereas in a system governed by *the* wave equation, all of the Fourier modes in a disturbance travel at the same wave speed, c. This explains the success of d'Alembert's solution. Consider an initial-value problem in which a disturbance of arbitrary shape is specified at $t = 0$. The natural way to find the shape of the disturbance a time τ later is to Fourier transform the initial condition and then track each Fourier mode individually. We then stop the clock at $t = \tau$ and use superposition to reconstruct the disturbance, which amounts to performing an inverse Fourier transform at $t = \tau$. In a non-dispersive system all the Fourier modes have translated the same distance, *i.e.* a distance $c\tau$ to the left or to the right, and so the disturbance does not change shape. By contrast, in a dispersive system, the various Fourier modes all travel at different speeds, and so the disturbance spreads out (disperses) at it propagates.

7.2 Two-dimensional Surface Gravity Waves of Small Amplitude

7.2.1 Surface Gravity Waves on Water of Arbitrary Depth

With this brief introduction to waves, we now turn to two-dimensional surface gravity waves of small amplitude. For simplicity, we shall assume that surface tension may be ignored (we shall rectify this later) and that the motion is irrotational and inviscid, so that (6.9) gives us

$$\mathbf{u} = \nabla \times (\psi \hat{\mathbf{e}}_z) = \nabla \varphi, \quad \nabla^2 \varphi = 0. \tag{7.7}$$

Consider the situation shown in Figure 7.4, in which the origin of coordinates lies at the undisturbed surface, $\eta(x, t)$ is the upward displacement of the free surface, and h_0 is the undisturbed depth. No restriction is placed on kh_0 and so we allow for both deep-water and shallow-water waves. Now, a fluid particle that starts out on the surface stays on the surface, and so the free surface is characterized by the statement that $y_s = \eta(x_s, t)$, where $\mathbf{x}_s$ is the position of a given fluid particle on the surface. Hence, in a time δt,

$$\delta y_s = \delta \eta = \frac{\partial \eta}{\partial t} \delta t + \frac{\partial \eta}{\partial x} \delta x = \frac{\partial \eta}{\partial t} \delta t + \frac{\partial \eta}{\partial x} u_x \delta t. \tag{7.8}$$

This then yields the non-linear, free-surface kinematic boundary condition,

$$u_y = \frac{\partial \eta}{\partial t} + u_x \frac{\partial \eta}{\partial x}, \quad \text{on} \quad y = \eta. \tag{7.9}$$

Since the waves are assumed to be of small amplitude, this may be linearized by dropping the quadratic term on the right and by applying (7.9) at $y = 0$, rather than at $y = \eta$. Our kinematic boundary condition is now

$$\frac{\partial \eta}{\partial t} = u_y = \frac{\partial \varphi}{\partial y}, \quad \text{on} \quad y = 0. \tag{7.10}$$

Next, we apply Bernoulli's equation to the free surface. Incorporating gravity into (6.4), absorbing $f(t)$ into the definition of φ, and setting the surface pressure to zero, gives

$$\frac{\partial \varphi}{\partial t} + g\eta + \frac{1}{2}\mathbf{u}^2 = 0, \quad \text{on} \quad y = \eta,$$

which linearizes to

$$\frac{\partial \varphi}{\partial t} + g\eta = 0, \quad \text{on} \quad y = 0. \tag{7.11}$$

Having established the free-surface boundary conditions, we now look for travelling-wave solutions of the form $\varphi = \hat{\varphi}(y)\sin(kx - \varpi t)$ and $\eta(x, t) = a\cos(kx - \varpi t)$, where a is an amplitude. Since $\nabla^2 \varphi = 0$, $\hat{\varphi}(y)$ must satisfy $\hat{\varphi}''(y) = k^2\hat{\varphi}$, whose general solution can be written as

$$\hat{\varphi} = B\cosh k(y + h_0) + C\sinh k(y + h_0). \tag{7.12}$$

The two constants, B and C, must be determined from the boundary conditions. For example, the vanishing of u_y on the lower boundary demands that $\hat{\varphi}'(y) = 0$ at $y = -h_0$, from which $C = 0$, while the linearized free-surface boundary condition (7.10) requires

$$kB\sinh kh_0 = \varpi a. \tag{7.13}$$

It remains to apply boundary condition (7.11), from which we obtain

$$\varpi B\cosh kh_0 = ga. \tag{7.14}$$

Eliminating B from these two expressions now yields the all-important dispersion relationship for surface gravity waves

$$\boxed{\varpi^2 = gk\tanh kh_0}. \tag{7.15}$$

The velocity potential, velocity components, and streamfunction are now given by

$$\varphi = \frac{\varpi a}{k}\frac{\cosh k(y + h_0)}{\sinh kh_0}\sin(kx - \varpi t), \tag{7.16}$$

$$u_x = \varpi a\frac{\cosh k(y + h_0)}{\sinh kh_0}\cos(kx - \varpi t), \tag{7.17}$$

$$u_y = \varpi a\frac{\sinh k(y + h_0)}{\sinh kh_0}\sin(kx - \varpi t), \tag{7.18}$$

$$\psi = \frac{\varpi a}{k}\frac{\sinh k(y + h_0)}{\sinh kh_0}\cos(kx - \varpi t). \tag{7.19}$$

7.2.2 Shallow-water and Deep-water Waves

It is instructive to consider the special cases of shallow-water and deep-water waves. We start with deep-water waves, which are formally defined by $|k|h_0 \gg 1$, although in practice $h_0 > \lambda/3$ is sufficient. Here, the dispersion relationship simplifies to $\varpi = \sqrt{g|k|}$, as noted in §7.1.2, while the velocity field and streamfunction are

$$\mathbf{u} = \varpi a e^{|k|y} \left(\pm \cos(kx - \varpi t), \; \sin(kx - \varpi t) \right), \tag{7.20}$$

$$\psi = \frac{\varpi a}{k} e^{|k|y} \cos(kx - \varpi t). \tag{7.21}$$

These are dispersive waves with a phase velocity of $c_p = \varpi/k = \pm\sqrt{g/|k|}$.

For shallow-water waves, on the other hand, we have $|k|h_0 \ll 1$ and the dispersion relationship simplifies to $\varpi = \sqrt{gh_0}\,|k|$, from which $c_p = \pm\sqrt{gh_0}$. These are non-dispersive waves and the velocity components and streamfunction are given by

$$u_x = \frac{\varpi}{k}\frac{a}{h_0}\cos(kx - \varpi t) + O(\varpi a k h_0), \tag{7.22}$$

$$u_y = \varpi \frac{a}{h_0}(y + h_0)\sin(kx - \varpi t), \tag{7.23}$$

$$\psi = \frac{\varpi}{k}\frac{a}{h_0}(y + h_0)\cos(kx - \varpi t). \tag{7.24}$$

Notice that, for shallow-water waves, $u_y \sim kh_0 u_x$ and so the motion is almost purely horizontal. Moreover, u_x is independent of depth, at least to leading order in kh_0.

When there is a mean horizontal motion in shallow water, say V, as well as waves, it is customary to make a sharp distinction between the case where $V < \sqrt{gh_0}$, which is called *subcritical flow*, and $V > \sqrt{gh_0}$, which is referred to as *supercritical flow*. In the latter case the mean motion is so strong that all waves are swept downstream, while waves can propagate upstream in a subcritical flow. There is a clear analogy here to the distinction between subsonic and supersonic flow in a compressible fluid, and indeed we shall see that the analogue of a shock wave in shallow-water flow is a hydraulic jump.

Finally, it is worth asking exactly what is meant by the caveat *small amplitude* in the context of surface gravity waves. In this respect we note that the linearization of (7.9) requires

$$u_x \frac{\partial \eta}{\partial x} \ll \frac{\partial \eta}{\partial t},$$

which is satisfied when $|u_x| \ll \varpi/|k|$. In the case of deep-water waves, (7.20) tells us that this linearization is valid whenever $|k|a \ll 1$, or equivalently $a \ll \lambda$. Of course, we could have anticipated that this is the case since, in deep water, λ is the only available length scale against which to compare a. For shallow-water waves, on the other hand, (7.22) tells us that we must satisfy the more stringent condition of $a \ll h_0$.

7.2.3 Particle Paths, Stokes Drift, and Energy Density in Deep-water Waves

It is of interest to calculate the particle paths induced by the passage of a wave. To keep the algebra simple, we shall focus on the simplest case of deep-water waves. Suppose that a fluid particle whose rest location is (x_0, y_0) has the instantaneous position

$$\mathbf{x}_p(t) = (x_p, y_p) = (x_0 + x', y_0 + y'), \tag{7.25}$$

so that the instantaneous velocity of the particle is $\mathbf{u}^{(p)} = \mathbf{u}(\mathbf{x}_p, t)$. Then (7.20) tells us that, for deep-water waves with a positive k,

$$\mathbf{u}^{(p)} = \varpi a e^{ky_0} e^{ky'} \left(\cos(kx_0 + kx' - \varpi t), \, \sin(kx_0 + kx' - \varpi t) \right), \tag{7.26}$$

which linearizes for small-amplitude waves to give

$$\mathbf{u}^{(p)} = \frac{d\mathbf{x}'}{dt} = \varpi a e^{ky_0} \left(\cos(kx_0 - \varpi t), \, \sin(kx_0 - \varpi t) \right). \tag{7.27}$$

On integration we find

$$\mathbf{x}' = a e^{ky_0} \left(-\sin(kx_0 - \varpi t), \, \cos(kx_0 - \varpi t) \right), \tag{7.28}$$

which yields $\mathbf{x}'^2 = a^2 e^{2ky_0}$. Evidently, in the linear approximation, the particle paths for deep-water waves are circular and with a radius R that decreases rapidly with depth in accordance with $R = a e^{ky_0}$. Moreover, it is simple to show that the sense of rotation of the fluid particles is clockwise for a forward-travelling wave. For a fluid of finite depth, the particle trajectories become elliptical, rather than circular, with an aspect ratio that becomes increasingly flat as kh_0 gets smaller (see Exercise 7.2).

If the calculation of the particle trajectories is taken to second order in amplitude, then an interesting phenomenon is revealed, known as *Stokes drift*. As before, our starting point is the particle velocity for *linear* (small-amplitude) waves travelling on deep water, i.e. (7.26). This time, however, we retain the terms involving $\mathbf{x}'$, though only to first order in $k\mathbf{x}'$. The horizontal speed of a fluid particle with rest position (x_0, y_0) is then

$$u_x^{(p)} = \varpi a e^{ky_0} \left[(1 + ky') \cos(kx_0 - \varpi t) - kx' \sin(kx_0 - \varpi t) \right] + O\left((k\mathbf{x}')^2 \right),$$

where, as before, we take $k > 0$. Approximating $\mathbf{x}'$ using (7.28) now yields

$$u_x^{(p)} = \varpi a e^{ky_0} \cos(kx_0 - \varpi t) + \varpi k a^2 e^{2ky_0} + O\left((ka)^2 \right),$$

and on averaging over a cycle we find that the mean horizontal velocity is non-zero:

$$\overline{u_x^{(p)}} = \varpi k a^2 e^{2ky_0}. \tag{7.29}$$

We conclude that there is a mean drift of the fluid particles in the direction of the phase velocity, which is known as Stokes drift. Thus, in each cycle, a fluid particle combines its orbital motion around a circle of radius $a e^{ky_0}$ with a forward movement of $2\pi k a^2 e^{2ky_0}$.

Equation (7.29) also suggests that deep-water gravity waves possess a mean horizontal momentum, and on integrating (7.29) through the depth of the wave we might anticipate that this net mean momentum is $M = \frac{1}{2} \rho \varpi a^2$ per unit plan area. This turns out to be correct, with the mean momentum concentrated close to the surface (Phillips, 1966).

It is interesting to compare this with the kinetic energy density carried by the waves, which from (7.20) is $\frac{1}{2}\rho\varpi^2 a^2 e^{2ky}$. Integrating through the depth of the wave and substituting for ϖ using the dispersion relationship yields a kinetic energy per unit plan area of $\frac{1}{4}\rho g a^2$. Moreover, the potential energy per unit plan area is $\frac{1}{2}\rho g \eta^2$ (see Exercise 7.3), which averages over a cycle to give $\frac{1}{4}\rho g a^2$. So we conclude that there is an equipartition of energy and that the total energy per unit plan area for a deep-water wave is $\frac{1}{2}\rho g a^2$. The ratio of mean energy to mean momentum is then $c_p = \varpi/k$, which is the deep-water version of a general rule satisfied by surface gravity waves (Phillips, 1966).

7.2.4 Wave Drag in Deep Water

The fact that waves carry with them energy leads directly to the phenomenon of *wave resistance*. Casual observation shows that a ship or boat passing through otherwise still water disturbs the surface of the water in two distinct ways. To the sides and the rear of the ship the surface exhibits signs of turbulence, which is the manifestation of a conventional turbulent boundary layer and wake. A similar wake appears behind a submerged body, such as a submarine, and within the wake the turbulence is very efficient at converting mechanical energy into heat. In addition, surface gravity waves are generated at the bow and stern of a ship or boat and these fan out to form a wedge-like wave field behind the ship. Because these waves are only mildly damped they can extend over large distances, as shown in Figure 7.5. Work must be done by the ship on the water in order to compensate for the energy dissipated by the turbulence in the wake, as well as the energy carried off to the far field by waves. This work is $F_D V$, where F_D is the net drag force acting on the ship and V is its speed. It follows that we can consider the drag force to be the sum of two distinct parts, a conventional viscous

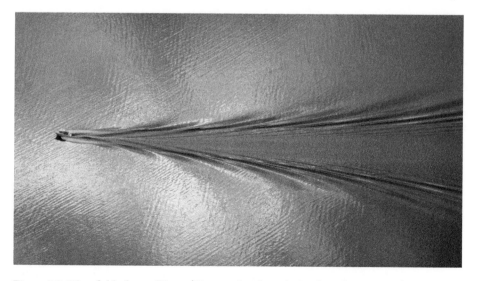

Figure 7.5 Wave field of a small boat. (Picture taken from the Preikestolen, Norway.)

and pressure drag associated with the turbulent boundary layer and wake, and a *wave drag* associated with the generation of surface gravity waves; in short $F_D = F_D^{(\text{turb})} + F_D^{(\text{wave})}$.

The wave pattern, and hence the wave drag, is found to be sensitive to the interaction of the bow and stern waves, with a particularly large drag arising when the bow and stern waves interfere constructively. This wave pattern is, in turn, controlled by the dimensionless *Froude number*, $\text{Fr} = V/\sqrt{gL}$, where L is the length of the ship. Indeed, for a given shape of ship or boat, it is found that the dimensionless wave drag $C_D^{(\text{wave})} = F_D^{(\text{wave})}/\frac{1}{2}\rho A_w V^2$, A_w being the wetted area of the hull, is effectively determined by Fr. This can be understood through dimensional analysis. For a *given shape of ship* we have $A_w \sim L^2$, while $F_D^{(\text{wave})}$ is controlled by ρ, V, L, and g. So, if we include $F_D^{(\text{wave})}$, there is a total of five parameters ($P = 5$) which contain between them the three dimensions of mass, length, and time ($D = 3$). The Buckingham Pi theorem now tells us that there are $G = P - D = 2$ independent dimensionless groups that can be formed from these five parameters, and these are Fr and $C_D^{(\text{wave})}$. Since one dimensionless group can depend only on another dimensionless group, we conclude that $C_D^{(\text{wave})} = f(\text{Fr})$ for some function f.

The curve $f(\text{Fr})$ depends on the shape of the ship or boat, but typically it rises as Fr increases, eventually reaching a broad peak, after which f declines. There are, however, occasional sharp peaks in the curve $f(\text{Fr})$, corresponding to the constructive interference of bow and stern waves. Cargo ships tend to operate in a regime which is a compromise between the need to maximize speed, which favours a large value of Fr, and the need to minimize drag and hence fuel costs, which calls for a low value of f and hence a lower value of Fr.

More generally, for a given shape of ship or boat, the *total* drag force, F_D, is a function of ρ, V, L, g, and ν. We now have $P = 6$ and $D = 3$, and so the number of independent dimensionless groups is $G = P - D = 3$. These groups are Fr, $\text{Re} = VL/\nu$, and $C_D = F_D/\frac{1}{2}\rho A_w V^2$, and so we conclude that $C_D = f(\text{Re}, \text{Fr})$. This is often approximated by the expression

$$C_D = C_D^{(\text{turb})}(\text{Re}) + C_D^{(\text{wave})}(\text{Fr}), \tag{7.30}$$

on the assumption that the wave radiation is insensitive to viscosity, while the boundary-layer drag and separation is insensitive to the wave emission. Often wave drag dominates over wake turbulence, in which case the laboratory testing of scaled models is carried out with a value of Fr, but not Re, which is characteristic of that encountered at full scale.

7.3 The General Theory of Dispersive Waves

7.3.1 Dispersion, Wave Packets, and the Group Velocity

We now temporarily set aside surface waves to discuss dispersive waves in general. These are waves for which $c_p = c_p(k)$, so that a wave pulse changes shape as it propagates. That is to say, if we Fourier decompose an initial disturbance, then each Fourier mode travels at a different speed and so propagates a different distance in a given time. So, on reconstructing the disturbance some time later (by summing all the Fourier modes), we obtain a shape which is different to that of the initial disturbance. Typically, a pulse starts out as localized (think of a hammer striking a beam), and then spreads out (disperses) as it propagates. Eventually the various Fourier modes become well separated in space and we talk of the disturbance taking the form of a *wave packet*, or a *slowly modulated wave train*. At a given location within such a wave train the disturbance looks locally like a single Fourier mode of given wavenumber and amplitude, a. However, as we move along the wave train both the wavenumber and amplitude slowly change. Typically, the amplitudes of the various Fourier modes within the wave train, $a(k)$, are narrow-banded and peak around the dominant wavenumber, k_0, of the initial disturbance (see Figure 7.6).

Dispersive waves have many counterintuitive properties. For example, the speed at which a wave packet propagates as a whole in a dispersive system, which is called the *group velocity*, is *not* the speed at which individual wave crests move, i.e. not the phase speed $c_p = \varpi/k_0$. Rather, it turns out to be $c_g = d\varpi/dk$. Consider, for example, Figure 7.7. If you strike a beam with a hammer, it turns out that the resulting wave crests travel at *half* the speed of the wave packet within which they sit, or, if you throw a stone into a deep pond, individual wave crests travel at *twice* the speed of the overall wave packet!

So how can we explain such a counterintuitive phenomenon? To focus our thoughts, consider the case of a stone striking the surface of a deep pond. The surface waves spread out in the form of an annulus, as shown in Figure 7.8(a), and after some time the waves become well dispersed. A cross-section through the wave train then looks something like that shown in Figure 7.8(b), where the dominant wavelength is set by the size of the stone. Now consider a particular wave crest within the wave packet. It will have an angular frequency of $\varpi = \sqrt{g|k|}$ and travel at a phase speed of $c_p = \sqrt{g/|k|}$. However, if you track the centre of the overall wave packet, then it turns out that it travels at the slower speed of $c_g = d\varpi/dk = \frac{1}{2}\sqrt{g/|k|}$. In order to accommodate this difference in speeds, wave crests appear out of nowhere at the rear of the wave packet, grow in amplitude as they ripple through the wave packet at a speed of $c_p = \sqrt{g/|k|}$, and then fade away as they approach the far end of the wave train. In the case of flexural vibrations of a beam, the wave crests travel at $c_p = \sqrt{IE/\rho_b}k$, in accordance with (7.6), but an overall wave packet moves at twice the speed of the wave crests, with a group velocity of $c_g = d\varpi/dk = 2\sqrt{IE/\rho_b}k$. Of course,

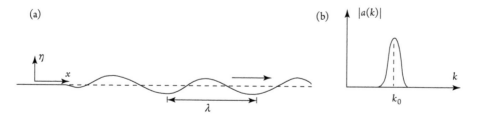

Figure 7.6 (a) A slowly modulated wave train. (b) The amplitudes of the Fourier modes.

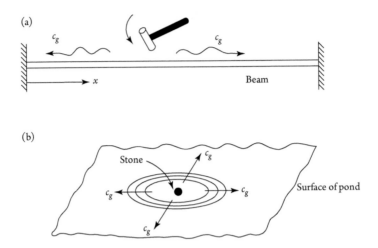

Figure 7.7 Waves generated in (a) a beam and (b) the surface of a deep pond.

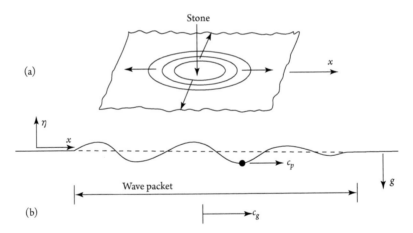

Figure 7.8 (a) Waves on the surface of a pond. (b) Cross-section through a wave packet.

the key question now is: why do wave packets travel at the speed $c_g = d\varpi/dk$, rather than the phase speed $c_p = \varpi/k$?

Let us try to understand the origins of the expression $c_g = d\varpi/dk$. Suppose that we have a slowly modulated wave train in a one-dimensional dispersive system, as shown in Figure 7.6. Then waves of wavenumber k propagate as

$$\eta(x, t) = a(k)\exp\left[j\left(kx - \varpi t\right)\right], \quad \varpi = \varpi(k), \tag{7.31}$$

where $\varpi = \varpi(k)$ is the dispersion relationship for the system at hand, $a(k)dk$ is the amplitude of the waves in the range $k \to k + dk$, and the real part is understood. Superposition now gives us

$$\eta(x, t) = \int_{-\infty}^{\infty} a(k) \exp\left[j\left(kx - \varpi t\right)\right] dk, \tag{7.32}$$

and we recognize that $a(k)$ is, in fact, the Fourier transform of the initial condition $\eta(x, 0)$. If $a(k)$ is narrow and centred on k_0, as shown in Figure 7.6, then we have

$$\varpi(k) = \varpi(k_0) + (k - k_0)\, c_g\,(k_0) + O\,(k - k_0)^2, \tag{7.33}$$

where c_g is *defined* as $c_g = d\varpi/dk$. We now substitute for ϖ in (7.32) while neglecting the quadratic term in expansion (7.33) on the grounds that $a(k) \approx 0$ when k differs significantly from k_0. After a little algebra we find

$$\eta(x, t) \approx \exp\left[j\left(k_0 x - \varpi(k_0)t\right)\right] \int_{-\infty}^{\infty} a(k) \exp\left[j\left((k - k_0)\left(x - c_g(k_0)t\right)\right)\right] d(k - k_0). \tag{7.34}$$

Treating $k - k_0$ as a dummy variable, we conclude that (7.34) has the general form

$$\eta(x, t) = F\left(x - c_g(k_0)t\right) \cdot \exp\left[j\left(k_0 x - \varpi(k_0)t\right)\right], \tag{7.35}$$

where the shape of the function F depends on the amplitude distribution, $a(k)$. This is a travelling wave of wavenumber k_0 and frequency $\varpi(k_0)$, which is modulated by the envelope $F(x - c_g(k_0)t)$ as shown in Figure 7.9. It is clear that the overall wave packet, as represented by $F(x - c_g(k_0)t)$, propagates at the speed $c_g(k_0) = (d\varpi/dk)_0$, which is the group velocity corresponding to the dominant wavenumber in the wave packet.

Actually, a more refined analysis yields the stronger results that:

(i) $c_g(k) = d\varpi/dk$ is the speed at which you must travel to continually observe waves of wavenumber k; and, by implication,

(ii) $c_g(k)$ is the speed at which waves of wavenumber k transport energy.

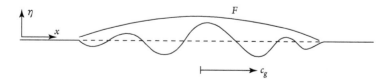

Figure 7.9 A traveling wave of dominant wavenumber k_0 sits inside the envelope $F\left(x - c_g(k_0)t\right)$.

Moreover, since there is always a range of wavenumbers in a given wave packet, this means that wave packets inevitably spread out as they propagate.

This analysis generalizes in an obvious way to three-dimensional waves. For example, solutions of the 3D wave equation such as acoustic or electromagnetic waves are non-dispersive, so everything just propagates at *the* wave speed—the speed of sound or the speed of light. However, for three-dimensional dispersive waves, such as internal gravity waves in a stratified fluid, one must look for travelling waves of the form

$$\eta(\mathbf{x}, t) \sim \exp\left[j\left(\mathbf{k}\cdot\mathbf{x} - \varpi t\right)\right], \tag{7.36}$$

where $\mathbf{k}$ is now a *wavevector*. Substituting this trial solution into the governing PDE for the system at hand yields the dispersion relationship $\varpi = \varpi(\mathbf{k})$. The group velocity is then defined as the vector $\mathbf{c}_g$ whose components are $(\mathbf{c}_g)_i = \partial\varpi/\partial k_i$. As in one dimension, $\mathbf{c}_g$ has the properties that:

(i) $\mathbf{c}_g(\mathbf{k}_0)$ is the velocity at which a wave packet dominated by the wavevector $\mathbf{k}_0$ propagates away from an initial disturbance;

(ii) $\mathbf{c}_g(\mathbf{k})$ is the velocity at which you must travel in order to keep seeing waves of wavevector, $\mathbf{k}$; and

(iii) $\mathbf{c}_g(\mathbf{k})$ is the velocity at which the energy of waves of wavevector $\mathbf{k}$ propagates through the system.

7.3.2 The Energy Flux in a Wave Packet

This last point suggests that the flux of wave energy per unit area in the direction of propagation of a wave packet is the product of the group velocity and the energy density of that wave. This is clearly so for an isolated wave packet in which the energy density is *spatially uniform*: that is to say, the packet as a whole travels at the velocity $\mathbf{c}_g(\mathbf{k}_0)$ and so, if we consider a plane orthogonal to $\mathbf{c}_g$, the rate at which energy is transported across that plane is the product of $\mathbf{c}_g$, the wave energy density, and the cross-sectional area of the packet. However, for surface gravity waves the energy density varies with depth and so it makes more sense to talk about the energy flux *per unit length of the crest*, and the energy density *per unit plan area*, i.e. the energy flux and energy density integrated up through the wave. The equivalent statement is then:

the energy flux per unit length of the crest (averaged over a cycle) is the group velocity times the energy density per unit plan area (also averaged over a cycle).

To be explicit, consider a surface gravity wave of the form $\eta(x, t) = a\cos(kx - \varpi t)$ propagating on water of depth h_0. The group velocity calculated from (7.15) is

$$c_g = \frac{\varpi}{2k}\left[1 + \frac{2kh_0}{\sinh 2kh_0}\right], \tag{7.37}$$

while the wave energy per unit plan area, averaged over a cycle, is shown in Exercise 7.3 to be $\frac{1}{2}\rho g a^2$. The flux of wave energy in the direction of propagation of the wave, measured per unit length of the wave crest and averaged over a cycle (see Exercise 7.4), is

$$\frac{\varpi}{2k}\left[1 + \frac{2kh_0}{\sinh 2kh_0}\right]\frac{1}{2}\rho g a^2 = c_g\left(\frac{1}{2}\rho g a^2\right), \tag{7.38}$$

exactly as expected.

7.4 The Dispersion of Small-amplitude Surface Gravity Waves

7.4.1 The Group Velocity and Energy Density for Waves on Water of Arbitrary Depth

Perhaps it is time to summarize all of our results relating to the group velocity and energy of surface gravity waves. We restrict ourselves to inviscid, small-amplitude waves of the form $\eta(x, t) = a\cos(kx - \varpi t)$ that propagate on water of depth h_0. From (7.15) and (7.37) we see that the dispersion relationship and group velocity are

$$\varpi^2 = gk\tanh kh_0, \qquad c_g = \frac{\partial \varpi}{\partial k} = \frac{\varpi}{2k}\left[1 + \frac{2kh_0}{\sinh 2kh_0}\right], \tag{7.39}$$

while Exercise 7.3 gives the kinetic and potential energies per unit horizontal area as

$$\frac{1}{2}\rho\int_{-h_0}^{0}\mathbf{u}^2 dy = \frac{1}{4}\rho g a^2\left[1 + \frac{2kh_0\left(\cos^2(kx - \varpi t) - \sin^2(kx - \varpi t)\right)}{\sinh(2kh_0)}\right], \tag{7.40}$$

$$\int_0^{\eta}\rho g y dy = \frac{1}{2}\rho g \eta^2 = \frac{1}{2}\rho g a^2\cos^2(kx - \varpi t). \tag{7.41}$$

When averaged over a cycle (either a wavelength or a period), the kinetic and potential energy densities are equal and given by $\frac{1}{4}\rho g a^2$, irrespective of the fluid depth. This yields a total energy density of $\frac{1}{2}\rho g a^2$ per unit horizontal area. Finally, from Exercise 7.4 we find that the wave energy flux, measured per unit length of the wave crests and averaged over a period, is

$$\frac{\varpi}{2k}\left[1 + \frac{2kh_0}{\sinh 2kh_0}\right]\frac{1}{2}\rho g a^2 = c_g\left(\frac{1}{2}\rho g a^2\right). \tag{7.42}$$

For deep-water waves, which are formally defined by $kh_0 \gg 1$, but in practice occur when $h_0 > \lambda/3$, these expressions simplify to

$$\varpi = \sqrt{g\,|k|}, \quad c_p = \sqrt{g/|k|}, \quad c_g = \frac{1}{2}\sqrt{g/|k|} = \frac{1}{2}c_p, \tag{7.43}$$

$$\frac{1}{2}\rho \int_{-h_0}^{0} \mathbf{u}^2 \, dy = \frac{1}{4}\rho g a^2, \quad \frac{1}{2}\rho g \eta^2 = \frac{1}{2}\rho g a^2 \cos^2(kx - \varpi t). \tag{7.44}$$

Crucially, wave packets and wave energy propagate at only half the speed of the wave crests. This lies behind the odd phenomenon illustrated in Figure 7.8, in which an annular wave packet spreads out from the point of impact of a stone. In order to accommodate the fact that the crests move at twice the speed of the wave packet, the crests first appear out of nowhere at the rear of the wave packet, grow in amplitude as they ripple through the wave packet, and then fade away as they approach the far end of the wave train.

By way of contrast, shallow-water waves, defined by $kh_0 \ll 1$, which in practice means $h_0 < \lambda/14$, are non-dispersive. They have the properties

$$\varpi = \sqrt{gh_0}\,|k|, \quad c_p = c_g = \sqrt{gh_0}, \tag{7.45}$$

$$\frac{1}{2}\rho \int_{-h_0}^{0} \mathbf{u}^2 \, dy = \frac{1}{2}\rho g \eta^2 = \frac{1}{2}\rho g a^2 \cos^2(kx - \varpi t). \tag{7.46}$$

In all cases, from shallow to deep water, the total energy density, averaged over a period, is $\frac{1}{2}\rho g a^2$ per unit plan area, while the averaged wave energy flux per unit length of the wave crests is $c_g(\frac{1}{2}\rho g a^2)$.

If the depth of undisturbed fluid changes slowly with x, then we can still use equations (7.39)–(7.42), but with h_0 replaced by $h(x)$. However, the relationship between ϖ and k, which is constrained to satisfy the local dispersion relationship (7.15), must now evolve as the depth slowly changes. The question then arises as to whether ϖ or k, or neither, is conserved as the wave moves to a different depth. Certainly *both* cannot be conserved, unless of course the water is deep, in which case $h(x)$ is irrelevant. To answer this question, it is necessary to embrace the theory of ray tracing, which is discussed in Lighthill (1978). However, for our purposes we need only know that it is the *frequency*, not the wavenumber, which is conserved as the fluid depth changes. The variation of wavenumber with depth can then be found by solving $gk \tanh kh = \varpi^2$ for a constant ϖ, while the relationship between phase velocity and depth is given by (7.15), rewritten in the form

$$\varpi c_p/g = \tanh(\varpi h/c_p), \quad \varpi = \text{const}. \tag{7.47}$$

The shape of the curve $\varpi c_p/g$ versus $\varpi^2 h/g$ is shown in Figure 7.10, with shallow water on the left and deep water on the right. Clearly, as waves move from deep to shallow water, the phase speed drops, which is consistent with, but extends, the trend set by the shallow-water equation $c_p = \sqrt{gh}$. Given that $c_p \sim \varpi \lambda$, any fall in c_p must be accompanied by a

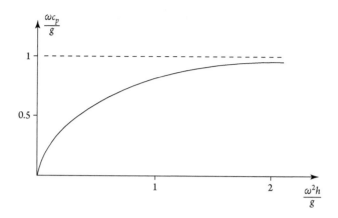

Figure 7.10 The shape of the curve $\varpi c_p/g$ versus $\varpi^2 h/g$ for surface gravity waves.

corresponding decline in the wavelength, λ. Thus, for example, according to (7.43) deep-water waves of wavelength 100 m have a phase speed of 12.5 m/s and period 8 s, but if these waves of period 8 s propagate into a region of shallow water, say 1 m deep, then according to (7.45) their wavelength drops to 25 m and their phase speed to 3.13 m/s, both of which are a factor of 4 smaller than their deep-water counterparts. The group velocity, by contrast, falls from 6.25 m/s to 3.13 m/s, *i.e.* by half.

7.4.2 Waves Approaching a Beach

It is noticeable that deep-water waves approaching a beach tend to align their wave crests with the shore, while the waves simultaneously steepen as the water depth falls. Both of these phenomena can be explained, at least in part, using the results of §7.4.1, on the assumption that the changes in depth are gradual. Consider, first, the tendency to align the wave crests with the depth contours of the beach. Suppose the deep-water wave crests are initially oblique to the beach. Then, as a particular wave crest approaches the shore, that part of the crest which first reaches shallow water slows down, while that further out retains its higher speed. In this way the wave crest automatically swings around to align with the beach. In effect, the waves are refracted.

The steepening of the waves is a little more complicated and arises from a combination of two processes. Suppose the wave crests are aligned with the beach. Then, as shown in Figure 7.10, the phase velocity falls as the beach is approached, and it is not difficult to show that the corresponding group velocity also falls. Now, the energy flux $c_g\left(\frac{1}{2}\rho g a^2\right)$ is conserved as the waves move from deep to shallow water, so any drop in c_g must be accompanied by a corresponding rise in amplitude, a. Moreover, the fall in phase velocity requires that the wavelength declines to conserve frequency, in accordance with $c_p \sim \varpi \lambda$. The steepness of a wave, $2\pi a/\lambda$, then increases partly from the increase in amplitude, a, and partly from the fall in λ, although the latter effect is always dominant.

It is instructive to examine a particular example. Let us consider, as Lighthill (1978) did, the case of deep-water waves of wavelength 100 m approaching shallow water 1 m deep, as discussed above. Then the phase velocity falls by a factor of 4, which requires the group velocity to decline by a factor of 2. (Remember that $c_g = c_p/2$ for deep-water waves, while $c_g = c_p$ in shallow water.) Conservation of $c_g a^2$ then requires that the wave amplitude a rises by a factor of $\sqrt{2}$. However, the wavelength also falls by a factor of 4, and so the rise in wave steepness, $2\pi a/\lambda$, is $4\sqrt{2}$.

7.4.3 The Influence of Surface Tension on Dispersion

So far we have ignored surface tension, which is important for waves of wavelength less than 1 cm or so. The Young–Laplace law tells us that, if a liquid surface is curved upward with air above it, the pressure in the liquid immediately adjacent to its surface exceeds atmospheric pressure by an amount

$$p_s = \gamma \left(\frac{1}{R} + \frac{1}{R'} \right), \tag{7.48}$$

where R and R' are the principal radii of curvature and γ is the surface tension coefficient. For two-dimensional surface waves of small amplitude we can approximate the surface curvature by $R^{-1} = -\partial^2\eta/\partial x^2$, where $\eta(x, t) = a\cos(kx - \varpi t)$, and so (7.48) becomes

$$p_s = -\gamma \frac{\partial^2\eta}{\partial x^2} = \gamma k^2 \eta. \tag{7.49}$$

The relative importance of surface tension is often measured using the *capillary length*, which is defined as $\ell_\gamma = \sqrt{\gamma/\rho g}$. As (7.49) shows, the ratio of capillary to gravitational forces in a surface wave is $(k\ell_\gamma)^2$, and so the two forces are of equal magnitude for a wavelength of $\lambda_\gamma = 2\pi\ell_\gamma$. The surface tension coefficient and capillary length for some common fluids are given in Table 7.1.

The primary effect of surface tension is to change the boundary condition (7.11). Instead of setting the surface pressure to zero in Bernoulli's equation, we must use (7.49), and so the linearized surface boundary condition becomes

$$\frac{\partial\varphi}{\partial t} + g\eta - \frac{\gamma}{\rho}\frac{\partial^2\eta}{\partial x^2} = 0, \text{ on } y = 0. \tag{7.50}$$

Table 7.1 Surface tension coefficients for some common liquids.

Liquid	Density (kg/m^3)	Surface tension coefficient (N/m)	ℓ_γ (cm)	λ_γ (cm)
Water	10^3	7.3×10^{-2}	0.27	1.7
Olive oil	920	3.2×10^{-2}	0.19	1.2
Mercury	13.5×10^3	0.47	0.19	1.2

For a travelling wave of the form $\eta(x, t) = a\cos(kx - \varpi t)$, the effect of this new boundary condition is simply to replace g in the old analysis by $g + (\gamma/\rho)k^2$. It follows immediately that the dispersion relationship for surface waves is modified to read

$$\varpi^2 = \left(g + \frac{\gamma k^2}{\rho}\right) k\tanh kh_0. \tag{7.51}$$

It is natural to restrict ourselves to deep-water waves at this point, as surface tension plays little role in most common liquids for wavelengths greater than a few cm. For deep-water waves with $k > 0$, the dispersion relationship above simplifies to

$$\varpi^2 = gk + \frac{\gamma k^3}{\rho}, \tag{7.52}$$

with phase and group velocities of

$$c_p = \sqrt{(g/k) + (\gamma k/\rho)}, \tag{7.53}$$

$$c_g = \frac{(g/k) + 3(\gamma k/\rho)}{2\sqrt{(g/k) + (\gamma k/\rho)}} = \frac{(g/k) + 3(\gamma k/\rho)}{2c_p}. \tag{7.54}$$

For wavelengths much less than λ_γ we have pure capillary waves, often called *ripples*, for which $c_p = \sqrt{\gamma k/\rho}$ and $c_g = \frac{3}{2}c_p$. These differ from pure gravity waves in two important respects. First, the *shortest* wavelengths travel fastest, and second, a wave packet (and hence also energy) travels *faster* than the wave crests. For mixed *capillary–gravity waves* the phase velocity exhibits a minimum at $\lambda = \lambda_\gamma = 2\pi\ell_\gamma$, as shown in Figure 7.11. For this wavelength (7.54) tells us that $c_g = c_p$. To the left of the minimum we find $c_g > c_p$, consistent with $c_g = \frac{3}{2}c_p$ for pure capillary waves, while to the right we have $c_g < c_p$, consistent with $c_g = \frac{1}{2}c_p$ for pure gravity waves. It turns out that for $\lambda < \lambda_\gamma/3$ gravitational effects are negligible, while for $\lambda > 4\lambda_\gamma$ surface tension may be ignored.

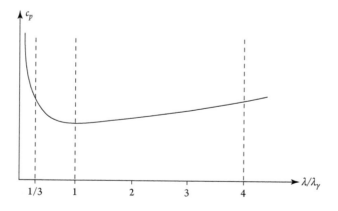

Figure 7.11 The phase velocity for mixed capillary–gravity waves as a function of λ.

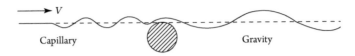

Figure 7.12 Flow past an object of scale λ_γ sets up a stationary wave pattern consisting of capillary waves upstream and gravity waves downstream.

The distinction between capillary and gravity waves can be most clearly seen when there is flow past an object of scale λ_γ, as shown in Figure 7.12. Provided the mean flow exceeds the minimum value of c_p in Figure 7.11 (which is 23 cm/s for water), a stationary wave pattern is set up in which the phase velocity of the waves is equal and opposite to the mean flow V, thus enabling the wave crests to adopt a fixed location relative to the object. From Figure 7.11 we see that, for a given V, there are two values of λ for which this is possible, and so both capillary and gravity-dominated waves are generated. It turns out that the gravity-dominated waves sit downstream of the object, while the capillary-dominated waves sit upstream. This is because the group velocity for the gravity-dominated waves is less than their phase velocity and so less than V, and hence all of their energy is swept downstream. On the other hand, the group velocity for the capillary-dominated waves is greater than their phase velocity and so greater than V, so the capillary waves can propagate upstream despite the mean flow. A twig breaking the surface of a stream exhibits just such a wave pattern, as do water beetles.

7.5 Finite-amplitude Waves in Shallow Water

7.5.1 The Inviscid Shallow-water Equations

We now turn to waves of finite amplitude, starting with the inviscid, shallow-water equations. Suppose that we have an inviscid, free-surface flow over a bed of constant height located at $y = 0$. The horizontal scale of the free-surface undulations is taken to be much larger than the depth of the fluid, $y = h(x, z, t)$, as shown in Figure 7.13. The flow is then predominantly horizontal and the vertical velocity is much smaller than the horizontal velocity, $u_y \ll |\mathbf{u}_H|$. If we ignore the vertical acceleration of the fluid then the horizontal and vertical components of Euler's equation are

$$\frac{D\mathbf{u}_H}{Dt} = -\nabla_H (p/\rho), \quad 0 = -\nabla_V (p/\rho) + \mathbf{g}, \tag{7.55}$$

where the subscripts H and V stand for horizontal and vertical. The second equation tells us that the pressure distribution is hydrostatic, with the gauge pressure related to the fluid depth h by $p = \rho g(h - y)$. The term $\nabla_H(p)$ is now independent of y, and so if $\mathbf{u}_H$ is initially independent of the depth, then it must stay that way. The momentum equation now reduces to

$$\frac{D\mathbf{u}_H}{Dt} = -\nabla (gh), \quad \mathbf{u}_H = \mathbf{u}_H(x, z, t). \tag{7.56}$$

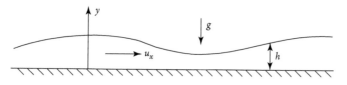

Figure 7.13 Shallow-water flow.

Of course, $\mathbf{u}_H$ and h are also related by mass conservation. Let $\mathbf{q} = \mathbf{u}_H h$ be the horizontal volumetric flow rate and consider mass conservation applied to the control volume $hdA = hdxdz$ whose vertical sides are $d\mathbf{S} = \pm hdz\hat{\mathbf{e}}_x$ and $d\mathbf{S} = \pm hdx\hat{\mathbf{e}}_z$. In *plan view* our control volume is a rectangle of sides $d\mathbf{S}_{2D} = \pm dz\hat{\mathbf{e}}_x$ and $d\mathbf{S}_{2D} = \pm dx\hat{\mathbf{e}}_z$, and so mass conservation yields

$$\frac{\partial}{\partial t}(hdA) = -\oint \mathbf{u}_H \cdot d\mathbf{S} = -\oint \mathbf{q} \cdot d\mathbf{S}_{2D} = -\int (\nabla \cdot \mathbf{q}) dA = -(\nabla \cdot \mathbf{q}) dA. \quad (7.57)$$

We conclude that

$$\frac{\partial h}{\partial t} = -\nabla \cdot \mathbf{q} = -\nabla \cdot (\mathbf{u}_H h). \quad (7.58)$$

For planar motion, independent of z, these shallow-water equations simplify to

$$\boxed{\frac{\partial u_x}{\partial t} + u_x \frac{\partial u_x}{\partial x} = -g \frac{\partial h}{\partial x}}, \quad (7.59)$$

$$\boxed{\frac{\partial h}{\partial t} = -\frac{\partial}{\partial x}(u_x h)}. \quad (7.60)$$

7.5.2 Finite-amplitude Waves and Non-linear Wave Steepening

In the linear regime of small-amplitude waves, the shallow-water phase velocity is $\sqrt{gh_0}$. It is natural, therefore, to replace h in (7.59) and (7.60) by $c = \sqrt{gh}$, which gives

$$\frac{\partial u_x}{\partial t} + u_x \frac{\partial u_x}{\partial x} = -2c \frac{\partial c}{\partial x}, \quad (7.61)$$

$$2\frac{\partial c}{\partial t} + 2u_x \frac{\partial c}{\partial x} = -c \frac{\partial u_x}{\partial x}. \quad (7.62)$$

Adding and subtracting these yields

$$\boxed{\left[\frac{\partial}{\partial t} + (u_x \pm c) \frac{\partial}{\partial x}\right](u_x \pm 2c) = 0}, \quad (7.63)$$

and it follows that $u_x + 2c$ is conserved along the trajectories defined by $dx/dt = u_x + c$ (so-called positive characteristics), while $u_x - 2c$ is conserved along the trajectories given by $dx/dt = u_x - c$ (negative characteristics). At any one location in space, we can think of the first of these as a *local* representation of a forward-propagating disturbance riding on the back of the local flow and with a *local* wave speed of $c = \sqrt{gh}$. Similarly, we may regard the second as a backward-propagating disturbance of local wave speed $-c$, also carried forward by the local flow, u_x. What is remarkable about these disturbances, however, is that we have placed no restriction on their amplitude. Moreover, the fact that the local wave speed increases with depth tells us that a wave front, such as that shown in Figure 7.14, will automatically steepen with time.

It is instructive to apply these equations to the formation of a bore, as shown in Figure 7.15. A wall initially located at $x = 0$ sits adjacent to undisturbed fluid of depth h_0, and at $t = 0$ the wall moves from left to right with constant acceleration, a. It pushes ahead of it a disturbance of finite extent and at the front of the disturbance the fluid velocity is zero and the wave speed $c_0 = \sqrt{gh_0}$. The front therefore moves at the speed c_0, and so at time t the entire disturbance occupies the region $\frac{1}{2}at^2 < x < c_0t$. Now backward-travelling waves emerging from the front progressively permeate the bore, requiring that $u_x - 2c = -2c_0$ at all points reached by those waves. Simultaneously, forward-travelling waves initiated at the wall push in from the left, governed by

$$\left[\frac{\partial}{\partial t} + (u_x + c)\frac{\partial}{\partial x}\right](u_x + 2c) = 0. \tag{7.64}$$

When these forward-travelling waves encounter the region permeated by the backward-travelling waves, where u_x and c are constrained to satisfy $u_x = 2(c - c_0)$, (7.64) becomes

$$\left[\frac{\partial}{\partial t} + (3c - 2c_0)\frac{\partial}{\partial x}\right](3c - 2c_0) = 0. \tag{7.65}$$

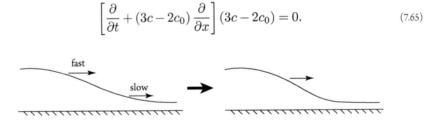

Figure 7.14 The non-linear steepening of a wave front.

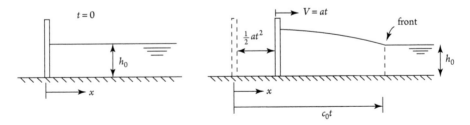

Figure 7.15 The non-linear formation of a bore. (a) Initial condition, (b) $t > 0$.

It is readily confirmed that this has the general solution

$$C = f(x - Ct), \quad C = 3c - 2c_0, \tag{7.66}$$

for some f. Crucially, this tells us that forward-travelling disturbances propagate at the speed $3\sqrt{gh} - 2\sqrt{gh_0}$, and so deep regions travel to the right faster than shallow regions. It is inevitable, therefore, that the front of the bore steepens with time, as suggested by Figure 7.14. Indeed, it may be shown that the front becomes *singular* within a *finite* time.

7.5.3 The Solitary Wave 1: Rayleigh's Solution

The discussion above suggests that the non-linearity associated with finite-amplitude waves is most readily handled in the shallow-water limit. This has been explored most exhaustively in the case of the *solitary wave*, which is a single elevation that travels large distances without change of shape (Figure 7.16), and the so-called *cnoidal wave*, which is periodic but non-sinusoidal. We shall explore solitary waves first.

Solitary waves were first studied experimentally by the Scottish engineer John Scott Russell. In 1844 he described his first encounter with such a wave in August of 1834. Having generated the wave by accident in a canal near Edinburgh he then: 'followed it on horseback, and overtook it still rolling on at a rate of some 8 or 9 miles per hour, preserving its original figure some 30 feet long and a foot to a foot and a half in height.' Russell's subsequent observations established the empirical law that the speed of propagation of the wave, V, is related to the depth of undisturbed water, h_0, and the wave amplitude, a, through the expression $V = \sqrt{g(h_0 + a)}$. He also confirmed the remarkable ability of solitary waves to retain their shape for long periods of time, something that is now attributed to a delicate balance between non-linear wave steepening, associated with a finite a/h_0, and the tendency for waves with a finite value of h_0/λ to disperse.

The first non-linear analysis of solitary waves was provided by Boussinesq in 1871, and then independently by Rayleigh in 1876. Similar results were later obtained using the so-called the KdeV equation. Let us start with Rayleigh, who assumed the flow to be inviscid, irrotational, of long wavelength ($\lambda \gg h_0$), and steady in a frame of reference moving with the wave. It is important to note, however, that Rayleigh makes no *explicit* assumption about the ratio a/h_0, although we shall see that there *is* an implicit limitation. We shall follow Lamb's 1932 account of Rayleigh's paper. Consider the complex potential

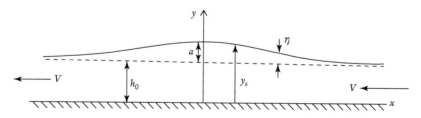

Figure 7.16 A solitary wave in a frame of reference in which the crest is stationary.

$$\varphi + j\psi = F(x + jy) = e^{jy\frac{d}{dx}} f(x),$$

which may be expanded as a power series in derivatives of x, as

$$\varphi = f - \tfrac{1}{2}y^2 f'' + \cdots, \quad \psi = yf' - \tfrac{1}{6}y^3 f''' + \cdots. \tag{7.67}$$

In a frame of reference moving with the crest, the velocity at large $|x|$ is $u_x = -V, u_y = 0$, and so Bernoulli's equation applied to the free surface, $y_s = h_0 + \eta$, requires that

$$u_x^2 + u_y^2 = V^2 - 2g(y_s - h_0) = V^2 - 2g\eta, \tag{7.68}$$

which combines with (7.67) to give

$$f'^2 - y_s^2 f' f''' + y_s^2 f''^2 + \cdots = V^2 - 2g\eta. \tag{7.69}$$

Moreover, (7.67) requires that the bottom streamline is $\psi = 0$, and so the free surface must be $\psi = -Vh_0$. It follows that

$$y_s f' - \tfrac{1}{6}y_s^3 f''' + \cdots = -Vh_0. \tag{7.70}$$

We now truncate (7.69) and (7.70) on the basis that horizontal gradients are small and then eliminate f from the two equations through successive approximations. Discarding terms containing four derivatives or more in $y_s(x)$ gives us the *Rayleigh–Lamb equation* (see Exercise 7.5):

$$\boxed{\frac{2y_s}{3}\frac{d^2\eta}{dx^2} - \frac{1}{3}\left(\frac{d\eta}{dx}\right)^2 = \frac{y_s^2}{h_0^2} - \frac{2y_s^2 g\eta}{h_0^2 V^2} - 1}. \tag{7.71a}$$

(Form 1)

Equation (7.71a) is the form given by Lamb (1932), which is slightly different to that which appears in Rayleigh's original paper. In any event, (7.71a) differentiates to give

$$(V^2 - c_0^2)\eta' = \frac{3c_0^2}{h_0}\eta\eta' + \frac{1}{3}V^2 h_0^2 \eta''' \quad, \quad c_0 = \sqrt{gh_0},$$

which then re-integrates to provide a simpler version of the Rayleigh–Lamb equation:

$$\boxed{(V^2 - c_0^2)\eta = \frac{3c_0^2}{2h_0}\eta^2 + \frac{1}{3}V^2 h_0^2 \eta''}. \tag{7.71b}$$

(Form 2)

We note in passing that, in a frame of reference in which the fluid is stationary at infinity, i.e. $\eta = \eta(x - Vt)$, this becomes

$$\frac{\partial^2 \eta}{\partial t^2} = c_0^2 \frac{\partial^2}{\partial x^2} \left[\eta + \frac{3}{2} \frac{\eta^2}{h_0} \right] + \frac{1}{3} h_0^2 \frac{\partial^4 \eta}{\partial t^2 \partial x^2},$$

which is often called the *regularized Boussinesq equation*. (Again, readers may wish to consult Exercise 7.5.) Returning to (7.71b), we multiply through by $d\eta/dx$, integrate, and set $d\eta/dx = 0$ at $\eta = 0$. This yields

$$\left(\frac{d\eta}{dx} \right)^2 = \frac{3\eta^2}{h_0^2} \left(1 - \frac{g(h_0 + \eta)}{V^2} \right). \tag{7.72a}$$

The same result was obtained by Rayleigh by integrating (7.71a). Since $\eta' = 0$ at the crest, we conclude that $V = \sqrt{g(h_0 + a)}$, exactly as observed by Russell. On rewriting (7.72a) as

$$\frac{d\eta}{dx} = \pm \frac{\eta}{b} \sqrt{1 - \eta/a} \quad \text{where} \quad b = \sqrt{\frac{h_0^2(h_0 + a)}{3a}}, \tag{7.72b}$$

one last integration yields

$$\boxed{\eta = a \operatorname{sech}^2 (x/2b)}, \quad \boxed{V = \sqrt{g(h_0 + a)}}. \tag{7.73}$$

It is a tribute to Rayleigh's powers as a mathematical physicist that he managed to find such a simple solution to a potentially complicated non-linear problem, although it should be emphasized that Boussinesq pre-empted many of Rayleigh's findings. For example, Boussinesq also obtained (7.73), but with $b^2 = h_0^3/3a$ (see Exercise 7.6). In any event, this vindication of Russell's empirical law for V was historically significant as Airy, who dominated the field of waves at the time, had been very dismissive of Russell's findings.

In a frame of reference in which the remote fluid is stationary, Rayleigh's solution gives the total kinetic energy of a solitary wave, per unit width into the page, as

$$\text{K.E.} = \frac{4\rho g a^2 h_0}{3\sqrt{3}} \sqrt{1 + \frac{h_0}{a}} \left[1 + \frac{2}{5} \frac{a}{h_0} - \frac{8}{35} \left(\frac{a}{h_0} \right)^2 + O\left((a/h_0)^3 \right) \right],$$

(see Exercise 7.8). By contrast, the potential energy per unit width is readily shown to be

$$\text{P.E.} = \frac{1}{2} \rho g \int_{-\infty}^{\infty} \eta^2 dx = \frac{4\rho g a^2 h_0}{3\sqrt{3}} \sqrt{1 + h_0/a}.$$

It is interesting that there is equipartition of energy *only* in the linear regime.

Lighthill (1978) introduced an *effective wavelength* for a solitary wave through the requirement that η/a drops to 0.03 when $|x| = \lambda_e/2$, which gives $\lambda_e = 9.76b$. From (7.72), this wavelength is related to a/h_0 and the maximum wave steepness, $(d\eta/dx)_{max}$, by

$$\frac{\lambda_e}{h_0} = 5.63\sqrt{1 + h_0/a}, \quad \left(\frac{d\eta}{dx}\right)_{max} = 3.76\frac{a}{\lambda_e}, \tag{7.74}$$

while the total wave energy per unit depth into the page is

$$\text{K.E.} + \text{P.E.} = 0.273\rho g a^2 \lambda_e \left[1 + \frac{a}{5h_0} - \frac{4a^2}{35h_0^2} + O\left((a/h_0)^3\right)\right].$$

Although Rayleigh made no explicit assumption about the magnitude of a/h_0, the fact that $a/h_0 \sim (h_0/\lambda_e)^2$, while h_0/λ_e is assumed small, implies an upper limit on a/h_0. In particular, Rayleigh's analysis retains terms up to $(h_0/\lambda_e)^3$, while discarding fourth-order terms in h_0/λ_e. It follows that Rayleigh's analysis is accurate up to order $(a/h_0)^{3/2}$ only. In practice, a/h_0 cannot exceed 0.78 due to instabilities (see Exercise 7.7). In any event, Rayleigh's prediction of the wave speed remains close to the measurements (within 2%) up to $a/h_0 = 0.6$, and it is closer than the predictions of the subsequent KdeV equation.

7.5.4 Solitary Waves 2: The KdeV Equation

In 1895 Korteweg and de Vries revisited Rayleigh's analysis and proposed the following equation for x-directed, shallow-water waves of small but finite amplitude:

$$\boxed{\frac{\partial \eta}{\partial t} + c_0\frac{\partial \eta}{\partial x} + \frac{3}{2}\frac{c_0\eta}{h_0}\frac{\partial \eta}{\partial x} + \frac{1}{6}c_0 h_0^2 \frac{\partial^3 \eta}{\partial x^3} = 0}, \tag{7.75}$$

where $c_0 = \sqrt{gh_0}$ and the other symbols are defined in Figure 7.16. The first two terms, taken by themselves, represent a non-dispersive, shallow-water wave in the limit of small amplitude, $a/h_0 \to 0$, and large wavelength, $h_0/\lambda \to 0$. The fourth term in this equation allows for a small but finite h_0/λ and rests on the fact that the low-k approximation to the dispersion relationship (7.15) is

$$\varpi = c_0 k \left(1 - (kh_0)^2/6\right) = c_0 k - \tfrac{1}{6}c_0 h_0^2 k^3.$$

So, for a single Fourier mode, the final term in (7.75) provides a first-order correction in h_0/λ and thus captures a small but finite amount of dispersion. The third term, on the other hand, comes from the non-linear, shallow-water equation for x-directed disturbances (7.65), rewritten as

$$\left[\frac{\partial}{\partial t} + c_0\left(3\sqrt{1 + \eta/h_0} - 2\right)\frac{\partial}{\partial x}\right]\eta = \left[\frac{\partial}{\partial t} + c_0\left(1 + \frac{3\eta}{2h_0} + \cdots\right)\frac{\partial}{\partial x}\right]\eta = 0.$$

This incorporates the effects of *weak* non-linearity in the small-amplitude limit.

Expression (7.75) is known as the *Korteweg–de Vries* (or KdeV) equation, although it was first written down by Boussinesq in 1877. It shares with Rayleigh's analysis an assumption of small h_0/λ, allowing for spatial derivatives in η up to third order only. Moreover, it limits the non-linearity to a first-order correction in a/h_0. We might expect, therefore, that this coincides with Rayleigh's equation in the limit of small a/h_0, and this is indeed the case (see Exercise 7.5). In fact, Korteweg and de Vries not only obtained a solution for solitary waves, but also for periodic travelling waves.

The KdeV equation has been much studied, as it crops up in several non-linear, dispersive systems in the appropriate limit. It is attractive as it allows the effects of weak non-linearity and dispersion to be isolated and then studied, particularly the competing influences of wave dispersion and non-linear wave steepening. In the case of a solitary wave, it turns out that these effects exactly cancel, allowing the wave to retain its shape. Another extraordinary feature of solitary waves is that two such waves can collide, interact in a non-linear manner, and then emerge with almost exactly the same properties as they had before the collision. Wave pulses of finite amplitude that retain their shape as they propagate, and can emerge unscathed from such collisions, are known as *solitons*. The discovery of the solitary wave provided the first hint that solitons exist (Miles, 1980).

Consider a solution of (7.75) of the form $\eta = f(x - Vt)$, so the KdeV equation becomes

$$(V - c_0)f' = \frac{3c_0}{4h_0}2ff' + \frac{1}{6}c_0 h_0^2 f'''. \tag{7.76}$$

Integrating, and demanding that the disturbance vanishes for $|x| \to \infty$, leads to

$$(V - c_0)f = \frac{3c_0}{4h_0}f^2 + \frac{1}{6}c_0 h_0^2 f'',$$

which should be compared with (7.71b). Next, we multiply by f', integrate again, and once more demand that $f_\infty = 0$. This yields

$$\frac{1}{6}c_0 h_0^2 f'^2 = (V - c_0)f^2 - \frac{c_0}{2h_0}f^3. \tag{7.77}$$

If the amplitude of the solitary wave is a, then the requirement that $f' = 0$ when $f = a$ fixes the wave speed at a *supercritical* value of

$$V = c_0\left(1 + a/2h_0\right). \tag{7.78}$$

This coincides with Russell's estimate (and Rayleigh's prediction) of $V = \sqrt{g(h_0 + a)}$ in the limit of small a/h_0, but yields a higher wave speed for finite a/h_0. Measurements of the wave speed show that the predictions of Rayleigh and the KdeV equation both overestimate V. However, the discrepancy in Rayleigh's estimate is only $\sim$2% at $a/h_0 = 0.6$, whilst that for KdeV is around 5%. In any event, substituting for V in (7.77) gives us

$$\frac{\partial \eta}{\partial x} = \pm\frac{\eta}{\hat{b}}\sqrt{1 - \eta/a}, \quad \text{where} \quad \hat{b} = \sqrt{h_0^3/3a}, \tag{7.79}$$

which integrates one last time to give

$$\eta = a\,\mathrm{sech}^2\left[(x - Vt)\big/2\hat{b}\right].\tag{7.80}$$

We have arrived back at Rayleigh's solution (7.73), except that b is now replaced by $\hat{b}$, the two being equal in the limit of small a/h_0. According to the KdeV solution, Lighthill's effective wavelength, $\lambda_e = 9.76\hat{b}$, and the maximum wave steepness, $(\partial\eta/\partial x)_{\mathrm{max}}$, are

$$\frac{\lambda_e}{h_0} = 5.63\sqrt{\frac{h_0}{a}}, \quad \left(\frac{\partial\eta}{\partial x}\right)_{\mathrm{max}} = \frac{2}{3}\left(\frac{a}{h_0}\right)^{3/2} = 3.76\frac{a}{\lambda_e},\tag{7.81}$$

which might be compared with (7.74). For $a/h_0 = 0.6$, the wavelength predicted by the KdeV solution is some 21% shorter than that predicted by Rayleigh's analysis.

7.5.5 More General Solutions of the KdeV Equation: Cnoidal Waves

The ratio of non-linearity to dispersion in the KdeV equation is of the order of $\eta_0\lambda_e^2/h_0^3$, where η_0 is an appropriately defined amplitude and λ_e the effective wavelength. Let us define η_0 to be one-half of the difference between the maximum and minimum values of η. In the case of a solitary wave we have $\eta_0 = a/2$, and so (7.81) gives us $\eta_0\lambda_e^2/h_0^3 \approx 16$, although the definition of λ_e here is a little bit arbitrary. Now, suppose we abandon the requirement that the disturbance vanishes for $|x| \to \infty$, while giving ourselves the freedom to vary $\eta_0\lambda_e^2/h_0^3$. Can we find other solutions of the KdeV equation in which dispersion and non-linear wave steepening balance, yielding a wave of fixed shape? It turns out that we can, as Korteweg and de Vries showed. Specifically, there exists a family of stable, periodic, travelling waves of non-sinusoidal shape for $\eta_0\lambda^2/h_0^3$ up to values of around $\eta_0\lambda^2/h_0^3 \approx 16$, with small $\eta_0\lambda^2/h_0^3$ corresponding to conventional linear gravity waves and $\eta_0\lambda^2/h_0^3 \approx 16$ to a periodic sequence of effectively decoupled solitary waves. These are called *cnoidal waves*, as their shape involves the Jacobian elliptic function normally denoted by cn. Exercise 7.9 outlines one particular derivation of cnoidal waves.

The surface profiles of these non-linear waves, taken from Lighthill (1978), are shown in Figure 7.17 for the range $0 < \eta_0\lambda^2/h_0^3 \leq 15$. Note that increasing the wave amplitude causes the wave crests to sharpen while the troughs flatten, something that is also exhibited by waves on deep water. As we approach a value of $\eta_0\lambda^2/h_0^3 = 15$, these periodic waves degenerate into a sequence of isolated elevations, or humps, with flat water between them. We may, therefore, regard the solitary wave as the limiting case of cnoidal waves as the non-linearity parameter $\eta_0\lambda^2/h_0^3$ approaches ~ 16. A single solitary wave is shown in Figure 7.17, below the cnoidal waves, for comparison.

7.5.6 The Hydraulic Jump Revisited

It turns out that, for solitary waves in which $\eta_0\lambda_e^2/h_0^3 > 16$, dispersion can no longer counter the effects of non-linear wave steepening in the KdeV equation (see Exercise 7.10

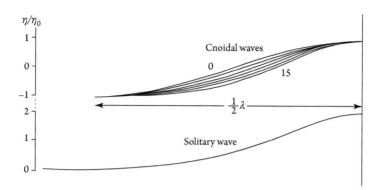

Figure 7.17 Surface profiles of cnoidal waves for $\eta_0 \lambda^2 / h_0^3 = 0$, 3, 6, 9, 12 and 15, with the corresponding limiting case of a solitary wave shown below. (From Lighthill, 1978.)

or Lighthill, 1978). The result is persistent wave steepening as shown in Figure 7.14, which eventually leads to a hydraulic jump or to a travelling bore.

The hydraulic jump was introduced in §1.4.2 as a phenomenon whose spatial location is fixed, say at the base of a dam spillway. Such stationary jumps represent an abrupt transition from supercritical flow (*i.e.* fast, shallow motion in which $\mathrm{Fr}_1 = V_1 / \sqrt{gh_1} > 1$) to a slow, deep flow which is subcritical, $\mathrm{Fr}_2 = V_2 / \sqrt{gh_2} < 1$. However, hydraulic jumps can equally surge through a fluid in the form of a tidal bore. In either case, in a frame of reference in which the jump is stationary, the change in height across the jump is related to the upstream Froude number by (1.63):

$$\mathrm{Fr}_1^2 = \frac{h_2 (h_1 + h_2)}{2h_1^2}. \tag{7.82}$$

Typically jumps are highly turbulent and so strongly dissipative. However, their structure varies somewhat with Fr_1. Very weak jumps, say $1 < \mathrm{Fr}_1 < 1.5$, are known as *undular jumps* and they are characterized by low levels of foaming (*i.e.* low levels of turbulence), a surface which rises up smoothly, and a wave pattern downstream of the jump which is stationary in the frame of reference of the jump (Figure 7.18). These waves are, in fact, cnoidal waves and they are stationary relative to the jump because their phase velocity is equal and opposite to V_2. Since the group velocity of cnoidal waves is less than their phase velocity, this means that wave energy is swept downstream of the jump, and indeed in undular jumps much of the energy at the front is carried off in the form of waves, rather than locally dissipated by turbulence. For larger values of Fr_1, say $\mathrm{Fr}_1 > 1.7$, the water surface rises rather abruptly and foaming (turbulence) is very evident. For $1.7 < \mathrm{Fr}_1 < 2.5$, the jump tends to be weak and steady-on-average, while in the range $2.5 < \mathrm{Fr}_1 < 4.5$ the turbulence levels are higher and the surface often displays irregular oscillations as waves randomly propagate downstream. This regime is sometimes called an oscillating jump. Finally, a fully developed hydraulic jump is formed for $\mathrm{Fr}_1 > 4.5$. This has a stable overall structure, is highly turbulent, and is strongly dissipative.

Figure 7.18 An undular bore with cnoidal waves. (By S. Bartsch-Winkler and D. K. Lynch.)

○ ○ ○

This concludes our brief discussion of surface gravity waves. Readers interested in more details might consult Faber (1995) for a gentle introduction, Acheson (1990) for a more detailed discussion, or Lighthill (1978) for the definitive account of water waves.

··

EXERCISES

7.1 *Standing surface gravity waves in a channel.* Show that two-dimensional standing waves in a channel of width, W, and depth, h_0, may be obtained by superimposing a forward-travelling wave of the form (7.19) with a similar backward-travelling wave of the same amplitude, to give

$$\psi = \frac{\varpi a}{k} \frac{\sinh k(y + h_0)}{\sinh k h_0} 2 \sin(kx) \sin(\varpi t), \quad k = n\pi/W.$$

7.2 *Particle trajectories for surface gravity waves on a fluid of finite depth.* Use expressions (7.17) and (7.18) to show that, in the linear approximation, the particle trajectories for a surface gravity wave are ellipses whose major and minor semi-axes are

$$\ell_{\text{major}} = a \frac{\cosh k(y_0 + h_0)}{\sinh k h_0}, \quad \ell_{\text{minor}} = a \frac{\sinh k(y_0 + h_0)}{\sinh k h_0},$$

where y_0 is the rest height of the particle and we consider the case of $k > 0$. Hence show that, for $k h_0 \gg 1$ (deep water), the major and minor semi-axes are both equal to $a e^{k y_0}$, while for $k h_0 \ll 1$ (shallow water), the ratio of the minor axis to the major axis is $k(y_0 + h_0) \ll 1$.

7.3 *The kinetic and potential energy of a surface gravity wave on a fluid of finite depth.* Use expressions (7.17) and (7.18), as well as the dispersion relationship (7.15), to show that the kinetic energy per unit plan area of a surface gravity wave is

$$\frac{1}{2}\rho \int_{-h_0}^{0} \mathbf{u}^2 dy = \frac{1}{4}\rho g a^2 \left[1 + \frac{2 k h_0 \left(\cos^2(kx - \varpi t) - \sin^2(kx - \varpi t) \right)}{\sinh(2 k h_0)} \right].$$

Confirm that for deep-water waves this reduces to $\frac{1}{4}\rho g a^2$, consistent with the discussion in §7.2.3, while for shallow-water waves we get $\frac{1}{2}\rho g a^2 \cos^2(kx - \varpi t)$, consistent with (7.22). Notice that, when averaged over a cycle, the kinetic energy per unit horizontal area becomes $\frac{1}{4}\rho g a^2$, irrespective of the water depth.

Now confirm that the excess potential energy per unit horizontal area for the same wave is

$$\int_{-h_0}^{\eta} \rho g y \, dy - \int_{-h_0}^{0} \rho g y \, dy = \frac{1}{2}\rho g \eta^2 = \frac{1}{2}\rho g a^2 \cos^2(kx - \varpi t),$$

which also averages over a cycle to give $\frac{1}{4}\rho g a^2$. So, when averaged over a cycle, there is an equipartition of energy and the total energy per unit horizontal area is $\frac{1}{2}\rho g a^2$.

7.4 *The flux of energy in a surface gravity wave on a fluid of finite depth.* The flux of wave energy per unit length of the crests in a surface gravity wave is discussed in Lighthill (1978). It can be calculated from the rate of working of the pressure forces acting on a vertical surface aligned with the wave crests, also of unit width. In Lighthill (1978) it is shown that the rate of working of these pressure forces, and hence the wave energy flux, is given by

$$\frac{\varpi}{k} \int_{-h_0}^{0} \rho u_x^2 \, dy = c_p \int_{-h_0}^{0} \rho u_x^2 \, dy.$$

Use expressions (7.15) and (7.17) to show that the flux of wave energy in the direction of propagation of the wave, measured per unit length of the wave crests and averaged over a cycle, is given by

$$c_p \int_{-h_0}^{0} \rho \overline{u_x^2} \, dy = \frac{\varpi}{2k} \left[1 + \frac{2 k h_0}{\sinh 2 k h_0} \right] \frac{1}{2}\rho g a^2 = c_g \left(\frac{1}{2}\rho g a^2 \right),$$

where the overbar denotes averaging over a cycle.

7.5 *The Rayleigh–Lamb solitary wave equation.* Show that, in the long wavelength limit, (7.70) gives us

$$f' = -\frac{V h_0}{y_s} + \frac{1}{6} y_s^2 f''' + \cdots = -V h_0 \left[\frac{1}{y_s} + \frac{1}{6} y_s^2 \left(\frac{1}{y_s} \right)'' + \cdots \right].$$

Use this to eliminate f from (7.69), discarding terms which contain four derivatives or more in $y_s(x)$. Hence deduce the Rayleigh–Lamb solitary wave equation

$$\frac{2 y_s}{3} \frac{d^2 \eta}{dx^2} - \frac{1}{3} \left(\frac{d\eta}{dx} \right)^2 = \frac{y_s^2}{h_0^2} - \frac{2 y_s^2 g \eta}{h_0^2 V^2} - 1.$$

Actually, this is the form in which Lamb expressed it. In Rayleigh (1876) the left-hand side is written as $(4 y_s^{3/2} / 3) d^2 y_s^{1/2} / dx^2$. (A particularly accessible account of Rayleigh's original analysis may be found in Basset, 1888.) Differentiate Rayleigh's equation to give

$$(V^2 - c_0^2) \eta' = \frac{3 c_0^2}{h_0} \eta \eta' + \frac{1}{3} V^2 h_0^2 \eta''', \qquad c_0 = \sqrt{g h_0},$$

and confirm that this is equivalent to the KdeV equation, (7.76), in the limit of small a/h_0. Now show that, if we change frame of reference so that $\eta = \eta\,(x - Vt)$, the Rayleigh–Lamb equation can be rewritten as

$$\frac{\partial^2 \eta}{\partial t^2} = c_0^2 \frac{\partial^2}{\partial x^2} \left[\eta + \frac{3}{2} \frac{\eta^2}{h_0} \right] + \frac{1}{3} h_0^2 \frac{\partial^4 \eta}{\partial t^2 \partial x^2}.$$

In recent years, this equation has been proposed as a refinement to Boussinesq's 1872 equation (as given in Exercise 7.6 below), although the connection to Rayleigh's theory appears not to have been noticed. It has acquired the name: *the regularized Boussinesq equation.*

7.6 *The Boussinesq solitary wave equation.* The original 1872 Boussinesq equation for solitary waves is

$$\frac{\partial^2 \eta}{\partial t^2} = c_0^2 \frac{\partial^2}{\partial x^2} \left[\eta + \frac{3}{2} \frac{\eta^2}{h_0} \right] + \frac{c_0^2}{3} h_0^2 \frac{\partial^4 \eta}{\partial x^4}$$

(see Miles, 1980). Confirm that this gives the same solitary wave solution as Rayleigh's analysis, *i.e.* (7.73), but with $b = \sqrt{h_0^3 / 3a}$. Also show that, in the absence of the non-linear term, both the original and regularized Boussinesq equations are consistent with the low-k approximation to the dispersion relationship (7.15),

$\varpi = c_0 k - \frac{1}{6} c_0 h_0^2 k^3 + \cdots$, but that the regularized equation has more accurate dispersion characteristics for $h_0 k \sim 1$.

7.7 *The highest solitary wave.* Use Bernoulli's equation in the form of (7.68), along with Rayleigh's prediction of $V = \sqrt{g(h_0 + a)}$, to show that the amplitude of a solitary wave, a, cannot exceed the undisturbed depth, h_0. In fact, the maximum amplitude of a solitary wave, corresponding to a cusped wave crest, turns out to be $a_{\max} = 0.83 h_0$ (Lenau, 1966), although in practice solitary waves are unstable for $a > 0.78 h_0$ (Tanaka, 1986).

7.8 *A challenge for the fearless reader: Rayleigh's predictions for the energy of a solitary wave.* Use the results of Exercise 7.5 to show that, in a frame of reference in which $\mathbf{u} = 0$ at infinity, Rayleigh's method predicts that the momentum flux and kinetic energy of a solitary wave are, at a given location,

$$\rho \int_0^{y_s} u_x^2 dy = \rho (f' + V)^2 y_s - \rho (f' + V) f''' \frac{1}{3} y_s^3 = \frac{\rho V^2 \eta^2}{y_s},$$

$$\frac{1}{2} \rho \int_0^{y_s} \mathbf{u}^2 dy = \frac{\rho}{2} (f' + V)^2 y_s + \frac{\rho}{6} \left[f''^2 - (f' + V) f''' \right] y_s^3 = \frac{\rho V^2}{2 y_s} \left[\eta^2 + \frac{1}{3} (h_0 \eta')^2 \right].$$

(To be consistent with Rayleigh's analysis, you must discard terms which contain four or more derivatives in $y_s(x)$.) Now use Rayleigh's solution for η, (7.73), along with the definite integral

$$\int_0^\infty \mathrm{sech}^n x\, dx = \int_0^1 \left(1 - t^2\right)^{(n/2)-1} dt,$$

to show that the total kinetic energy is

$$\mathrm{K.E.} = \frac{4\rho g a^2 h_0}{3\sqrt{3}} \sqrt{1 + h_0/a} \left[1 + \frac{2}{5} \frac{a}{h_0} - \frac{8}{35} \frac{a^2}{h_0^2} + O\left((a/h_0)^3\right) \right],$$

while the potential energy is

$$\mathrm{P.E.} = \frac{1}{2} \rho g \int_{-\infty}^\infty \eta^2 dx = \frac{4\rho g a^2 h_0}{3\sqrt{3}} \sqrt{1 + h_0/a}.$$

Evidently, there is equipartition of energy *only* in the linear regime. It so happens that an *exact*, non-linear property of solitary waves is

$$\frac{d}{da} \left(\frac{\mathrm{K.E.}}{V^2} \right) = \frac{1}{V^2} \frac{d}{da} (\mathrm{P.E.})$$

(see Longuet-Higgins, 1974). Confirm that the estimates above satisfy this to within the order suggested. Also, starting with Rayleigh's estimate for the potential energy, show that this exact result suggests the more general expression

$$\text{K.E.} = \frac{4\rho g a h_0^2}{3\sqrt{3}}\sqrt{1 + h_0/a}\left[3 + 2\frac{a}{h_0} - 3\sqrt{1 + h_0/a}\sinh^{-1}\sqrt{a/h_0}\right],$$

although this differs from the Rayleigh's estimate by no more than 1% for $a/h_0 < 0.5$. Finally, show that the Rayleigh–Lamb equation, (7.71), but not the KdeV equation, yields

$$\text{P.E.} = \frac{1}{3}\left(V^2 - c_0^2\right)\rho\int_{-\infty}^{\infty} \eta\, dx,$$

which is another exact, non-linear property of solitary waves (Longuet-Higgins, 1974).

7.9 *Cnoidal waves.* The existence of cnoidal waves can be established using either the KdeV equation, as originally done by Korteweg and de Vries (1895), or else using the method of Rayleigh. We consider the latter, following the development in Lamb (1932). In the derivation of the solitary wave equation, (7.72a), we twice took advantage of the assumption that the surface disturbance disappears at infinity: first, in writing Bernoulli's equation as (7.68), and second, when evaluating the constant of integration in (7.72a). If we now drop the assumption of a localized disturbance then Bernoulli's equation becomes $u_x^2 + u_y^2 = C - 2g\eta$, where C is to be determined. We then have two free parameters at our disposal: C plus one constant of integration. Show that, on repeating the build-up to (7.72a), but without the assumption of a localized disturbance, we obtain

$$\left(\frac{dy_s}{dx}\right)^2 = -\frac{3g}{V^2 h_0^2}(y_s - h_1)(y_s - h_2)(y_s - \ell), \quad \ell = \frac{V^2 h_0^2}{gh_1 h_2},$$

where h_1 and h_2 are our two free parameters and Vh_0 is defined via (7.70) as the volumetric flow rate. If we interpret h_1 and h_2 to be the smallest and largest values of y_s, with ℓ assumed less than h_1, then this has the periodic solution

$$y_s = h_1 + (h_2 - h_1)\text{cn}^2\left(\frac{x}{\beta} : m\right)$$

(see, Lamb, 1932), where

$$\beta = \sqrt{\frac{4h_1 h_2 \ell}{3(h_2 - \ell)}}, \quad m = \frac{h_2 - h_1}{h_2 - \ell}, \quad \frac{(h_2 - \ell)}{(h_0 - \ell)} = \frac{K(m)}{E(m)}.$$

Here $K(m)$ and $E(m)$ are complete elliptic integrals and $\text{cn}(z; m)$ is the Jacobian elliptic function usually represented by cn, which looks a bit like a cosine with a period of $4K(m)$. Evidently, this solution represents a wave of wavelength $\lambda = 2\beta K(m)$ and amplitude $\eta_0 = (h_2 - h_1)/2$. A solitary wave formally corresponds to the limiting case of $m \to 1$ and $K(m) \to \infty$, which yields $\ell = h_1 = h_0$, $\lambda \to \infty$ and $V = \sqrt{g(h_0 + a)}$.

7.10 *Unsteady wave steepening of a solitary wave of large amplitude.* In practice, any cnoidal wave for which $m > 0.9$ behaves like a solitary wave, with successive crests effectively decoupled. When $m = 0.9$ we have $\lambda = 5.16\beta$. Use the results of Exercise 7.9 to show that the ratio of non-linearity to dispersion, as measured by $\eta_0 \lambda^2 / h_0^3$, is then

$$\frac{\eta_0 \lambda^2}{h_0^3} = 26.6 \frac{\eta_0 \beta^2}{h_0^3} = 17.7 m \frac{V^2}{gh_0} = 16.0 \frac{V^2}{gh_0} \approx 16.$$

This coincides with the estimate of $\eta_0 \lambda_e^2 / h_0^3$ for a solitary wave given in §7.5.5, based on Lighthill's definition of the effective wavelength of a solitary wave, λ_e. Korteweg and de Vries (1895) show that, if the amplitude of a solitary wave exceeds $\eta_0 \approx 16 h_0^3 / \lambda_e^2$, its shape cannot remain fixed, because the effects of non-linearity outweigh those of dispersion. Non-linear wave steepening then leads to a bore or a hydraulic jump.

....

REFERENCES

Acheson, D.J., 1990, *Elementary fluid dynamics*, Oxford University Press.
Basset, A.B., 1888, *Hydrodynamics*, Vol. II, Deighton, Bell, and Co.
Faber, T.E., 1995, *Fluid dynamics for physicists*, Cambridge University Press.
Korteweg, D.J. & de Vries, G., 1895, On the change of form of long waves advancing in a rectangular channel and on a new type of long stationary waves. *Phil Mag* (series 5), **39**, 422.
Lamb, H., 1932, *Hydrodynamics*. 6th Ed., Cambridge University Press.
Lenau, C.W., 1966, The solitary wave of maximum amplitude. *J. Fluid Mech.*, **26**, 309.
Lighthill, J., 1978, *Waves in fluids*, Cambridge University Press.
Longuet-Higgins, M.S., 1974, On the mass, momentum, energy and circulation of a solitary wave. *Proc. Roy. Soc. A*, **337**, 1–13.
Miles, J.W., 1980, Solitary waves. *Ann. Rev. Fluid Mech.*, **12**, 11–43. (This contains all the original references to Boussinesq's work on solitary waves.)
Phillips, O.M., 1966, *The dynamics of the upper ocean*, Cambridge University Press.
Rayleigh, Lord, 1876, On waves. *Phil. Mag.* (series 5), **1**(4), 257–79.
Tanaka, M, 1986, The stability of solitary waves. *Phys. Fluids*, **29**, 650–55.

8

. . . . .

Vortex Dynamics: Classical Theory and Illustrative Examples

Vortex dynamics has its origins in a seminal paper in 1858 by Hermann von Helmholtz (left, 1821–1894). It was further developed by Lord Kelvin in 1869 (right, 1824–1907).

8.1 Vorticity and its Evolution Equation (Revisited)

We have already suggested that motion that is entirely irrotational is a rather rare occurrence, and that most real flows have non-zero vorticity, *i.e.* $\boldsymbol{\omega} = \nabla \times \mathbf{u} \neq 0$. Sometimes this vorticity is generated by a rotational body force, such as buoyancy or electromagnetic forces, but more commonly it is generated in thin boundary layers and then redistributed across the fluid following separation of the boundary layer.

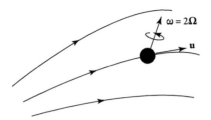

Figure 8.1 The vorticity at location **x** and at time t is twice the intrinsic angular velocity of a fluid element passing through the point **x** at time t.

As noted in §2.2.3, vorticity represents the intrinsic spin of fluid particles; that is to say, the vorticity at location **x** and time t is twice the angular velocity of a fluid particle passing through that point at that instant (Figure 8.1). Note that this angular velocity is measured about the centre of the particle. We also observed, in §2.7.1, that the expression $\boldsymbol{\omega} = \nabla \times \mathbf{u}$ can be inverted using the Biot–Savart law of magnetostatics. In particular, in an infinite domain the velocity associated with a given distribution of $\boldsymbol{\omega}$ is

$$\mathbf{u}(\mathbf{x},t) = \frac{1}{4\pi} \int \frac{\boldsymbol{\omega}(\mathbf{x}',t) \times \mathbf{r}}{r^3} \, d\mathbf{x}', \quad \mathbf{r} = \mathbf{x} - \mathbf{x}', \tag{8.1}$$

where $r = |\mathbf{r}|$. Finally, in §2.7.2, we showed that the curl of the Navier–Stokes equation yields an evolution equation for the vorticity field in the form of

$$\boxed{\frac{D\boldsymbol{\omega}}{Dt} = (\boldsymbol{\omega} \cdot \nabla)\mathbf{u} + \nu\nabla^2\boldsymbol{\omega}}. \tag{8.2}$$

When both terms on the right of (8.2) are zero, as occurs in a two-dimensional inviscid flow, then we have $D\boldsymbol{\omega}/Dt = 0$ and each fluid particle holds on to its vorticity as it moves around. On the other hand, the viscous term on the right represents a redistribution of vorticity through diffusion, while the term $(\boldsymbol{\omega} \cdot \nabla)\mathbf{u}$ represents the intensification (or reduction) of vorticity through vortex-line stretching (or compression), in accordance with angular momentum conservation. (Again, see the discussion in §2.7.2.)

8.2 Inviscid Vortex Dynamics

When Helmholtz wrote his seminal paper in 1858, he was unaware of Stokes' 1845 work incorporating viscous stresses into the equation of motion for a fluid. As a consequence, he did not know how to handle viscosity and so restricted himself to an inviscid fluid, although he was fully aware of the importance of friction in practice. So let us start by considering the classical inviscid theories of Helmholtz and Kelvin, postponing our discussion of viscous effects until later. Our evolution equation is then

$$\frac{D\boldsymbol{\omega}}{Dt} = (\boldsymbol{\omega} \cdot \nabla)\mathbf{u}, \tag{8.3}$$

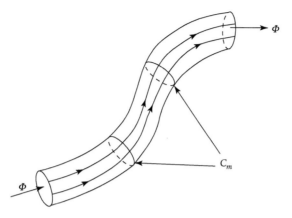

Figure 8.2 A vortex tube constructed from a bundle of vortex lines carries a vorticity flux Φ. According to Helmholtz's second law, Φ is the same at all cross-sections of the vortex tube, which follows from $\nabla \cdot \boldsymbol{\omega} = 0$, and also independent of time.

or, equivalently,

$$\frac{\partial \boldsymbol{\omega}}{\partial t} = \nabla \times (\mathbf{u} \times \boldsymbol{\omega}). \tag{8.4}$$

8.2.1 The Classical Theories of Helmholtz and Kelvin

Helmholtz's laws are couched in terms of *vortex lines*, lines drawn in space that are every-where parallel to $\boldsymbol{\omega}$, and *vortex tubes*, imaginary tubes whose boundaries are composed of an aggregate of vortex lines, much like a streamtube is composed of an aggregate of streamlines. The laws apply to an inviscid fluid, free of rotational body forces, and come in two parts:

(1) *Vortex lines* are frozen into the fluid like dye lines, so fluid particles that lie on a vortex line at some initial instant continue to lie on that vortex line for all time.

(2) The flux of vorticity along a *vortex tube*, say Φ, is:

    (a) the same at all cross-sections of the tube (see Figure 8.2); and

    (b) independent of time.

As we shall see, the origin of the first law, as well as part (a) of the second law, is straight forward. However, the ideas behind part (b) of the second law, which turn out to be closely related to Kelvin's circulation theorem of §6.3.3, are somewhat more involved.

The simplest way to see where Helmholtz's first law comes from is to invoke (6.41). Let $d\mathbf{l}$ be a short material line that moves with the fluid, like a dyeline, with $\mathbf{x}$ and $\mathbf{x} + d\mathbf{l}$ locating the ends of $d\mathbf{l}$. Then the change in $d\mathbf{l}$ in a time δt is $[\mathbf{u}(\mathbf{x} + d\mathbf{l}) - \mathbf{u}(\mathbf{x})]\,\delta t$, from which we obtain

$$\frac{D}{Dt}(d\mathbf{l}) = \mathbf{u}(\mathbf{x} + d\mathbf{l}) - \mathbf{u}(\mathbf{x}) = (d\mathbf{l} \cdot \nabla)\mathbf{u}, \tag{8.5}$$

which is (6.41). Comparing this with (8.3) we see that $\boldsymbol{\omega}$ and $d\mathbf{l}$ obey the same evolution equation and hence they must evolve in identical ways under the influence of $\mathbf{u}$. We conclude that vortex lines move with the fluid, like dye lines.

Turning now to part (a) of Helmholtz's second law, the fact that the vorticity flux is constant along a vortex tube follows directly from $\nabla \cdot \boldsymbol{\omega} = 0$, which in turn is a direct consequence of the definition $\boldsymbol{\omega} = \nabla \times \mathbf{u}$. (The divergence of the curl of any vector field is zero.) However, part (b) of law (2), the temporal invariance of the vorticity flux Φ, is altogether less obvious. One way to see where it comes from is to invoke the kinematic equation

$$\frac{d}{dt}\int_S \mathbf{G} \cdot d\mathbf{S} = \int_S \frac{\partial \mathbf{G}}{\partial t} \cdot d\mathbf{S} - \oint_{C_m} (\mathbf{u} \times \mathbf{G}) \cdot d\mathbf{r}, \tag{8.6}$$

where $\mathbf{G}(\mathbf{x}, t)$ is any solenoidal vector field that links the open surface S, which in turn spans a closed *material* curve, C_m (Figure 8.3). The idea behind (8.6) is the following. The flux of $\mathbf{G}$ through S may change for one of two reasons. First, even if C_m and S are both fixed in space there may be a change in flux through S whenever $\mathbf{G}$ is time dependent. This is the surface integral on the right of (8.6). Second, if the curve C_m moves it may expand at points to include additional flux, or perhaps contract at other points to exclude flux. This is the reason for the second integral on the right of (8.6).

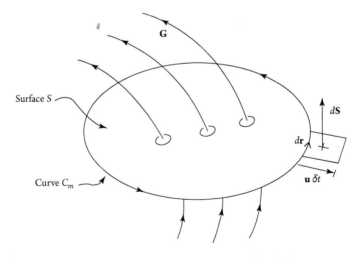

Figure 8.3 A solenoidal vector field $\mathbf{G}$ links an open surface S, which in turn spans a closed material curve C_m. Note that the surface element $d\mathbf{S}$ swept out by a line element $d\mathbf{r}$ in a time δt is $d\mathbf{S} = (\mathbf{u}\delta t) \times d\mathbf{r}$.

The origin of the line integral in (8.6) is, perhaps, not immediately obvious, but it arises as follows. The surface element, $d\mathbf{S}$, that is swept out by a line element, $d\mathbf{r}$, of the curve C_m in a time δt has sides $d\mathbf{r}$ and $\mathbf{u}_\perp \delta t$, where $\mathbf{u}_\perp$ is the component of $\mathbf{u}$ perpendicular to $d\mathbf{r}$. This vector area, which is normal to both $d\mathbf{r}$ and $\mathbf{u}$, can then be written as $d\mathbf{S} = (\mathbf{u}\delta t) \times d\mathbf{r}$. Thus, the change in the flux of $\mathbf{G}$ through S by virtue of movement of the boundary element $d\mathbf{r}$ is $\mathbf{G} \cdot d\mathbf{S} = \mathbf{G} \cdot (\mathbf{u} \times d\mathbf{r})\delta t$. Rearranging the scalar triple product gives us $\mathbf{G} \cdot d\mathbf{S} = -[(\mathbf{u} \times \mathbf{G}) \cdot d\mathbf{r}]\delta t$, and summing all such contributions around the closed curve C_m then yields the required result.

We now apply the kinematic equation, (8.6), to the vorticity field, using Stokes' theorem to rewrite the line integral as a surface integral. The evolution equation, (8.4), then gives us

$$\frac{d}{dt}\int_S \boldsymbol{\omega} \cdot d\mathbf{S} = \int_S \left(\frac{\partial \boldsymbol{\omega}}{\partial t} - \nabla \times (\mathbf{u} \times \boldsymbol{\omega}) \right) \cdot d\mathbf{S} = 0, \tag{8.7}$$

and so the flux of vorticity through any closed material curve C_m is an invariant. Now consider a vortex tube, such as that shown in Figure 8.2. From Helmholtz's first law the vortex tube moves with the fluid, and so we may draw a closed curve on the surface of the tube encircling it, which is a material curve, C_m. Part (b) of Helmholtz's second law, the invariance of the flux, Φ, now follows directly from applying (8.7) to just such a closed curve.

Note that, when combined with Stokes' theorem, (8.7) can be rewritten as

$$\frac{d\Phi}{dt} = \frac{d\Gamma}{dt} = \frac{d}{dt}\oint_{C_m} \mathbf{u} \cdot d\mathbf{r} = 0, \tag{8.8}$$

where Γ is the circulation of $\mathbf{u}$ around C_m. This is, of course, Kelvin's circulation theorem, (6.43). This confirms that, as suggested above, there is a close relationship between Kelvin's theorem and Helmholtz's second law. It also provides an alternative proof of Kelvin's theorem, that is somewhat different to that offered in §6.3.3.

Helmholtz's two laws place strong constraints on the evolution of an inviscid fluid. Consider, for example, two thin, interlinked vortex tubes, as shown in Figure 8.4. Each tube creates a velocity field via the Biot–Savart law (8.1), and so they will advect each other, as well as themselves, in some complex way. However, no matter how complicated the motion becomes, Helmholtz's first law tells us that the two tubes must remain linked in the same manner for all time, because to break the linkage would require distinct fluid particles to occupy the same region of space, which is not possible. Moreover, the second law tells us that the strength of each vortex tube, Φ, is preserved.

8.2.2 Helicity and its Conservation

The conservation of vortex-line topology in an inviscid fluid, illustrated in Figure 8.4, is captured by an integral invariant called *helicity*. This is defined as

$$h = \int_{V_m} \mathbf{u} \cdot \boldsymbol{\omega}\, dV, \tag{8.9}$$

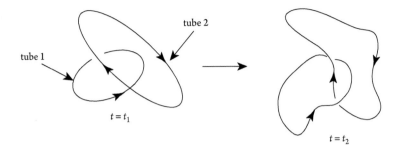

Figure 8.4 Two inviscid vortex tubes preserve their topology and flux strengths as they mutually advect each other.

where V_m is any material volume for which $\boldsymbol{\omega} \cdot d\mathbf{S} = 0$ on its surface S_m. We may confirm that h is indeed an invariant by combining Euler's equation with the inviscid vorticity equation, (8.3), to give

$$\frac{D}{Dt}(\mathbf{u} \cdot \boldsymbol{\omega}) = \frac{D\mathbf{u}}{Dt} \cdot \boldsymbol{\omega} + \frac{D\boldsymbol{\omega}}{Dt} \cdot \mathbf{u} = -\nabla\left(\frac{p}{\rho}\right) \cdot \boldsymbol{\omega} + (\boldsymbol{\omega} \cdot \nabla\mathbf{u}) \cdot \mathbf{u}. \qquad (8.10)$$

Since $\boldsymbol{\omega}$ is solenoidal, this becomes

$$\frac{D}{Dt}(\mathbf{u} \cdot \boldsymbol{\omega}) = \nabla \cdot \left[\left(\frac{1}{2}\mathbf{u}^2 - \frac{p}{\rho}\right)\boldsymbol{\omega}\right], \qquad (8.11)$$

which may be integrated over a set of *material* volume elements, δV_m, to give

$$\frac{dh}{dt} = \frac{d}{dt}\int_{V_m}(\mathbf{u} \cdot \boldsymbol{\omega})dV_m = \int_{V_m}\frac{D(\mathbf{u} \cdot \boldsymbol{\omega})}{Dt}dV_m = \oint_{S_m}\left(\frac{1}{2}\mathbf{u}^2 - \frac{p}{\rho}\right)\boldsymbol{\omega} \cdot d\mathbf{S} = 0, \qquad (8.12)$$

where the surface integral vanishes by virtue of the boundary condition $\boldsymbol{\omega} \cdot d\mathbf{S} = 0$ on S_m. Equation (8.12) confirms that h, as defined by (8.9), is indeed an integral invariant.

The connection between the invariance of h and Helmholtz's two laws may be illustrated by returning to Figure 8.4. Suppose that vortex tube 1 has volume V_1, flux Φ_1, and centreline C_1, while vortex tube 2 has volume V_2, flux Φ_2, and centreline C_2. Integrating over all space, we find that the net helicity of this configuration is

$$\int_{V_1}\mathbf{u} \cdot \boldsymbol{\omega}dV + \int_{V_2}\mathbf{u} \cdot \boldsymbol{\omega}dV = \oint_{C_1}\mathbf{u} \cdot (\Phi_1 d\mathbf{l}) + \oint_{C_2}\mathbf{u} \cdot (\Phi_2 d\mathbf{l})$$

$$= \Phi_1\oint_{C_1}\mathbf{u} \cdot d\mathbf{l} + \Phi_2\oint_{C_2}\mathbf{u} \cdot d\mathbf{l}, \qquad (8.13)$$

where we have used $\boldsymbol{\omega}dV = \Phi d\mathbf{l}$ as well as part (a) of Helmholtz's second law. However, provided that the loops are interlinked once, Stokes' theorem tells us that the circulation around loop 1 is $\pm\Phi_2$, where the positive sign corresponds to a right-handed linkage and the negative sign to a left-handed linkage. Similarly, the circulation around loop 2 is $\pm\Phi_1$ and it follows that

$$h = h_1 + h_2 = \pm 2\Phi_1\Phi_2. \tag{8.14}$$

We conclude that the invariance of h in this simple example is a direct consequence of the conservation of the linkage of the vortex tubes and of their fluxes Φ_1 and Φ_2.

Of course, one can construct more complex combinations of knotted and linked vortex tubes, wound around each other multiple times, and the conservation of helicity in each case will reflect the preservation of the vortex-line topology. Kelvin was so impressed by the *immutable* nature of inviscid vortex-line topology that he proposed a theory of matter based on it, in which the atoms of various elements are composed of linked, inviscid vortices of varying complexity! While Kelvin's vortex theory of matter is now seen as a bizarre footnote in the history of science, Helmholtz's laws of vortex dynamics remain a powerful tool, though of course one must be ever vigilant as to the effects of viscosity in a real fluid.

8.2.3 Steady, Axisymmetric Flows and the Squire–Long Equation

We now turn to the special case of *steady, axisymmetric motion*. In §3.3.6 we saw that, for axisymmetric flows written in terms of cylindrical polar coordinates, it is natural to decompose the velocity field into its azimuthal, $\mathbf{u}_\theta$, and poloidal, $\mathbf{u}_p = (u_r, 0, u_z)$, components, each of which are individually solenoidal. The vorticity may be likewise decomposed into two solenoidal fields, $\boldsymbol{\omega}_\theta$ and $\boldsymbol{\omega}_p$, with $\boldsymbol{\omega}_\theta = \nabla \times \mathbf{u}_p$ and $\boldsymbol{\omega}_p = \nabla \times \mathbf{u}_\theta$. It is also natural to introduce the Stokes streamfunction, $\Psi(r, z)$, defined in (3.35) by $\mathbf{u}_p = \nabla \times [(\Psi/r)\hat{\mathbf{e}}_\theta]$, and the angular momentum density, $\Gamma = ru_\theta$. Evidently, the entire velocity field is uniquely determined by these two scalar functions. The vorticity is likewise uniquely determined by Γ and Ψ according to (3.61) and (3.62),

$$\boldsymbol{\omega}_p = \left(-\frac{1}{r}\frac{\partial\Gamma}{\partial z}, 0, \frac{1}{r}\frac{\partial\Gamma}{\partial r}\right), \quad \omega_\theta = -\frac{1}{r}\nabla_*^2\Psi, \tag{8.15}$$

where ∇_*^2 is the Stokes operator defined by (3.31).

So far we have simply performed a kinematic decomposition. For an inviscid flow this decomposition may be supplemented by the axisymmetric Euler equation which yields evolution equations for both Γ and ω_θ in the form of (3.70):

$$\boxed{\frac{D\Gamma}{Dt} = 0, \quad \frac{D}{Dt}\left(\frac{\omega_\theta}{r}\right) = \frac{\partial}{\partial z}\left(\frac{\Gamma^2}{r^4}\right).} \tag{8.16}$$

If the flow is also *steady* we have $\mathbf{u}_p \cdot \nabla\Gamma = 0$ and so Γ is constant on a streamline. Since the definition of Ψ ensures that $\mathbf{u}_p \cdot \nabla\Psi = 0$, we may now write $\Gamma = \Gamma(\Psi)$ and the azimuthal vorticity equation becomes

$$\mathbf{u} \cdot \nabla \left(\frac{\omega_\theta}{r} \right) = \frac{2\Gamma\Gamma'(\Psi)}{r^4} \frac{\partial \Psi}{\partial z} = -\Gamma\Gamma'(\Psi) \frac{2u_r}{r^3},$$

or equivalently,

$$\mathbf{u} \cdot \nabla \left(\frac{\omega_\theta}{r} - \frac{\Gamma\Gamma'(\Psi)}{r^2} \right) = 0.$$

This, in turn, gives us

$$r\omega_\theta = \Gamma\Gamma'(\Psi) - r^2 H'(\Psi), \tag{8.17}$$

for some function $H(\Psi)$.

Combining (8.15) with (8.17), we conclude that a steady, axisymmetric, inviscid flow is governed by

$$\boxed{\nabla_*^2 \Psi = r \frac{\partial}{\partial r} \frac{1}{r} \frac{\partial \Psi}{\partial r} + \frac{\partial^2 \Psi}{\partial z^2} = r^2 H'(\Psi) - \Gamma\Gamma'(\Psi)}. \tag{8.18}$$

This is often called the *Squire–Long equation*. In principle, it can be solved for Ψ provided that $\Gamma(\Psi)$ and $H(\Psi)$ are specified, say at some upstream location. In fact, it is not difficult to show that $H(\Psi)$ is Bernoulli's function, $H = \frac{1}{2}\mathbf{u}^2 + p/\rho$. This is most readily seen from the axial component of the Euler equation, rewritten as

$$\frac{\partial H}{\partial z} = H'(\Psi) \frac{\partial \Psi}{\partial z} = (\mathbf{u} \times \boldsymbol{\omega})_z = \omega_\theta u_r - u_\theta \omega_r = -\frac{\omega_\theta}{r} \frac{\partial \Psi}{\partial z} + \frac{\Gamma}{r^2} \frac{\partial \Gamma}{\partial z}.$$

Dividing through by $\partial \Psi / \partial z$ yields

$$H'(\Psi) = -\frac{\omega_\theta}{r} + \frac{\Gamma\Gamma'(\Psi)}{r^2}, \tag{8.19}$$

which is consistent with (8.17).

Perhaps the most important case in practice is where the upstream conditions take the form of rigid body rotation, $u_\theta = \Omega r$, combined with a uniform axial motion, $u_z = V$. In such a case the upstream distributions of Ψ, Γ, and p/ρ are $\Psi = \frac{1}{2}Vr^2$, $\Gamma = \Omega r^2$, and, to within a constant, $p/\rho = \frac{1}{2}\Omega^2 r^2$. After a little algebra, this gives $\Gamma = (2\Omega/V)\Psi$ and $H'(\Psi) = 2\Omega^2/V$. We now let F represent the departure of Ψ from its upstream value, defined through the expression $\Psi = \frac{1}{2}Vr^2 + rF(r, z)$. Substituting for Ψ in the Squire–Long equation then gives

$$\frac{\partial^2 F}{\partial z^2} + \frac{1}{r} \frac{\partial}{\partial r} r \frac{\partial F}{\partial r} + \left[\frac{4\Omega^2}{V^2} - \frac{1}{r^2} \right] F = 0. \tag{8.20}$$

It is remarkable that, for these particular upstream conditions, our flow is governed by a simple linear equation. A multitude of inviscid, swirling flows are governed by (8.20).

By way of an example, consider swirling flow in a pipe of radius b with upstream conditions $u_\theta = \Omega r$ and $u_z = V$. A sphere of radius R sits on the axis of the pipe, as shown in Figure 8.5, and we look for axisymmetric solutions *downstream* of the sphere which are periodic in z and steady relative to the object. These periodic solutions are governed by (8.20) and it is readily shown that they take the form $F = AJ_1(k_r r)\cos(k_z z)$, where A is an amplitude, J_1 is the usual Bessel function, and k_r and k_z are related by $k_r^2 + k_z^2 = (2\Omega/V)^2$. In a frame of reference moving at speed V, such periodic solutions may be regarded as progressive waves of arbitrary magnitude propagating *upstream* with a phase speed of V. The admissible values of k_r are set by the requirement that the radial velocity disappears at $r = b$, and this in turn demands that $k_r b = \gamma_n$, where γ_n are the zeros of J_1. It follows that these waves have axial wavenumbers governed by

$$k_z^2 b^2 = (2\Omega b/V)^2 - \gamma_n^2. \tag{8.21}$$

Evidently, such waves exist only when $2\Omega b/V > \gamma_1 = 3.83$, or equivalently $V/\Omega b < 0.52$, and it is gratifying that they are observed in experiments downstream of an axisymmetric

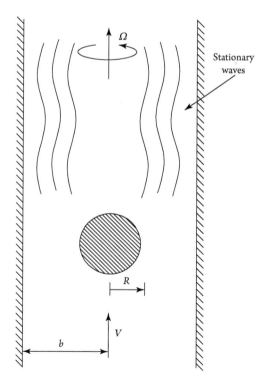

Figure 8.5 Stationary waves generated by rotating flow past a sphere. (From Davidson, 2013.)

blockage when the *Rossby number*, $\mathrm{Ro} = V/\Omega b$, is less than ~ 0.4. These finite-amplitude waves are examples of so-called *inertial waves*, which are maintained by the background rotation. We shall discuss inertial waves in Chapter 10, but one point to note now is that, in this geometry, their group velocity is smaller than their phase speed, V, which is why the waves triggered by the sphere cannot propagate upstream against the oncoming flow.

8.2.4 Viscous versus Inviscid Vortex Dynamics

Perhaps it is time to consider the relationship between inviscid vortex dynamics and the vortex dynamics of real fluids. In §2.7.2 we emphasized that three physical processes are embedded in the vorticity equation

$$\frac{D\boldsymbol{\omega}}{Dt} = (\boldsymbol{\omega}\cdot\nabla)\mathbf{u} + \nu\nabla^2\boldsymbol{\omega}. \tag{8.22}$$

On the left we have advection, the tendency for fluid particles to cling onto their vorticity, while on the right we have vortex stretching and diffusion. Away from boundaries, advection and diffusion merely act to redistribute vorticity, rather than create new vorticity, and of course vortex stretching can only occur if there is some vorticity on which to act. So these three processes cannot, by themselves, introduce vorticity into a fluid otherwise devoid of it. In the absence of rotational body forces, the vorticity gets into a fluid by diffusing in from the boundaries, as emphasized in §2.7.3. Indeed, one way to think of boundary layers is that they are diffusion layers for vorticity. This is illustrated in Figure 8.6, where there is no vorticity upstream of the cylinder, yet clearly there is vorticity on the downstream side. This vorticity diffuses out from the surface of the cylinder and into the upstream boundary layers, and the Kármán vortices form when these boundary layers separate.

In inviscid vortex dynamics, Helmholtz's laws capture very nicely the processes of advection and vortex stretching, with advection closely related to the first law (fluid particles want to cling on to their vorticity), and vortex stretching captured by the second law. Indeed, the second law gives us an extremely convenient way of quantifying vortex stretching. Consider a short element of the thin vortex tube shown in Figure 8.2, of length $d\ell$ and cross-sectional area A. According to Helmholtz, the element conserves its volume, $A d\ell$, as well as its flux, $\Phi = \omega A$. Taking the ratio of these tells us that $\omega/d\ell$ is conserved by the element, and hence $\omega \sim d\ell$. So, for example, if the short element is stretched by a factor of two, its vorticity will be doubled.

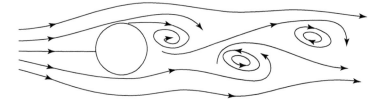

Figure 8.6 Vortex shedding behind a cylinder.

Helmholtz also emphasized the usefulness of the analogy between the kinematics of vorticity, $\nabla \times \mathbf{u} = \boldsymbol{\omega}$ and $\nabla \cdot \mathbf{u} = 0$, and Ampère's law of magnetostatics, $\nabla \times \mathbf{B} = \mu_0 \mathbf{J}$ and $\nabla \cdot \mathbf{B} = 0$, in which $\mathbf{B}$ is the magnetic field induced by the current density $\mathbf{J}$. Clearly, in this analogy, we have $\mathbf{B} \leftrightarrow \mathbf{u}$ and $\mu_0 \mathbf{J} \leftrightarrow \boldsymbol{\omega}$. Not only does this gives us the Biot–Savart law (8.1), which relates $\mathbf{u}$ to $\boldsymbol{\omega}$, but it also gives us a simple way of visualizing the velocity field associated with a given distribution of vorticity. For example, Figure 8.7 not only shows the magnetic field lines associated with a circular loop of current, but, using Helmholtz's analogy, it also gives us the velocity field associated with a thin vortex ring. We shall see in the next section that this analogy can be very helpful.

So far, so good. Of course, what is missing in Helmholtz's laws are the viscous effects, and in particular:

(i) the redistribution of vorticity by diffusive spreading; and

(ii) the generation of vorticity within boundary layers.

The neglect of boundary layers is particularly important, as is clear from Figure 8.6 where the entire phenomenon rests on the vorticity generated within the boundary layers. Moreover, even when the vorticity is supplied by some other means, boundary layers can still play an important role. Consider the configuration shown in Figure 8.5. Since the vorticity in the inviscid theory is supplied by the upstream rotation, with $\omega_z = 2\Omega$, it might be hoped that the neglect of boundary layers in the theory is not so critical. However, in order obtain a good quantitative comparison between the inviscid predictions and laboratory experiments, it is

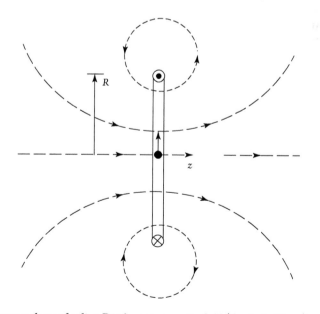

Figure 8.7 A current loop of radius R induces a magnetic field (the dashed lines).

necessary to minimize the effects of the boundary layers on the sphere. To that end, typically the sphere and fluid are made to co-rotate to eliminate any boundary layer associated with differential rotation, while a cone is often attached to the rear of the sphere to minimize the effect of boundary-layer separation associated with the axial velocity.

8.3 A Qualitative Overview of some Simple Isolated Vortices

We now consider some simple examples of isolated vortices, chosen to illustrate the ideas developed in §8.2, starting with the simplest configuration of all, that of the line vortex.

8.3.1 The Interaction of Line Vortices

A line vortex is a long, thin, straight tube of vorticity whose strength can be measured either by the circulation around the vortex, Γ, or the flux of vorticity along the vortex tube, Φ, the two being the same by virtue of Stokes' theorem:

$$\Gamma = \oint \mathbf{u} \cdot d\mathbf{r} = \int \boldsymbol{\omega} \cdot d\mathbf{S} = \Phi. \tag{8.23}$$

From the magnetostatic analogy to a long, straight wire carrying a current, it is clear that the streamlines are concentric circles centred on the vortex tube. Moreover, if the tube sits on the z-axis, then (8.23) gives us

$$\mathbf{u} = \frac{\Phi}{2\pi r} \hat{\mathbf{e}}_\theta, \tag{8.24}$$

in cylindrical polar coordinates. Since the motion is two-dimensional, attention is often focussed on the x–y plane, in which case a line vortex of vanishingly small cross-section is often called a point vortex.

If the line vortex finds itself in a two-dimensional crossflow, also independent of z, and the motion is considered inviscid, then the point vortex is carried along by the crossflow in accordance with Helmholtz's first law, or equivalently, in accordance with $D\omega_z/Dt = 0$. The vortex also conserves its strength, Φ, as demanded by Helmholtz's second law. This leads to the idea that two or more point vortices can interact, with each vortex caught up in the velocity field of the others. Two simple cases are:

(a) two point vortices of equal strength, Φ; and

(b) two vortices of equal but opposite strength, Φ and $-\Phi$.

These two cases are illustrated in Figure 8.8. If the instantaneous separation of the two point vortices is ℓ, then it is readily confirmed that two point vortices of like sign rotate about their midpoint with an angular velocity of $\Omega = \Phi/\pi\ell^2$. This follows directly from (8.24). On the other hand, the point vortices of opposite sign exhibit no rotation, but rather propagate as a pair at the speed $V = \Phi/2\pi\ell$, while maintaining their separation fixed at ℓ.

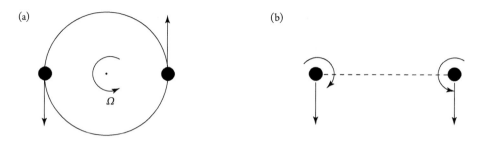

Figure 8.8 The relative motion of two point vortices. (a) Vortices of equal strength. (b) Vortices of equal but opposite strength.

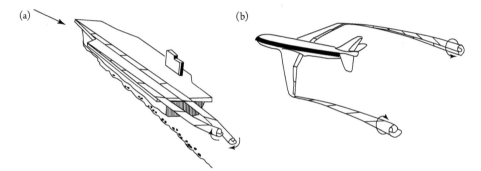

Figure 8.9 Examples of trailing vortices of: (a) equal strength, and (b) opposite strength. (Reproduced from Lugt, 1983.)

These two simple cases are not quite as idealized as they might at first seem. For example, the case of line vortices of equal but opposite strength occurs naturally with wing-tip vortices, as shown in Figure 8.9(b). These vortices propagate as a pair towards the ground. Moreover, an example of the relative rotation of like-signed vortices is shown in Figure 8.9 (a).

The case of two point vortices of the same sign but different strengths, say Φ_1 and Φ_2, is considered in Exercise 8.2. Here the distance between the vortices remains fixed as they both move in circular paths about the same fixed point with the same angular velocity, $(\Phi_1 + \Phi_2)/2\pi\ell^2$. Their centre of rotation lies on the line joining them, a distance $\Phi_2\ell/(\Phi_1 + \Phi_2)$ from vortex 1 and $\Phi_1\ell/(\Phi_1 + \Phi_2)$ from vortex 2.

It is of some interest to consider the dynamics of multiple point vortices. The three-vortex problem turns out to be integrable though complex, while four or more vortices results in *chaotic* motion (a sort of precursor to turbulence), although there is order within that chaos. For example, it turns out that a cloud of point vortices has a *centre of vorticity* defined by

$$x_c = \frac{\int x\omega dA}{\int \omega dA}, \quad y_c = \frac{\int y\omega dA}{\int \omega dA}, \tag{8.25}$$

which is an *invariant* of the motion (see Exercise 8.6). This invariant is closely related to the principle of linear momentum conservation. One can also define a dispersion, D^2, for the cloud of vortices through the expression

$$D^2 = \frac{\int \left[(x - x_c)^2 + (y - y_c)^2 \right] \omega dA}{\int \omega dA},$$
(8.26)

and it too turns out to be an invariant, related to angular momentum conservation. We shall not pause to prove the invariance of D, but instead refer the reader to Batchelor (1967).

Note the subtle interplay between the velocity and vorticity fields in these simple examples. On the one hand, the instantaneous vorticity distribution sets up a velocity field via the Biot–Savart law, and in this sense **u** is slave to ω. On the other hand, the instantaneous velocity field redistributes the vorticity by advection, so that ω is now slave to **u**. It is this non-linear interaction which characterizes vortex dynamics and also underlies the phenomenon of turbulence.

8.3.2 A Glimpse at Vortex Rings

Perhaps the next simplest geometry to consider is the axisymmetric vortex ring, which turns out to be a remarkably robust structure. The generic geometry of a vortex ring is shown in Figure 8.10. In effect, it is a vortex tube bent into a circular loop, with the vorticity running around the ring in the azimuthal direction. We shall use cylindrical polar coordinates centred on the axis of the ring, with R denoting the radius of the centreline and a the radius of the cross-section. The magnetostatic analogy to a current loop tells us that the velocity field associated with the ring is poloidal, and has the general shape shown in Figure 8.7. The precise details of the velocity field depend somewhat on how the vorticity is distributed across the cross-section of the ring, with two common assumptions being uniform vorticity and a Gaussian distribution peaked on the centreline. However, if we restrict ourselves to the

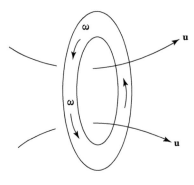
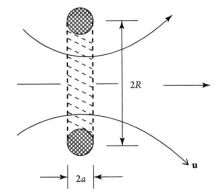

Figure 8.10 A vortex ring.

thin-ring approximation, $R \gg a$, then the dominant behaviour is more or less independent of the vorticity distribution within the ring.

In the thin-ring approximation, the velocity field induced by the ring is readily calculated using the Biot–Savart law (see §8.6.2) and its magnetic analogue can be found in many textbooks on electromagnetism. The velocity on the centreline of the ring's cross-section is in the axial direction and, by symmetry, the same at all points around the ring. Thus, like a pair of line vortices of opposite sign, a vortex ring will propagate without change of shape through otherwise still fluid. Such rings are surprisingly stable and indeed vortex rings produced by volcanic eruptions, which may be 150 m in diameter, can remain coherent for as long as ten minutes. The speed of propagation turns out to be

$$\mathbf{V} = \frac{\Phi}{4\pi R}\left[\ln\frac{8R}{a} - \alpha\right]\hat{\mathbf{e}}_z,$$

(8.27)

where Φ is the flux of vorticity around the ring and α depends on the distribution of vorticity within the ring, being equal to $1/4$ for a uniform distribution. Expression (8.27) was attributed to Kelvin by Tait in his 1867 translation of Helmholtz's 1858 paper.

Note that the speed of a vortex ring varies more or less inversely with its radius, R. This leads to a surprising phenomenon known as leap-frog. Suppose that two vortex rings are created, one behind the other, as shown schematically in Figure 8.11. Then the front ring (ring 1) finds itself in the velocity field of the rear ring (ring 2), and so starts to expand. Similarly, the rear ring feels the velocity field of the front ring and so contracts. This causes the front ring to slow down while the rear ring speeds up, and so the rear ring passes through the front ring. The axial position of the two rings in now reversed, and so ring 2 now expands under the influence of ring 1, which lies behind it, while ring 1 contracts due to the velocity field of ring 2, which lies ahead of it. The two rings then return to their original radii, having exchanged position. This is illustrated in Figure 8.12 for the case of smoke rings in air. The equations governing this process are discussed in Exercise 8.3.

In the laboratory, vortex rings are usually generated using a piston and cylinder. The piston pushes fluid through a submerged tube, causing vorticity to diffuse out from the inner surface of the tube to form an internal boundary layer. The mean flow then extrudes this boundary-layer vorticity out past the end of the tube to form an annular vortex sheet, as shown by the dye in Figure 8.13 (a). This annular vortex sheet now rolls up into a spiralled ring, while necking down (Figure 8.13 (b)). Finally, the vortex ring pinches off with the spiralled internal structure shown in Figure 8.13 (c).

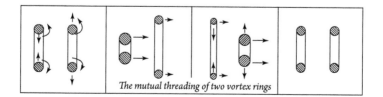

The mutual threading of two vortex rings

Figure 8.11 Schematic illustration of the leap-frog of two vortex rings.

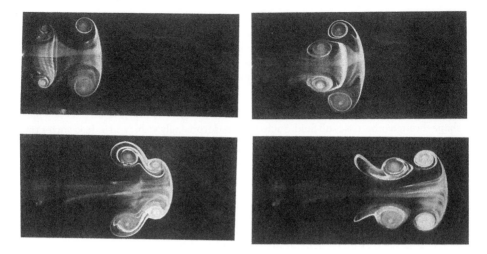

Figure 8.12 The leap-frog of two smoke rings in air. (From Yamada and Matsui, 1978.)

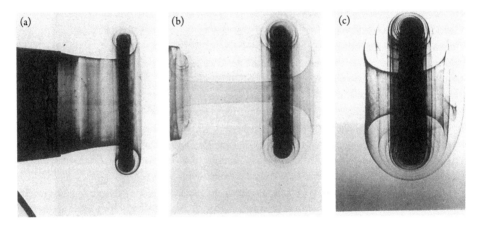

Figure 8.13 A vortex ring is formed by pushing water through a submerged tube. (a) Boundary-layer vorticity is extruded out of the tube to form an annular vortex sheet. (b) The vortex sheet necks down as it rolls up into a ring. (c) Final ring. (From Didden, 1977.)

8.3.3 Vortices due to Boundary-layer Separation

The manner in which vortex rings are generated is just one example of vortex formation through boundary-layer separation. The planar analogue of vortex ring formation is shown in Figure 8.14. Here, water is suddenly driven at constant speed past a wedge of $30°$ semi angle. The boundary layer which forms of the front face of the wedge separates at the apex to form a vortex sheet. This then rolls up into a starting vortex, not unlike that of an aerofoil.

A more dramatic example of vortex-sheet roll-up is provided by the scroll vortices on the upper surface of a delta-wing, as shown in Figure 8.15. Here, high-pressure air on

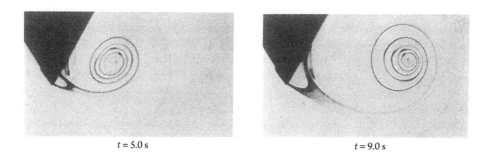

$t = 5.0$ s $t = 9.0$ s

Figure 8.14 The starting vortex on a wedge. (From Pullin and Perry, 1980.)

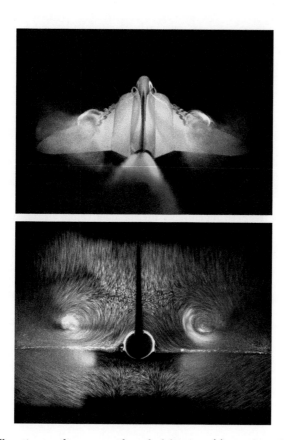

Figure 8.15 Scroll vortices on the upper surface of a delta wing: (a) view from above, (b) rear view. (Photographs by H. Werlé of ONERA, courtesy of J. Delery.)

the underside of the wing streams past the leading edge of the wing to get to the low pressure on the upper surface. This produces boundary-layer vorticity at the leading edge, which immediately separates and then rolls up to form the scroll vortices shown. These vortices have the beneficial effect of ensuring high lift and large stall angles, as illustrated in Figure 6.21.

We have already discussed the vortices shed by a circular cylinder in §2.6.2, but perhaps it is worth closing this section with two revealing images. Figure 8.16 shows the shed vortex from an impulsively started cylinder at a Reynolds number of 2000. This provides us with a beautiful example of how boundary-layer separation introduces vorticity into a flow. Finally, Figure 8.17 shows the Kármán vortex street behind a circular cylinder at a Reynolds number

Figure 8.16 Shed vortex of an impulsively started cylinder. (From Visualised Flow, 1988.)

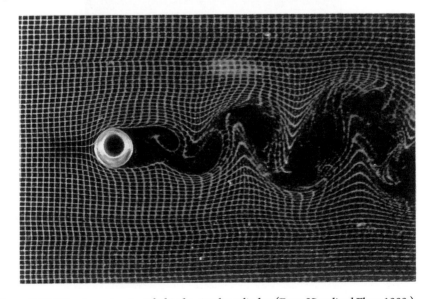

Figure 8.17 Kármán vortex street behind a circular cylinder. (From Visualised Flow, 1988.)

of 170. This kind of periodic vortex shedding also occurs behind cylinders of other cross-sections, and it gives rise to cyclic forces on the cylinder. In a high wind this can lead to oscillations of suspended cables, tall chimneys, or even suspension bridges, and large amplitudes can develop if resonance occurs.

8.3.4 Columnar Vortices in the Atmosphere and Oceans

Perhaps the most dramatic of the naturally occurring vortices are tropical cyclones and tornados. Since the rotation of the Earth is important for tropical cyclones, we shall defer our discussion of cyclones until Chapter 10. Here, we focus on tornadoes and related vortices. Tornadoes generate some of the highest natural wind speeds on Earth, typically around 50 m/s, although particularly intense tornadoes can generate winds of $\sim$150 m/s. They are around 100 m in diameter and 1–2 km high, stretching from the ground to cumulonimbus cloud above. Their sense of rotation is cyclonic (*i.e.* in the same direction as the Earth), yet they are too small to feel the Earth's rotation directly. This suggests that they are generated by larger-scale atmospheric motions and indeed tornadoes are often associated with so-called supercells (a class of thunderstorm). When generated over water, tornadoes are called waterspouts, an example of which is shown in Figure 8.18.

To leading order, tornadoes are just vortex tubes, with the axial vorticity giving rise to a horizontal swirling motion. However, superimposed on this swirl there is a complicated secondary flow in the vertical plane, as shown in Figure 8.19. The origin of this secondary flow, which is called Ekman pumping, is rather subtle, and we shall defer our discussion

Figure 8.18 A water spout off the coast of Florida. (Image by J. Golden, NOAA.)

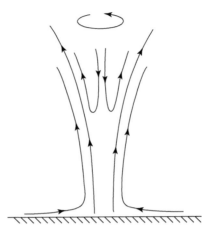

Figure 8.19 The secondary flow in a tornado or a dust devil.

of it until Chapter 10. In brief, it turns out that, whenever a fluid rotates over a stationary horizontal boundary, in this case the ground, a secondary motion is produced in which the fluid spirals radially in towards the axis of the vortex, as well as up and away from the boundary. When superimposed on the primary rotation, this means that each fluid particle spirals radially inward along the ground (or sea surface) and then spirals radially upward. The spiralled motion near the ground can be seen in the pattern of felled trees or, in the case of waterspouts, surface waves. In addition to the Ekman pumping, some researchers report a region of downward vertical motion near the axis of the tornado, also shown in Figure 8.19. As yet, there is no clear explanation for this reversed flow.

Dust devils have a structure not unlike tornadoes, although they are smaller and weaker, with wind speeds less than 20 m/s. They occur in deserts under a clear sky, and so their generation mechanism is entirely different to that of tornadoes. When the radiation from the Sun is particularly intense, radiative heating of the ground can produce an unstable thermal stratification, which gives rise to convection. This provides the energy for dust devils, though not their angular momentum. However, if a shear flow is simultaneously created by, say, wind flowing around an object, then the combination of thermal convection and a shear flow can give rise to a dust devil. An example of a dust devil, in this case on the surface of Mars, is shown in Figure 8.20.

The oceanic analogue of a tornado is a whirlpool or *tidal vortex*, sometimes given the Norwegian name of *maelstrom*. These have proved popular in fiction, from Homer's *Odyssey* to Edgar Allan Poe's *Decent into the Maelstrom*. In effect, they are large bathtub-like vortices, perhaps 10 m across, driven by strong tides. They tend to appear in narrow straights where flow rates are high and are often associated with the roll-up of tidal shear layers, or with wakes behind submerged or partially submerged rock formations. Figure 8.21 shows the Corryvreckan whirlpool located off the Scottish island of Jura.

Figure 8.20 A dust devil on the surface of Mars. (Source: JPL/MSSS/NASA)

Figure 8.21 Corryvreckan whirlpool off the island of Jura. (Image: Walter Baxter.)

8.4 Viscous Vortex Dynamics I: the Prandtl–Batchelor Theorem

We now return to quantitative analysis and focus on the role of viscosity. Our discussion of viscous effects comes in three parts, starting with the Prandtl–Batchelor theorem.

8.4.1 The Physical Origins of the Prandtl–Batchelor Theorem

The Prandtl–Batchelor theorem is an important result in the theory of two-dimensional laminar flows, which would be difficult to rationalize without the aid of vortex dynamics. In effect, it says the following. Suppose that we have a two-dimensional flow whose Reynolds number is large, but not so large that the flow becomes turbulent. Suppose also that the flow is steady, devoid of rotational body forces, and contains a region of closed streamlines, as in the cavity flow shown in Figure 8.22. Then the theorem says that, within the region of closed streamlines, but outside the boundary layers (or any other highly sheared region), the vorticity is perfectly uniform.

The physical explanation for this result is the following. Since the Reynolds number is large, it is natural to drop the viscous term from the vorticity equation, (2.48), at least outside the boundary layers. For a steady, two-dimensional flow this yields $\mathbf{u} \cdot \nabla \omega = 0$, which tells us that the vorticity is constant along a streamline, but yields no information about the variation of vorticity across the streamlines. The problem seems to be underspecified, in that we are missing some information. In the case of open streamlines, say a flow in which the streamlines come in from the left and depart on the right, this dilemma is readily resolved; we simply specify the distribution of vorticity across the streamlines at some upstream location, say as $\omega = \omega_0(\psi)$, where ψ is the streamfunction defined by (1.15). This cross-stream variation of ω is preserved as the fluid passes from left to right, and so we may combine this with the kinematic result $\omega = -\nabla^2 \psi$ to give

$$\nabla^2 \psi = -\omega_0(\psi). \tag{8.28}$$

In principle, (8.28) can be solved for ψ, and hence ω, although solutions may be difficult to find when $\omega_0(\psi)$ is non-linear. Crucially, however, for flows with closed streamlines there is no upstream location at which we may specify the cross-stream variation in vorticity.

In order to resolve this impasse, we must recall that the fluid particles in a region of closed streamlines remain trapped there forever. Moreover, when we approximate the two-dimensional vorticity equation by $\mathbf{u} \cdot \nabla \omega = 0$, we ignore the diffusion of vorticity on the grounds that, outside the boundary layers, the diffusive flux of ω is small whenever ν is small. However, since the trapped fluid is not going anywhere, the cumulative effects of very weak diffusion can, and do, build up over time to produce an order-one effect. In short, in regions of closed streamlines we cannot ignore the long-term influence of weak diffusion. Typically, when setting up a flow such as that shown in Figure 8.22, the vorticity is soon found to be constant along streamlines in accordance with $\mathbf{u} \cdot \nabla \omega = 0$, yet varies across

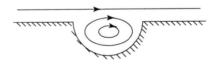

Figure 8.22 Laminar flow in a cavity at large Re. The vorticity is uniform outside the boundary layers and in the region of closed streamlines.

the streamlines. However, this is not the final steady-state. Weak, cross-stream diffusion of vorticity now slowly but progressively acts to eradicate cross-stream gradients in ω, and this continues until all such gradients have been eliminated. The true steady-state is then one in which the vorticity is uniform outside the boundary layers. It is this slow, cross-stream diffusion of vorticity that underpins the Prandtl–Batchelor theorem.

The theorem is given added weight by the fact that it is physically robust. For example, it generalizes to axisymmetric flows, as discussed in §8.4.2, and it seems to hold, at least approximately, for the time-averaged vorticity distribution in a statistically steady turbulent flow, where presumably turbulent diffusion replaces molecular diffusion in the eradication of cross-stream gradients in vorticity. The theorem also extends to the temperature field of a steady, two-dimensional flow with closed streamlines, provided that the Peclet number is large, *i.e.* $u\ell/\alpha \gg 1$ where α is the thermal diffusivity. The physical interpretation of the thermal case is similar to that for vorticity. Suppose that the solid surface in Figure 8.22 is held at temperature T_{hot}, the free stream is at the cooler temperature of T_{cold}, and we ask how the temperature varies across the region of closed streamlines. One first argues that, outside the thermal boundary layers, a large Peclet number means that the advection–diffusion equation for heat, (2.49), reduces to $\mathbf{u} \cdot \nabla T \approx 0$ in a steady flow. This tells us that T is approximately conserved along streamlines, but says nothing about the variation of temperature across the streamlines. However, in regions where the streamlines are closed, a weak cross-stream diffusion of temperature eventually eradicates all gradients in T. In short, outside the thermal boundary layers (and a thin transition layer across the top of the cavity), the temperature in the cavity will be uniform and take a value that is intermediate between T_{hot} and T_{cold}.

8.4.2 A Proof of the Theorem

To prove the Prandtl–Batchelor theorem we proceed as follows. According to (1.15), the velocity and vorticity fields of a two-dimensional flow can be expressed in terms of the streamfunction ψ as

$$u_x = \frac{\partial \psi}{\partial y}, \quad u_y = -\frac{\partial \psi}{\partial x}, \quad \omega = -\nabla^2 \psi, \tag{8.29}$$

while the steady vorticity equation is

$$\mathbf{u} \cdot \nabla \omega = \nu \nabla^2 \omega. \tag{8.30}$$

We now consider the situation where ν is very small but nonetheless finite, *i.e.* Re $\gg 1$. Outside the boundary layers we may then drop the diffusion term in (8.30), at least to leading order in Re^{-1}. We then have $\mathbf{u} \cdot \nabla \omega \approx 0$ and the vorticity is more or less constant along streamlines:

$$\omega = \omega(\psi) + O(\text{Re}^{-1}). \tag{8.31}$$

This is as far as a quasi-inviscid analysis will get us. We now need to engage with the fact that there is a weak but persistent diffusion of vorticity across streamlines, and we do this through an integral constraint. Let us rewrite (8.30) as

$$\nabla \cdot (\omega \mathbf{u}) = \nabla \cdot (\nu \nabla \omega), \tag{8.32}$$

and integrate over a two-dimensional volume bounded by a closed streamline. Gauss' theorem allows us to convert both volume integrals into surface integrals and we obtain

$$\nu \oint \nabla \omega \cdot d\mathbf{S} = 0, \tag{8.33}$$

$\omega \mathbf{u} \cdot d\mathbf{S}$ being zero as we are integrating over a streamline. We now combine (8.31) with (8.33), dropping the Re^{-1} correction in (8.31) on the understanding that $\mathrm{Re} \gg 1$. This yields

$$\nu \frac{d\omega}{d\psi} \oint \nabla \psi \cdot d\mathbf{S} = 0, \tag{8.34}$$

where we have taken advantage of the fact that $d\omega/d\psi$ is constant on a streamline to take it outside the integral sign. This is our integral constraint. Crucially, (8.34) must be satisfied no matter how small we make ν, and for a small but finite viscosity this demands that

$$\frac{d\omega}{d\psi} \oint \nabla \psi \cdot d\mathbf{S} = 0. \tag{8.35}$$

However, the line integral in (8.35) cannot be zero because the theorems of Gauss and Stokes applied to the area enclosed by the streamline yield

$$\oint \nabla \psi \cdot d\mathbf{S} = \int \nabla^2 \psi \, dA = -\int \omega \, dA = -\oint \mathbf{u} \cdot d\mathbf{r} \neq 0. \tag{8.36}$$

We are therefore forced to the conclusion that $d\omega/d\psi = 0$, so there is no cross-stream variation in vorticity outside the boundary layers. This is the Prandtl–Batchelor theorem. To determine the value of the vorticity in the region of closed streamlines it is necessary to analyse the boundary layers, as discussed in Batchelor (1956).

The theorem extends in an obvious way to axisymmetric flow. Suppose that $\mathbf{u}(r, z) = (u_r, 0, u_z)$ in cylindrical polar coordinates. Then the vorticity is purely azimuthal and governed by (3.69):

$$\frac{D}{Dt}\left(\frac{\omega_\theta}{r}\right) = \nu \nabla \cdot [r^{-2} \nabla (r\omega_\theta)] = \frac{\nu}{r^2} \nabla \cdot [r^2 \nabla (\omega_\theta/r)]. \tag{8.37}$$

Evidently, ω_θ/r is conserved along streamlines in an inviscid, steady flow. Moreover, when ω_θ/r is uniform over some extended region of space, the diffusive term in (8.37) is zero, and so diffusion shuts down. It should come as no surprise, therefore, that the axisymmetric version of the Prandtl–Batchelor theorem says that, in a steady, laminar flow at high Re, ω_θ/r is uniform in a region of closed streamlines, as long as the streamlines lie outside the boundary layers (Batchelor, 1956).

8.5 Viscous Vortex Dynamics II: Burgers' Vortex

8.5.1 A Dilemma in Turbulence: Finite Energy Dissipation for Vanishing Viscosity

In §1.4.1 we calculated the rate of conversion of mechanical energy into heat caused by turbulence in the shoulder of a sudden pipe expansion, as shown in Figure 8.23. On the assumption that the Reynolds number is large, we showed the rate of energy loss to be:

$$\text{rate of loss of mechanical energy} = \frac{1}{2}\dot{m}\left(V_1 - V_2\right)^2. \tag{8.38}$$

Curiously, this is independent of the value of the viscosity of the fluid.

At the time we noted that the absence of ν in (8.38) appears paradoxical, as viscous stresses are responsible for the conversion of mechanical energy into heat. Actually, this is not an isolated result: it turns out that one of the most fundamental properties of turbulence is that the rate of dissipation of energy is finite, order one, and asymptotically independent of the Reynolds number in the limit of Re $\rightarrow \infty$. The implication is that there is a fundamental distinction between setting $\nu = 0$, corresponding to zero dissipation, and taking the limit of $\nu \rightarrow 0$, which leads to order unity dissipation. This empirical (but well-established) law

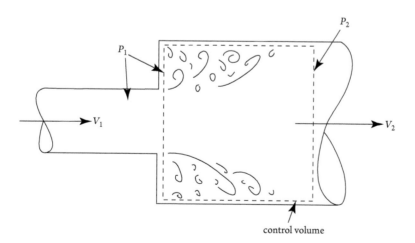

Figure 8.23 Flow behind the sudden expansion in a pipe.

underpins nearly all theories of turbulence and its paradoxical nature has exercised the minds of many a theorist, including Burgers.

In 1948 Burgers reasoned that, if he could find a simple, exact solution of the Navier–Stokes equation which possesses the property of a finite energy dissipation rate in the limit of Re $\to \infty$, then this might provide some hints as to what occurs in a turbulent flow at large Reynolds number. In this respect, he was remarkably successful, as we now discuss.

8.5.2 Burgers' Axisymmetric Vortex

Consider the axisymmetric vortex

$$\boldsymbol{\omega} = \frac{\Phi_0}{\pi \delta^2} \exp\left[-r^2/\delta^2\right] \hat{\mathbf{e}}_z, \tag{8.39}$$

written in cylindrical polar coordinates (r, θ, z). Here, the constant Φ_0 is the net flux of vorticity along the vortex tube and δ is the characteristic radius of the tube. If left to itself, this vortex tube will diffuse radially outward, while the peak vorticity on its axis will fall as the flux Φ_0 is spread over an ever-larger area. So, in order to obtain a steady solution, Burgers placed the vortex tube in an imposed, axisymmetric, irrotational strain field,

$$\mathbf{u}^{(\alpha)} = (u_r, 0, u_z) = \left(-\tfrac{1}{2}\alpha r, 0, \alpha z\right), \tag{8.40}$$

where α is the imposed strain rate. The total velocity field is then the sum of $\mathbf{u}^{(\alpha)}$ and that associated, via the Biot–Savart law, with the vortex itself,

$$\mathbf{u}^{(\omega)} = \frac{\Phi_0}{2\pi r} \left[1 - \exp\left(-r^2/\delta^2\right)\right] \hat{\mathbf{e}}_\theta, \tag{8.41}$$

as shown in Figure 8.24. (Sceptical readers may wish to confirm that the curl of (8.41) does indeed yield (8.39).) Note that the radial component of $\mathbf{u}^{(\alpha)}$ sweeps the vorticity back towards the axis while the axial component stretches the vortex lines.

It is readily confirmed that, provided we choose the magnitude of the strain rate, α, to satisfy the expression $\delta = \sqrt{4\nu/\alpha}$, (8.39) and (8.40) represent a steady solution of the axial component of the vorticity equation,

$$\frac{D\omega_z}{Dt} = \omega_z \frac{\partial u_z}{\partial z} + \nu \nabla^2 \omega_z. \tag{8.42}$$

In fact, Burgers' vortex turns out to represent a delicate balancing act. By carefully matching the magnitude of the strain rate to the radius of the vortex, the inward sweeping motion represented by the left of (8.42) balances perfectly the tendency for vorticity to diffuse radially outward. Simultaneously, the axial stretching of the vortex lines, represented by the first term on the right of (8.42), exactly counters the tendency for the centre-line vorticity to fall by diffusion. In short, by choosing $\delta = \sqrt{4\nu/\alpha}$, the advection, stretching, and diffusion of vorticity are all perfectly balanced, yielding a steady flow.

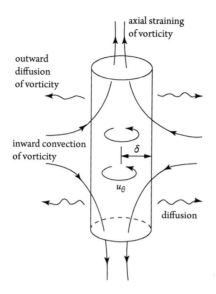

Figure 8.24 Burgers' vortex.

Crucially, Burgers showed that, in the limit of $\nu \to 0$, the net viscous dissipation per unit length of the vortex tube is finite and independent of ν, equal to $\rho \alpha \Phi_0^2 / 8\pi$ (see Exercise 8.5). It was precisely this feature of the steady solution that interested Burgers since, as noted above, a finite dissipation rate in the limit of $\nu \to 0$ is exactly what is observed in turbulent motion at high Reynolds number.

8.5.3 The Robust Nature of Burgers' Vortex

One reason why Burgers' vortex is conceptually appealing is that it is physically robust, in the following sense. Suppose that we allow the strain rate and vortex radius in (8.39) and (8.40) to be functions of time, i.e. $\alpha(t)$ and $\delta(t)$. Then it is readily confirmed that we still have an exact solution of (8.42), but one in which $\delta(t)$ is governed by

$$\frac{d\delta^2}{dt} + \alpha(t)\delta^2 = 4\nu \tag{8.43}$$

(see Davidson, 2015). For example, if α is a constant, but the initial value of δ does not satisfy $\delta = \sqrt{4\nu/\alpha}$, then (8.43) yields

$$\delta^2 = \delta_0^2 e^{-\alpha t} + (4\nu/\alpha)\left[1 - e^{-\alpha t}\right], \quad \delta_0 = \delta(0), \tag{8.44}$$

which always asymptotes to the steady solution. Thus, if the initial radius exceeds the steady-state value of $\delta = \sqrt{4\nu/\alpha}$, diffusion is weaker than advection and the vortex core shrinks due to the inward advection of vorticity, until eventually we reach $\delta = \sqrt{4\nu/\alpha}$ and the

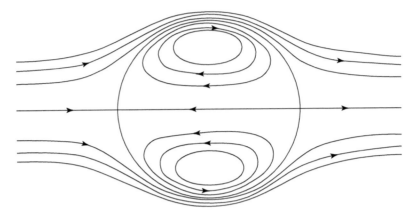

Figure 8.25 Schematic of Hill's spherical vortex.

steady solution is recovered. Conversely, if the initial vortex radius is smaller than the steady-state value, then diffusion dominates over advection and the vortex core grows by diffusion, until once again we reach $\delta = \sqrt{4\nu/\alpha}$.

Burgers' hope that this simple vortex might provide insights into the nature of turbulence turns out to be rather prophetic. It is now known that the small scales in a turbulent flow, which act as the centres of dissipation, are characterized by thin vortex tubes, and these tubes are highly reminiscent of an unsteady Burgers' vortex, albeit curved rather than straight.

8.6 More Axisymmetric Vortices (both Viscous and Inviscid)

8.6.1 Hill's Spherical Vortex

As another example of an axisymmetric vortex, let us consider *Hill's spherical vortex*, as shown in Figure 8.25. Although this has several interpretations, it is usually portrayed as an inviscid, steady, spherical region of finite vorticity embedded within an otherwise irrotational cross-flow that is uniform at large distances from the vortex. Such a spherical vortex is sometimes used as a model for a large drop of immiscible fluid falling through another liquid.

In order to describe Hill's vortex, we first note that the Stokes streamfunction for irrotational flow over a sphere of radius R, centred on the origin (see Exercise 6.10), is

$$\Psi = \frac{1}{2}Vr^2\left(1 - R^3\big/|\mathbf{x}|^3\right),$$

in cylindrical polar coordinates. In particular, it is readily confirmed that the velocity field corresponding to this streamfunction has zero curl, is equal to $V\hat{\mathbf{e}}_z$ far from the sphere, and is tangential to the surface of the sphere, as is evident from the fact that $\Psi = 0$ at $|\mathbf{x}| = R$.

Turning to the flow within the sphere, it is natural to ask if we can find a poloidal velocity field that satisfies $\Psi = 0$ at $|\mathbf{x}| = R$ and also matches the external velocity (and hence pressure) at the surface of the sphere. Rather remarkably, it turns out that we can, and this velocity field corresponds to the azimuthal vorticity distribution $\omega_\theta/r = C$, where C is a constant. The Stokes streamfunction for such a flow is given by (8.15), rewritten as

$$\nabla_*^2 \Psi = -r^2 C, \tag{8.45}$$

and a comparison with (8.18) shows that Bernoulli's function, $H(\Psi) = \frac{1}{2}\mathbf{u}^2 + p/\rho$, is related to C by $H = -C\Psi$. The general structure of (8.45) suggests that we seek a solution for the internal flow of the form

$$\Psi = r^2 F(\eta), \quad \eta = |\mathbf{x}|/R.$$

Substituting this into (8.45) yields

$$\frac{1}{\eta^4}\frac{d}{d\eta}\eta^4\frac{dF}{d\eta} = -CR^2,$$

which integrates to give

$$\Psi = \frac{1}{10}Cr^2\left(R^2 - |\mathbf{x}|^2\right).$$

Clearly this satisfies $\Psi = 0$ at $|\mathbf{x}| = R$ and so it simply remains to match the external velocity at the surface of the sphere. It is left as an exercise for the reader to show that the internal and external velocity fields do indeed match at $|\mathbf{x}| = R$, provided that we take $C = -15V/2R^2$. The net flux of vorticity around the axis of symmetry is then $\Phi = -5VR$.

It is instructive to change the frame of reference so that the fluid at infinity is stationary. In such a frame, we have a spherical vortex which propagates through otherwise still fluid with a speed of $V = |\Phi|/5R$, in much the same way that a vortex ring propagates through still fluid. Indeed, in some sense, Hill's vortex may be thought of as the limiting case of a vortex ring that has been shrunk down onto its axis of symmetry.

Finally, we note that, although Hill's vortex is often thought of as an inviscid flow, it actually represents an exact solution of the viscous vorticity equation, (8.37), with the vorticity distribution $\omega_\theta/r = C$ being precisely that expected from the axisymmetric version of the Prandtl–Batchelor theorem, as discussed in §8.4.2. However, in the viscous case there is a discontinuity in shear stress at the surface of the sphere, and this complicates matters, giving rise to a boundary layer at the spherical interface.

8.6.2 The Velocity Field and Kinetic Energy of a Thin Vortex Ring

We now turn to the velocity field of a thin vortex ring. Recall that, since $\mathbf{u}$ is solenoidal, we may introduce a vector potential for the velocity field, which is uniquely defined by

$$\nabla \times \mathbf{A} = \mathbf{u}, \quad \nabla \cdot \mathbf{A} = 0, \quad |\mathbf{A}|_\infty = 0. \tag{8.46}$$

(A vector field is uniquely determined if its curl and divergence are specified, along with suitable boundary conditions.) The curl of (8.46) now yields $\nabla^2 \mathbf{A} = -\boldsymbol{\omega}$, which may be inverted using Green's inversion integral (see Appendix 2) to give

$$\mathbf{A}(\mathbf{x}) = \frac{1}{4\pi} \int \boldsymbol{\omega}(\mathbf{x}') \frac{d\mathbf{x}'}{|\mathbf{x}' - \mathbf{x}|}, \tag{8.47}$$

where the integral is taken over all space. In the case of a vortex ring of radius R and vorticity flux Φ, this becomes

$$\mathbf{A}(\mathbf{x}) = \frac{\Phi}{4\pi} \int_0^{2\pi} \frac{\hat{\mathbf{e}}'_\theta R d\theta'}{|\mathbf{x} - \mathbf{x}'|}, \quad \mathbf{x}' = R\hat{\mathbf{e}}'_r, \tag{8.48}$$

where we use cylindrical polar coordinates centred on the ring and the fact that $\omega' d\mathbf{x}' = \Phi R d\theta'$. Given that

$$\left(\mathbf{x} - R\hat{\mathbf{e}}'_r\right)^2 = \mathbf{x}^2 + R^2 - 2(r\hat{\mathbf{e}}_r) \cdot (R\hat{\mathbf{e}}'_r) = \mathbf{x}^2 + R^2 - 2rR\cos(\theta - \theta'),$$

(8.48) yields

$$\mathbf{A}(\mathbf{x}) = \frac{\Psi(r, z)}{r}\hat{\mathbf{e}}_\theta = \frac{R\Phi}{4\pi} \int_0^{2\pi} \frac{\cos\tilde{\theta}\, d\tilde{\theta}}{\sqrt{z^2 + r^2 + R^2 - 2rR\cos\tilde{\theta}}}\hat{\mathbf{e}}_\theta, \tag{8.49}$$

where Ψ is the Stokes steamfunction, $\tilde{\theta} = \theta - \theta'$, and we have used $\hat{\mathbf{e}}_\theta \cdot \hat{\mathbf{e}}'_\theta = \cos\tilde{\theta}$ as well as the fact that $\mathbf{A}$ is azimuthal. This can be tidied up through the use of elliptic integrals. In particular, the denominator in the integrand can be rewritten as

$$\sqrt{z^2 + r^2 + R^2 - 2rR\cos\tilde{\theta}} = \frac{2\sqrt{Rr}}{k}\sqrt{1 - k^2\cos^2\left(\tilde{\theta}/2\right)}, \quad k^2 = \frac{4Rr}{z^2 + (r + R)^2},$$

from which

$$\Psi = \frac{\Phi\sqrt{Rr}}{4\pi} \int_0^\pi \left\{ \left(\frac{2}{k} - k\right) \frac{1}{\sqrt{1 - k^2\cos^2\left(\tilde{\theta}/2\right)}} - \frac{2}{k}\sqrt{1 - k^2\cos^2\left(\tilde{\theta}/2\right)} \right\} d\tilde{\theta}.$$

This then yields

$$\boxed{\Psi(r, z) = \frac{\Phi\sqrt{Rr}}{2\pi} \left\{ \left(\frac{2}{k} - k\right) K(k) - \frac{2}{k} E(k) \right\},} \tag{8.50}$$

where $K(k)$ and $E(k)$ are the complete elliptic integrals of the first and second kinds.

Two cases of particular interest are the flow in the far field, $|\mathbf{x}| \gg R$, and the velocity on the axis, $r = 0$, both of which correspond to $k \ll 1$. The low-k forms of $K(k)$ and $E(k)$ are

$$K(k) = \frac{\pi}{2}\left\{1 + \frac{k^2}{4} + \frac{9}{64}k^4 + \cdots\right\}, \quad E(k) = \frac{\pi}{2}\left\{1 - \frac{k^2}{4} - \frac{3}{64}k^4 + \cdots\right\},$$

and so for small k we have

$$\Psi(r, z) = \frac{\Phi R^2 r^2}{4\left(z^2 + (r + R)^2\right)^{3/2}} \quad (k \ll 1), \tag{8.51}$$

from which

$$\mathbf{u} = \frac{\Phi R^2 \left[3rz\hat{\mathbf{e}}_r + \left(2z^2 - r^2 + rR + 2R^2\right)\hat{\mathbf{e}}_z\right]}{4\left(z^2 + (r + R)^2\right)^{5/2}} \quad (k \ll 1).$$

The kinetic energy of a vortex ring is most readily found by noting that $\mathbf{u}^2$ and $\boldsymbol{\omega} \cdot \mathbf{A}$ differ by $\nabla \cdot (\mathbf{A} \times \mathbf{u})$, and so

$$\frac{1}{2}\int \mathbf{u}^2 dV = \frac{1}{2}\int \boldsymbol{\omega} \cdot \mathbf{A} dV = \frac{1}{2}\Phi \oint \mathbf{A} \cdot d\mathbf{r} = \frac{1}{2}\Phi \int \mathbf{u} \cdot d\mathbf{S},$$

where the line integral is along the centreline of the ring, the surface integral is the flux of $\mathbf{u}$ through the ring, and we have used $\boldsymbol{\omega} dV = \Phi d\mathbf{r}$. The flux of $\mathbf{u}$ through the ring is readily found in the magnetostatic literature, where the equivalent calculation for $\mathbf{B}$ is commonplace (see, for example, Jackson, 1999, problem 5.32). For a ring of circular cross-section and uniform vorticity, this yields

$$\frac{1}{2}\int \mathbf{u}^2 dV = \frac{1}{2}\Phi^2 R\left[\ln\left(8R/a\right) - 7/4\right]. \tag{8.52}$$

8.7 Viscous Vortex Dynamics III: the Impulse of Localized Vorticity Fields

We are used to the idea that a current loop generates a magnetic field $\mathbf{B}$ that pervades all of space, as shown in Figure 8.7. In fact, the magnetic field shown in this figure falls off as a power law at large distances, as $|\mathbf{B}|_\infty \sim |\mathbf{x}|^{-3}$. The analogy between magnetostatics and vortex dynamics, discussed in §8.2.4, tells us that the velocity field of a vortex ring also falls as $|\mathbf{x}|^{-3}$ in the far field, as is evident from (8.51), and indeed we shall see that this is typical of most localized vorticity distributions. The fact that the far-field velocity falls off slowly as $|\mathbf{u}|_\infty \sim |\mathbf{x}|^{-3}$ has a number of important consequences for vortex dynamics, not the least

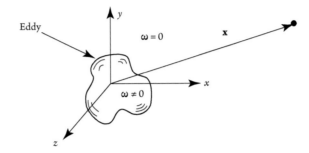

Figure 8.26 A localized distribution of vorticity.

of which is that certain integrals taken over all space, such as the global linear momentum $\int \rho \mathbf{u} dV$, are only conditionally convergent. We now explore these issues. In particular, we introduce two important integral quantities, called the *linear and angular impulse*, which allow us to invoke global linear and angular momentum conservation in a way that sidesteps the lack of absolute convergence of integrals like $\int \rho \mathbf{u} dV$. Let us start, however, with the far-field velocity induced by a localized vorticity distribution, as shown in Figure 8.26.

8.7.1 The Far field of a Localized Vorticity Distribution

Consider a vorticity field that is localized near the origin in an unbounded fluid that is motionless at infinity. The vector potential for $\mathbf{u}$ is given by (8.47). To find this vector potential at large distances from the vorticity field we expand $|\mathbf{x}' - \mathbf{x}|^{-1}$ as a Taylor series in $\mathbf{x}'$:

$$\frac{1}{|\mathbf{x} - \mathbf{x}'|} = \frac{1}{|\mathbf{x}|} - \frac{\partial}{\partial x_i}\left(\frac{1}{|\mathbf{x}|}\right)x_i' + C_{ij}x_i'x_j' + \cdots, \quad C_{ij}(\mathbf{x}) = \frac{1}{2}\frac{\partial^2}{\partial x_i \partial x_j}\left(\frac{1}{|\mathbf{x}|}\right).$$

(8.53)

On substituting this expansion into integral (8.47), we obtain

$$\mathbf{A}(\mathbf{x}) = \frac{1}{4\pi |\mathbf{x}|}\int \boldsymbol{\omega}(\mathbf{x}')d\mathbf{x}' - \frac{1}{4\pi}\frac{\partial}{\partial x_i}\left(\frac{1}{|\mathbf{x}|}\right)\int x_i'\boldsymbol{\omega}(\mathbf{x}')d\mathbf{x}'$$

$$+ \frac{1}{4\pi}C_{ij}\int x_i'x_j'\boldsymbol{\omega}(\mathbf{x}')d\mathbf{x}' + \cdots.$$

(8.54)

However, since the vorticity field is localized, the first term on the right integrates to zero,

$$\int_{V_\infty} \omega_i d\mathbf{x} = \int_{V_\infty} \boldsymbol{\omega} \cdot \nabla(x_i)\, d\mathbf{x} = \int_{V_\infty} \nabla \cdot (\boldsymbol{\omega} x_i)\, d\mathbf{x} = \oint_{S_\infty} x_i \boldsymbol{\omega} \cdot d\mathbf{S} = 0, \quad (8.55)$$

where V_∞ is a large volume whose surface, S_∞, recedes to infinity. Also, the second integral can be transformed using the integral relationship

$$\int_{V_\infty} [x_i\omega_j + x_j\omega_i]\, d\mathbf{x} = \int_{V_\infty} \nabla\cdot(x_i x_j \boldsymbol{\omega})d\mathbf{x} = \oint_{S_\infty} x_i x_j \boldsymbol{\omega}\cdot d\mathbf{S} = 0, \qquad (8.56)$$

to give

$$\frac{\partial}{\partial x_i}\left(\frac{1}{|\mathbf{x}|}\right)\int 2x_i'\omega_j'\,d\mathbf{x}' = \frac{\partial}{\partial x_i}\left(\frac{1}{|\mathbf{x}|}\right)\int (x_i'\omega_j' - x_j'\omega_i')\,d\mathbf{x}'$$

$$= -\nabla\left(\frac{1}{|\mathbf{x}|}\right)\times\int (\mathbf{x}'\times\boldsymbol{\omega}')\,d\mathbf{x}'. \qquad (8.57)$$

Our expansion for $\mathbf{A}$ now yields

$$\boxed{\mathbf{A}(\mathbf{x}) = \frac{1}{4\pi}\nabla\left(\frac{1}{|\mathbf{x}|}\right)\times\mathbf{L} + \frac{1}{4\pi}C_{ij}\int x_i' x_j'\boldsymbol{\omega}'d\mathbf{x}' + \cdots}, \qquad (8.58)$$

where

$$\boxed{\mathbf{L} = \frac{1}{2}\int (\mathbf{x}\times\boldsymbol{\omega})d\mathbf{x}}. \qquad (8.59)$$

The integral $\mathbf{L}$ is called the *linear impulse* of the vorticity distribution and we shall see shortly that it is a measure of the net linear momentum introduced into the fluid by virtue of the presence of the vorticity. We shall also show that $\mathbf{L}$ is an invariant of the motion. Taking the curl of (8.58) gives the far-field velocity as

$$\boxed{\mathbf{u}(\mathbf{x}) = \frac{1}{4\pi}(\mathbf{L}\cdot\nabla)\nabla\left(\frac{1}{|\mathbf{x}|}\right) + \frac{1}{4\pi}(\nabla C_{ij})\times\int x_i' x_j'\boldsymbol{\omega}'d\mathbf{x}' + \cdots}. \qquad (8.60)$$

Evidently, the velocity in the far field is $O(|\mathbf{x}|^{-3})$ if $\mathbf{L}$ is finite, but $O(|\mathbf{x}|^{-4})$ when $\mathbf{L} = 0$. We can find the corresponding far-field scalar potential for $\mathbf{u}$ by rewriting (8.60) as

$$\mathbf{u} = \nabla\varphi = \frac{1}{4\pi}\nabla\left[\mathbf{L}\cdot\nabla\left(\frac{1}{|\mathbf{x}|}\right)\right] + \cdots,$$

which tells us that the scalar potential for $\mathbf{u}$ in the far field is $\varphi_\infty = (4\pi)^{-1}\mathbf{L}\cdot\nabla(1/|\mathbf{x}|)$.

8.7.2 The Spontaneous Redistribution of Momentum in Space

It is important to note that, even if a velocity field starts out as localized in space, it does not stay that way for long. Rather, it spontaneously develops a power-law tail in the far field. An inviscid, axisymmetric example of this process is given in Exercise 8.1, where the initial velocity field, in cylindrical polar coordinates, is $u_\theta = \Omega r\exp\left[-|\mathbf{x}|^2\big/\delta^2\right]$, and zero poloidal velocity. Although $\mathbf{u}_p = 0$ at $t = 0$, it immediately grows, since (8.16)

$$\frac{D}{Dt}\left(\frac{\omega_\theta}{r}\right) = \frac{\partial}{\partial z}\left(\frac{u_\theta^2}{r^2}\right),$$

(8.61)

ensures the growth of azimuthal vorticity and this then generates a poloidal flow through

$$\nabla_*^2 \Psi = -r\omega_\theta, \quad \mathbf{u}_p = \nabla \times \left[(\Psi/r)\,\hat{\mathbf{e}}_\theta\right].$$

(8.62)

Solving (8.62) for Ψ, and hence $\mathbf{u}_p$, yields, in the case of Exercise 8.1, a far-field velocity of $|\mathbf{u}_p| \sim |\mathbf{x}|^{-4}$ for $t > 0$, the linear impulse being zero because of symmetry. This spontaneous redistribution of momentum throughout all space is achieved through the non-local pressure field, which is governed by the divergence of the Navier–Stokes equation,

$$\nabla^2 p = -\rho \nabla \cdot (\mathbf{u} \cdot \nabla \mathbf{u}).$$

Notice that velocity and vorticity fields behave very differently in this respect. Vorticity can be redistributed in space only by advection or diffusion, and this means that vorticity fields that start out as localized, stay localized. Indeed, this is the hallmark of all of the examples discussed in §8.3. So, typically, vorticity distributions are localized in space, while their corresponding velocity fields fall off as power laws. One consequence of this is that the integral moments of a vorticity field are usually well behaved, whereas the integral moments of the corresponding velocity field either diverge or are only conditionally convergent. For example, the global linear momentum, $\int \rho \mathbf{u} dV$, can be shown to be conditionally convergent for a localized vorticity distribution, in the sense that its value depends on the way in which the volume of integration is allowed to recede to infinity. Suppose, for example, that we have a vorticity field that is localized near the origin in an unbounded fluid. Let V_∞ be a large control volume whose surface S_∞ is sufficiently far from the vorticity for (8.58) to apply on S_∞. Then

$$\int_{V_\infty} \mathbf{u} dV = \int_{V_\infty} \nabla \times \mathbf{A} dV = -\oint_{S_\infty} \mathbf{A} \times d\mathbf{S} = \frac{1}{4\pi} \oint_{S_\infty} \left[\mathbf{L} \times \nabla\left(\frac{1}{|\mathbf{x}|}\right)\right] \times d\mathbf{S},$$

(8.63)

and it may be shown that the surface integral on the right depends on the shape of S_∞. In order to avoid such ambiguities, it is conventional to work with linear impulse, rather than linear momentum, the two integrals being closely related, as we now show.

8.7.3 Conservation of Linear Impulse and its Relationship to Linear Momentum

Let us now return to the linear impulse $\mathbf{L}$, defined by (8.59). It is readily confirmed that $\mathbf{L}$ is an invariant of the motion associated with a localized distribution of vorticity. Somewhat surprisingly, this means that the leading-order contribution to the far-field velocity is steady, even when the flow itself is unsteady. To show that $\mathbf{L}$ is an invariant we note that (8.2) yields

$$\frac{d\mathbf{L}}{dt} = \frac{1}{2}\int \mathbf{x} \times [\nabla \times (\mathbf{u} \times \boldsymbol{\omega} - \nu \nabla \times \boldsymbol{\omega})]\,dV. \tag{8.64}$$

However, the identity

$$\mathbf{x} \times (\nabla \times \mathbf{G}) = 2\mathbf{G} + \nabla(\mathbf{x} \cdot \mathbf{G}) - \partial(G_i x_j)/\partial x_j$$

tells us that any localized vector field, say $\mathbf{G}$, satisfies

$$\frac{1}{2}\int_{V_\infty} \mathbf{x} \times (\nabla \times \mathbf{G})\,dV = \int_{V_\infty} \mathbf{G}\,dV. \tag{8.65}$$

It follows that

$$\frac{d\mathbf{L}}{dt} = \int_{V_\infty} (\mathbf{u} \times \boldsymbol{\omega} - \nu \nabla \times \boldsymbol{\omega})\,dV = \int_{V_\infty} (\mathbf{u} \times \boldsymbol{\omega})\,dV + \nu \oint_{S_\infty} \boldsymbol{\omega} \times d\mathbf{S}$$

$$= \int_{V_\infty} (\mathbf{u} \times \boldsymbol{\omega})\,dV. \tag{8.66}$$

However, $\mathbf{u} \times \boldsymbol{\omega} = \nabla(u^2/2) - \mathbf{u} \cdot \nabla \mathbf{u}$ can be written as a divergence, and so the final integral of (8.66) converts to a surface integral whose integrand falls off as $O(|\mathbf{x}|^{-6})$ for large $|\mathbf{x}|$. We conclude that $\int \mathbf{u} \times \boldsymbol{\omega}\,dV = 0$ and hence $\mathbf{L}$ is conserved.

The invariance of $\mathbf{L}$ is related to linear momentum conservation, and indeed it may be shown that $\mathbf{L} \sim \int \mathbf{u}\,dV$. We proceed as follows. Let V_R be *any* spherical volume of radius R that is centred on the origin and encloses the vorticity field (Figure 8.27). Then

$$\int_{V_R} \mathbf{u}\,d\mathbf{x} = \int_{V_R} \nabla \times \mathbf{A}\,d\mathbf{x} = -\oint_{S_R} \mathbf{A} \times d\mathbf{S} = -\frac{1}{4\pi}\oint_{S_R}\left[\int_{V_R} \boldsymbol{\omega}(\mathbf{x}')\frac{d\mathbf{x}'}{|\mathbf{x}' - \mathbf{x}|}\right] \times d\mathbf{S}, \tag{8.67}$$

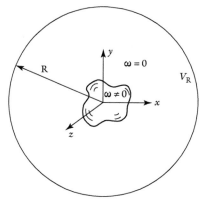

Figure 8.27 A spherical control volume V_R which encloses the vorticity field.

where $\mathbf{x}$ locates a point on the surface S_R and $\mathbf{x}'$ is a point within the interior of V_R. Exchanging the order of integration, we find

$$\int_{V_R} \mathbf{u}\,d\mathbf{x} = \frac{1}{4\pi} \int_{V_R} \left[\oint_{S_R} \frac{d\mathbf{S}}{|\mathbf{x}'-\mathbf{x}|} \right] \times \boldsymbol{\omega}\,(\mathbf{x}')\,d\mathbf{x}' = \frac{1}{3} \int (\mathbf{x}' \times \boldsymbol{\omega}')d\mathbf{x}', \qquad (8.68)$$

since the surface integral in (8.68) is readily shown to be equal to $(4\pi/3)\,\mathbf{x}'$. (See, for example, Jackson, 1999, §5.6.) We conclude that

$$\int_{V_R} \mathbf{u}\,dV = \frac{2}{3}\mathbf{L} = \frac{1}{3} \int (\mathbf{x} \times \boldsymbol{\omega})d\mathbf{x}, \qquad (8.69)$$

for any spherical control volume that encloses the vorticity field. In particular, this holds for the asymptotic case of $R \to \infty$. It is tempting to assume that there is no linear momentum outside V_R in this limit, but this is not the case, as noted by Batchelor (1967). In fact, no matter how large we make V_R, there is always some linear momentum outside V_R. When this external integral is evaluated we find that, for $R \to \infty$, the contribution to $\int \mathbf{u}\,dV$ from outside V_R is also proportional to $\mathbf{L}$, (Batchelor, 1967). In summary, then,

$$\boxed{\mathbf{L} = \frac{1}{2} \int (\mathbf{x} \times \boldsymbol{\omega})d\mathbf{x} = \frac{3}{2} \int_{V_R} \mathbf{u}\,d\mathbf{x} = \text{constant}}. \qquad (8.70)$$

8.7.4 Conservation of Angular Impulse and its Relationship to Angular Momentum

We shall now show that a localized vorticity distribution has a second invariant, which is closely related to the conservation of angular momentum. Our starting point is the identity

$$6\,(\mathbf{x} \times \mathbf{u}) = 2\mathbf{x} \times (\mathbf{x} \times \boldsymbol{\omega}) + 3\nabla \times (x^2\mathbf{u}) - \boldsymbol{\omega} \cdot \nabla\,(x^2\mathbf{x}). \qquad (8.71)$$

Integrating (8.71) over a spherical control volume of radius R which encloses all of the vorticity, we obtain

$$\int_{V_R} \mathbf{x} \times \mathbf{u}\,dV = \frac{1}{3} \int_{V_R} \mathbf{x} \times (\mathbf{x} \times \boldsymbol{\omega})\,d\mathbf{x} - \frac{1}{2}R^2 \oint_{S_R} \mathbf{u} \times d\mathbf{S} = \frac{1}{3} \int \mathbf{x} \times (\mathbf{x} \times \boldsymbol{\omega})\,d\mathbf{x}, \qquad (8.72)$$

where the surface integral converts to a volume integral, which is zero by virtue of (8.55). We now apply conservation of angular momentum in the form (see Exercise 1.8)

$$\frac{d}{dt} \int_{V_R} \mathbf{x} \times \mathbf{u}\,dV = -\oint_{S_R} (\mathbf{x} \times \mathbf{u})\,\mathbf{u} \cdot d\mathbf{S} - \nu \int_{V_R} \mathbf{x} \times (\nabla \times \boldsymbol{\omega})dV = -\oint_{S_R} (\mathbf{x} \times \mathbf{u})\,\mathbf{u} \cdot d\mathbf{S}, \qquad (8.73)$$

the viscous torque acting on V_R being zero because of (8.65) applied to $\boldsymbol{\omega}$, combined with (8.55). In the limit of $R \to \infty$, the surface integral in (8.73) vanishes, leaving $\int \mathbf{x} \times \mathbf{u}dV$ conserved. It follows that

$$\boxed{\mathbf{H} = \frac{1}{3} \int \mathbf{x} \times (\mathbf{x} \times \boldsymbol{\omega})\, d\mathbf{x} = \int_{V_R} \mathbf{x} \times \mathbf{u}dV = \text{constant}}. \qquad (8.74)$$

The integral $\mathbf{H}$ is known as the *angular impulse* of the vorticity distribution. It is sometimes rewritten, using the identity

$$\boldsymbol{\omega} \cdot \nabla \left(\mathbf{x}^2 \mathbf{x}\right) = 3\mathbf{x}^2\boldsymbol{\omega} + 2\mathbf{x} \times (\mathbf{x} \times \boldsymbol{\omega}),$$

as

$$\boxed{\mathbf{H} = -\frac{1}{2} \int \mathbf{x}^2 \boldsymbol{\omega}dV = \text{constant}}. \qquad (8.75)$$

Of course, for an *inviscid* fluid, we can add to $\mathbf{L}$ and $\mathbf{H}$ two other invariants: helicity and kinetic energy. Noting that $\mathbf{u}^2$ and $\boldsymbol{\omega} \cdot \mathbf{A}$ differ by $\nabla \cdot (\mathbf{A} \times \mathbf{u})$, the latter can be written as

$$\boxed{\frac{1}{2} \int \mathbf{u}^2 dV = \frac{1}{2} \int \boldsymbol{\omega} \cdot \mathbf{A}dV = \frac{1}{8\pi} \iint \frac{\boldsymbol{\omega}(\mathbf{x}) \cdot \boldsymbol{\omega}(\mathbf{x}')}{|\mathbf{x}' - \mathbf{x}|} d\mathbf{x}d\mathbf{x}' = \text{constant}}. \qquad (8.76)$$

(inviscid fluid only)

8.7.5 Axisymmetric Examples of Impulse and Vortex Rings Revisited

The expressions above for linear and angular impulse simplify somewhat in axisymmetric flows, $\mathbf{u}(r, z) = \mathbf{u}_\theta + \mathbf{u}_p$, described in terms of cylindrical polar coordinates, (r, θ, z).

Consider first the azimuthal contribution to the velocity field, $\mathbf{u}_\theta$. The corresponding vorticity is poloidal, $\boldsymbol{\omega}_p = \nabla \times \mathbf{u}_\theta$, and since $\mathbf{x} \times \boldsymbol{\omega}_p$ is azimuthal, it makes no contribution to $\mathbf{L}$ in (8.70). On the other hand, (8.75) yields

$$\mathbf{H} = -\frac{1}{2} \int \left(r^2 + z^2\right) \omega_z dV \hat{\mathbf{e}}_z = -\frac{1}{2} \int \left(r^2 + z^2\right) \frac{1}{r}\frac{\partial \Gamma}{\partial r} dV \hat{\mathbf{e}}_z = \int \Gamma dV \hat{\mathbf{e}}_z, \quad (8.77)$$

where $\Gamma = ru_\theta$ is the angular momentum density. The conservation of $\mathbf{H}$ then follows from integrating (3.63) in the form

$$\frac{D\Gamma}{Dt} = \nu\nabla_*^2\Gamma = \nu\nabla \cdot \left[r^2\nabla \left(\frac{\Gamma}{r^2}\right)\right],$$

applying Gauss' theorem, and assuming that Γ, and hence $\boldsymbol{\omega}_p$, is spatially localized.

Conversely, the poloidal contribution to the velocity field has azimuthal vorticity, $\boldsymbol{\omega}_\theta = \nabla \times \mathbf{u}_p$, and so (8.75) tells us that it makes no contribution to $\mathbf{H}$. The linear impulse, on the other hand, is finite and given by

$$\mathbf{L} = \frac{1}{2} \int r\omega_\theta dV \hat{\mathbf{e}}_z. \tag{8.78}$$

For example, it is readily confirmed by direct integration that the linear impulse of Hill's spherical vortex is

$$\mathbf{L} = \frac{1}{2} \int r\omega_\theta dV \hat{\mathbf{e}}_z = -2\pi R^3 V \hat{\mathbf{e}}_z, \tag{8.79}$$

a result that can also be obtained directly from (8.69) applied in a reference frame for which $\mathbf{u} = 0$ in the far field. The conservation of $\mathbf{L}$ in axisymmetric flows follows either from (8.66), or from integrating (3.69) in the form

$$\frac{Dr\omega_\theta}{Dt} = r^2 \frac{D}{Dt}\left(\frac{\omega_\theta}{r}\right) + 2u_r\omega_\theta = \nabla \cdot \left[\nu r^2 \nabla\left(\frac{\omega_\theta}{r}\right) + \left(u_r^2 + u_\theta^2 - u_z^2\right)\hat{\mathbf{e}}_z - 2u_r u_z \hat{\mathbf{e}}_r\right], \tag{8.80}$$

applying Gauss' theorem to the divergence, and assuming that ω_θ is spatially localized.

Perhaps the simplest example of a localized azimuthal vorticity field is the thin-cored vortex ring shown in Figure 8.28. Here the linear impulse is

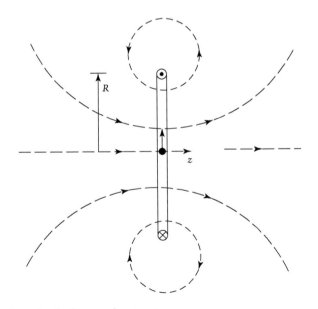

Figure 8.28 Schematic of a thin-cored vortex ring.

$$\mathbf{L} = \frac{1}{2} \int r\omega_\theta dV \hat{\mathbf{e}}_z = (\pi R^2)\Phi\hat{\mathbf{e}}_z, \qquad (8.81)$$

where R is the radius of the centreline of the vortex ring and Φ the flux of vorticity around the ring. Actually, (8.81) is a special case of a more general result. It turns out that the linear impulse of any closed, filamentary vortex loop with flux Φ can be written as

$$\mathbf{L} = \Phi \int_S d\mathbf{S}, \qquad (8.82)$$

where the surface S spans the centreline C of the vortex filament. (This follows from combining the result $\mathbf{x} \times (\boldsymbol{\omega} dV) = \mathbf{x} \times (\Phi d\mathbf{r})$, where $d\mathbf{r}$ is part of C, with the geometrical observation that $\oint_C \mathbf{x} \times d\mathbf{r} = 2\int_S d\mathbf{S}$.) It is sometimes convenient to rewrite (8.81) in terms of the velocity of translation of the vortex ring, say $\mathbf{V}$, as the two properties of a ring that are most readily observed are $\mathbf{V}$ and R. Combining (8.27) with (8.81) yields

$$\mathbf{L} = \frac{4\pi^2 R^3}{\ln(8R/a) - \alpha}\mathbf{V}. \qquad (8.83)$$

The manner in which two coaxial vortex rings exchange linear impulse, leading to the leap-frog shown in Figure 8.12, is discussed in Exercise 8.3.

Sea salps, jellyfish, and squid all take advantage of the finite linear impulse of vortex rings to propel themselves through still water by intermittently expelling such rings. For salps, but perhaps not squid, this is a particularly efficient form of propulsion. However, probably the most dramatic example of naturally occurring vortex rings are those produced by active volcanoes, such as the ring shown in Figure 8.29. Sometimes appearing as precursors to more dramatic activity, such rings can exceed 150 m across.

∘ ∘ ∘

This concludes our discussion of vortex dynamics. Readers seeking more background could do worse than consult Acheson (1990) or Lugt (1983) for an introduction, Batchelor (1967) for a fuller discussion, or Saffman (1992) for a comprehensive overview.

···

EXERCISES

8.1 *The centrifugal bursting of a vortex.* Consider an inviscid, axisymmetric flow whose initial velocity in cylindrical polar coordinates is $u_\theta = \Omega r \exp\left[-(r^2 + z^2)/\delta^2\right]$ and $\mathbf{u}_p = 0$ (zero poloidal velocity). Use (8.16),

$$\frac{D}{Dt}\left(\frac{\omega_\theta}{r}\right) = \frac{\partial}{\partial z}\left(\frac{\Gamma^2}{r^4}\right),$$

Figure 8.29 A vortex ring over Mount Etna. (Image courtesy of Jürg Alean.)

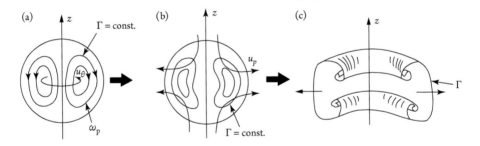

Figure 8.30 The bursting of an axisymmetric vortex. (a) The initial condition has $\mathbf{u}_p = 0$. (b) The poloidal flow sweeps the angular momentum contours radially outward. (c) The asymptotic state takes the form of a poloidal vortex sheet which thins exponentially fast.

to show that, for $t > 0$, we have $\omega_\theta < 0$ for $z > 0$ and $\omega_\theta > 0$ for $z < 0$. Hence show that the poloidal velocity has the form shown in Figure 8.30 (b). This poloidal flow sweeps the angular momentum, $\Gamma = r u_\theta$, radially outward, in accordance with $D\Gamma/Dt = 0$.

Now confirm that

$$\frac{d}{dt} \int_{z<0} (\omega_\theta/r) \, dV = \frac{d}{dt} \int_{z>0} (|\omega_\theta|/r) \, dV = 2\pi \int_0^\infty \left(\Gamma_0^2/r^3\right) dr > 0,$$

where $\Gamma_0(r, t)$ is the angular momentum on the plane $z = 0$. Clearly, the integral of $|\omega_\theta|/r$ rises monotonically. This causes the energy of the poloidal flow to increase steadily at the expense of the kinetic energy of u_θ, which in turn must decline as Γ is swept outward.

Clearly, the vortex is centrifuged radially outward and it may be shown that the angular momentum, Γ, is swept into a thin axisymmetric sheet, which is in fact a poloidal vortex sheet ω_p (see, also, Figure 11.22). This vortex sheet thins exponentially fast and propagates radially outward at a constant speed, as shown in Davidson et al. (2007).

8.2 *The interaction of two line vortices.* Consider two line vortices of the same sign but different strengths, say Φ_1 and Φ_2. Show that the centre of vorticity (8.25) lies on the line joining them and is a distance $\Phi_2\ell/(\Phi_1 + \Phi_2)$ from vortex 1 and $\Phi_1\ell/(\Phi_1 + \Phi_2)$ from vortex 2, where ℓ is the separation of the vortices. Now show that they both move about the centre of vorticity in circular paths with the same angular velocity, $(\Phi_1 + \Phi_2)/2\pi\ell^2$, and that consequently the distance between the vortices remains constant. Sketch the trajectory of the vortices. Find an expression for the dispersion, defined by (8.26), in terms of $\Phi_1, \Phi_2,$ and ℓ and show that it is conserved. What happens when the vortices are of opposite sign?

8.3 *The exchange of linear impulse between vortices.* Consider two non-overlapping vortices of scale ℓ, $\omega^{(1)}$ and $\omega^{(2)}$, whose centres (suitably defined) are separated by the displacement vector $\mathbf{r}$, with $|\mathbf{r}| \gg \ell$. Let $\mathbf{L}^{(1)}$ and $\mathbf{L}^{(2)}$ be the linear impulses of the two vortices and $\mathbf{u}^{(1)}$ and $\mathbf{u}^{(2)}$ the two velocity fields associated with $\omega^{(1)}$ and $\omega^{(2)}$ via the Biot–Savart law. Thus $\nabla \times \mathbf{u}^{(1)} = \omega^{(1)}$ and $\nabla \times \mathbf{u}^{(2)} = \omega^{(2)}$. Also, let $\mathbf{x}$ be the position vector of a point measured in a coordinate system whose origin lies at the centre of vortex 1, and $\mathbf{x}'$ the position of the same point measured in a coordinate system whose origin lies at the centre of vortex 2, with $\mathbf{x}' = \mathbf{x} + \mathbf{r}$. Use the identity $\mathbf{u} \times \omega = \nabla\left(u^2/2\right) - \mathbf{u} \cdot \nabla\mathbf{u}$ to show that

$$\int \mathbf{u}^{(1)} \times \omega^{(1)} dV = \int \mathbf{u}^{(2)} \times \omega^{(2)} dV = 0,$$

and hence deduce

$$\frac{d\mathbf{L}^{(1)}}{dt} = \int \mathbf{u}^{(2)} \times \omega^{(1)} dV, \qquad \frac{d\mathbf{L}^{(2)}}{dt} = \int \mathbf{u}^{(1)} \times \omega^{(2)} dV,$$

with the help of (8.66). Since $\omega_i = \nabla \cdot (\omega x_i)$ ensures that $\int \omega^{(1)} dV = 0$, we can rewrite the first of these expressions as

$$\frac{d\mathbf{L}^{(1)}}{dt} = \int \left(\mathbf{u}^{(2)} - \mathbf{u}_0^{(2)}\right) \times \omega^{(1)} dV,$$

where $\mathbf{u}_0^{(2)}$ is the value of $\mathbf{u}^{(2)}$ at the centre of vortex 1. Write down the equivalent expression for the rate of change of $\mathbf{L}^{(2)}$. Next, recalling that $|\mathbf{r}| \gg \ell$, and that $\mathbf{u}^{(2)}$ is irrotational throughout vortex 1 while $\mathbf{u}^{(1)}$ is irrotational throughout vortex 2, show that

$$\frac{d\mathbf{L}^{(1)}}{dt} = -\left[\left(\mathbf{L}^{(1)} \cdot \nabla\right)\mathbf{u}^{(2)}\right]_{|\mathbf{x}|=0}, \qquad \frac{d\mathbf{L}^{(2)}}{dt} = -\left[\left(\mathbf{L}^{(2)} \cdot \nabla'\right)\mathbf{u}^{(1)}\right]_{|\mathbf{x}'|=0},$$

(Davidson, 2015, p. 209). The significance of these expressions is that a vortex with a finite linear impulse casts a long shadow, so that distant vortices exchange linear impulse.

By way of an example, consider the case of two well-separated, coaxial vortex rings of fluxes Φ_1 and Φ_2 and areas A_1 and A_2. Noting that the velocity on the axis of such a ring is

$$u_z(z) = \frac{\Phi R^2}{2\left(R^2 + z^2\right)^{3/2}} = \frac{|\mathbf{L}|}{2\pi\left(R^2 + z^2\right)^{3/2}}, \quad |\mathbf{L}| = A\Phi = \pi R^2 \Phi,$$

show that

$$\frac{dL^{(2)}}{dt} = -\frac{dL^{(1)}}{dt} = -\frac{3L^{(1)}L^{(2)}}{2\pi}\frac{s}{(s^2)^{5/2}}, \quad \frac{ds}{dt} = V_1 - V_2,$$

where (1) indicates the lead ring, $s(t)$ is the axial separation of the rings, and V_1 and V_2 are the speeds of propagation of the two rings. Hence explain the early stages of behaviour shown in Figures 8.11 and 8.12, in which the rings go on to perform leap-frog. Of course, the equations above break down as the rings get close, as is clear from the s^{-4} behaviour for dL/dt. Confirm that an alternative formulation of the problem, using (8.51) to find u_r, and hence dR_1/dt and dR_2/dt, leads to the more robust set of equations:

$$\frac{dL^{(2)}}{dt} = -\frac{dL^{(1)}}{dt} = -\frac{3L^{(1)}L^{(2)}}{2\pi}\frac{s}{\left((R_1 + R_2)^2 + s^2\right)^{5/2}}, \quad \frac{ds}{dt} = V_1 - V_2.$$

Show, by direct computation, that these equations capture the leap-frog shown in Figure 8.12.

8.4 *Burgers' vortex sheet*. The Cartesian analogue of Burgers' tubular vortex is the vortex sheet,

$$\boldsymbol{\omega} = \frac{\Phi_0}{\sqrt{\pi}\delta}\exp\left[-x^2/\delta^2\right]\hat{\mathbf{e}}_z,$$

subject to the irrotational straining flow $\mathbf{u}^{(\alpha)} = (u_x, 0, u_z) = (-\alpha x, 0, \alpha z)$. (Here Φ_0 is the vorticity flux per unit length of the sheet.) Show that, when α is a constant, and provided we choose $\delta = \sqrt{2\nu/\alpha}$, this represents a steady solution of the vorticity equation in which the advection, stretching, and diffusion of vorticity are all perfectly balanced. However, as Burgers noticed, the net dissipation per unit area of the sheet goes to zero as $\nu \to 0$, and so such vortex sheets cannot represent the centres of dissipation in a turbulent flow.

8.5 *Dissipation in a Burgers vortex.* Confirm the identity

$$2\rho\nu S_{ij}S_{ij} = \rho\nu\omega^2 + 2\rho\nu\nabla\cdot(\mathbf{u}\cdot\nabla\mathbf{u})$$

and show that, for a tubular Burgers vortex,

$$\rho\nu\int_0^\infty \omega^2 2\pi r\,dr = \frac{\rho\alpha\Phi_0^2}{8\pi},$$

which is independent of ν. This is Burgers' expression for the energy dissipation per unit length of the vortex. Now show that, in the far field of the vortex,

$$2\rho\nu\nabla\cdot(\mathbf{u}\cdot\nabla\mathbf{u}) = 3\rho\nu\alpha^2.$$

Confirm that, when integrated over a large cylindrical control volume centred on the vortex, this represents the far-field rate of working of the viscous stresses acting on the surface of the control volume.

8.6 *The centre of vorticity for an inviscid, two-dimensional vorticity field.* Show that, for a localized, two-dimensional vorticity field $\boldsymbol{\omega}(x,y) = \omega\hat{\mathbf{e}}_z$, the Biot–Savart law (8.1) may be integrated over z to give

$$\mathbf{u}(\mathbf{x}) = \frac{1}{2\pi}\int\frac{\boldsymbol{\omega}(\mathbf{x}')\times\mathbf{r}}{r^2}\,dA', \quad \mathbf{r} = \mathbf{x} - \mathbf{x}',$$

where $\mathbf{x}$ and $\mathbf{x}'$ are now two-dimensional position vectors. Now show that the inviscid vorticity equation, $D\omega/Dt = 0$, ensures that $\int\omega\,dA$ is an invariant, and also that

$$\frac{d}{dt}\int x\omega\,dA = -\int x(\mathbf{u}\cdot\nabla\omega)\,dA = \int u_x\omega\,dA.$$

Substitute for u_x using the Biot–Savart law, to show that $\int x\omega\,dA = $ constant, as is $\int y\omega\,dA$. (This is the two-dimensional version of conservation of linear impulse, as discussed in §8.7.3.) It follows that

$$x_c = \int x\omega\,dA \bigg/ \int\omega\,dA, \quad y_c = \int y\omega\,dA\bigg/\int\omega\,dA,$$

are invariants of the motion. This defines the *centre of vorticity*.

8.7 *A generalization of Hill's spherical vortex to include swirl.* Consider an inviscid, axisymmetric flow with vorticity which is confined to the sphere $|\mathbf{x}| = R$ and governed by (8.18). Like Hill's spherical vortex, this is to be matched to an external irrotational cross flow $V\hat{\mathbf{e}}_z$ whose Stokes streamfunction is

$$\Psi = \frac{1}{2}Vr^2\left(1 - R^3\big/|\mathbf{x}|^3\right).$$

Hill's spherical vortex corresponds to $H = -C\Psi$, where C is a constant, and to $\Gamma = 0$ (or equivalently, $\boldsymbol{\omega}_p = 0$). Consider the generalized case in which $H = -C\Psi$ and $\Gamma = \pm\alpha\Psi$ (or equivalently, $\boldsymbol{\omega}_p = \pm\alpha\mathbf{u}_p$), where α is a constant. Show that (8.18) takes the form

$$\nabla^2_*\Psi = -Cr^2 - \alpha^2\Psi,$$

and that this has the general solution

$$\Psi = -\frac{Cr^2}{\alpha^2}\left[1 - \frac{R^{3/2}}{|\mathbf{x}|^{3/2}}\frac{J_{3/2}(\alpha|\mathbf{x}|)}{J_{3/2}(\alpha R)}\right].$$

Find the relationship between V/CR^2 and αR that matches this solution to the external irrotational flow.

....

REFERENCES

Acheson, D.J., 1990, *Elementary fluid dynamics*, Oxford University Press.

Batchelor, G.K., 1956, On steady laminar flow with closed streamlines at large Reynolds number. *J. Fluid Mech.*, **1**(2), 177–90.

Batchelor, G.K., 1967, *An introduction to fluid dynamics*, Cambridge University Press.

Burgers, J.M., 1948, A mathematical model illustrating the theory of turbulence. *Adv. Appl. Mech.*, **1**, 171.

Davidson, P.A., 2013, *Turbulence in rotating, stratified and electrically conducting fluids*, Cambridge University Press.

Davidson, P.A., 2015, *Turbulence: an introduction for scientists and engineers*, 2nd Ed., Oxford University Press.

Davidson, P.A., Sreenivasan, B., & Aspen, A.J., 2007, Evolution of localised blobs of swirling or buoyant fluid. *Physical Review E*, **75**(2), 026304.

Didden, N., 1977, Mitt. Max-Plank-Inst. Stromungsforsch. Aerodyn. Versuchsanst, 64.

Helmholtz, H., 1858, On integrals of the hydrodynamical equations which express vortex motion. *J. für die reine and angewandte Mathematik*, **55**, 25–55. (Translated into English by P.G. Tait in: *Phil Mag.* (series 4), 1867, **33**, 485–512.)

Jackson, J.D., 1999, *Classical electrodynamics*, 3rd Ed., Wiley.

Japan Society of Mechanical Engineers, 1988, *Visualised flow*, Pergamon Press.

Lugt, H.J., 1983, *Vortex flows in nature and technology*, Wiley.

Pullin, D.I. & Perry, A.E., 1980, Some flow visualisation experiments on the starting vortex. *J. Fluid Mech.*, **97**, 239–55.

Saffman, P.G., 1992, *Vortex dynamics*, Cambridge University Press.

Yamada, H. & Matsui, T., 1978, Preliminary study of mutual slip-through of a pair of vortices. *Phys. Fluids*, **21**, 292–94.

9

· · • · ·

Waves and Flow in a Stratified Fluid

Vilhelm Bjerknes (1862–1951) was a Norwegian physicist who established the Bergen school of meteorology. Highly influential, he was one of the first to promote the possibility of numerical weather forecasting and to appreciate the significance of internal gravity waves. Many of his students, such as the Swedish oceanographer Vagn Walfrid Ekman and meteorologist Carl-Gustaf Rossby, also made major contributions to the field.

9.1 The Boussinesq Approximation and a Second Definition of the Froude Number

We now consider motion in a fluid whose density varies with height, as is the case in the atmosphere and in the oceans. We shall restrict the discussion to incompressible fluids and to a uniform, stable stratification, in which the density decreases uniformly with height, z. We shall also adopt the so-called *Boussinesq approximation*, in which the variations in density are sufficiently small that they can be ignored in the momentum equation, except to the extent that they introduce a buoyancy force. (In fact, it was the German physicist Anton Oberbeck, rather than Boussinesq, who first introduced this approximation, in 1879.)

Let the density of the unperturbed fluid be $\rho_0(z)$, where $d\rho_0/dz$ is uniform and negative and the corresponding hydrostatic pressure gradient is $\nabla p_0 = \rho_0(z)\mathbf{g}$. The density of the perturbed fluid is then $\rho(\mathbf{x}, t) = \rho_0(z) + \rho'$, where $\rho' \ll \rho_0$, and incompressibility in the form of $D\rho/Dt = 0$ gives us

$$\frac{D\rho'}{Dt} + u_z\frac{d\rho_0}{dz} = 0. \tag{9.1}$$

As usual, the mass conservation equation, (1.13), applied to an incompressible fluid requires $\nabla \cdot \mathbf{u} = 0$, while the Navier–Stokes equation in the Boussinesq limit can be written as

$$\bar{\rho}\frac{D\mathbf{u}}{Dt} = -\nabla p + \rho'\mathbf{g} + \bar{\rho}\nu\nabla^2\mathbf{u}, \tag{9.2}$$

where $\bar{\rho}$ is a constant representative of the mean density and it is understood that p represents the departure from the hydrostatic pressure, p_0. The corresponding vorticity equation is, of course,

$$\frac{D\boldsymbol{\omega}}{Dt} = (\boldsymbol{\omega} \cdot \nabla)\mathbf{u} + \nabla(\rho'/\bar{\rho}) \times \mathbf{g} + \nu\nabla^2\boldsymbol{\omega}. \tag{9.3}$$

Note that the buoyancy force does not directly influence the vertical component of vorticity, which is a point to which we shall return.

We now introduce the *Väisälä–Brunt frequency*, N, defined by

$$N^2 = -\frac{g}{\bar{\rho}}\frac{d\rho_0}{dz}, \tag{9.4}$$

so that (9.1) can be rewritten as

$$\frac{D\rho'}{Dt} = \frac{\bar{\rho}N^2}{g}u_z. \tag{9.5}$$

Note that N, which is a measure of the strength of the stratification, has the dimensions of a frequency. Indeed, we shall see that the frequency of internal gravity waves is of order N, so the speed of propagation of such waves is of order $N\lambda$, where λ is their wavelength.

If u and ℓ are typical velocity and length scales, then $u/N\ell$ is clearly dimensionless, and it is called the *internal Froude number* by analogy with the definition $\mathrm{Fr} = u/\sqrt{gh}$ for free-surface flows. That is to say, for both stratified and shallow-water flow, the appropriate Froude number represents the ratio of u to a wave speed. (See §1.4.2 for a discussion of Fr in free-surface flow.) By convention, the same symbol Fr is used for *both* dimensionless groups.

It is common to provide an alternative interpretation of Fr in terms of a ratio of forces. In particular, a rough estimate of the ratio of inertial to buoyancy forces in (9.2), in which (9.5) is used to estimate ρ', suggests that $(u/N\ell)^2$ is a measure of the ratio of these forces. However, great care must be made in making such estimates. The point is that, as we shall see shortly, the hallmark of a strongly stratified flow is that motion in the vertical direction tends to be supressed relative to the horizontal motion. This has three immediate consequences for any scaling analysis. First, we need to distinguish carefully between the *intrinsic* length scales of the motion in the vertical and horizontal directions, say $\ell_\parallel$ and $\ell_\perp$, where we might anticipate that $\ell_\parallel \ll \ell_\perp$ in a strongly stratified flow. (The subscripts here indicates directions parallel and perpendicular to $\mathbf{g}$.) Second, since $\nabla \cdot \mathbf{u} = 0$ requires $u_\perp/\ell_\perp \sim u_\parallel/\ell_\parallel$, we must distinguish between the characteristic speeds in the vertical and horizontal directions, with $u_\parallel \ll u_\perp$ in the case of a strong stratification. Third, we must distinguish between $\ell_\parallel$ and $\ell_\perp$, the *intrinsic scales of the motion*, and L, an imposed *geometric length scale*.

All of this suggests that we must be careful as to which velocity and length scale we use in the definition of the Froude number. It turns out to be convenient to introduce two different definitions:

$$\boxed{\text{Fr} = u_\perp/NL}, \tag{9.6}$$

which is based on the imposed geometric length scale, and its cousin

$$\text{Fr}_\parallel = u_\perp/N\ell_\parallel, \tag{9.7}$$

which is the Froude number based on the intrinsic vertical length scale of the motion. We shall see shortly that Fr $\ll 1$ constitutes the case of strong stratification, in the sense that this

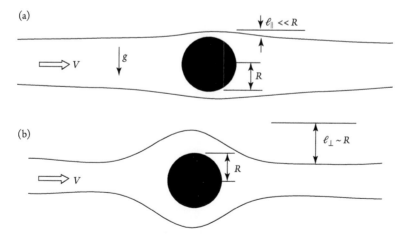

Figure 9.1 Strongly stratified flow past a sphere: (a) side view; (b) plan view. Most of the fluid flows around the sides of the sphere in horizontal planes and very little passes above or below it. Evidently $\ell_\parallel \ll R$ and $u_z \ll V$. (Adapted from Davidson, 2013.)

is usually sufficient to ensure $\ell_\| \ll \ell_\perp$. Moreover, for three-dimensional flow we typically find $\ell_\perp \sim L$, so that $\text{Fr} \sim u_\perp / N\ell_\perp$, though this *not* always true in two dimensions.

This disparity in length scales is illustrated in Figure 9.1, which shows strongly stratified flow past a sphere of radius R. The motion here is steady and $\text{Fr} = V/NR \ll 1$, V being the oncoming speed of the flow. Most of the fluid flows around the sides of the sphere and very little passes above or below it, with the flow resembling a stack of partially decoupled layers of almost horizontal motion. In this example, then, $\ell_\perp$ is set by the imposed geometric length scale, $\ell_\perp \sim R$, while $u_\perp$ is set by the imposed velocity, $u_\perp \sim V$. However, $\ell_\|$ and $u_\|$ are determined by the internal dynamics, with $\ell_\| \ll R$ and $u_z \ll V$.

9.2 The Suppression of Vertical Motion: a Simple Scaling Analysis

Let us explore this suppression of vertical motion in a little more detail. If we ignore the viscous stresses on the grounds that Re is sufficiently large, then outside the boundary layer the steady flow shown in Figure 9.1 is governed by

$$(\mathbf{u} \cdot \nabla)\rho' = \left(\frac{\bar{\rho}N^2}{g} \right) u_z, \tag{9.8}$$

and

$$(\mathbf{u} \cdot \nabla)\boldsymbol{\omega} = (\boldsymbol{\omega} \cdot \nabla)\mathbf{u} + \nabla(\rho'/\bar{\rho}) \times \mathbf{g}. \tag{9.9}$$

Since continuity requires $u_\perp / \ell_\perp \sim u_\| / \ell_\|$, these yield the estimates

$$\rho' \sim \left(\frac{\bar{\rho}N^2}{g} \right) \ell_\|, \tag{9.10}$$

and

$$\frac{u_\perp}{\ell_\perp}\omega_\perp \sim \frac{\omega_\perp}{\ell_\perp}u_\perp + \frac{g\rho'}{\bar{\rho}\ell_\perp}, \tag{9.11}$$

and given that $\omega_\perp \sim u_\perp / \ell_\|$, (9.11) can be rewritten as

$$\frac{u_\perp^2}{\ell_\|} \sim \frac{u_\perp^2}{\ell_\|} + \frac{g\rho'}{\bar{\rho}}. \tag{9.12}$$

We now *define* strongly stratified flow as that motion for which the gravitational term in estimate (9.12) plays a leading-order role and in which $\ell_\| \ll \ell_\perp$. For such flows, the elimination of ρ' from (9.10) and (9.12) gives us the scaling $u_\perp \sim N\ell_\|$, and with the help of $u_\perp / \ell_\perp \sim u_\| / \ell_\|$ we find

$$\boxed{\frac{u_{\parallel}}{u_{\perp}} \sim \frac{\ell_{\parallel}}{\ell_{\perp}} \sim \frac{u_{\perp}}{N\ell_{\perp}} \sim \text{Fr}}\,, \tag{9.13}$$

(strongly stratified 3D flow with negligible viscous stresses)

where we have assumed that $\ell_{\perp} \sim R$. Expression (9.13) tells us that, when $\text{Fr} \ll 1$, we do indeed have strong stratification, in the sense that $\ell_{\parallel} \ll \ell_{\perp}$. It also tells us that $\text{Fr}_{\parallel} = u_{\perp}/N\ell_{\parallel} \sim 1$, irrespective of the value of Fr, something the internal dynamics ensures by fixing the vertical scale of the motion in accordance with $\ell_{\parallel} \sim \ell_{\perp}\text{Fr}$.

It is important to note the caveat in (9.13) of negligible viscous stresses. If we retrospectively determine the ratio of inertia forces to viscous stresses in the vorticity equation, then we find that

$$\frac{\text{inertial forces}}{\text{viscous forces}} \sim \frac{u_{\perp}/\ell_{\perp}}{\nu/\ell_{\parallel}^2} \sim \frac{u_{\perp}\ell_{\perp}}{\nu}\frac{\ell_{\parallel}^2}{\ell_{\perp}^2} \sim \text{Re}_{\perp}\text{Fr}^2, \tag{9.14}$$

where $\text{Re}_{\perp} = u_{\perp}\ell_{\perp}/\nu$ and we have used (9.13) to estimate the ratio of the length scales. The quantity $\Re = \text{Re}_{\perp}\text{Fr}^2$ is known as the *buoyancy Reynolds number*, and the requirement that viscous forces are negligible in a strongly stratified flow is evidently

$$\Re = \text{Re}_{\perp}\text{Fr}^2 \gg 1.$$

This is a more stringent requirement than $\text{Re}_{\perp} \gg 1$.

Let us now consider the opposite case, in which viscous forces dominate over inertia. By balancing buoyancy against viscous forces in the vorticity equation, rather than against inertia, we find

$$\frac{g\rho'}{\bar{\rho}\ell_{\perp}} \sim \nu\frac{\omega_{\perp}}{\ell_{\parallel}^2} \sim \nu\frac{u_{\perp}/\ell_{\parallel}}{\ell_{\parallel}^2}.$$

Eliminating ρ' using (9.10) now yields an alternative estimate for $\ell_{\parallel}/\ell_{\perp}$,

$$\frac{\ell_{\parallel}^4}{\ell_{\perp}^4} \sim \frac{\nu u_{\perp}}{N^2\ell_{\perp}^3} \sim \frac{\text{Fr}^2}{\text{Re}_{\perp}}, \tag{9.15}$$

where, as before, we take $\ell_{\perp} \sim R$ and $\text{Re}_{\perp} = u_{\perp}\ell_{\perp}/\nu$. So, for highly viscous flow we have

$$\boxed{\frac{\ell_{\parallel}}{\ell_{\perp}} \sim \frac{u_{\parallel}}{u_{\perp}} \sim \frac{\text{Fr}^{1/2}}{\text{Re}_{\perp}^{1/4}}}\,. \tag{9.16}$$

(strongly stratified 3D flow with dominant viscous stresses)

Now the assertion that the viscous forces dominate over inertia in the vorticity equation requires

$$\nu \frac{\omega_\perp}{\ell_\parallel^2} \gg \frac{u_\perp \omega_\perp}{\ell_\perp},$$

which combines with (9.16) to give

$$\Re = \mathrm{Re}_\perp \mathrm{Fr}^2 \ll 1. \tag{9.17}$$

Once again, it is the buoyancy Reynolds number, rather than $\mathrm{Re}_\perp$, which is important. Estimate (9.17) suggests that, in a strongly stratified flow with $\mathrm{Fr} \ll 1$, we can have dominant viscous forces even when $\mathrm{Re}_\perp > 1$. We conclude that great care must be exercised if it is proposed to neglect viscous forces. (See, also, Exercise 9.1 on this point.)

9.3 The Phenomenon of Blocking

Suppose the sphere in Figure 9.1 is replaced by a circular cylinder whose axis is horizontal and which extends right across the flow. The motion is then two dimensional, say in the x–z plane, and the fluid is no longer free to flow around the object in horizontal planes. Since vertical motion is, to a large extent, suppressed by a strong stratification, one might wonder how the fluid manages to pass over the cylinder. The answer, which is a little surprising, is shown in Figure 9.2. It turns out that an extended region of stagnant fluid forms ahead of the cylinder and the fluid particles slowly drift upwards (or downwards) until they are able to pass over the top (or bottom) of the cylinder. This phenomenon is known as *blocking*.

   Let us see if we can estimate the length of the blocked region, L_b. As before, (9.8) provides us with an estimate of ρ', in this case

$$(V/L_b)\rho' \sim \left(\frac{\bar{\rho} N^2}{g}\right) u_z \sim \left(\frac{\bar{\rho} N^2}{g}\right) \frac{VR}{L_b},$$

from which we find

$$\rho' \sim \left(\frac{\bar{\rho} N^2}{g}\right) R. \tag{9.18}$$

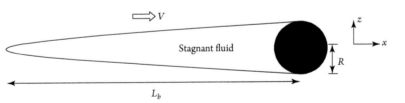

Figure 9.2 Stratified flow past a long cylinder, showing the upstream blocked region. (Adapted from Davidson, 2013.)

Moreover, the two-dimensional version of the steady vorticity equation is

$$(\mathbf{u} \cdot \nabla) \omega_y = \frac{g}{\bar{\rho}} \frac{\partial \rho'}{\partial x} + \nu \nabla^2 \omega_y, \quad \omega_y \sim V/R, \tag{9.19}$$

which combines with (9.18) to give the force balance

$$\frac{V}{L_b} \frac{V}{R} \sim \frac{N^2 R}{L_b} + \frac{\nu}{R^2} \frac{V}{R}. \tag{9.20}$$

Dividing through by $N^2 R$ gives us

$$\frac{\mathrm{Fr}^2}{L_b} \sim \frac{1}{L_b} + \frac{1}{R} \frac{\mathrm{Fr}^2}{\mathrm{Re}}, \tag{9.21}$$

where $\mathrm{Fr} = V/NR$ and $\mathrm{Re} = VR/\nu$. When the stratification is strong we have $\mathrm{Fr} \ll 1$ and so, according to (9.21), inertia cannot balance the buoyancy force. The leading-order force balance is then between buoyancy and the viscous stresses, which requires

$$L_b \sim R \frac{\mathrm{Re}}{\mathrm{Fr}^2}. \tag{9.22}$$

Evidently, for small Fr and large Re, the blocked region can become extremely elongated. Indeed, in the limit of $\mathrm{Fr} \ll 1$ and $\mathrm{Re} \to \infty$, the stagnant region extends out to infinity and the flow upstream of the cylinder is characterized by $\partial u_x / \partial x = 0$. If we now change frame of reference, so that the cylinder moves to the left at the speed V while the fluid at infinity is quiescent, we come to the surprising conclusion that a cylinder translating slowly through a strongly stratified fluid will push an extended layer of fluid ahead of it, and for $\mathrm{Re} \to \infty$ that layer extends out to infinity. It is natural to ask how the fluid at large distances from the cylinder knows about the existence of the moving cylinder, and here we get a second surprise: we shall see that the information is transmitted out to the far field by internal gravity waves.

9.4 Lee Waves

We now turn to waves in a stratified fluid, which of course arise because fluid parcels displaced from their equilibrium position want to return to that height, with inertia ensuring an overshoot and subsequent oscillations. Perhaps the simplest type of internal gravity wave to analyse are two-dimensional *lee waves*.

9.4.1 Linear Lee Waves in Two Dimensions

Lee waves, sometimes called mountain waves, are commonly observed on the *downwind* side of a mountain range. The most striking cases occur when the wave crests cause condensation and hence oscillations in the cloud pattern, thus allowing the waves to be visualized

Figure 9.3 Lee waves made visible by the oscillations in the cloud pattern. (Image courtesy of NOAA and the Comet Program.)

(see Figure 9.3). Triggered by the wind crossing the mountain ridge, these waves are stationary relative to the ground, which means that their phase velocity (the velocity of the wave crests relative to the fluid) is directed upstream, and equal but opposite to the wind speed.

Before discussing lee waves in the atmosphere, let us consider how such waves might be generated in the laboratory. The steady, two-dimensional motion shown in Figure 9.4 consists of the initially uniform flow V of a stratified fluid in a channel of height H. A small, two-dimensional object is placed on the base of the channel, of length b and height profile $h(x)$, where $h \ll H$ and x is the stream-wise coordinate. If the Froude number based on h exceeds unity, $V/Nh > 1$, then there will be negligible blocking of the upstream flow, and if $\mathrm{Fr} = V/NH < \pi^{-1}$, then stationary lee waves of small amplitude develop downstream of the object. We shall focus on the regime where $h \ll H$ and $V/Nh > 1$, but with $V/NH \ll 1$.

Ignoring viscous stresses, these steady waves are governed by the two-dimensional versions of (9.8) and (9.9). These are

$$(\mathbf{u} \cdot \nabla)\rho' = \left(\frac{\bar{\rho} N^2}{g} \right) u_z \quad \text{and} \quad (\mathbf{u} \cdot \nabla)\omega_y = \frac{g}{\bar{\rho}} \frac{\partial \rho'}{\partial x}, \tag{9.23}$$

which, for small-amplitude waves, linearize to give

$$V \frac{\partial \rho'}{\partial x} = \left(\frac{\bar{\rho} N^2}{g} \right) u_z \quad \text{and} \quad V \frac{\partial \omega_y}{\partial x} = \frac{g}{\bar{\rho}} \frac{\partial \rho'}{\partial x}. \tag{9.24}$$

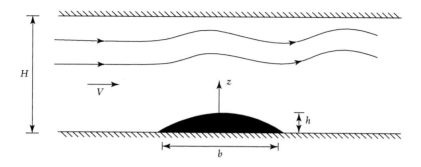

Figure 9.4 Two-dimensional flow over an obstruction in a channel generates lee waves.

Eliminating ρ' now yields

$$\frac{\partial \omega_y}{\partial x} = \frac{\partial}{\partial x}\left(\frac{\partial u_x}{\partial z} - \frac{\partial u_z}{\partial x}\right) = \frac{N^2}{V^2} u_z, \tag{9.25}$$

and $\nabla \cdot \mathbf{u} = 0$ allows us to rewrite (9.25) as

$$\nabla^2 u_z + (N/V)^2 u_z = 0. \tag{9.26}$$

This supports solutions of the form $u_z \sim \cos(k_x x)\sin(k_z z)$ provided that $k_z = n\pi/H$, which ensures that $u_z = 0$ at the top $(z = H)$ and bottom $(z = 0)$ of the channel, n being an integer. Moreover, (9.26) requires

$$k_x^2 + k_z^2 = N^2/V^2, \tag{9.27}$$

and so the *free modes* in a channel must satisfy

$$(k_x H)^2 = \mathrm{Fr}^{-2} - (n\pi)^2, \tag{9.28}$$

where $\mathrm{Fr} = V/NH$. We conclude that stationary waves can form provided $\mathrm{Fr} < 1/\pi$, with only one allowable stream-wise wavenumber in the range $1/2\pi < \mathrm{Fr} < 1/\pi$, but multiple values of k_x for lower values of Fr. (Waves can also form for $\mathrm{Fr} > 1/\pi$, but in such cases V is too high to allow stationary waves, and instead the disturbances are swept downstream.)

It is instructive to consider these waves in a frame of reference moving with the mean flow. In such a frame we have

$$u_z \sim \sin(k_z z)\cos\left(k_x(x + Vt)\right) = \sin(k_z z)\cos\left(k_x x - \varpi t\right)), \tag{9.29}$$

where V is now the phase speed in the negative x direction and ϖ the angular frequency, which is positive by convention. It follows that

$$\varpi = -V k_x = -\frac{N k_x}{\sqrt{k_x^2 + k_z^2}}, \quad k_x < 0. \tag{9.30}$$

The stream-wise group velocity (the velocity at which wave energy propagates relative to the fluid) is then

$$c_{g,x} = \frac{\partial \varpi}{\partial k_x} = -\frac{Nk_z^2}{(k_x^2 + k_z^2)^{3/2}} = -V\frac{k_z^2}{k_x^2 + k_z^2}, \tag{9.31}$$

which is also in the negative x direction relative to the fluid. Crucially, $c_{g,x}$ has a magnitude which is less than V, so when we return to a frame of reference attached to the obstruction in the channel, the wave energy is swept downstream. This is why lee waves cannot form upstream of the object that triggers them.

Thus far we have considered only the free modes. The question now arises as to which combination of free modes will be generated by a given obstruction. This is an altogether more complicated question, as it requires the imposition of the linearized boundary condition $u_z = V dh/dx$ at $z = 0$ in the vicinity of the obstruction. (Because we have linearized the problem, this boundary condition is applied at $z = 0$, rather than $z = h$.) We shall not pause to solve this problem but merely note that, perhaps not surprisingly, the wave amplitude is largest when the length of the object is comparable with the stream-wise wavelength of one of the allowable free modes, $k_x b \sim \pi$, so that a near-resonance occurs.

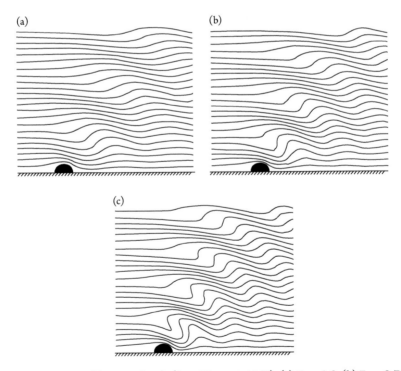

Figure 9.5 Lee waves of finite amplitude (from Huppert, 1968). (a) Fr $= 1.0$, (b) Fr $= 0.79$, and (c) Fr $= 2/3$, where Fr $= V/NR$ and R is the radius of the cylinder.

We now turn to lee waves in the atmosphere, where there is no abrupt upper boundary. Here, wave energy travels upward as well as in the horizontal direction, as illustrated in Figure 9.5. If allowance is made for the variation in wind speed with height, then the background vorticity, dV/dz, leads to an additional term on the left of the linearized vorticity equation, which now reads

$$V\frac{\partial \omega_y}{\partial x} + u_z \frac{d^2 V}{dz^2} = \frac{g}{\bar{\rho}}\frac{\partial \rho'}{\partial x}. \tag{9.32}$$

Stationary lee waves are then governed by a generalization of (9.26):

$$\nabla^2 u_z + \left(\frac{N^2}{V^2} - \frac{1}{V}\frac{d^2 V}{dz^2}\right)u_z = 0. \tag{9.33}$$

Clearly, something singular must occur for heights at which the wind reverses direction, i.e. $V = 0$. It turns out that, on approaching such a critical altitude, k_z becomes extremely large so the Laplacian in (9.33) can continue to balance the second term on the left (Turner, 1973). This rise in k_z causes the vertical component of the group velocity to fall to zero, and so wave energy approaching a critical layer from below cannot pass through that layer. Instead, large wave amplitudes build up in the vicinity of a critical layer, and this can cause the waves to break and turbulence to form.

9.4.2 Finite-amplitude Lee Waves in Two Dimensions

Somewhat surprisingly, there exist exact analytical solutions for finite-amplitude lee waves in an inviscid fluid, provided that those waves are steady and two-dimensional. We proceed as follows. If the motion is two-dimensional then $\nabla \cdot \mathbf{u} = 0$ allows us to introduce a stream-function, ψ, defined by $\mathbf{u} = \nabla \times (\psi \hat{\mathbf{e}}_y)$. Moreover, incompressibility gives us $\mathbf{u} \cdot \nabla \rho = 0$ in a steady flow, and so the *total density* is constant along streamlines. It follows that we can write $\rho = \rho(\psi)$ since, by definition, $\mathbf{u} \cdot \nabla \psi = 0$. We now introduce

$$\rho^* = \rho - \bar{\rho} = (\rho_0(z) - \bar{\rho}) + \rho' = \frac{d\rho_0}{dz}z + \rho' = \rho' - \frac{\bar{\rho}N^2}{g}z, \tag{9.34}$$

where we have taken $\bar{\rho} = \rho_0(z = 0)$. This also satisfies $\mathbf{u} \cdot \nabla \rho^* = 0$, and so $\rho^* = \rho^*(\psi)$.

Next, we note that the steady, inviscid version of the momentum equation, (9.2), can be written as

$$\bar{\rho}(\mathbf{u} \cdot \nabla)\mathbf{u} = -\nabla(p + p_0) + \rho \mathbf{g}, \tag{9.35}$$

where $p + p_0$ is the total pressure, including the hydrostatic contribution, and p is the departure from p_0. This, in turn, can be rewritten with the aid of the identities

$$(\mathbf{u} \cdot \nabla)\mathbf{u} = \nabla\left(\tfrac{1}{2}u^2\right) - \mathbf{u} \times \boldsymbol{\omega}, \quad \text{and} \quad \mathbf{g} = -\nabla(gz),$$

as

$$\bar{\rho}\boldsymbol{\omega} \times \mathbf{u} = -\nabla \left(p + p_0 + \frac{1}{2}\bar{\rho}\mathbf{u}^2 \right) - \rho\nabla(gz),$$

or, equivalently,

$$\bar{\rho}\boldsymbol{\omega} \times \mathbf{u} = -\nabla \left(p + p_0 + \rho gz + \frac{1}{2}\bar{\rho}\mathbf{u}^2 \right) + gz\nabla(\rho^*). \tag{9.36}$$

We now introduce a Bernoulli-like function defined by

$$H^* = (p_0 + \rho_0 gz) + p + \rho' gz + \frac{1}{2}\bar{\rho}\mathbf{u}^2, \tag{9.37}$$

so that (9.36) becomes

$$\bar{\rho}\boldsymbol{\omega} \times \mathbf{u} = -\nabla (H^*) + gz\nabla(\rho^*). \tag{9.38}$$

Since $\mathbf{u} \cdot \nabla\rho^* = 0$, we conclude that $\mathbf{u} \cdot \nabla H^* = 0$ and so H^* is also constant along the streamlines, $H^* = H^*(\psi)$. Noting that $\nabla p_0 = \rho_0(z)\mathbf{g}$ integrates to give

$$p_0 + \rho_0 gz = -\frac{1}{2}\bar{\rho}N^2 z^2,$$

we can rewrite H^* as

$$H^*(\psi) = p + \frac{1}{2}\bar{\rho}\mathbf{u}^2 + \rho' gz - \frac{1}{2}\bar{\rho}N^2 z^2. \tag{9.39}$$

The final step is to observe that the z-component of (9.38) gives us

$$-\bar{\rho}\omega_y u_x = \bar{\rho}\omega_y \frac{\partial\psi}{\partial z} = -\frac{dH^*}{d\psi}\frac{\partial\psi}{\partial z} + gz\frac{d\rho^*}{d\psi}\frac{\partial\psi}{\partial z},$$

and dividing through by u_x yields

$$\bar{\rho}\omega_y = -\frac{dH^*}{d\psi} + gz\frac{d\rho^*}{d\psi}, \tag{9.40}$$

or, since $\omega_y = -\nabla^2\psi$,

$$\boxed{\bar{\rho}\nabla^2\psi = \frac{dH^*}{d\psi} - gz\frac{d\rho^*}{d\psi}.} \tag{9.41}$$

This is the key result: if $\rho^* = \rho^*(\psi)$ and $H^* = H^*(\psi)$ are specified at some upstream location, then (9.41) may be solved for ψ at all downstream points, at least in principle. Note that this is a fully non-linear result, reminiscent of the Squire–Long equation, (8.18). It is sometimes called Yih's equation, after the Chinese-American engineer Chia-Shun Yih.

In the case of two-dimensional lee waves triggered by uniform flow over an obstruction, the upstream conditions require $\psi = -Vz$ and also

$$H^*(\psi) = \frac{1}{2}\bar{\rho}V^2 - \frac{1}{2}\bar{\rho}\frac{N^2}{V^2}\psi^2, \quad \rho^* = -\frac{\bar{\rho}N^2}{g}z = \frac{\bar{\rho}N^2}{gV}\psi. \tag{9.42}$$

It follows that, downstream of an obstruction or mountain ridge, we have

$$\frac{dH^*}{d\psi} = -\bar{\rho}\frac{N^2}{V^2}\psi, \quad \frac{d\rho^*}{d\psi} = \frac{\bar{\rho}N^2}{gV}. \tag{9.43}$$

If we write the downstream streamfunction as $\psi = -Vz + \psi'$, where ψ' is the departure from the upstream value, then (9.41) and (9.43) combine to yield

$$\nabla^2\psi' + (N/V)^2\,\psi' = 0, \tag{9.44}$$

which is the same as the governing equation for linear lee waves, (9.26). Crucially, however, (9.44) is not subject to any limitations in wave amplitude.

There are, of course, many points of contact between the discussion above and that offered in §8.2.3, where we deduced the governing equation for finite-amplitude inertial waves in a fluid subject to background rotation. This is no coincidence. We shall see in Chapter 10 that there are many analogies, both exact and approximate, between stratified and rapidly rotating fluids, with the Coriolis force in a rotating fluid often playing a role analogous to that of the buoyancy force in a stratified fluid.

9.5 Internal Gravity Waves of Small Amplitude

9.5.1 Linear Theory and Simple Examples

We now generalize our discussion of internal gravity waves to unsteady, three-dimensional wave motion. In order to keep the discussion brief, we shall assume that there is no background flow, the waves are of small amplitude, the fluid is inviscid, and N is uniform. In this section we establish the governing equations and focus on waves radiating from a localized source, while in §9.5.2 we discuss the reflection of plane waves from boundaries.

Our governing equations are the linearized versions of (9.3) and (9.5) in which $\mathbf{u}$ and ρ' are assumed small:

$$\frac{\partial\boldsymbol{\omega}}{\partial t} = \nabla\left(\rho'/\bar{\rho}\right) \times \mathbf{g} \quad \text{and} \quad \frac{\partial\rho'}{\partial t} = \frac{\bar{\rho}N^2}{g}u_z. \tag{9.45}$$

Note that (9.45) requires $\partial\omega_z/\partial t = 0$, so that linear gravity waves cannot transport ω_z. Eliminating ρ' from these equations gives us

$$\frac{\partial^2\boldsymbol{\omega}}{\partial t^2} = N^2\hat{\mathbf{e}}_z \times \nabla u_z, \tag{9.46}$$

whose curl is

$$\frac{\partial^2}{\partial t^2}\nabla^2\mathbf{u} + N^2\left[(\nabla^2 u_z)\hat{\mathbf{e}}_z - \nabla(\partial u_z/\partial z)\right] = 0. \tag{9.47}$$

Of particular importance is the z component of (9.47), which yields the wave-like equation

$$\boxed{\frac{\partial^2}{\partial t^2}\nabla^2 u_z + N^2\nabla_\perp^2 u_z = 0}\,, \tag{9.48}$$

where $\nabla_\perp^2$ is the horizontal Laplacian in which derivatives in z are supressed.

When confronted with a wave-like equation such as (9.48), it is natural to look for plane-wave solutions of the form

$$\mathbf{u}(\mathbf{x}, t) = \hat{\mathbf{u}}\exp\left[j\left(\mathbf{k}\cdot\mathbf{x} - \varpi t\right)\right] \tag{9.49}$$

(real part understood), where $\mathbf{k}$ is the wavevector, ϖ the angular frequency, $\hat{\mathbf{u}}$ is an amplitude, and we adopt the convention that $\varpi > 0$, as in Chapter 7. Substituting the z component of this trial solution into (9.48) yields the dispersion relationship,

$$\boxed{\varpi(\mathbf{k}) = Nk_\perp/k = N\left|\cos\theta\right|}\,, \quad k_\perp = \sqrt{k_x^2 + k_y^2}, \tag{9.50}$$

where $k = |\mathbf{k}|$ and θ is the angle $\mathbf{k}$ makes to the horizontal plane. Evidently, the frequency of internal gravity waves is restricted to the range $0 < \varpi < N$. Returning to (9.47), our trial solution combined with (9.50) now demands

$$k_z k_\perp^2\hat{\mathbf{u}} = \left(k^2\mathbf{k}_\parallel - k_z^2\mathbf{k}\right)\hat{u}_z = \left(k_\perp^2\mathbf{k}_\parallel - k_z^2\mathbf{k}_\perp\right)\hat{u}_z = \left[\mathbf{k}\times(\mathbf{k}_\parallel\times\mathbf{k})\right]\hat{u}_z, \tag{9.51}$$

where $\mathbf{k}_\parallel = k_z\hat{\mathbf{e}}_z$ and $\mathbf{k}_\perp = \mathbf{k} - \mathbf{k}_\parallel$. Note that $\mathbf{k}\cdot\hat{\mathbf{u}} = 0$, which is a consequence of $\nabla\cdot\mathbf{u} = 0$.

The phase velocity (the velocity of the wave crests) is, by definition, in the direction of $\mathbf{k}$ and has magnitude ϖ/k:

$$\boxed{\mathbf{c}_p = \frac{Nk_\perp\mathbf{k}}{k^3}}\,. \tag{9.52}$$

The group velocity, however, which is the velocity at which wave energy propagates in the form of wave packets, has components $(\mathbf{c}_g)_i = \partial\varpi/\partial k_i$, as discussed in §7.3.1. A little algebra confirms that

$$\boxed{\mathbf{c}_g = \frac{N}{k^3 k_\perp}\mathbf{k}\times(\mathbf{k}\times\mathbf{k}_\parallel) = \frac{N}{k^3 k_\perp}\left[k_\parallel^2\mathbf{k}_\perp - k_\perp^2\mathbf{k}_\parallel\right]}\,. \tag{9.53}$$

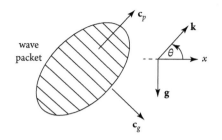

Figure 9.6 The relative orientation of the phase and group velocities for a two-dimensional wave packet. Notice that the wave energy travels *along* the wave crests.

We have reached the extraordinary conclusion that the phase and group velocities are perpendicular! That is to say, the direction of propagation of a wave packet is normal to the velocity of the wave crests which sit within that wave packet. It follows that wave energy travels *along* the wave crests, rather than normal to the crests, as shown in Figure 9.6. This is the hallmark of three-dimensional dispersive waves; not only is the *speed* of propagation of the wave packets and wave crests different, as is the case for one-dimensional dispersive waves, but the *direction* of propagation is also different. Note that (9.51) tells us that $\mathbf{c}_g$ is aligned with $\hat{\mathbf{u}}$. Note also that $|\mathbf{c}_g| = (N/k)|\sin\theta|$, so the fastest wave packets travel horizontally, with $\mathbf{k}$ vertical.

From (9.50) and (9.53) we see that low-frequency waves, $\varpi \ll N$, have $\mathbf{k}$ vertical and a horizontal group velocity of magnitude N/k. These constitute the fastest wave packets. Conversely, high-frequency waves, $\varpi \to N$, have a wavevector that is horizontal and a group velocity that is vertical and of vanishingly small magnitude. Of course, waves of intermediate frequency propagate at an oblique angle.

It is instructive to consider the special case of a two-dimensional radiation field, say confined to the x–z plane. Then (9.52) and (9.53) become

$$\mathbf{c}_p = \pm\frac{N\cos\theta}{k}(\cos\theta\hat{\mathbf{e}}_x + \sin\theta\hat{\mathbf{e}}_z), \quad \mathbf{c}_g = \pm\frac{N\sin\theta}{k}(\sin\theta\hat{\mathbf{e}}_x - \cos\theta\hat{\mathbf{e}}_z), \quad (9.54)$$

where the upper (lower) signs correspond to positive (negative) k_x and θ is the angle $\mathbf{k}$ makes to the x-axis, as shown in Figure 9.6. Notice that the horizontal components of $\mathbf{c}_p$ and $\mathbf{c}_g$ have the same sense, whereas the vertical components are anti-parallel.

These expressions for $\mathbf{c}_p$ and $\mathbf{c}_g$ are illustrated in Figure 9.7, where a circular cylinder of radius R, aligned with the y-axis, oscillates horizontally at some intermediate frequency, say at $\varpi = N/2$. Because there is only one frequency in the problem, and $|\cos\theta|$ is fixed by (9.50), there are only two allowable values of θ. Given the $\pm$ signs in (9.54), we conclude that there are four directions in which wave energy can radiate away from the cylinder. It follows that the internal gravity waves are confined to four sheets, two directed to the right of the cylinder and two directed to the left. These sheets are bounded at large $|\mathbf{x}|$ by wave fronts which travel outward at the group velocity $|\mathbf{c}_g| = (N/k)|\sin\theta| \sim NR|\sin\theta|$. Since $\mathbf{c}_p$ and $\mathbf{c}_g$ are mutually perpendicular, the wave crests are aligned with the sheets, with

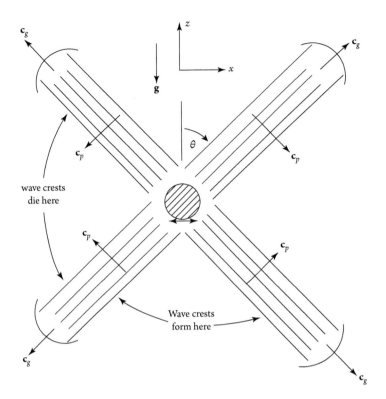

Figure 9.7 The dispersion pattern for gravity waves generated by an oscillating cylinder.

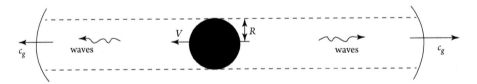

Figure 9.8 A cylinder moving slowly to the left generates low-frequency gravity waves which propagate horizontally at a group velocity of $c_g \sim NR$. These waves rapidly establish the blocked region shown in Figure 9.2. (Adapted from Davidson, 2013.)

crests appearing spontaneously on one side of each sheet, propagating across the sheet in the direction indicated, and then disappearing into thin air on the far side of the sheet!

It is of interest to consider the case of very slow oscillations of the cylinder. Here, low-frequency gravity waves propagate horizontally outward from the cylinder, carrying the information out to the far field that the cylinder is moving. In the limit of $\varpi \to 0$, the cylinder at any instant travels either to the left or to the right, say at the speed V, generating a slab of gravity waves which radiate horizontally away from the cylinder at the fast speed of $c_g \sim NR$, the term 'fast' indicating that $NR \gg V$. The situation is as shown in Figure 9.8.

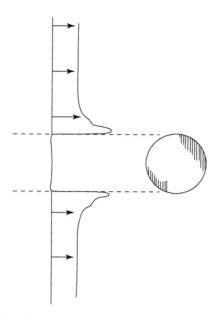

Figure 9.9 Bretherton's (1967) inviscid solution for stratified flow past a cylinder. This confirms that the blocked region is established by horizontally propagating gravity waves.

This provides the answer to the question posed in §9.3, as to how the blocked region ahead of the cylinder in Figure 9.2 initially forms. The information about the relative movement of the cylinder and the fluid is transmitted out to the far field by low-frequency gravity waves. This was confirmed by Bretherton (1967), who calculated the wave pattern for the configuration shown in Figure 9.8. Bretherton's solution is shown in Figure 9.9 in a frame of reference attached to the cylinder.

9.5.2 The Reflection of Internal Gravity Waves

We now turn to a topic which is of considerable importance in oceanography: the reflection of internal gravity waves off an inclined lower boundary, such as the continental shelf. Because gravity waves are dispersive, their reflection properties are more complicated than those encountered in, say, geometrical optics. In order to keep the discussion brief, we shall restrict ourselves to inviscid, two-dimensional, plane waves in the x–z plane striking a flat surface, also located in the x–z plane. The incident wave has wavevector $\mathbf{k}$ and group velocity $\mathbf{c}_g$, which in turn make angles of θ and $\phi = \pi/2 - \theta$ to the horizontal, as shown in Figure 9.10 (a). The boundary from which the waves reflect subtends an angle of β to the horizontal and we shall see that the nature of the reflection depends critically on whether $\phi > \beta$ (a shallow slope), or $\phi < \beta$ (a steep slope). In particular, when $\phi > \beta$, the group velocity of the reflected wave, $\mathbf{c}'_g$, is directed up the slope, while for $\phi < \beta$ we find that $\mathbf{c}'_g$ is directed down the slope (Figure 9.10 (b)). Crucially, for $\phi \to \beta$ the group velocity of the

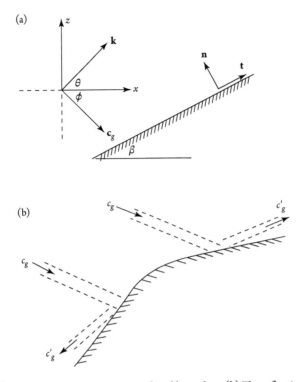

Figure 9.10 (a) A plane wave encounters an inclined boundary. (b) The reflections from a shallow slope ($\phi > \beta$) and a steep slope ($\phi < \beta$) are different.

reflected wave goes to zero and so its energy density tends to infinity, since the reflected energy cannot escape. In the oceans this leads to turbulence and mixing.

If we use a prime to denote the reflected wave, then the velocity field associated with the combined incident and reflected wave is

$$\mathbf{u} = \hat{\mathbf{u}}\exp j(\mathbf{k}\cdot\mathbf{x} - \varpi t) + \hat{\mathbf{u}}'\exp j(\mathbf{k}'\cdot\mathbf{x} - \varpi' t), \tag{9.55}$$

and our task is to choose the reflected wave so as to ensure $\mathbf{u}\cdot\mathbf{n} = 0$ at the boundary, where $\mathbf{n}$ is a unit normal. Since (9.55) must be satisfied at all times and at all points on the surface, we require that the arguments in the two exponentials are identical, which in turn demands that $\varpi = \varpi'$ and $\mathbf{k}\cdot\mathbf{t} = \mathbf{k}'\cdot\mathbf{t}$, where $\mathbf{t}$ is a unit tangential vector to the surface. The first of these restrictions tells us that $\cos\theta = \pm\cos\theta'$, while the second dictates k'. When both of these prerequisites are satisfied, the boundary condition simplifies to $(\hat{\mathbf{u}} + \hat{\mathbf{u}}')\cdot\mathbf{n} = 0$.

Let us now determine the orientation of $\mathbf{k}'$ and $\mathbf{c}'_g$ for the reflected wave. Since $\varpi = \varpi'$, there are three possible angles for $\mathbf{k}'$, and these are shown in Figure 9.7. To find the appropriate angle we note that $\mathbf{c}'_g$ must point away from the boundary, and $\mathbf{k}'$ must be directed up the slope, since we have $\mathbf{k}\cdot\mathbf{t} > 0$ and $\mathbf{k}\cdot\mathbf{t} = \mathbf{k}'\cdot\mathbf{t}$. When $\phi > \beta$ (a shallow slope), only the orientation shown in the top right of Figure 9.7 satisfies these conditions, while for $\phi < \beta$ (a steep slope), we must choose the configuration shown on the bottom left. This is summarized in Figure 9.11. Note that, since $\varpi/N = \cos\theta = \sin\phi$, the reflected wave is directed up the slope when $\varpi/N > \sin\beta$, and down the slope for $\varpi/N < \sin\beta$. Notice also that $\cos\theta = \pm\cos\theta'$ while $\sin\theta = \mp\sin\theta'$, where the upper sign corresponds to shallow slopes and the lower one to steep slopes.

We must now satisfy $(\hat{\mathbf{u}} + \hat{\mathbf{u}}') \cdot \mathbf{n} = 0$ on the boundary. To that end, we note that (9.51) gives us

$$\hat{\mathbf{u}} = \left(\hat{\mathbf{e}}_z - \frac{\sin\theta}{\cos\theta}\hat{\mathbf{e}}_x\right)\hat{u}_z, \quad \hat{\mathbf{u}}' = \left(\hat{\mathbf{e}}_z - \frac{\sin\theta'}{\cos\theta'}\hat{\mathbf{e}}_x\right)\hat{u}'_z,$$

and since $\sin\theta'/\cos\theta' = -\sin\theta/\cos\theta$ for both shallow and steep slopes, we conclude that

$$\hat{\mathbf{u}} + \hat{\mathbf{u}}' = (\hat{u}'_z + \hat{u}_z)\hat{\mathbf{e}}_z + \frac{\sin\theta}{\cos\theta}(\hat{u}'_z - \hat{u}_z)\hat{\mathbf{e}}_x. \tag{9.56}$$

Moreover, $\hat{\mathbf{e}}_x \cdot \mathbf{n} = -\sin\beta$ and $\hat{\mathbf{e}}_z \cdot \mathbf{n} = \cos\beta$, and so (9.56) yields

$$\cos\theta\,(\hat{\mathbf{u}} + \hat{\mathbf{u}}') \cdot \mathbf{n} = (\hat{u}'_z + \hat{u}_z)\cos\theta\cos\beta - (\hat{u}'_z - \hat{u}_z)\sin\theta\sin\beta. \tag{9.57}$$

The boundary condition $(\hat{\mathbf{u}} + \hat{\mathbf{u}}') \cdot \mathbf{n} = 0$ now provides a relationship between $\hat{u}_z$ and $\hat{u}'_z$,

$$\hat{u}_z \cos(\theta - \beta) + \hat{u}'_z \cos(\theta + \beta) = 0, \tag{9.58}$$

which is best rewritten in term of the complementary angle $\phi = \pi/2 - \theta$, as

$$\hat{u}'_z = -\frac{\sin(\phi + \beta)}{\sin(\phi - \beta)}\hat{u}_z. \tag{9.59}$$

Notice that $\hat{u}'_z \to \infty$ at the critical slope of $\beta = \phi$, which corresponds to the frequency $\varpi/N = \sin\beta$.

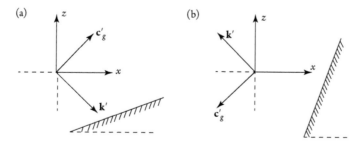

Figure 9.11 The orientation of $\mathbf{k}'$ and $\mathbf{c}'_g$ for the reflected wave. (a) The case of $\phi > \beta$ (a shallow slope). (b) The case of $\phi < \beta$ (a steep slope).

It remains to explore the implications of $\mathbf{k} \cdot \mathbf{t} = \mathbf{k'} \cdot \mathbf{t}$. Since $\hat{\mathbf{e}}_x \cdot \mathbf{t} = \cos\beta$ and $\hat{\mathbf{e}}_z \cdot \mathbf{t} = \sin\beta$, we have

$$\mathbf{k} \cdot \mathbf{t} = k \left(\cos\theta \hat{\mathbf{e}}_x + \sin\theta \hat{\mathbf{e}}_z \right) \cdot \mathbf{t} = k \cos\left(\theta - \beta\right).$$

Similarly,

$$\mathbf{k'} \cdot \mathbf{t} = k' \cos\left(\theta' - \beta\right) = \pm k' \cos\left(\theta + \beta\right),$$

where the upper sign corresponds to shallow slopes $(\phi > \beta)$ and the lower one to steep slopes $(\phi < \beta)$. It follows that

$$k \cos\left(\theta - \beta\right) \mp k' \cos\left(\theta + \beta\right) = 0, \tag{9.60}$$

which is reminiscent of (9.58). Again, this is best rewritten in term of the complementary angle $\phi = \pi/2 - \theta$, this time as

$$k' = \pm \frac{\sin\left(\phi + \beta\right)}{\sin\left(\phi - \beta\right)} k = \frac{\sin\left(\phi + \beta\right)}{\left|\sin\left(\phi - \beta\right)\right|} k. \tag{9.61}$$

Notice that $k' \to \infty$ at the critical slope of $\beta = \phi$. Since $\left|\mathbf{c}'_g\right| = (N/k')\left|\sin\theta'\right|$, this corresponds to the group velocity of the reflected wave going to zero. It is no coincidence that $\left|\mathbf{c}'_g\right| \to 0$ and $\hat{u}'_z \to \infty$ at the same critical angle: as the group velocity of the reflected wave gets smaller and smaller, its energy density must rise to compensate, so as to avoid an accumulation of energy at the boundary.

The near-critical reflection of internal waves from the continental shelf is thought to be an important process in the oceans, resulting in intense dissipation and mixing. This, in turn, has important consequences for biological species and sediment transport. These waves are typically generated by tidal flow over bottom topography, say seamounts, and have long wavelengths and low frequencies. The near-critical reflection of such waves on the gentle slope of the continental shelf can result in turbulence in one of two ways; either the large amplitude of the reflected wave causes wave-breaking, or the reflected wave drives an intense boundary-layer flow along the slope, which in turn becomes unstable. Lamb (2014) provides a convenient review of these (and related) processes.

9.6 Generalized Vortex Dynamics: Bjerknes' Theorem and Ertel's Potential Vorticity

We now consider the vortex dynamics of a fluid of variable density. This will lead us to *Bjerknes' theorem*, which is a generalization of Kelvin's theorem, and to the related concept of *Ertel's potential vorticity*. This theory applies to any inviscid flow and in particular it is not restricted to a Boussinesq fluid. So, in this final section, we set aside the Boussinesq approximation and place no restriction on ρ, other than mass conservation. Our governing equations are then mass conservation and Euler's equation:

$$\frac{D\rho}{Dt} + \rho\nabla \cdot \mathbf{u} = 0, \qquad \frac{D\mathbf{u}}{Dt} = -\frac{1}{\rho}\nabla p + \mathbf{g}. \qquad (9.62)$$

Recall that the identity $\nabla\left(\frac{1}{2}\mathbf{u}^2\right) = (\mathbf{u}\cdot\nabla)\mathbf{u} + \mathbf{u}\times\boldsymbol{\omega}$ allows us to rewrite Euler's equation as

$$\frac{\partial\mathbf{u}}{\partial t} = \mathbf{u}\times\boldsymbol{\omega} - \frac{1}{\rho}\nabla p - \nabla\left(\frac{1}{2}\mathbf{u}^2\right) + \mathbf{g},$$

whose curl is evidently

$$\frac{\partial\boldsymbol{\omega}}{\partial t} = \nabla\times(\mathbf{u}\times\boldsymbol{\omega}) - \nabla(1/\rho)\times\nabla p. \qquad (9.63)$$

We shall now show how to obtain a transport equation for the vorticity field from (9.63), which is not unlike (8.3). Expanding the first term on the right of (9.63) we have

$$\frac{\partial\boldsymbol{\omega}}{\partial t} = (\boldsymbol{\omega}\cdot\nabla)\mathbf{u} - (\mathbf{u}\cdot\nabla)\boldsymbol{\omega} - \boldsymbol{\omega}(\nabla\cdot\mathbf{u}) - \nabla(1/\rho)\times\nabla p,$$

and using mass conservation to substitute for $\nabla\cdot\mathbf{u}$ yields

$$\frac{D\boldsymbol{\omega}}{Dt} - \frac{\boldsymbol{\omega}}{\rho}\frac{D\rho}{Dt} = (\boldsymbol{\omega}\cdot\nabla)\mathbf{u} - \nabla(1/\rho)\times\nabla p.$$

Dividing by ρ now gives us

$$\frac{1}{\rho}\frac{D\boldsymbol{\omega}}{Dt} + \boldsymbol{\omega}\frac{D}{Dt}\left(\frac{1}{\rho}\right) = \frac{1}{\rho}(\boldsymbol{\omega}\cdot\nabla)\mathbf{u} - \frac{1}{\rho}\nabla(1/\rho)\times\nabla p,$$

from which

$$\boxed{\frac{D}{Dt}\left(\frac{\boldsymbol{\omega}}{\rho}\right) = \frac{\boldsymbol{\omega}}{\rho}\cdot\nabla\mathbf{u} - \frac{1}{\rho}\nabla(1/\rho)\times\nabla p}. \qquad (9.64)$$

This is our generalization of the vorticity transport equation, (8.3). It differs from the case of uniform density in two respects. First, we must replace $\boldsymbol{\omega}$ by $\boldsymbol{\omega}/\rho$ in the convective derivative and second, gradients in density can act as a source of vorticity. Note that for the case of a *baratropic* fluid, for which $\rho = \rho(p)$, this source term disappears.

Let us now return to (9.63) and see if we can generalize Kelvin's theorem. To that end we need the kinematic equation, (8.6), which, when applied to the vorticity field, becomes

$$\frac{d}{dt} \oint_{C_m} \mathbf{u} \cdot d\mathbf{r} = \frac{d}{dt} \int_S \boldsymbol{\omega} \cdot d\mathbf{S} = \int_S \left(\frac{\partial \boldsymbol{\omega}}{\partial t} - \nabla \times (\mathbf{u} \times \boldsymbol{\omega}) \right) \cdot d\mathbf{S}. \qquad (9.65)$$

Here S is any surface that spans the closed, material curve, C_m, i.e. a closed loop that moves with the fluid. Combining (9.63) with (9.65) yields

$$\boxed{\frac{d}{dt} \oint_{C_m} \mathbf{u} \cdot d\mathbf{r} = \frac{d}{dt} \int_S \boldsymbol{\omega} \cdot d\mathbf{S} = - \int_S [\nabla (1/\rho) \times \nabla p] \cdot d\mathbf{S} = - \oint_{C_m} \frac{1}{\rho} \nabla p \cdot d\mathbf{r}}.$$

$$(9.66)$$

This is Bjerknes' theorem (Bjerknes, 1898) which, for a baratropic or constant density fluid, reverts to Kelvin's theorem.

Next, suppose that ϑ is any materially conserved quantity, $D\vartheta/Dt = 0$. Then our generalized vorticity equation, (9.64), gives us the somewhat uninspiring result

$$\rho \frac{D}{Dt} \left(\frac{\boldsymbol{\omega}}{\rho} \cdot \nabla \vartheta \right) = \boldsymbol{\omega} \cdot \frac{D}{Dt} (\nabla \vartheta) + (\boldsymbol{\omega} \cdot \nabla \mathbf{u}) \cdot \nabla \vartheta - [\nabla (1/\rho) \times \nabla p] \cdot \nabla \vartheta.$$

However, since we have specified $D\vartheta/Dt = 0$, the first two terms on the right cancel; i.e.

$$\frac{D}{Dt} \nabla \vartheta = \frac{\partial}{\partial x_i} \frac{\partial \vartheta}{\partial t} + u_j \frac{\partial^2 \vartheta}{\partial x_i \partial x_j} = - \frac{\partial u_j}{\partial x_i} \frac{\partial \vartheta}{\partial x_j},$$

from which,

$$\boldsymbol{\omega} \cdot \frac{D}{Dt} (\nabla \vartheta) = - (\boldsymbol{\omega} \cdot \nabla \mathbf{u}) \cdot \nabla \vartheta.$$

We conclude, therefore, that

$$\rho \frac{D}{Dt} \left(\frac{\boldsymbol{\omega}}{\rho} \cdot \nabla \vartheta \right) = - [\nabla (1/\rho) \times \nabla p] \cdot \nabla \vartheta. \qquad (9.67)$$

If we further restrict ϑ to be a prescribed function of p and ρ, $\vartheta = \vartheta(p, \rho)$, then we obtain the remarkably simple result

$$\boxed{\frac{D}{Dt} \left(\frac{\boldsymbol{\omega}}{\rho} \cdot \nabla \vartheta \right) = 0}. \qquad (9.68)$$

The materially conserved quantity

$$\boxed{\Pi = \frac{\boldsymbol{\omega}}{\rho} \cdot \nabla \vartheta} \qquad (9.69)$$

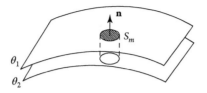

Figure 9.12 Bjerknes' theorem applied to the surface element S_m tells us that $\boldsymbol{\omega} \cdot (S_m \mathbf{n})$ is conserved, as it the mass of the cylinder, $\rho h S_m$. The invariance of $(\boldsymbol{\omega}/\rho) \cdot \nabla \vartheta$ follows.

is known as Ertel's potential vorticity (Ertel, 1942); a rather obscure name. For the case of an incompressible fluid, where $D\rho/Dt = 0$, the obvious choice for ϑ is $\vartheta = \vartheta(\rho)$, which ensures $D\vartheta/Dt = 0$. In such a case it is convenient to write Π as $\Pi = \boldsymbol{\omega} \cdot \nabla \rho$.

The conservation of Π is readily interpreted in terms of Bjerknes' theorem, as Ertel himself noted. Consider two adjacent material surfaces, ϑ_1 and ϑ_2, separated by a short distance $h(\mathbf{x})$. Then the normal $\mathbf{n}$ to the surfaces, which is parallel to $\nabla \vartheta$, must be perpendicular to $\nabla \rho \times \nabla p$, because $\vartheta = \vartheta(p, \rho)$. Now consider a small, cylindrical, material volume element located between the two surfaces, of cross-section S_m and mass $\rho h S_m$, as shown in Figure 9.12. Since $\mathbf{n}$ is perpendicular to $\nabla \rho \times \nabla p$, Bjerknes' theorem applied to the material surface element $S_m \mathbf{n}$ tells us that $\boldsymbol{\omega} \cdot (S_m \mathbf{n})$ is materially conserved. Moreover, the mass of the material element, $\rho h S_m$, is also conserved, and so the ratio of the two, $\boldsymbol{\omega} \cdot \mathbf{n}/\rho h$, is a material invariant. The final step is to note that

$$\mathbf{n}/h = \nabla \vartheta / (\vartheta_1 - \vartheta_2),$$

which tells us that $(\boldsymbol{\omega}/\rho) \cdot \nabla \vartheta$ is conserved for our small material element. We have arrived back at Ertel's potential vorticity.

Potential vorticity plays a central role in much of geophysical fluid dynamics. For a Boussinesq fluid with uniform stratification, the conservation of $\Pi = \boldsymbol{\omega} \cdot \nabla \rho$ leads to the material invariant

$$\Pi = \boldsymbol{\omega} \cdot \nabla \rho' - \left(\bar{\rho} N^2 / g\right) \omega_z.$$

In the linear regime of small-amplitude waves this simplifies to $\Pi = -\left(\bar{\rho} N^2/g\right) \omega_z$, which is consistent with the fact that linear gravity waves cannot transport ω_z.

○ ○ ○

This concludes our discussion of stratified fluids. Readers seeking additional introductory material could do worse than consult Tritton (1988). A more advanced treatment may be found in Turner (1973) and Sutherland (2014), while Lighthill (1978) provides the definitive account of internal gravity waves.

....

EXERCISES

9.1 *The importance of viscosity for buoyancy-driven flows in confined domains.* We have often found it convenient to ignore viscosity in this chapter. However, for natural convection in a confined domain, the neglect of viscosity can be problematic, as we now show. Consider a steady, two-dimensional flow in a confined domain in which $\mathbf{u} = 0$ on the boundary. The streamlines are necessarily closed. Show that integrating (9.2) around a closed streamline yields

$$\oint (\rho' \mathbf{g}) \cdot d\mathbf{r} + \bar{\rho}\nu \oint (\nabla^2 \mathbf{u}) \cdot d\mathbf{r} = 0.$$

This says that the work done by the buoyancy force on a fluid particle as it passes once around a streamline must be perfectly matched by the work done by the viscous forces. Viscosity is therefore essential for reaching a steady state. Moreover, show that the dot product of (9.2) with $\mathbf{u}$, integrated over the domain, yields

$$\int \rho' \mathbf{g} \cdot \mathbf{u} \, dV = \bar{\rho}\nu \int \omega^2 dV.$$

This is also an energy balance, this time between the net rate of working of the buoyancy force and the rate of viscous dissipation. Once again, viscosity is essential to achieving a steady state.

9.2 *A relationship between the phase and group velocities of internal gravity waves.* Use equations (9.52) and (9.53) to show that

$$\mathbf{c}_p + \mathbf{c}_g = (N/k)(\mathbf{k}_\perp/k_\perp) \quad \text{and} \quad c_p^2 + c_g^2 = N^2/k^2.$$

Evidently $\mathbf{c}_p$ and $\mathbf{c}_g$ have equal and opposite vertical components.

9.3 *A relationship between the group velocity and fluid velocity for internal gravity waves.* Use equations (9.51) and (9.53) to show that

$$\mathbf{c}_g = -\frac{Nk_\perp k_z}{k^3 \hat{u}_z}\hat{\mathbf{u}} = \varpi \sin^2 \theta \frac{\hat{\mathbf{u}}}{\mathbf{k}_\perp \cdot \hat{\mathbf{u}}_\perp},$$

so that $\mathbf{c}_g$ is aligned with $\hat{\mathbf{u}}$.

Give a physical explanation for why $\mathbf{c}_g$ must be aligned with $\mathbf{u}$.

9.4 *Evanescent gravity waves.* Suppose that an inviscid, stratified fluid is bounded from below by an undulating sinusoidal surface whose surface displacement is

$$\eta(x,t) = \eta_0 \cos(k_x x) \cos \varpi t, \quad k_x > 0.$$

For $\varpi < N$, this generates a standing wave whose vertical wave number is chosen to satisfy the dispersion relationship (9.50), *i.e.* $\varpi = N k_x / |\mathbf{k}|$. Now suppose that $\varpi > N$. Clearly, standing-wave solutions are forbidden as they violate the dispersion relationship. However, there must still be some form of disturbance localized around the lower boundary, it just cannot take the form of a standing wave. It is natural to look for a disturbance of the form $u_z = \hat{u}_z(z) \cos(k_x x) \sin \varpi t$. Show that this leads to the solution

$$u_z \sim e^{-\gamma z} \cos(k_x x) \sin \varpi t, \quad \gamma = k_x \sqrt{1 - N^2/\varpi^2}.$$

A wave-like motion whose amplitude decays exponentially with distance from its source is known as an *evanescent wave*. An intriguing feature of evanescent gravity waves is that $\gamma \to 0$ as ϖ approaches N from above, and this creates highly elongated disturbances.

9.5 *The disturbance energy of a stratified fluid.* Consider the rescaled perturbed density field $b' = g\rho'/N\bar{\rho}$, sometimes called the *buoyancy field*. Use (9.2) and (9.5) to show that, in an inviscid fluid,

$$\frac{D}{Dt}\left(\tfrac{1}{2}\bar{\rho}\mathbf{u}^2 + \tfrac{1}{2}\bar{\rho}b'^2\right) = -\nabla \cdot (p\mathbf{u}).$$

Clearly this is an energy equation, but we need to explain the origin of the term $\tfrac{1}{2}\bar{\rho}b'^2$. Consider slowly lifting a small parcel of fluid a short distance z_p through a quiescent, stratified fluid. Confirm that the change in potential energy (per unit volume) of the parcel is

$$\frac{1}{2}\left|\frac{d\rho_0}{dz}\right|g z_p^2 = \frac{1}{2}\bar{\rho}N^2 z_p^2,$$

where z_p is its height relative to its equilibrium position. Now show that a solution of (9.5) is $\rho'(\mathbf{x}_p, t) = (\bar{\rho}N^2/g)z_p$ and hence confirm that $\tfrac{1}{2}\bar{\rho}b'^2$ is the potential energy per unit volume of a disturbance.

....

REFERENCES

Bjerknes, V., 1898, On the generation of circulation and vortices in inviscid fluids. *Skr. Nor. Vidensk.-Akad. 1: Mat.-Naturvidensk. kl.*, **5**, 3–29.

Bretherton, F.P., 1967, The time-dependent motion due to a cylinder moving in an unbounded rotating or stratified fluid. *J. Fluid Mech.*, **28**, 545–70.

Davidson, P.A., 2013, *Turbulence in rotating, stratified and electrically conducting fluids*, Cambridge University Press.

Ertel, H., 1942, Ein neuer hydrodynamischer Wirbelsatz. (A new hydrodynamic eddy theorem), *Meteorol. Z.*, **59**, 277–81.

Huppert, H.E., 1968, Appendix to a paper by J.W. Miles. *J. Fluid Mech.*, **33**(4), 811–13.

Lamb, K.G., 2014, Internal wave breaking and dissipation mechanisms on the continental slope/shelf. *Annual Rev. Fluid Mech.*, **46**, 231–54.

Lighthill, J., 1978, *Waves in fluids*, Cambridge University Press.

Oberbeck, A., 1879, On the thermal conduction of liquids taking into account flows due to temperature differences. *Ann. Phys. Chem.*, **7**, 271–92.

Sutherland, B.R., 2014, *Internal gravity waves*, Cambridge University Press.

Tritton, D.J., 1988, *Physical fluid dynamics*, Oxford University Press.

Turner, J.S., 1973, *Buoyancy effects in fluids*, Cambridge University Press.

10

· · • · ·

Waves and Flow in a Rotating Fluid

The Swedish-US meteorologist Carl-Gustaf Rossby (left, 1898–1957) and the Swedish oceanographer Vagn Walfrid Ekman (right, 1874–1954) were protégés of Vilhelm Bjerknes. They were both central to the early development of geophysical fluid mechanics.

10.1 Rayleigh's Stability Criterion for Inviscid, Swirling Flow

We start our discussion of rotating fluids by introducing Rayleigh's celebrated stability criterion for an inviscid, swirling flow of the form $\mathbf{u}_0 = (0, u_\theta(r), 0)$ in (r, θ, z) coordinates. We do so, not because we want to discuss instability per se (we shall discuss this at length in Chapter 11), but rather because Rayleigh's analysis introduces a number of important ideas about rotating flows, ideas that will recur many times in this chapter.

Rayleigh was interested in *axisymmetric* instabilities of the rotating flow $\mathbf{u}_0$. So let us start by recalling the governing equations for inviscid, axisymmetric flow of a fluid of uniform density. These are

$$\frac{D\Gamma}{Dt} = 0, \quad \Gamma = ru_\theta, \tag{10.1}$$

$$\frac{Du_r}{Dt} = -\frac{1}{\rho}\frac{\partial p}{\partial r} + \frac{\Gamma^2}{r^3}, \tag{10.2}$$

$$\frac{Du_z}{Dt} = -\frac{1}{\rho}\frac{\partial p}{\partial z}, \tag{10.3}$$

(see §3.3.1). Rayleigh (1916) noticed that there is an exact analogy between (10.1)–(10.3) and the axisymmetric (non-rotating) motion of an incompressible Boussinesq fluid driven by density gradients in a hypothetical *radial* gravitational field. To see that this is so, we write $\rho = \bar{\rho} + \rho'$ for the density of the Boussinesq fluid, $\bar{\rho}$ being a mean density, and $\mathbf{u}_p(r, z, t) = (u_r, 0, u_z)$ for the poloidal motion driven by the gravitational field. The axisymmetric motion of our Boussinesq fluid is then governed by $\nabla \cdot \mathbf{u} = 0$ and

$$\frac{D\rho'}{Dt} = 0, \tag{10.4}$$

$$\frac{D\mathbf{u}_p}{Dt} = -\nabla (p/\bar{\rho}) + (\rho'/\bar{\rho})\,\mathbf{g}, \tag{10.5}$$

where $\mathbf{g}$ is the (radial) gravitational acceleration. If we choose for $\mathbf{g}$ the *irrotational* vector $\widehat{g}\hat{\mathbf{e}}_r/r^3$, $\widehat{g}$ being a constant, and substitute Γ^2 for $\widehat{g}\rho'/\bar{\rho}$ in (10.4) and (10.5), then these equations do indeed revert to (10.1)–(10.3). Note that $\mathbf{g} = -\nabla(\widehat{g}/2r^2)$, and so the potential energy per unit mass associated with ρ' is $\widehat{g}\rho'/2\bar{\rho}r^2$. Under the analogy $\Gamma^2 \leftrightarrow \widehat{g}\rho'/\bar{\rho}$, this becomes $\Gamma^2/2r^2$, which corresponds to the kinetic energy of the swirling flow, $u_\theta^2/2$.

Several interesting observations follow directly from this analogy. First, since $\mathbf{g}$ points radially outward, the Boussinesq fluid has a stable, static equilibrium of the form $\rho_0 = \bar{\rho} + \rho'(r)$, provided that $\rho'(r)$ increases monotonically with radius (light fluid lying within heavy fluid). Conversely, the equilibrium is unstable if there exists a radius at which $\rho'(r)$ decreases with r, because potential energy can then be released by exchanging the radial position of two adjacent rings of fluid. It follows that the necessary and sufficient condition for the stability of the swirling flow $\mathbf{u}_0 = (0, u_\theta(r), 0)$ to inviscid, axisymmetric disturbances is that Γ^2 increases monotonically with radius. This is *Rayleigh's centrifugal stability criterion*, which is usually expressed in terms of *Rayleigh's discriminant*, defined as

$$\boxed{\Phi(r) = \frac{1}{r^3}\frac{d\Gamma^2}{dr}}. \tag{10.6}$$

That is to say, an inviscid swirling flow is stable to axisymmetric disturbances if and only if $\Phi(r) \geq 0$ at all points in the flow. This criterion is usually arrived at by a more conventional perturbation analysis, and we shall do just that in Chapter 11.

The second point to note is that the stable equilibrium $\rho_0 = \bar{\rho} + \rho'(r)$ must support internal gravity waves. Of course, these waves are not exactly the same as those discussed in Chapter 9, because we have rotated $\mathbf{g}$ to point radially outward. Nevertheless, there must be many common features. In any event, this tells us that a rotating fluid also supports internal waves and such waves are referred to as *inertial waves*, so-called because all of the wave energy is kinetic (see the discussion in §8.2.3).

We note in passing that Rayleigh's argument is a simple example of a more general technique in Hamiltonian dynamics, known as *Routh's procedure*, which can be used to reduce the number of degrees of freedom of a system with symmetries. For those readers particularly interested in Hamiltonian dynamics, we might summarize this procedure as follows. When a conservative system possesses a symmetry, say axial symmetry, there is a corresponding (so-called) cyclic coordinate whose generalized velocity appears in the Lagrangian, $L = T - V$, which in our case is u_θ. (In Hamiltonian dynamics it is conventional to use T for kinetic energy and V for potential energy.) Moreover, for each cyclic coordinate there is a corresponding conserved generalized momentum, which in our case is Γ. Routh's procedure says that we can create a new Lagrangian (the *Routhian*), in which the kinetic energy associated with the cyclic coordinates, written in terms of their conserved momenta, is transferred from T to V. The resulting Lagrangian then involves only the non-cyclic coordinates and the conserved momenta (see Goldstein, 1980). This reduces the number of degrees of freedom. In our case, the transfer of the kinetic energy $\Gamma^2/2r^2$ to a potential energy associated with some external force, $\rho'\mathbf{g}$, allows us to convert a problem involving (u_r, u_θ, u_z) into one involving only $(u_r, 0, u_z)$ and the materially conserved quantity ρ'.

Rayleigh also invoked a more conventional energy argument in support of his stability criterion. Consider two thin, circular hoops of fluid in the base flow $\mathbf{u}_0$. One has radius r_1, angular momentum Γ_1, volume δV, and kinetic energy $\frac{1}{2}\rho\delta V \left(\Gamma_1^2/r_1^2\right)$, while the other has radius $r_2 = r_1 + \delta r$, angular momentum $\Gamma_2 = \Gamma_1 + \delta\Gamma$, and the same volume δV. We now perturb the flow in such a way that these two fluid rings exchange position while retaining their angular momenta. The change in kinetic energy is then

$$\delta T = \tfrac{1}{2}\rho\delta V \left(\Gamma_2^2 - \Gamma_1^2\right)\left(\frac{1}{r_1^2} - \frac{1}{r_2^2}\right) = \rho\delta V \left(\Gamma_2^2 - \Gamma_1^2\right)\frac{\delta r}{r_1^3} = \Phi(r)\rho\delta V \left(\delta r\right)^2. \quad (10.7)$$

When $\Phi(r) < 0$, the energy associated with u_θ falls as a result of the perturbation and this releases energy to the disturbance, yielding an instability. Of course, in our analogue system of a radially stratified Boussinesq fluid, this is equivalent to releasing potential energy by interchanging two adjacent hoops of fluid in a region that is gravitationally unstable.

As noted at the start of this section, we are less interested in stability here than in the general properties of rotating fluids, so perhaps we should summarize what Rayleigh's analysis tells us about rotating flows. The first point to note is that the analogy to buoyancy-driven flow is often helpful. Second, a stable base flow, $\Gamma(r)$, supports axisymmetric oscillations whose properties are identical to internal gravity waves in a radially stratified, Boussinesq fluid. As we shall see, these inertial waves exist even for non-axisymmetric disturbances, and they have many similarities to the internal gravity waves described in Chapter 9. Third, in the remainder of this chapter we shall be particularly concerned with fluids subject to a uniform background rotation Ω, where the base flow is $\mathbf{u}_0 = (0, \Omega r, 0)$. In such cases we have $\Phi = 4\Omega^2$, and so Rayleigh's analysis tells us that rigid-body rotation is stable to inviscid, axisymmetric disturbances and supports internal wave motion.

10.2 The Equations of Motion in a Rotating Frame of Reference

10.2.1 The Coriolis Force and the Rossby Number

We shall now focus on fluids which are subject to a uniform, steady, background rotation, Ω. In such cases it is natural to transfer to a non-inertial frame of reference rotating with the background flow. Of course, Newton's laws of motion do not apply in a non-inertial frame, but we can remedy this as follows. Time derivatives in the rotating and non-rotating frames are related by

$$(d/dt)_{\text{inertial}} = (d/dt)_{\text{rotating}} + \Omega \times \qquad (10.8)$$

(see, for example, Goldstein, 1980). It follows that

$$\mathbf{u}_{\text{inertial}} = \mathbf{u}_{\text{rotating}} + \Omega \times \mathbf{x}, \qquad (10.9)$$

where $\mathbf{u}_{\text{rotating}}$ is measured in the rotating frame. Applying (10.8) to (10.9) now gives

$$(d\mathbf{u}/dt)_{\text{inertial}} = (d\mathbf{u}/dt)_{\text{rotating}} + 2\Omega \times \mathbf{u}_{\text{rotating}} + \Omega \times (\Omega \times \mathbf{x}), \qquad (10.10)$$

and for a particle of mass m that is subject to a force $\mathbf{F}$, this implies

$$\mathbf{F} = m \left(d\mathbf{u}/dt\right)_{\text{inertial}} = m \left(d\mathbf{u}/dt\right)_{\text{rotating}} + 2m\Omega \times \mathbf{u}_{\text{rotating}} + m\Omega \times (\Omega \times \mathbf{x}),$$

from which

$$m \left(d\mathbf{u}/dt\right)_{\text{rotating}} = \mathbf{F} - 2m\Omega \times \mathbf{u}_{\text{rotating}} - m\Omega \times (\Omega \times \mathbf{x}). \qquad (10.11)$$

We conclude that, although Newton's second law does not apply in a rotating frame, we can pretend that it does by introducing two fictitious forces per unit mass, $-2\Omega \times \mathbf{u}_{\text{rotating}}$ and $-\Omega \times (\Omega \times \mathbf{x})$. These are known as the Coriolis and centrifugal forces, respectively. It follows that the Navier–Stokes equation written in a rotating frame of reference is

$$\frac{D\mathbf{u}}{Dt} = -\nabla \left(p/\rho\right) - \Omega \times (\Omega \times \mathbf{x}) + 2\mathbf{u} \times \Omega + \nu \nabla^2 \mathbf{u}, \qquad (10.12)$$

where we have assumed uniform density and dropped the subscript 'rotating' on the understanding that everything is measured in the rotating frame. However, the centrifugal force is irrotational,

$$\mathbf{\Omega} \times (\mathbf{\Omega} \times \mathbf{x}) = -\frac{1}{2}\nabla\left[(\mathbf{\Omega} \times \mathbf{x})^2\right],$$

and so it is conventional to rewrite (10.12) as

$$\frac{D\mathbf{u}}{Dt} = 2\mathbf{u} \times \mathbf{\Omega} - \nabla(p/\rho) + \nu\nabla^2\mathbf{u}, \qquad (10.13)$$

where it is understood that p represents the reduced pressure, $p_{\text{true}} - \frac{1}{2}\rho(\mathbf{\Omega} \times \mathbf{x})^2$. Notice that the Coriolis force does no work, $(\mathbf{u} \times \mathbf{\Omega}) \cdot \mathbf{u} = 0$, as seems appropriate for a fictitious force.

From now on we shall take $\mathbf{\Omega}$ to point in the z direction, $\mathbf{\Omega} = \Omega\hat{\mathbf{e}}_z$, so that, in (r, θ, z) coordinates,

$$2\mathbf{u} \times \mathbf{\Omega} = 2\Omega u_\theta\hat{\mathbf{e}}_r - 2\Omega u_r\hat{\mathbf{e}}_\theta. \qquad (10.14)$$

Evidently, a fluid particle moving radially outward will be subject to a negative azimuthal Coriolis force, and hence tend to reduce its azimuthal velocity in the rotating frame. This is consistent with angular momentum conservation in the inertial frame since, in the absence of other forces, particles moving radially outward reduce their angular velocity.

Note that the ratio of the inertial force $\mathbf{u} \cdot \nabla\mathbf{u}$ to the Coriolis force $2\mathbf{u} \times \mathbf{\Omega}$ is of the order of $u/\Omega\ell$, where ℓ is a characteristic length scale. It is convenient, therefore, to introduce the *Rossby number*,

$$\mathrm{Ro} = \frac{u}{\Omega\ell}, \qquad (10.15)$$

which is analogous to the internal Froude number in a stratified fluid. Flows for which $\mathrm{Ro} \ll 1$ are said to be subject to rapid rotation.

10.2.2 Rapid Rotation: the Taylor–Proudman Theorem and Drifting Taylor Columns

Consider an inviscid, rapidly rotating flow. Since $\mathrm{Ro} \ll 1$, we may neglect the inertial force $\mathbf{u} \cdot \nabla\mathbf{u}$ and our equation of motion reduces to

$$\frac{\partial\mathbf{u}}{\partial t} = 2\mathbf{u} \times \mathbf{\Omega} - \nabla(p/\rho), \qquad (10.16)$$

whose corresponding vorticity equation is evidently

$$\boxed{\frac{\partial \boldsymbol{\omega}}{\partial t} = 2(\boldsymbol{\Omega} \cdot \nabla)\mathbf{u}}.$$

(10.17)

A steady (or quasi-steady) flow therefore satisfies $(\boldsymbol{\Omega} \cdot \nabla)\mathbf{u} = 0$, and so a steady velocity field in a rapidly rotating fluid must be strictly two-dimensional, $\mathbf{u} = \mathbf{u}(x, y)$. In particular, the flow must satisfy $\partial u_z / \partial z = 0$, so that columns of fluid orientated with the z-axis cannot change their length. Although first derived by Hough in 1897, the constraint $(\boldsymbol{\Omega} \cdot \nabla)\mathbf{u} = 0$ is known as the *Taylor–Proudman theorem* and it has many counterintuitive consequences.

Perhaps the most famous illustration of the Taylor–Proudman theorem is an experiment conceived by the physicist G.I. Taylor and shown schematically in Figure 10.1. A tank of water is placed on a turntable and a small object is *slowly* traversed across the base of the tank. The entire process is then observed in the rotating frame of reference. One might have expected to see the fluid ahead of the object to move up and over the object to make way for it, as would occur in the non-rotating case. However, the flow is quasi-steady and must satisfy $(\boldsymbol{\Omega} \cdot \nabla)\mathbf{u} = 0$, which excludes such a three-dimensional motion. In particular, water rising up and over the object would require vertical columns of fluid to contract, which is forbidden. Instead, what happens is that the water sitting above the object drifts across the tank keeping pace with the object. It is as if the fluid sitting inside the imaginary cylinder which circumscribes the object is rigidly attached to the object. This cylinder of drifting fluid is known as a *Taylor column*, and the fluid outside the column flows around it as if the column were rigid, thus maintaining a two-dimensional motion.

This striking behaviour can be confirmed by placing dye at the points A and B shown in Figure 10.1 (a). The dye at A is seen to drift across the tank, always centred above the

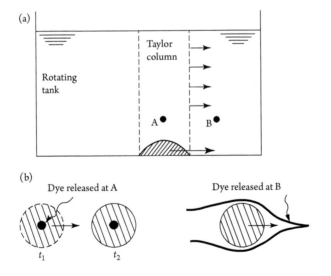

Figure 10.1 A small object is slowly dragged across the base of a rotating tank of water, producing a Taylor column. (a) Side view. (b) Plan view. (From Davidson, 2015.)

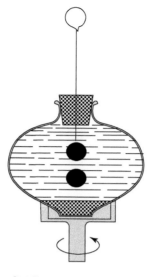

Figure 10.2 Kelvin's experiment of 1868.

object, while that at B splits into two as the object passes below it, making way for the Taylor column. This is shown in Figure 10.1 (b).

A less well-known experiment, which predates Taylor's visualization of columnar motion by 55 years, was performed by Kelvin in 1868. His apparatus is shown in Figure 10.2. It impressed Kelvin's brother, James Thomson, who wrote in a letter to Lord Kelvin:

> 'I saw Professor Tait in Belfast last week and he mentioned to me an experiment which you had been making in which you put a cork into the axis of a whirlpool of water having the same angular velocity at all parts, and got the result that the cork stayed down without floating up to the top; and he mentioned that if you put two corks at different levels in the axis …and that if you pushed the upper one down a little the lower one would go down too, as if there was an almost rigid mass of water between them.' (1868)

This is a particularly elegant demonstration of the power of the constraint $\partial u_z / \partial z = 0$.

One curious feature of Taylor's experiment is the movement of the dye at point A. How does the fluid at A know the instantaneous location of the object below? Here we get a surprise: the information is transmitted upward by wave motion, and in particular by inertial waves. In fact, just as the blocked region ahead of the cylinder in Figure 9.2 is established by low-frequency gravity waves propagating horizontally outward from the cylinder, as shown in Figure 9.8, so we shall see that Taylor columns are established and maintained by low-frequency inertial waves travelling along the rotation axis. In short, it is not strictly correct to think of the flow in Figure 10.1 as quasi-steady. Rather, the object at the base of the tank is acting like a radio antenna, continually emitting low-frequency inertial waves which travel upward, carrying the information that the object is moving. To show that this is indeed the case, we first need to establish the basic properties of inertial waves.

10.3 Inertial Waves of Small Amplitude

10.3.1 Their Dispersion Relationship, Group Velocity, and Spatial Structure

The application of the operator $\nabla \times (\partial/\partial t)$ to (10.17) yields the wave-like equation

$$\boxed{\frac{\partial^2}{\partial t^2}\nabla^2 \mathbf{u} + (2\mathbf{\Omega} \cdot \nabla)^2 \mathbf{u} = 0}, \tag{10.18}$$

which might be compared with the governing equation for gravity waves, (9.48). This supports plane waves of the form

$$\mathbf{u}(\mathbf{x}, t) = \hat{\mathbf{u}}\exp\left[j\left(\mathbf{k} \cdot \mathbf{x} - \varpi t\right)\right], \tag{10.19}$$

where the real part is understood, $\mathbf{k}$ is the wavevector, ϖ the angular frequency, $\nabla \cdot \mathbf{u} = 0$ demands $\mathbf{k} \cdot \hat{\mathbf{u}} = 0$, and we adopt the convention that $\varpi > 0$, as in Chapter 7. Substituting (10.19) into (10.18) we obtain $\varpi^2 |\mathbf{k}|^2 = (2\mathbf{k} \cdot \mathbf{\Omega})^2$, which restricts ϖ to $0 < \varpi < 2\Omega$ and yields the dispersion relationship

$$\boxed{\varpi = \pm 2(\mathbf{k} \cdot \mathbf{\Omega})/k = 2\Omega|\sin\theta|}, \tag{10.20}$$

where $k = |\mathbf{k}|$ and θ is the angle $\mathbf{k}$ makes to the *horizontal*. By definition, the phase velocity (the velocity of the wave crests) is in the direction of $\mathbf{k}$ and has magnitude ϖ/k, and so

$$\mathbf{c}_p = \varpi\mathbf{k}/k^2 = \pm 2(\mathbf{k} \cdot \mathbf{\Omega})\mathbf{k}/k^3. \tag{10.21}$$

By way of contrast, the group velocity $\mathbf{c}_g = \partial\varpi/\partial k_i$, which is the velocity at which wave energy disperses in the form of wave packets, is readily shown to be

$$\boxed{\mathbf{c}_g = \pm 2\frac{\mathbf{k} \times (\mathbf{\Omega} \times \mathbf{k})}{k^3} = \pm 2\frac{k^2\mathbf{\Omega} - (\mathbf{k} \cdot \mathbf{\Omega})\mathbf{k}}{k^3}}, \tag{10.22}$$

from which $|\mathbf{c}_g| = 2(\Omega/k)|\cos\theta|$. Perhaps some comments are in order at this point. First, inertial waves, like internal gravity waves, have the extraordinary property that their group velocity is perpendicular to their phase velocity. That is to say, the direction of propagation of a wave packet is normal to the velocity of the wave crests which sit within that wave packet. Thus wave energy travels *along* the wave crests, rather than normal to the crests. Second, (10.22) gives

$$\mathbf{c}_g \cdot \mathbf{\Omega} = \pm 2k^{-3}\left[k^2\Omega^2 - (\mathbf{k} \cdot \mathbf{\Omega})^2\right], \tag{10.23}$$

so the positive (negative) sign corresponds to wave energy travelling upward (downwards).

Third, $\mathbf{c}_g$ and $\mathbf{c}_p$ are related by

$$\mathbf{c}_g + \mathbf{c}_p = \pm 2\Omega/k, \quad \mathbf{c}_p^2 + \mathbf{c}_g^2 = 4\Omega^2/k^2, \tag{10.24}$$

so $\mathbf{c}_g$ is co-planar with $\boldsymbol{\Omega}$ and $\mathbf{k}$. Fourth, from (10.22) we see that wave packets of intermediate frequency, say $\varpi \sim \Omega$, travel obliquely. This is illustrated in Figure 10.3 (a), where the waves are generated by an oscillating disc and spread to fill two conical annuli. Low-frequency waves, by contrast, have $\mathbf{k} \cdot \boldsymbol{\Omega} \approx 0$ and a group velocity of $\mathbf{c}_g = \pm 2\Omega/k$, so $\mathbf{k}$ is horizontal and energy disperses up and down the rotation axis, as shown in Figure 10.3 (b). These are the fastest wave packets. (It turns out that for a disc of radius R the dominant wavenumber is $k \approx \pi/R$, and so the group velocity in Figure 10.3 (b) has a magnitude of $c_g \approx (2/\pi)\Omega R$.) Conversely, according to (10.20) and (10.22), high-frequency waves in which $\varpi \to 2\Omega$ have $\mathbf{k}$ vertical and a group velocity which is horizontal and vanishingly small.

An important feature of inertial waves is that they are helical, *i.e.* they have a finite helicity, $h = \mathbf{u} \cdot \boldsymbol{\omega}$. To see why, note that (10.17) requires $\varpi \hat{\boldsymbol{\omega}} = -2(\boldsymbol{\Omega} \cdot \mathbf{k})\hat{\mathbf{u}}$, which combines with (10.20) to yield

$$\hat{\boldsymbol{\omega}} = \mp k\hat{\mathbf{u}}, \tag{10.25}$$

where $\hat{\boldsymbol{\omega}}$ is the amplitude of the vorticity, $\hat{\boldsymbol{\omega}} = j\mathbf{k} \times \hat{\mathbf{u}}$. It follows that the velocity and vorticity fields are in phase and parallel, so that inertial waves have maximal helicity. Moreover, the $+$ sign in (10.23) corresponds to negative helicity, and the $-$ sign to positive helicity, so a wave packet with negative helicity travels upward ($\mathbf{c}_g \cdot \boldsymbol{\Omega} > 0$), while a wave packet with positive helicity travels downward ($\mathbf{c}_g \cdot \boldsymbol{\Omega} < 0$), as illustrated in Figure 10.4.

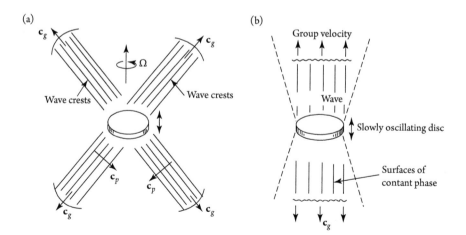

Figure 10.3 (a) An oscillating disc radiates inertial waves at $\varpi \approx \Omega$. The radiation pattern consists of two conical annuli, one above the disc and one below. (b) Low-frequency waves.

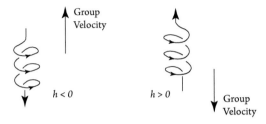

Figure 10.4 A wave packet with negative helicity travels upward, while one with positive helicity travels downward. The helical lines shown here happen to be the streamlines within the wave packets given by solutions (10.32a) and (10.32c). (From Davidson, 2013.)

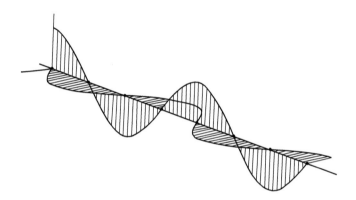

Figure 10.5 The instantaneous velocity field for an inertial wave with negative helicity.

In order to determine the spatial structure of a monochromatic inertial wave, it is convenient to adopt a set of coordinates, (X, Y, Z), with Z parallel to $\mathbf{k}$. Then $\mathbf{k} \cdot \hat{\mathbf{u}} = 0$ demands that $\hat{\mathbf{u}} = (\hat{u}_X, \hat{u}_Y, 0)$, while (10.25) in the form $j\mathbf{k} \times \hat{\mathbf{u}} = \mp k\hat{\mathbf{u}}$ requires $\hat{u}_Y = \mp j\hat{u}_X$. It follows that $\hat{\mathbf{u}} = \hat{u}_X (1, \mp j, 0)$, and if we choose $\hat{u}_X$ to be real then

$$\mathbf{u} = \hat{u}_X (\cos(kZ - \varpi t), \pm \sin(kZ - \varpi t), 0). \qquad (10.26)$$

This is a circularly polarized wave with a velocity of constant magnitude that rotates about the Z-axis as the wave propagates. Note that the sense of rotation is determined by the helicity. If the wave has negative helicity, so that the upper sign in (10.26) applies, then the velocity distribution at $t = 0$ is as shown in Figure 10.5, with $\mathbf{u}$ rotating about the Z-axis in a right-handed fashion as Z increases. Conversely, if we fix attention on the point $Z = 0$, then $\mathbf{u}$ rotates about the Z-axis with an angular velocity that is antiparallel to $\mathbf{k}$. Of course, a wave in which $h > 0$ behaves in the opposite manner.

10.3.2 The Formation of Transient Taylor Columns by Low-frequency Waves

It is instructive to return to the disc shown in Figure 10.3, which co-rotates with the fluid. Suppose that, instead of slowly oscillating the disc, it is suddenly set into slow, steady motion at $t = 0$, moving along the z-axis with a speed V. Provided that $V \ll \Omega R$, R being the radius of the disc, low-frequency waves will be excited at the surface of the disc for $t > 0$, and these will propagate up and down the rotation axis with a group velocity of $\mathbf{c}_g = \pm 2\boldsymbol{\Omega}/k$. Since the dominant wavenumber is $k \approx \pi/R$, the group velocity has a magnitude of $c_g \approx (2/\pi)\Omega R$. It follows that, at a given time t, there will be a cylindrical region of space above and below the disc of length $\ell \approx (2/\pi)\Omega Rt$ that is filled with inertial waves. The situation is as shown in Figure 10.6.

All of space is now divided into one of two regions. Within a cylinder of overall length $\ell \approx (4/\pi)\Omega Rt$ there exist waves, and the fluid inside this cylindrical region knows that the disc is moving because the waves have carried that information from the disc to the fluid. Outside this growing cylinder, however, there are no waves, and so the fluid there does not know the disc has been set in motion, there being no mechanism for transferring that information. The fluid outside the cylinder is therefore quiescent in the rotating frame of reference. Now, all of this may seem rather unremarkable until we look to see what happens to the fluid within the cylindrical wave region. Here we get a surprise: all of the fluid within the cylinder, both above and below the disc, moves upward with the speed V, shadowing the disc (see, for example, Figure 4.2 in Greenspan, 1968). Moreover, if we were to move the cylinder slowly to the right, instead of upwards, then we would obtain essentially the same result, in the sense that waves fill a cylindrical region of space of length $\ell \approx (4/\pi)\Omega Rt$, and within that region the fluid mimics the behaviour of the disc, drifting to the right. Of course, what we are doing is generating a Taylor column.

It is now clear how the fluid at A in Figure 10.1 knows that the object below is moving. The information is transmitted by low-frequency inertial waves that propagate upward from the surface of the object, with the object itself acting like a radio antenna. In short, the Taylor column is continually formed and reformed by the constant emission of low-frequency inertial waves from the object below.

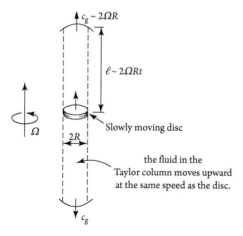

Figure 10.6 A slowly moving disc generates low-frequency inertial waves which fill a cylindrical region of space above and below the disc of length $\ell \approx (2/\pi)\Omega Rt$.

Note that the columnar structure shown in Figure 10.6 is a consequence of the fact that we chose to move the disc *slowly*, and so only low-frequency waves are generated, and of course such waves propagate along the rotation axis. If we were to oscillate the disc at some intermediate frequency, say $\varpi \approx \Omega$, then waves would propagate at an oblique angle determined by the dispersion relationship (10.20); that is to say, given $\varpi/2\Omega \approx 1/2$, the dispersion relationship fixes the orientation of $\mathbf{k}$ through $k_z/k = \pm\varpi/2\Omega \approx \pm 1/2$, and then $\mathbf{c}_g$ is chosen to be normal to $\mathbf{k}$. The net result is that waves propagate away from the disc in the form of two conical annuli, one directing waves upwards and one downwards, as shown in Figure 10.3 (a). The resulting dispersion pattern is not unlike that shown in Figure 9.7 for two-dimensional gravity waves.

10.3.3 The Spontaneous Focussing of Inertial Waves and the Formation of Columnar Vortices

So far, we have considered waves generated by a moving or oscillating boundary, where we can control the direction of wave propagation through the choice of frequency. We might classify such cases as boundary-value problems. However, this is not the only possibility.

One common situation is where the waves are excited not by a moving boundary, but by some initial perturbation in the velocity field, for example by a localized vortex in the rotating frame of reference. Typically, one solves such *initial-value* problems by taking a three-dimensional Fourier transform of the initial velocity field, $\mathbf{u}_0(\mathbf{x})$. (Readers not familiar with three-dimensional Fourier transforms may wish to consult Appendix 4 at this point.) This decomposes $\mathbf{u}_0(\mathbf{x})$ into a sum of Fourier modes of amplitude $\hat{\mathbf{u}}_0(\mathbf{k})$, with each mode looking like a plane wave whose orientation is fixed by its wavevector, $\mathbf{k}$. The transform also attributes a certain amount of kinetic energy to each mode, or we might say to each $\mathbf{k}$, of order $\hat{\mathbf{u}}_0(\mathbf{k}) \cdot \hat{\mathbf{u}}_0^{\dagger}(\mathbf{k})$, where $\dagger$ indicates a complex conjugate. These modes now propagate as plane waves according to (10.22), carrying their energy with them, and the disturbance field at later times is found by summing the modes, which amounts to performing an inverse Fourier transform on $\hat{\mathbf{u}}(\mathbf{k}, t)$. In short, the initial transform creates a continuous spectrum of $\mathbf{k}$ vectors and attributes a certain proportion of the initial kinetic energy to each $\mathbf{k}$. That energy is then carried off in the form of plane waves and the dispersion pattern at later times is found using superposition.

Now suppose that our initial disturbance, $\mathbf{u}_0(\mathbf{x})$, is localized near the origin. Clearly, we can vary $\hat{\mathbf{u}}_0(\mathbf{k})$, and hence the way that energy is distributed across the $\mathbf{k}$ vectors, through the choice of $\mathbf{u}_0(\mathbf{x})$. So we control the way in which wave energy disperses from the origin through our choice of the $\mathbf{k}$ vectors excited at $t = 0$, rather than by maintaining a fixed frequency at a boundary. Moreover, since $\mathbf{u}_0(\mathbf{x})$ is constrained only by $\nabla \cdot \mathbf{u}_0 = 0$, one might anticipate that, as in a boundary-value problem, we can orientate the wave energy flux at will. Somewhat surprisingly, it turns out that this is *not* the case. Rather, for almost any localized disturbance, the dispersion pattern is such that the wave energy flux is largest on the rotation axis. This ability to focus the wave energy flux in a particular direction (along the z-axis), irrespective of the initial condition, makes inertial waves unique amongst internal waves, and also explains why inertial-wave dispersion patterns are so often dominated by columnar structures aligned with the rotation axis.

In order to understand why the radiation of inertial waves from an isolated vortex tends to be dominated by dispersion along the rotation axis, it is convenient to follow an angular momentum argument given in Davidson et al. (2006). Consider an initial condition consisting of a vortex (a bob of vorticity) of scale δ, that is located near the origin and is circumscribed by an imaginary cylinder of radius $\delta/2$. The cylinder, V_c, is orientated with the rotation axis and infinitely long, as shown in Figure 10.7 (a). If we treat the fluid as inviscid, the angular momentum density of the fluid is governed by the cross product of $\mathbf{x}$ with (10.16);

$$\frac{\partial}{\partial t}(\mathbf{x} \times \mathbf{u}) = 2\mathbf{x} \times (\mathbf{u} \times \mathbf{\Omega}) + \nabla \times (p\mathbf{x}/\rho). \tag{10.27}$$

The z component of this may be rewritten using the relationship

$$\nabla \cdot \left[\left(\mathbf{x}^2 - z^2\right)\mathbf{\Omega}\mathbf{u}\right] = 2(\mathbf{x} \cdot \mathbf{u})\mathbf{\Omega} - 2(z\mathbf{\Omega})u_z = -\left[2\mathbf{x} \times (\mathbf{u} \times \mathbf{\Omega})\right]_z,$$

to give

$$\frac{\partial}{\partial t}(\mathbf{x} \times \mathbf{u})_z = -\nabla \cdot \left[r^2\mathbf{\Omega}\mathbf{u}\right] + \left[\nabla \times (p\mathbf{x}/\rho)\right]_z, \tag{10.28}$$

in (r, θ, z) coordinates. We now integrate (10.28) over the cylinder, V_c, and note that the pressure term converts to a surface integral which is readily shown to be zero, reflecting the fact that there is no axial torque acting on V_c arising from pressure. The net result is

$$\frac{dH_z}{dt} = \frac{d}{dt}\int_{V_c}(\mathbf{x} \times \mathbf{u})_z\, dV = -(\delta/2)^2\mathbf{\Omega}\oint \mathbf{u} \cdot d\mathbf{S} = 0, \tag{10.29}$$

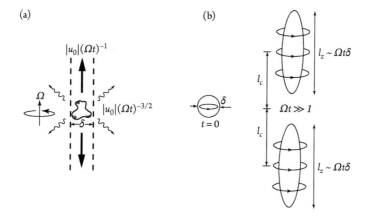

Figure 10.7 (a) The radiation of inertial waves from a localized vortex. The solid arrows represent the dispersion of angular momentum and the wiggly arrows the dispersion of energy. (b) The dispersion pattern after some time is dominated by two columnar vortices.

where $\mathbf{H}$ is the angular momentum within the cylinder and the integral on the right is zero by virtue of continuity.

We conclude that, despite the radiation of energy in the form of inertial waves, the axial component of angular momentum within V_c is conserved. Now, after a time t, inertial waves will have travelled a distance of order $\Omega \delta t$, and so the initial kinetic energy fills a volume of order $V_{3D} \sim (\Omega \delta t)^3$. Conservation of energy now requires that a characteristic velocity at time t satisfies $u^2 (\Omega \delta t)^3 \sim u_0^2 \delta^3$, where u_0 is a typical velocity at $t = 0$. It follows that, as the energy disperses, the velocity falls off as $u \sim u_0 (\Omega t)^{-3/2}$. This is true everywhere except within V_c, where conservation of angular momentum places a constraint on the decline in u. That is to say, as the inertial waves disperse, the angular momentum is spread over a cylindrical volume of the order of $V_{1D} \sim (\Omega \delta t) \delta^2$, and so conservation of H_z requires that, within V_c, $(u\delta)(\Omega \delta t) \delta^2 \sim (u_0 \delta) \delta^3$. It follows that a typical velocity near the rotation axis declines no faster than $u \sim u_0 (\Omega t)^{-1}$. This is indicated in Figure 10.7 (a).

These scaling arguments, and in particular $u \sim u_0 (\Omega t)^{-1}$ for on-axis radiation versus $u \sim u_0 (\Omega t)^{-3/2}$ for off-axis waves, are supported by an asymptotic analysis of the *dispersion integral*, i.e. the inverse Fourier transform obtained by Fourier analysis. Thus, we conclude that the energy density is always highest on the rotation axis. A typical dispersion pattern is shown schematically in Figure 10.7 (b). While there is some off-axis radiation, this tends to fall off quickly and the dominant pattern consists of two lobes propagating along the rotation axis, one upwards and one downwards. The centres of the lobes propagate at the group velocity corresponding to the dominant wavenumber $c_g = 2\Omega/k_{dom} \sim \Omega \delta$, and so are located at $|z| = \ell_c \sim \Omega \delta t$. However, since there is some spread of wavenumbers in the initial condition, say $\Delta k \sim 1/\delta$, and $c_{g,z} = 2\Omega/k$ for each wavenumber, the lobes elongate as they propagate, growing in length as $\ell_z \sim \Omega \delta t$. In short, radiation from a localized eddy spontaneously generates a pair of columnar vortices.

It is instructive to consider the simple example of axisymmetric dispersion initiated by the Gaussian vortex

$$\mathbf{u}_0 = \Lambda r \exp\left[-\left(r^2 + z^2\right)/\delta^2\right] \hat{\mathbf{e}}_\theta, \tag{10.30}$$

in (r, θ, z) coordinates. For axisymmetric flows the θ component of $\nabla^2 \mathbf{u}$ is

$$r\left(\nabla^2 \mathbf{u}\right)_\theta = r\frac{\partial}{\partial r}\frac{1}{r}\frac{\partial \Gamma}{\partial r} + \frac{\partial^2 \Gamma}{\partial z^2} = \nabla_*^2 \Gamma, \quad \Gamma = r u_\theta,$$

(see §3.3.1), and so the azimuthal component of our wave equation, (10.18), can be written in the form

$$\frac{\partial^2}{\partial t^2}\left[r\frac{\partial}{\partial r}\frac{1}{r}\frac{\partial \Gamma}{\partial r} + \frac{\partial^2 \Gamma}{\partial z^2}\right] + (2\Omega)^2 \frac{\partial^2 \Gamma}{\partial z^2} = 0. \tag{10.31}$$

This may be solved using a Hankel-cosine transform, as described in Exercise 10.1. (For axisymmetric problems a 3D Fourier transform often reduces to a Hankel-cosine transform in (r, z). See Bracewell, 1986.) The inverse transform then yields the dispersion integral

$$u_\theta \approx \Lambda\delta \int_0^\infty \kappa^2 e^{-\kappa^2} J_1\left(\frac{2\kappa r}{\delta}\right) \left\{ \exp\left[-\left(\frac{z}{\delta} - \frac{\Omega t}{\kappa}\right)^2\right] + \exp\left[-\left(\frac{z}{\delta} + \frac{\Omega t}{\kappa}\right)^2\right] \right\} d\kappa,$$

(10.32a)

from which we find

$$\omega_z \approx 2\Lambda \int_0^\infty \kappa^3 e^{-\kappa^2} J_0\left(\frac{2\kappa r}{\delta}\right) \left\{ \exp\left[-\left(\frac{z}{\delta} - \frac{\Omega t}{\kappa}\right)^2\right] + \exp\left[-\left(\frac{z}{\delta} + \frac{\Omega t}{\kappa}\right)^2\right] \right\} d\kappa,$$

(10.32b)

$$u_z \approx \Lambda\delta \int_0^\infty \kappa^2 e^{-\kappa^2} J_0\left(\frac{2\kappa r}{\delta}\right) \left\{ -\exp\left[-\left(\frac{z}{\delta} - \frac{\Omega t}{\kappa}\right)^2\right] + \exp\left[-\left(\frac{z}{\delta} + \frac{\Omega t}{\kappa}\right)^2\right] \right\} d\kappa,$$

(10.32c)

where $\kappa = k_r \delta/2$, and J_0 and J_1 are the usual Bessel functions. (Again, the details are spelt out in Exercise 10.1). The presence of the term $\kappa^2 e^{-\kappa^2}$ tells us that the integrands are dominated by $\kappa = O(1)$. Consequently, in order to keep the argument in one of the other two exponentials of order unity for $\Omega t \gg 1$, we require $z \sim \pm\Omega\delta t$. This tells us that the energy is centred at $|z| \sim \Omega\delta t$, exactly as expected. Figure 10.8 shows the corresponding energy distribution at various times in the upper half of the r–z plane. The dominance of the on-axis radiation is clear.

The form of (10.32a) for $\Omega t \gg 1$ may be found by demanding that the argument in one of the exponentials remains of order one as $\Omega t \to \infty$. At $z = \pm\delta\Omega t$ this requires $\kappa \to 1$, which gives $u_\theta \sim \Lambda\delta J_1(2r/\delta)(\Omega t)^{-1}$, as shown in Exercise 10.1. This, in turn, yields $u_\theta \sim \Lambda r (\Omega t)^{-1}$ near the axis and $u_\theta \sim \Lambda\delta (\Omega t)^{-3/2} (z/r)^{1/2}$ off the axis, consistent with the scalings above.

A second, more geometrical way of understanding why wave energy concentrates on the rotation axis is given in Davidson (2013). Consider a wave packet of dominant wavevector

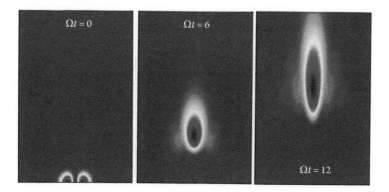

Figure 10.8 The radiation pattern for inertial waves dispersing from a Gaussian vortex.

k which leaves the source at $t = 0$. If the packet propagates upward, then at time t it reaches the point

$$\mathbf{x} = \mathbf{c}_g t = \frac{2\Omega t}{k^3}\left[k^2\hat{\mathbf{e}}_z - k_z\mathbf{k}\right].$$ (10.33)

Evidently, **x** is normal to **k** and coplanar with **k** and **Ω**. Now, for any **x** which lies off the axis, such as point B in Figure 10.9, there is one and only one orientation of **k** that takes wave energy to that point (the inclined **k** in Figure 10.9). However, if **x** is located on the rotation axis, such as at point A in Figure 10.9, then any **k** that lies in the horizontal plane transports energy to that location. It follows that all of the energy that lies within a thin horizontal disc in **k**-space is folded up into a narrow cylinder in real space. It is this process of channelling energy from a two-dimensional object in **k**-space into a one-dimensional object in real space that is responsible for the concentration of energy on the rotation axis.

This second explanation for the focussing of energy onto the rotation axis has the advantage of simplicity although, unlike the angular momentum argument, it does not yield the scaling laws $u \sim u_0\,(\Omega t)^{-1}$ and $u \sim u_0\,(\Omega t)^{-3/2}$ for waves initiated by a vortex. Another advantage of the geometrical argument is that it is independent of how the waves are excited, and requires only that the excitation is local and has no imposed time scale. For example, Figure 10.10 shows inertial wave packets radiating from a thin layer of buoyant blobs that are drifting horizontally under the influence of gravity through a rotating, Boussinesq fluid at low Ro. Yet again, we see the formation of columnar vortices.

Columnar vortices are ubiquitous in the numerical simulations and laboratory experiments of rotating turbulence. The only prerequisite it that Ro $\sim O(1)$, or smaller, where Ro is based on the rms fluctuating velocity and a characteristic transverse scale of the large eddies. A snapshot of a typical numerical simulation is shown in Figure 10.11. Notice the helical particle trajectories as well as the columnar nature of the large vortices. Notice also that the small-scale structures, whose small size means that they feel the rotation less strongly, are much more three-dimensional.

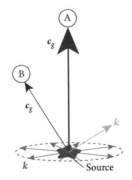

Figure 10.9 The group velocity, $\mathbf{c}_g$, and the associated orientation of **k** for on-axis (red) and off-axis (blue) wave packets. (Figure by Oliver Bardsley.)

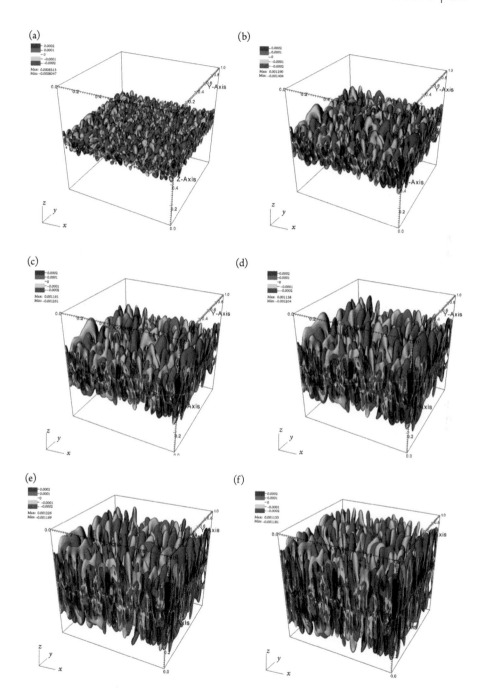

Figure 10.10 The radiation of inertial waves packets from a layer of buoyant blobs slowly drifting through a rotating Boussinesq fluid under the influence of gravity, with $\mathrm{Ro} \ll 1$ and $\mathbf{g} = -g\hat{\mathbf{e}}_x$. The panels show surfaces of u_z, with blue for $u_z < 0$ and red for $u_z > 0$, at the times $\Omega t = 2, 4, 6, 8, 10, 12$. (From Davidson and Ranjan, 2015.)

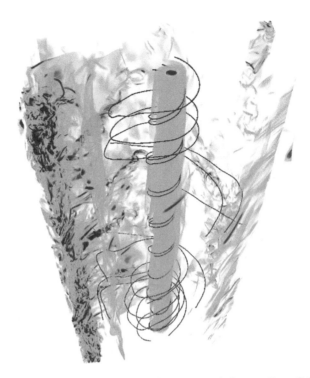

Figure 10.11 The numerical simulation of rapidly rotating turbulence at Ro ~ 0.06, visualized by thresholding on the vorticity. (Courtesy of Pablo Mininni and Annick Pouquet.)

Similar behaviour is seen in physical experiments. For example, the laboratory experiment of Davidson et al. (2006) is shown schematically in Figure 10.12. Here, turbulence is created in the upper part of a tank of water by dragging a planar mesh of orthogonal bars part way through the tank and then removing the mesh altogether. This generates turbulence whose initial value of Ro is large, which ensures that no inertial waves are generated by the mesh used to create the turbulence. However, with time, Ro drifts downwards as the energy of the turbulence decays, and eventually we enter the regime of Ro $\sim O(1)$. When Ro reaches a value of ~ 0.4, columnar vortices start to emerge from the turbulent cloud, propagating in the axial direction, as shown schematically in Figure 10.12. Note that the wider eddies elongate faster than the thinner vortices, which is exactly what would be expected from the expression for group velocity for inertial waves.

Measurements show that these columnar vortices elongate at a constant rate, as shown in Figure 10.13 (a). Moreover, the growth rate is proportional to Ω and to the transverse scale of the vortices, δ, for which the mesh size, M, acts as a proxy (see Figure 10.13 (b)). Thus we have $\ell_z \sim \delta \Omega t \sim M \Omega t$, which confirms that the columnar vortices are nothing more than inertial wave packets.

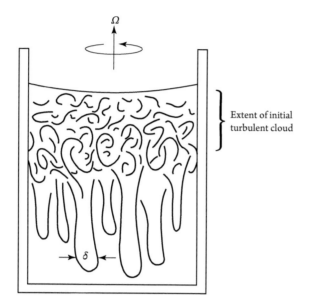

Figure 10.12 Schematic of the experiment of Davidson et al. (2006). Columnar vortices emmerge from a turbulent cloud.

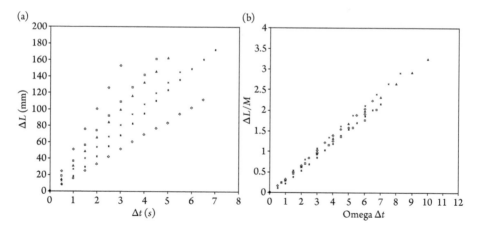

Figure 10.13 The variation of length, ΔL, versus time of the dominant columnar vortex in six experiments which had different rotation rates and different mesh sizes, M. (a) ΔL versus t shows that each columnar vortex grows at a constant rate. (b) $\Delta L/M$ versus Ωt confirms that the columnar vortices grow as $\ell_z \sim M\Omega t$.

10.3.4 Helicity Generation and Helicity Segregation by Inertial Waves

Equation (10.25) tells us that a monochromatic inertial wave possesses a helicity $h = \mathbf{u} \cdot \boldsymbol{\omega}$ which is maximal, $|h| = |\mathbf{u}|\,|\boldsymbol{\omega}|$, with $h < 0$ when the wave energy propagates upward $(\mathbf{c}_g \cdot \boldsymbol{\Omega} > 0)$ and $h > 0$ for the downward propagation of wave energy $(\mathbf{c}_g \cdot \boldsymbol{\Omega} < 0)$. It does not necessarily follow that a *wave packet*, which contains many overlapping wavenumbers, behaves in the same way, but it turns out that wave packets do (more or less) exhibit the same behaviour, with the helicity of the packet close to maximal and $h < 0$ when $\mathbf{c}_g \cdot \boldsymbol{\Omega} > 0$ and $h > 0$ for $\mathbf{c}_g \cdot \boldsymbol{\Omega} < 0$ (see Davidson and Ranjan, 2015). It follows that packets of inertial waves are a particularly efficient means of transporting helicity across a fluid.

This is illustrated in Figure 10.14, which shows the influence of background rotation on a slab of turbulence. The initial condition is shown in Figure 10.14 (a) and consists of a random collection of eddies located near the central plane of the computational domain, with Ro ~ 0.1 based on the rms fluctuating velocity and the size of the vortices. The flow is then allowed to evolve, which it does by radiating wave packets up and down the rotation axis, and by $\Omega t = 6$ it has reached the state shown in Figure 10.14 (b). As expected, the wave packets take the form of columnar vortices aligned with the rotation axis. The key point about Figure 10.14, however, is that the vortices are coloured according to their helicity, with red for $h < 0$ and green for $h > 0$. Clearly, the upward propagating wave packets carry negative helicity and the downward propagating packets positive helicity.

Essentially the same thing is seen if the initial condition consists of a layer of buoyant blobs slowly drifting through a rotating Boussinesq fluid at low Ro under the influence of gravity. The velocity field generated by just such a random distribution of buoyant blobs is shown in Figure 10.10. Here, the gravitational field is orthogonal to $\boldsymbol{\Omega}$, $\mathbf{g} = -g\hat{\mathbf{e}}_x$, and the drifting buoyant blobs are located near the central plane of the computational domain. As in

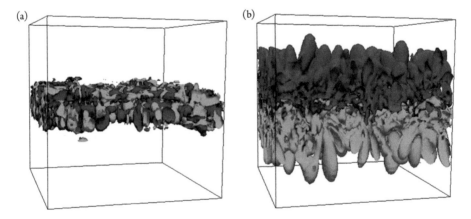

Figure 10.14 The influence of background rotation on a slab of turbulence. (a) The initial condition consists of a random collection of eddies confined to the central plane of the computational domain, with Ro ~ 0.1. (b) The flow at $\Omega t = 6$ coloured by helicity, with red for $h < 0$ and green for $h > 0$. (From Davidson, 2013.)

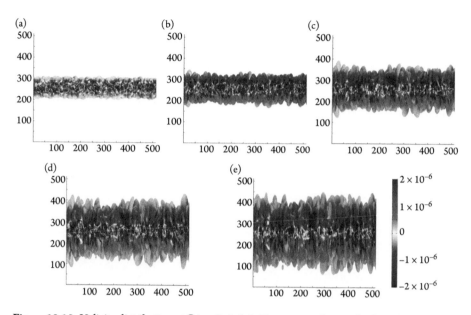

Figure 10.15 Helicity distribution at $\Omega t = 2, 4, 6, 8, 10$ corresponding to the flow shown in Figure 10.10, with blue for $h < 0$ and red for $h > 0$. (From Davidson and Ranjan, 2015.)

Figure 10.14, the flow evolves by radiating wave packets up and down the rotation axis. The same buoyancy-induced flow is reproduced in Figure 10.15, only this time it is coloured by helicity, with blue for $h < 0$ and red for $h > 0$. Once again, we see that the upward propagating wave packets carry negative helicity and the downward propagating packets positive helicity. Evidently, inertial waves triggered by a localized source not only disperse helicity throughout the fluid, but also segregate that helicity, with left-handed spirals above the source and right-handed spirals below.

This spatial segregation of helicity can be important in certain geophysical flows. For example, numerical simulations of the flow in the fluid core of a rapidly rotating planet, such as the Earth, show that the basic flow structure consists of a multitude of thin columnar vortices aligned with the rotation axis (Figure 10.16 (a)), more or less arranged as alternating cyclones ($\omega_z > 0$) and anti-cyclones ($\omega_z < 0$). These columnar eddies come and go in a chaotic fashion and span much of the fluid core. They are also strongly helical and planetary dynamo theory shows that this helicity is crucial to the maintenance of the planetary magnetic field. Moreover, for an efficient planetary dynamo, it turns out that one needs a statistically persistent segregation of the helicity, with helicity predominantly of one sign in the north, and the opposite sign in the south. The numerical simulations show just that, with the azimuthally averaged helicity distribution predominantly negative in the north and positive in the south (Figure 10.16 (b)). The similarity of this flow to that shown in Figures 10.10 and 10.15 is striking.

This similarity has led to the speculation that the columnar vortices in these dynamo simulations are generated by buoyant blobs in and around the equatorial plane, blobs that

(a)
(b)

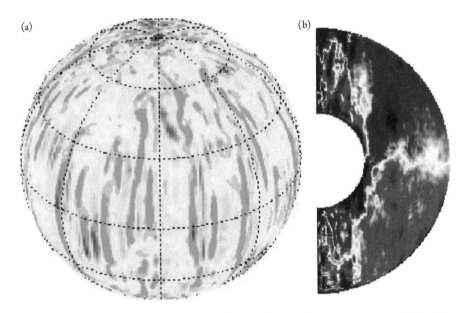

Figure 10.16 (a) Alternating cyclonic–anticyclonic columnar vortices in a numerical simulation of a planetary dynamo. (b) Azimuthally averaged helicity in a vertical plane, with blue for $h < 0$, red for $h > 0$. (From Ranjan et al., 2018.)

(a)
(b)
(c)

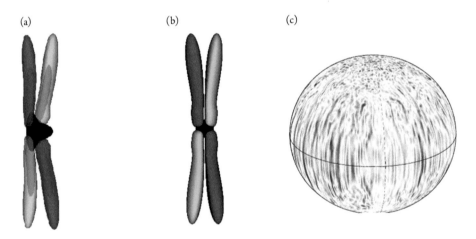

Figure 10.17 (a) A buoyant blob drifts out of the page while emitting inertial waves ($u_z < 0$ is blue, $u_z > 0$ red). The waves are organized as four columnar vortices: a cyclone–anticyclone pair above the blob matched to a similar pair below. Ro = 0.1. (b) As for (a), but with Ro = 0.01. (c) Alternating cyclones–anticyclones in a planetary core. (Davidson and Ranjan, 2015.)

drift radially outward towards the mantle (Davidson and Ranjan, 2018). This is consistent with the observation that, near the equatorial plane, there is an intense outward flow of hot, buoyant material. The dispersion pattern for waves generated by a *single* buoyant blob is discussed in Exercise 10.2 and illustrated in Figure 10.17 (a) and (b). It consists of a pair of cyclonic–anti-cyclonic columnar vortices above the blob matched to a similar pair below, which is highly reminiscent of the flow observed in the numerical dynamos.

10.3.5 Finite-amplitude Inertial Waves

Axisymmetric inertial waves of *small amplitude* are governed by (10.31),

$$\frac{\partial^2}{\partial t^2}\left[r\frac{\partial}{\partial r}\frac{1}{r}\frac{\partial \Gamma}{\partial r} + \frac{\partial^2 \Gamma}{\partial z^2}\right] + (2\Omega)^2\frac{\partial^2 \Gamma}{\partial z^2} = 0, \tag{10.34}$$

where $\Gamma = ru_\theta$ and we use (r,θ,z) coordinates. This supports travelling waves of the form

$$\Gamma = A\mathrm{J}_1(k_r r)\cos(k_z(z \pm Vt)), \quad V > 0, \tag{10.35}$$

where A is an amplitude, V an axial phase velocity, J_1 the usual Bessel function, and k_r and k_z are related by $k_r^2 + k_z^2 = (2\Omega/V)^2$ (Kelvin, 1880). Moreover, the radial velocity can be obtained from Γ through the azimuthal component of (10.16), and so it too has a radial structure of the form $u_r \sim \mathrm{J}_1(k_r r)$. It follows that, if the waves are confined to a tube of radius b, then the admissible values of k_r are set by the requirement that $u_r(r = b) \sim \mathrm{J}_1$ $(k_r b) = 0$, and this in turn demands that $k_r b = \gamma_n$, where γ_n are the zeros of J_1. Hence these waves have axial wavenumbers governed by

$$k_z^2 b^2 = (2\Omega b/V)^2 - \gamma_n^2, \tag{10.36}$$

subject to the restriction that $2\Omega b/V > \gamma_1 = 3.83$. We conclude that there is an upper limit on the phase velocity of confined axisymmetric waves, of $V/\Omega b < 0.522$.

We have been here before. Suppose that, *in the laboratory frame of reference*, the fluid has a uniform axial velocity of U and an axial rotation rate of Ω. Suppose also that we take the waves to propagate upstream, against the mean flow. If $U = V$, then the waves will appear stationary in the laboratory frame, and so we have precisely the situation shown in Figure 8.5. Moreover, it is readily confirmed that the solution derived above for u_r is identical in form to that of §8.2.3, with (10.38) equivalent to (8.21). The key point, however, is that the steady analysis of §8.2.3 is fully non-linear. If we now revert to a frame of reference that moves with the mean flow, we conclude that finite-amplitude inertial waves of axisymmetric form can propagate through a fluid that is stationary in the rotating frame of reference.

Of course, there is a direct analogy between these non-linear inertial waves and the finite-amplitude lee waves discussed in §9.4.2. Indeed, much of the non-linear analysis of §8.2.3 is similar to that used to obtain finite-amplitude lee waves in §9.4.2, the main difference being that the inertial waves are axisymmetric while the lee waves are two-dimensional. Yet again, we see that the analogy between swirling and stratified fluids can be suggestive.

10.4 Rossby Waves

In this section we continue our discussion of dynamics at low Ro by introducing a second type of wave motion, known as *Rossby waves*. These are commonly observed in the atmosphere and in the oceans, typically at large scales. For example, atmospheric Rossby waves may have wavelengths as long as 10^3 km, and are sometimes called planetary waves. Although Rossby waves have some connection to standing (as opposed to progressive) inertial waves, it probably is best to think of them as a completely new class of wave.

A steady motion in which the Coriolis force is balanced by a pressure gradient is called a *geostrophic flow*. Such flows are necessarily two-dimensional in the sense that (10.17) demands $(\mathbf{\Omega} \cdot \nabla)\mathbf{u} = 0$, from which $\mathbf{u} = \mathbf{u}(x, y)$. Large-scale geophysical flows are often represented as small, slow departures from a geostrophic force balance and such flows are referred to as *quasi-geostrophic*. Rossby waves are an example of just such a motion, being *low-frequency* perturbations of a geostrophic base state.

Perhaps the simplest configuration in which Rossby waves arise is that shown in Figure 10.18. Here, the fluid is confined between two planes, say $z = 0$ and $z = h(y)$, which are almost, but not quite, parallel. We represent this by $h = h_0 - sy$, where s is a small, positive constant. Now, in a geostrophic flow we have $\partial u_z / \partial z = 0$, so a fluid column cannot change its height and individual particles must follow contours of constant depth. If a fluid column is displaced laterally in the configuration below, then this geostrophic constraint is broken, and as the fluid returns to its original geostrophic height it overshoots, triggering oscillations. The governing equations for these oscillations are often deduced by performing a formal perturbation analysis about the geostrophic state, using s as a small parameter. However, we start with a simpler, more intuitive approach.

Let us set aside any background flow, ignore viscosity, and assume that the wave motion is predominantly planar, so that $|\omega_z| \gg |\omega_x|, |\omega_y|$ in our rotating frame of reference. Since ω_x and ω_y are small, (10.17) yields

$$\frac{\partial u_x}{\partial z} = \frac{1}{2\Omega} \frac{\partial \omega_x}{\partial t} \approx 0, \quad \frac{\partial u_y}{\partial z} = \frac{1}{2\Omega} \frac{\partial \omega_y}{\partial t} \approx 0, \tag{10.37}$$

$$\frac{\partial \omega_z}{\partial t} = 2\Omega \frac{\partial u_z}{\partial z}. \tag{10.38}$$

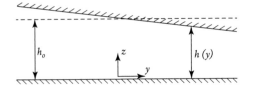

Figure 10.18 A simple configuration in which Rossby waves arise.

Now, (10.37) combined with the continuity equation demands $\partial^2 u_z / \partial z^2 = 0$, and since $u_z(z = h) = -s u_y$, we conclude that $\partial u_z / \partial z = -s u_y / h_0$. Our governing equations are then

$$\frac{\partial u_x}{\partial x} + \frac{\partial u_y}{\partial y} = \frac{s u_y}{h_0}, \qquad (10.39)$$

and

$$\frac{\partial \omega_z}{\partial t} = -\frac{2\Omega s}{h_0} u_y. \qquad (10.40)$$

We now look for solutions which are independent of y, i.e. $u_x = u_x(x, t), u_y = u_y(x, t)$. Equations (10.39) and (10.40) then simplify to

$$\frac{\partial u_x}{\partial x} = \frac{s u_y}{h_0}, \quad \frac{\partial^2 u_y}{\partial x \partial t} = -\frac{2\Omega s}{h_0} u_y, \qquad (10.41)$$

which support progressive waves of the form $u_y \sim \exp\left[j\left(k_x x - \varpi t\right)\right]$ with the dispersion relationship

$$\varpi = -2\Omega s / k_x h_0, \quad k_x < 0. \qquad (10.42)$$

The phase and group velocities are then

$$\mathbf{c}_p = -\frac{2\Omega s}{h_0 k_x^2} \hat{\mathbf{e}}_x, \quad \mathbf{c}_g = \frac{2\Omega s}{h_0 k_x^2} \hat{\mathbf{e}}_x. \qquad (10.43)$$

Perhaps some comments are appropriate at this point. First, we have $\varpi \sim s\Omega \ll \Omega$ and so these are *low-frequency* waves. Second, Rossby waves are dispersive as $\mathbf{c}_g$ is a function of k_x. Third, since $k_x < 0$, these are progressive waves that can propagate in *one direction only*, which is quite different to travelling inertial waves. Fourth, we have $|\mathbf{c}_g| \sim s\Omega / |k_x| \ll \Omega / |k_x|$ and so Rossby waves have a group velocity that is much smaller than that of inertial waves. Fifth, $\mathbf{c}_g$ and $\mathbf{c}_p$ are of *opposite* signs, and so a wave packet propagates in the opposite direction to the wave crests that sit within that packet! Sixth, suppose that such a wave is superimposed on a uniform geostrophic motion whose velocity is chosen to be equal and opposite to the phase velocity. Then the undulating wave pattern will appear steady, rather like that of lee waves downwind of a mountain chain, and indeed this is what tends to happen when Rossby waves are triggered by flow over an obstacle.

The analysis above might be considered somewhat ad hoc, and so we close this section by indicating how a more formal perturbation analysis might proceed. We start by writing $\mathbf{u} = \mathbf{u}^{(0)} + \mathbf{u}^{(1)} + \cdots$, where (0) indicates a quasi-steady geostrophic motion corresponding to $s = 0$ and (1) indicates a small perturbation to the geostrophic state caused by a small but finite s. Next, we note that $(\mathbf{\Omega} \cdot \nabla)\mathbf{u}^{(0)} = 0$ requires $\mathbf{u}^{(0)} = \mathbf{u}_\perp^{(0)}(x, y, t)$ and hence

$\boldsymbol{\omega}^{(0)} = \omega_z^{(0)}(x, y, t)\hat{\mathbf{e}}_z$, where $\perp$ indicates the horizontal components of $\mathbf{u}$. The first-order perturbation is now governed by

$$\frac{\partial \boldsymbol{\omega}^{(0)}}{\partial t} = 2(\boldsymbol{\Omega} \cdot \nabla)\mathbf{u}^{(1)}, \quad \nabla \cdot \mathbf{u}^{(1)} = 0, \tag{10.44}$$

which allows the geostrophic motion to evolve *slowly* by relaxing the constraint of strict z-independence. This now yields

$$\frac{\partial u_x^{(1)}}{\partial z} = \frac{\partial u_y^{(1)}}{\partial z} = 0, \tag{10.45}$$

$$\frac{\partial \omega_z^{(0)}}{\partial t} = 2\Omega \frac{\partial u_z^{(1)}}{\partial z}, \tag{10.46}$$

which are, in effect, (10.37) and (10.38). Next, (10.45) combined with $\nabla \cdot \mathbf{u}^{(1)} = 0$ demands $\partial^2 u_z^{(1)}/\partial z^2 = 0$, and since $u_z^{(1)}(z = h_0) = -s u_y^{(0)}$, we conclude that $\partial u_z^{(1)}/\partial z = -s u_y^{(0)}/h_0$. Our governing equation is now

$$\frac{\partial \omega_z^{(0)}}{\partial t} = -\frac{2\Omega s}{h_0} u_y^{(0)}, \tag{10.47}$$

which is just (10.40). If we introduce a streamfunction $\psi^{(0)}(x, y, t)$ for the two-dimensional motion $\mathbf{u}^{(0)}$ then (10.47) becomes

$$\frac{\partial}{\partial t} \nabla_\perp^2 \psi^{(0)} = -\frac{2\Omega s}{h_0} \frac{\partial \psi^{(0)}}{\partial x}, \tag{10.48}$$

which supports two-dimensional waves of the form $\psi^{(0)} \sim \exp[j(\mathbf{k} \cdot \mathbf{x} - \varpi t)]$ whose dispersion relationship is

$$\varpi = -2\Omega s k_x/k^2 h_0, \quad k_x < 0. \tag{10.49}$$

Finally, this generalization of (10.42) gives the components of the group velocity as

$$c_{g,x} = \frac{2\Omega s}{h_0 k^4}(k_x^2 - k_y^2), \quad c_{g,y} = \frac{2\Omega s}{h_0 k^4}(2k_x k_y), \tag{10.50}$$

which revert to (10.43) when $k_y = 0$.

This is all we shall say about Rossby waves, other than to note that they arise, not only from non-parallel boundaries, but also from the variation with latitude of the component of $\boldsymbol{\Omega}$ normal to the surface of the Earth: the so-called β-effect. There is a vast literature on Rossby waves, reflecting their importance in large-scale meteorological and oceanic dynamics. Readers seeking more details will find a gentle introduction in Tritton (1988), while Pedlosky (1987) and Greenspan (1968) provide a detailed analysis.

10.5 Ekman Boundary Layers and Ekman Pumping

10.5.1 Confined Swirling Flows: the Solutions of Kármán, Bödewadt, and Ekman

We now turn to boundary layers in a rotating fluid. To that end, it is convenient to revert to an inertial frame of reference and relax the assumption that Ro is small. We start with Kármán's (1921) solution for the flow generated by a rotating disc of *infinite* radius.

Suppose that a disc rotates with angular velocity Ω in an otherwise still fluid. A boundary layer forms on the surface of the disc and the associated flow is completely determined by the two parameters ν and Ω, from which we can construct the length scale $\hat{\delta} = \sqrt{\nu/\Omega}$. Since there is no geometric length scale associated with an infinite disc, the boundary-layer thickness must scale on $\hat{\delta}$. In addition, the no-slip boundary condition suggests that $u_\theta \sim \Omega r$ in (r, θ, z) coordinates centred on the disc surface, and so Kármán proposed a solution of the form

$$u_r = \Omega r F(z/\hat{\delta}), \quad u_\theta = \Omega r G(z/\hat{\delta}), \quad u_z = \Omega \hat{\delta} H(z/\hat{\delta}), \tag{10.51}$$

with pressure a function of z only. The dimensionless functions F, G, and H can now be determined from continuity, which demands $2F + H' = 0$, and from the radial and azimuthal components of the Navier–Stokes equation, which require

$$F^2 + F'H - G^2 = F'', \quad 2FG + HG' = G''. \tag{10.52}$$

The corresponding boundary conditions are no-slip at surface of the disc and $u_r = u_\theta = 0$ in the far field, which translate to

$$F(0) = 0, \ G(0) = 1, \ H(0) = 0, \quad F(\infty) = G(\infty) = 0. \tag{10.53}$$

The integration of (10.52) subject to these boundary conditions is routine and the solutions for F, G, and $-H$ are shown in Figure 10.19. Note that $H(\infty) = -0.884$ and $G'(0) = -0.616$.

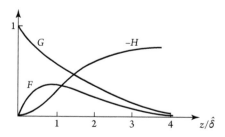

Figure 10.19 Kármán's solution of F, G, and $-H$ for the flow induced by a rotating disc.

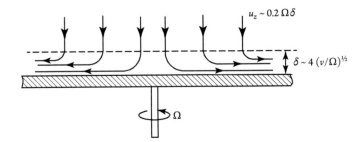

Figure 10.20 The secondary flow in Kármán's problem of flow induced by a rotating disc.

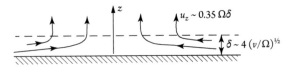

Figure 10.21 The secondary flow induced by a rotating fluid above a stationary surface.

In effect, we have a centrifugal fan, with a flow which spirals radially outward within a boundary layer of thickness $\delta \approx 4\hat{\delta}$. The outward mass flux within the boundary layer is fed by a weak axial flow towards the disc of $u_z(z \to \infty) \approx -0.884\Omega\hat{\delta}$, with the resulting secondary flow in the r–z plane shown in Figure 10.20. The azimuthal surface stress is $|\tau_{\theta z}| = 0.616\rho\Omega^2 r\hat{\delta}$, from which the frictional torque on the disc can be calculated.

The reverse problem, in which the disc is stationary and the fluid remote from the disc rotates at the rate Ω, was studied by Bödewadt (1940). Here, the azimuthal velocity adjusts from $u_\theta = \Omega r$ in the far field to $u_\theta = 0$ at $z = 0$ across a thin boundary layer. As before, the characteristic boundary-layer thickness is $\hat{\delta} = \sqrt{\nu/\Omega}$, and so Kármán's assumption that $\mathbf{u}$ takes the form (10.51) remains valid. However, there is now a radial pressure gradient in the far field, which adds a new term to the radial component of the Navier–Stokes equation. In particular, in order to satisfy the boundary conditions $F(\infty) = 0, G(\infty) = 1$, the term G^2 in (10.52) has to be replaced by $G^2 - 1$, so the radial equation becomes

$$F^2 + F'H - G^2 + 1 = F''. \tag{10.54}$$

Integration of the resulting equations, subject to the new boundary conditions of $G(0) = 0$ and $G(\infty) = 1$, yields a boundary-layer thickness of $\delta \approx 4\hat{\delta}$, as in Kármán's problem. However, the resulting flow differs from Kármán's solution in that the secondary motion in the r–z plane is reversed, with the fluid particles spiralling radially inward within the boundary layer, eventually drifting up and out of the layer (Figure 10.21). This results in a weak axial flow *away* from the disc, of $u_z(z \to \infty) \approx 1.35\Omega\hat{\delta}$.

The reason for the radial inflow within the boundary layer is a force imbalance. Outside the boundary layer there is a radial force balance between $\partial p/\partial r$ and $\rho u_\theta^2/r$, resulting in a low pressure on the axis. This radial pressure gradient is imposed on the fluid within the boundary layer where the centrifugal force is smaller, and the resulting imbalance between $\partial p/\partial r$ and $\rho u_\theta^2/r$ drives the fluid radially inward.

A third common situation is where both the disc and the fluid rotate, but at slightly different rates. In such cases it pays to return to a rotating frame in which the fluid at infinity is stationary, and solve the low-Ro version of the Navier–Stokes equation:

$$\nabla\left(p/\rho\right) = 2\mathbf{u} \times \mathbf{\Omega} + \nu\nabla^2\mathbf{u}. \tag{10.55}$$

Boundary layers which are driven by small differences in rotation are called *Ekman layers* and their primary features are similar to the two flows already discussed. In particular, when the fluid rotates faster than the boundary, we find something resembling Bödewadt's problem, with the fluid spiralling radially inward, and when the boundary rotates faster than the fluid, the flow resembles Kármán's solution, with a radial outflow.

The structure of these Ekman layers is readily found. Suppose that, in an inertial frame, the fluid remote from the disc rotates at Ω and the disc at $\Omega + \Delta\Omega$, with $\Delta\Omega \ll \Omega$. Then we take

$$u_r = \Delta\Omega r F(z/\hat{\delta}), \quad u_\theta = \Delta\Omega r G(z/\hat{\delta}), \quad u_z = \Delta\Omega\hat{\delta}H(z/\hat{\delta}), \tag{10.56}$$

in the *rotating frame*, with $\hat{\delta} = \sqrt{\nu/\Omega}$, $\eta = z/\hat{\delta}$ and $z = 0$ at the surface of the disc. The radial and azimuthal components of (10.55) then yield

$$F''(\eta) = -2G, \quad G''(\eta) = 2F, \tag{10.57}$$

the reduced pressure in (10.55) being a function of z only. Solving for F and G, subject to the boundary conditions $F = 0$ and $G = 1$ at $\eta = 0$ and $F = G = 0$ at $\eta \to \infty$, gives

$$F = \exp(-\eta)\sin\eta, \quad G = \exp(-\eta)\cos\eta. \tag{10.58}$$

It follows that u_r is mostly positive for $\Delta\Omega > 0$, and negative for $\Delta\Omega < 0$, as suggested above, although there are weak oscillations in sign at large η. The net velocity in a horizontal plane is then

$$\mathbf{u}_\perp = \Delta\Omega r \exp(-\eta)\left(\sin\eta\hat{\mathbf{e}}_r + \cos\eta\hat{\mathbf{e}}_\theta\right), \tag{10.59}$$

whose orientation varies with height in a pattern known as the *Ekman spiral*. Finally, the axial velocity can be found from u_r using the continuity equation in the form $2F + H' = 0$.

In the geophysical literature it has become common to refer to *any* boundary layer established by a difference in rotation rate between the fluid and a solid boundary as an Ekman layer, including the non-linear problems of Kármán and Bödewadt. The associated secondary flow in the r–z plane, particularly the axial velocity in the far field, is then called *Ekman pumping*.

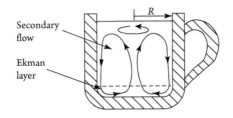

Figure 10.22 Spin-down of a stirred cup of tea.

10.5.2 Ekman Layers as a Mechanism for Energy Dissipation

When the Reynolds number is large, the far-field axial velocity induced by Ekman pumping is much weaker than the boundary-layer flow, by a factor of δ/R, where R is the radius of the disc. Nevertheless, if the fluid is confined, this weak axial flow inevitably has important consequences for the flow as a whole. This is often illustrated by *spin-down* in a stirred cup of tea, a problem investigated by Kelvin's brother in 1857 and Einstein in 1926.

Suppose that, with the help of a spoon, the tea is set into a state of rotation. The spoon is then removed and we ask: how long does it take for the tea to stop spinning? To answer this, we note that an Ekman layer of the Bödewadt type is established on the bottom of the teacup, inducing a radial inflow along the base of the cup. Continuity then requires that the fluid drifts up and out of the boundary layer, where it is recycled through a layer on the side of the cup. The end result is that a secondary flow is established as shown in Figure 10.22, with the fluid spiralling radially inwards through the Ekman layer, then up into the core of the cup, and finally back down through the side-wall boundary layer. Crucially, as each fluid particle passes through the Ekman layer, it gives up a significant fraction of its kinetic energy. So the tea comes to rest when all of the contents of the cup have been flushed (once or twice) through the Ekman layer. It follows that the spin-down time is of the order of the turn-over time of the secondary flow, $\tau \sim R/u_z \sim R/1.4\sqrt{\nu\Omega}$, which is often rewritten in terms of the *Ekman number*, $\mathrm{Ek} = \nu/\Omega R^2$, as $\tau \sim \mathrm{Ek}^{-1/2}\Omega^{-1}$. This might be compared with spin-down in a *long* cylinder where there is no Ekman pumping and so the spin-down is controlled by the time taken for the centreline vorticity to diffuse to the wall, which is $\tau \approx R^2/\gamma_1^2\nu$ where $\gamma_1 = 3.83$ (see Exercise 3.3). Suppose, for example, that $R = 4\,\mathrm{cm}$, $\Omega = 4\,\mathrm{s}^{-1}$ and $\nu = 10^{-6}\,\mathrm{m}^2/\mathrm{s}$. Then $\tau \sim 15\,\mathrm{s}$ in a teacup, as compared to $\tau \sim 2$ minutes in a long cylinder where there is no Ekman pumping. The moral is: Ekman pumping provides a particularly efficient mechanism for destroying mechanical energy.

10.6 Tropical Cyclones

10.6.1 The Anatomy of a Tropical Cyclone

Tropical cyclones, sometimes called hurricanes or typhoons, are vast storm systems typically 300–1000 km in radius, that form over warm tropical seas. They extend from the ocean surface up to the tropopause, which has an altitude of $H \sim 15$ km, and their primary energy source is the warm moisture that evaporates from the sea surface to heat the air, moisture that condenses at cooler altitudes, forming clouds and releasing latent heat. The rotation of the Earth also plays an important role in the dynamics of cyclones, acting as a source of angular momentum. Thus, cyclones form *near* the equator, where the seas are warm, but not *at* the equator, as they require a finite component of rotation normal to the surface of the Earth. Indeed, most tropical cyclones form within a colatitude of $5°$–$25°$.

The overall flow in a tropical cyclone is roughly axisymmetric and is dominated by azimuthal motion, u_θ in (r, θ, z) coordinates centred on the cyclone and rotating with the Earth. A Bödewadt-like Ekman layer forms at the sea surface and so the air spirals radially inward along that surface, picking up speed as it goes through angular momentum conservation. Before reaching the centre of the cyclone the air turns, typically at a radius of $R_{\text{eye}} = 15$–30 km, to spiral vertically upwards towards the tropopause. Finally, the flow spirals radially outward along the tropopause, losing speed to conserve angular momentum. There is a quiescent region at the centre of the cyclone known as the *eye*, which is striking because of the lack of cloud cover (Figure 10.23). Here the vertical motion is reversed, consisting of subsiding air (Figure 10.24). The outer edge of the eye is conical and that part of the cyclone that lies adjacent to the eye is known as the *eyewall*. The eyewall is the location of the largest azimuthal velocities, typically in the range 30–80 m/s. Since the θ component of the Coriolis force is $-2\Omega u_r$, Ω being the component of the Earth's rotation normal to the sea surface, the motion is cyclonic ($u_\theta > 0$) near the sea surface where $u_r < 0$, and it remains cyclonic as it turns up into the tropopause. However, it becomes

Figure 10.23 Hurricane Isabel as seen from the International Space Station (NASA).

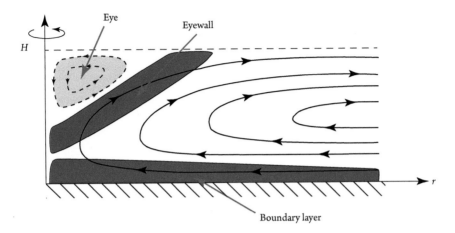

Figure 10.24 Sketch of the eye, eyewall, and boundary layer. Blue indicates $\omega_\theta < 0$.

anti-cyclonic ($u_\theta < 0$) within the tropopause at larger radii. The Rossby number, Ro, based on H and a local value of u_θ, $\mathrm{Ro} = |u_\theta|/\Omega H$, is of the order of unity in the bulk of the cyclone, so the Earth's rotation is important there, but rises to Ro $\sim$ 20–80 at the eyewall, so the background rotation of the Earth is unimportant near the eyewall. Note that, when $r \gg R_{\mathrm{eye}}$, an inflow along the sea surface and an outflow at the tropopause means that $\omega_\theta > 0$, except in the bottom boundary layer where $\omega_\theta < 0$, because $\partial u_r/\partial z < 0$ due to frictional drag. However, we have $\omega_\theta < 0$ within the eye and in the eyewall (Figure 10.24), with ω_θ particularly intense in the eyewall and the boundary layer. As we shall see, the eyewall probably acquires its negative azimuthal vorticity from the boundary layer, as ω_θ is stripped off the sea surface and advected upward. The eye is then a passive response to the eyewall, with vorticity diffusing across the streamlines from the eyewall into the eye.

10.6.2 A Simple Model of a 'Dry' Cyclone

Tropical cyclones are extremely complicated objects, and in many respects they are still poorly understood. Difficulties arise from the complex thermodynamics of moist convection and latent heat release, the continual evolution of cyclones through changes in their environment, ill-defined boundary conditions such as a poorly understood air–sea interaction involving wave generation and sea spray, and the fact that the air flow itself is three-dimensional and turbulent. However, Oruba et al. (2017) and Atkinson et al. (2019) have explored a simple model problem that captures some of the relevant fluid dynamics, at least in a zero-order sense, while sidestepping the complexities of moist convection, turbulence, and a poorly understood air–sea interaction. This toy problem is steady and axisymmetric, consisting of flow in a rotating cylindrical container whose base is heated (Figure 10.25). In effect, it replaces turbulent moist convection by laminar dry convection.

Perhaps it is worth taking a moment to describe the analysis of Oruba et al. (2017), who were interested particularly in how eyes form in cyclones. Consider the steady, laminar flow

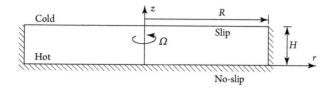

Figure 10.25 Flow domain and boundary conditions in a 'toy' model of a cyclone.

of a Boussinesq fluid in a shallow, rotating, cylindrical domain of height H, and radius R, where the lower surface, $z = 0$, and the outer radius, $r = R$, are no-slip boundaries and the upper surface is stress free. The motion is driven by buoyancy with a fixed upward heat flux maintained between the surfaces $z = 0$ and $z = H$, while the outer radial boundary is thermally insulating. In static equilibrium there is a uniform temperature gradient of $dT_0/dz = -\gamma$, but when in motion the density and temperature distributions are

$$\rho = \rho_0(z) + \rho' = \rho_0(z) - \bar{\rho}\beta\vartheta, \quad T = T_0(z) + \vartheta, \tag{10.60}$$

where $\beta = -(\partial\rho/\partial T)/\bar{\rho}$ is the thermal expansion coefficient, $\bar{\rho}$ a mean density, and ϑ the perturbation in temperature. In a reference frame that rotates with the boundaries, the momentum and heat equations are

$$\frac{D\mathbf{u}}{Dt} = -\nabla(p/\bar{\rho}) + 2\mathbf{u} \times \boldsymbol{\Omega} + \nu\nabla^2\mathbf{u} - \beta T\mathbf{g}, \tag{10.61}$$

$$\frac{D\vartheta}{Dt} = \alpha\nabla^2\vartheta + \gamma u_z, \tag{10.62}$$

where $\mathbf{u}$ is the (solenoidal) velocity in the rotating frame, α the thermal diffusivity, and $-\beta T\mathbf{g}$ the buoyancy force. For axisymmetric motion, the azimuthal component of (10.61) becomes an evolution equation for the angular momentum density $\Gamma = ru_\theta$,

$$\frac{D\Gamma}{Dt} = -2r\Omega u_r + \nu\nabla_*^2\Gamma, \tag{10.63}$$

while the curl of the poloidal components yields an evolution equation for ω_θ,

$$\frac{D}{Dt}\left(\frac{\omega_\theta}{r}\right) = \frac{\partial}{\partial z}\left(\frac{\Gamma^2}{r^4}\right) + \frac{2\Omega}{r^2}\frac{\partial\Gamma}{\partial z} - \frac{\beta g}{r}\frac{\partial\vartheta}{\partial r} + \frac{\nu}{r^2}\nabla_*^2(r\omega_\theta), \tag{10.64}$$

where ∇_*^2 is the Stokes operator (3.31) and we have added the curl of the buoyancy and Coriolis forces to (3.69) to give (10.64). Our governing equations are (10.62)–(10.64).

The heat flux is now chosen so that the Rossby number, $\mathrm{Ro} = |u_\theta|/\Omega H$, is of the order unity at large radius, $r \sim R$, but is large near the eyewall, $r \sim H$, which is typical of a tropical cyclone. Moreover, the viscosity is chosen so that a suitably defined Reynolds number is large, though not so large that the laminar flow becomes unsteady. (We shall see that a moderately large Reynolds number is crucial to eye formation in this toy problem as it is essential that the boundary-layer vorticity can be advected upward to form an eyewall.) Under these conditions the flow in the model problem resembles that of a tropical cyclone, including an eye and eyewall in which $\omega_\theta < 0$. The fluid spirals radially inward along the lower boundary and outward near $z = H$, as shown in Figure 10.26. Moreover, the Coriolis force in (10.63) ensures that Γ rises as the fluid spirals inward along the bottom boundary, but falls as it spirals back out along the upper surface. As a result, particularly high levels of $|u_\theta|$ build up near the axis, with a correspondingly large value of Ro near the eyewall.

While the global flow pattern in a tropical cyclone is established and shaped by the buoyancy and Coriolis forces, the large value of Ro near the eye means that the Coriolis force is locally negligible near the eye and eyewall, and it turns out that this is true also of the buoyancy force in this model problem. In short, the very forces that establish the global flow pattern in the model problem play no role in the local dynamics of the eye and eyewall, whose behaviour is controlled by the simplified equations

$$\frac{D\Gamma}{Dt} \approx \nu \nabla_*^2 \Gamma, \tag{10.65a}$$

$$\frac{D}{Dt}\left(\frac{\omega_\theta}{r}\right) \approx \frac{\partial}{\partial z}\left(\frac{\Gamma^2}{r^4}\right) + \frac{\nu}{r^2}\nabla_*^2(r\omega_\theta). \tag{10.65b}$$

Now the eye is characterized by anti-clockwise motion in the r–z plane ($\omega_\theta < 0$), in contrast to the global vortex which is clockwise ($\omega_\theta > 0$). The eye is also characterized by low levels of Γ. A natural question to ask, therefore, is where the negative azimuthal vorticity in the eye comes from. Since Γ is small in the eye, ω_θ/r is locally governed by a simple advection–diffusion equation with no source term. It follows that the negative azimuthal vorticity must diffuse into the eye from the eyewall in a process reminiscent of the Prandtl–Batchelor theorem (see §8.4), and in this sense the eye is a passive response to the accumulation of negative ω_θ in the eyewall. So the key to eye formation is the generation of significant levels of negative ω_θ in the eyewall, and an important question is how, and under what conditions, the eyewall acquires intense negative azimuthal vorticity. It is tempting to assume that this is generated locally by axial gradients in Γ via the first term on the right of (10.65b), and indeed this mechanism has been proposed in some meteorological papers. However, in Oruba et al. (2017) it is shown that the first term on the right of (10.65b) takes the form of a flux that inevitably generates as much positive as negative ω_θ in the eyewall (see Exercise 10.8). Rather, in this model problem, the eyewall vorticity comes directly from the bottom boundary layer, carried upward by advection.

Figure 10.26 A typical solution of the model problem showing ω_θ/r (blue is negative, red positive) overlaid on streamlines. The eyewall and eye are evident. (Figure by J. Atkinson.)

○ ○ ○

This concludes our discussion of rotating flows. Readers seeking more details will find gentle introductions in Batchelor (1967), Tritton (1988), and Guyon et al. (2015), while Greenspan (1968) fills in many of the gaps. Montgomery and Smith (2017) review cyclones.

. . . .

EXERCISES

10.1 *The inertial-wave dispersion integral for a Gaussian vortex.* The axisymmetric inertial waves that radiate from the Gaussian initial condition (10.30) are governed by

$$\frac{\partial^2}{\partial t^2}\left[r\frac{\partial}{\partial r}\frac{1}{r}\frac{\partial \Gamma}{\partial r} + \frac{\partial^2 \Gamma}{\partial z^2}\right] + (2\Omega)^2\frac{\partial^2 \Gamma}{\partial z^2} = 0.$$

This PDE is most readily solved using the Hankel-cosine transform pair

$$\hat{u}_\theta(k_r, k_z) = \frac{1}{2\pi^2}\int_0^\infty\int_0^\infty u_\theta J_1(k_r r)\cos(k_z z)r\,dr\,dz,$$

$$u_\theta(r, z) = 4\pi\int_0^\infty\int_0^\infty \hat{u}_\theta J_1(k_r r)\cos(k_z z)k_r\,dk_r\,dk_z,$$

where J_1 is the usual Bessel function. Show that the transform of our PDE yields

$$\partial^2 \hat{u}_\theta/\partial t^2 + \varpi^2\hat{u}_\theta = 0,$$

where $\varpi = 2\Omega k_z/k$ and $k^2 = k_r^2 + k_z^2$. With the initial conditions $\hat{u}_\theta = \hat{u}_\theta^{(0)}$ and $\hat{u}_r = \hat{u}_z = 0$, which requires $\partial\hat{u}_\theta/\partial t = 0$ at $t = 0$, we find that $\hat{u}_\theta = \hat{u}_\theta^{(0)}\cos\varpi t$. Confirm that the inverse transform now yields

$$u_\theta = 2\pi\int_0^\infty\int_0^\infty \hat{u}_\theta^{(0)}k_r J_1(k_r r)\left[\cos\left(k_z(z - 2\Omega t/k)\right) + \cos\left(k_z(z + 2\Omega t/k)\right)\right]$$

$$dk_r\,dk_z.$$

Now show that the Gaussian initial condition (10.30) requires

$$\hat{u}_\theta^{(0)} = \frac{\Lambda\delta^5}{16\pi^{3/2}}k_r\exp\left(-k^2\delta^2/4\right),$$

and hence u_θ is given by

$$u_\theta = \frac{\Lambda\delta^5}{8\pi^{1/2}}\int_0^\infty k_r^2\exp\left[-k_r^2\delta^2/4\right]J_1(k_r r)I(k_r, z, t)dk_r,$$

where

$$I(k_r, z, t) = \int_0^\infty \exp\left[-k_z^2\delta^2/4\right] \left[\cos\left(k_z\left(z - 2\Omega t/k\right)\right) + \cos\left(k_z\left(z + 2\Omega t/k\right)\right)\right] dk_z.$$

The power spectrum for $\hat{u}_\theta^{(0)}$ is dominated by wavevectors in the vicinity of $k_z \approx 0$ and $k_r \sim \delta^{-1}$, so a good approximation to the integral above is obtained by putting $k_z/k \approx k_z/k_r$ in the argument of the cosines. Confirm that $I(k_r)$ can then be evaluated exactly to give

$$u_\theta \approx \Lambda\delta \int_0^\infty \kappa^2 e^{-\kappa^2} J_1(2\kappa r/\delta) \left[\exp\left[-\left(\frac{z}{\delta} - \frac{\Omega t}{\kappa}\right)^2\right] + \exp\left[-\left(\frac{z}{\delta} + \frac{\Omega t}{\kappa}\right)^2\right]\right] d\kappa,$$

where $\kappa = k_r\delta/2$. In order to keep the argument in one or other of the two exponentials small, we require $\kappa z \approx \pm\Omega\delta t$, which tells us that the wave energy is centred at $|z| \sim \Omega\delta t$. Find the corresponding expression for ω_z and use (10.17) to show that u_z is given by (10.32c).

The asymptotic form of this dispersion integral for $\Omega t \gg 1$ may be found by demanding that the argument in one or other of the two exponentials remains of order unity as $\Omega t \to \infty$. At location $z = \delta\Omega t$ this requires $\kappa \to 1$. Show that this leads to

$$u_\theta(z = \delta\Omega t, \Omega t \to \infty) \approx \Lambda\delta\left[\sqrt{\pi}/e\right] J_1(2r/\delta)(\Omega t)^{-1}.$$

Finally, noting that $J_1(x \to 0) = x/2$ and $J_1(x \to \infty) \sim x^{-1/2}$, use this expression to confirm that $u_\theta \sim \Lambda r(\Omega t)^{-1}$ near the rotation axis while $u_\theta \sim \Lambda\delta(\Omega t)^{-3/2}(z/r)^{1/2}$ away from the axis.

10.2 *The inertial-wave dispersion pattern created by a buoyant blob in a rotating, Boussinesq fluid.* Consider an isolated blob of buoyant material of scale δ sitting in a rotating, incompressible, Boussinesq fluid, governed by $\nabla \cdot \mathbf{u} = 0$ and

$$\frac{\partial\mathbf{u}}{\partial t} = 2\mathbf{u} \times \mathbf{\Omega} - \nabla\left(p/\bar{\rho}\right) + b\mathbf{g}, \quad \mathbf{g} = -g\hat{\mathbf{e}}_x,$$

where $\rho = \bar{\rho} + \rho'$, $b = \rho'/\bar{\rho}$ and $\mathrm{Ro} = u/\Omega\delta \ll 1$. The corresponding vorticity equation is

$$\frac{\partial\boldsymbol{\omega}}{\partial t} = 2(\mathbf{\Omega} \cdot \nabla)\mathbf{u} + \nabla b \times \mathbf{g}.$$

Since ρ' is governed by an advection equation it evolves on a slow time scale set by $\mathbf{u}$, while inertial wave packets evolve on the fast timescale of Ω^{-1}. At low Ro, then, we may take ρ' as quasi-steady as far as the initiation of inertial waves is concerned. Treating ρ' as independent of time, show that $\mathbf{u}$ is governed by the inhomogeneous wave-like equation

$$\frac{\partial^2}{\partial t^2}\left(\nabla^2 \mathbf{u}\right) + (2\mathbf{\Omega} \cdot \nabla)^2 \mathbf{u} = (2\mathbf{\Omega} \cdot \nabla)(\mathbf{g} \times \nabla b).$$

Thus buoyancy acts as a local source of inertial waves whose dispersion pattern is necessarily dominated by low-frequency waves propagating up and down the rotation axis, in the sense that the radiation density is highest on the axis. (This focussing of radiated energy onto the rotation axis is a geometrical property of the dispersion relationship, as discussed in §10.3.3.) These low-frequency wave packets propagate in the $\pm\mathbf{\Omega}$ directions at a speed of $\Omega\delta$, creating columnar vortices that carry negative (positive) helicity upwards (downwards). We can determine the structure of these columnar vortices by considering the *vertical jump conditions* across the buoyant blob. Since ρ' is quasi-steady and the inertial waves in question are of low frequency, within the buoyant blob we have

$$2\left(\mathbf{\Omega} \cdot \nabla\right)\mathbf{u} + \nabla b \times \mathbf{g} \approx 0,$$

whose curl is

$$2\left(\mathbf{\Omega} \cdot \nabla\right)\boldsymbol{\omega} \approx \mathbf{g}\nabla^2 b - (\mathbf{g} \cdot \nabla)\nabla b.$$

Show that the integrated vertical jump condition in vertical vorticity across the blob is $\Delta\omega_z \approx 0$, while the vertical jump conditions in velocity are

$$\Delta u_x \approx 0, \quad \Delta u_y \approx 0, \quad \Delta u_z \approx -\frac{g}{2\Omega}\int_{-\infty}^{\infty}(\partial b/\partial y)\, dz.$$

Now show that, for the Gaussian buoyant anomaly $b = -|b_0|\exp\left(-\mathbf{x}^2/\delta^2\right)$, Δu_z is positive for $y < 0$ and negative for $y > 0$. Hence show that u_z, which is antisymmetric about the plane $z = 0$, diverges from the plane $z = 0$ for $y < 0$ and converges to $z = 0$ for $y > 0$. Moreover, from the jump condition for ω_z we see that a cyclonic ($\omega_z > 0$) columnar vortex below the buoyant blob must correspond to a cyclonic vortex above the blob, while an anti-cyclonic ($\omega_z < 0$) columnar vortex below corresponds to an anti-cyclonic vortex above. Finally, use the fact that upward propagating inertial waves have negative helicity ($u_z\omega_z < 0$), while downward propagating waves have positive helicity ($u_z\omega_z > 0$), to show that the inertial-wave dispersion pattern consists of a pair of cyclonic and anti-cyclonic columnar vortices above the blob, matched to a pair of cyclonic and anti-cyclonic vortices below the blob, with the anti-cyclones located at negative y, and the cyclones at positive y. This is precisely the dispersion pattern shown in Figure 10.17 (a) and (b), where x points out of the page and the cyclones (anti-cyclones) are on the right (left) of the buoyant blob.

10.3 *The reflection of a plane inertial wave at a flat horizontal boundary.* Let ϕ be the polar angle measured from the rotation axis to the wavevector $\mathbf{k}$ of a plane inertial wave, with $0 \le \phi \le \pi$. Show that $|\mathbf{c}_g| = (2\Omega/k)\sin\phi$ and $\varpi = \pm 2\Omega\cos\phi$. Now consider a plane wave whose group velocity is directed upward. It reflects off a flat, horizontal boundary. We wish to find the form of the reflected wave. Equation (10.24) tells us that the group velocity of the incident wave is co-planar with $\boldsymbol{\Omega}$ and $\mathbf{k}$ and so we consider the plane containing $\mathbf{c}_g$, $\boldsymbol{\Omega}$, and $\mathbf{k}$. If we use a prime to denote the reflected wave, then the velocity field associated with the combined incident and reflected wave is

$$\mathbf{u} = \hat{\mathbf{u}} \exp j(\mathbf{k} \cdot \mathbf{x} - \varpi t) + \hat{\mathbf{u}}' \exp j(\mathbf{k}' \cdot \mathbf{x} - \varpi' t),$$

and we must choose the reflected wave so as to ensure $\mathbf{u} \cdot \mathbf{n} = 0$ at the boundary, where $\mathbf{n}$ is a unit normal to the surface. Since this must be satisfied at all times and at all points on the surface, we require that the arguments in the two exponentials are identical, which in turn demands that $\varpi = \varpi'$ and $\mathbf{k} \cdot \mathbf{t} = \mathbf{k}' \cdot \mathbf{t}$, where $\mathbf{t}$ is a unit tangential vector to the surface. Show that $\varpi = \varpi'$ requires $\cos \phi = -\cos \phi'$, from which we have $\phi' = \pi - \phi$ and $\sin \phi = \sin \phi'$. Sketch the relative orientations of $\boldsymbol{\Omega}$, $\mathbf{c}_g$, $\mathbf{k}$, $\mathbf{c}'_g$, and $\mathbf{k}'$, noting that all vectors lie in the same plane and that (10.24) requires the horizontal components of the phase and group velocities of the incident (and reflected) wave to be equal but opposite. Now show that, for a horizontal boundary, $\mathbf{k} \cdot \mathbf{t} = \mathbf{k}' \cdot \mathbf{t}$ requires $k \sin \phi = k' \sin \phi'$, from which we deduce that $k = k'$ and $|\mathbf{c}_g| = |\mathbf{c}'_g|$. Hence show that $\mathbf{c}_g$ and $\mathbf{c}'_g$ have equal horizontal components but opposite vertical components.

10.4 *Standing inertial waves in a circular cylinder.* Consider inertial waves in a rotating cylindrical container of radius R and length L. Use (10.35) to show that axisymmetric standing waves have a frequency of

$$\varpi / 2\Omega = \left[1 + (\gamma_n / |R)^2 (L/m\pi)^2 \right]^{-1/2},$$

where γ_n are the zeros of J_1 and $m = 1, 2, 3 \ldots$

10.5 *Secondary flow in a curved duct.* Water is pumped through a helical duct of square cross-section. The pitch of the helix can be considered as almost flat and so the mean flow along the axis of the duct is more or less azimuthal, $u_{\text{mean}} \approx u_\theta \approx \Omega r$ in cylindrical polar coordinates. Explain the origin of the secondary flow shown in Figure 10.27. Why does this secondary flow greatly enhance the energy dissipation? (Hint: consult Figure 10.22.)

10.6 *The secondary flow in a meandering river.* The cross-section through a meandering river is shown in Figure 10.28, where the inside of the curve is on the left and the outside on the right. Use the lower half of Figure 10.27 to explain the origin of the secondary flow in the river cross-section. This secondary flow induces outer-bank scour (erosion) and inner-bank deposition of sediment, causing the meandering of the river to increase with time.

10.7 *Evanescent inertial waves.* Suppose that an inviscid, rotating fluid is bounded internally by a cylindrical surface whose axis is aligned with the rotation axis. The surface of the cylinder undulates radially, according to $\eta(z,t) \sim \cos(k_z z) \cos \varpi t$. For $\varpi < 2\Omega$ this generates a standing wave pattern whose radial structure is given by (10.35), where

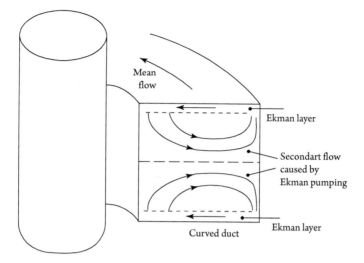

Figure 10.27 Water flows along a helical duct and a secondary flow is generated as shown.

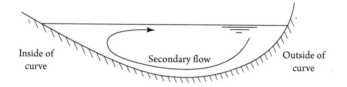

Figure 10.28 The secondary flow in the cross-section of a meandering river.

$$k_r = k_z \sqrt{(2\Omega)^2/\varpi^2 - 1}.$$

Now suppose that $\varpi > 2\Omega$. Clearly standing-wave solutions are forbidden as they violate the dispersion relationship. However, there must still be some form of disturbance localized around the cylindrical boundary. It is natural to look for a disturbance of the form $u_r = \hat{u}_r(r)\cos(k_z z)\sin\varpi t$. Show that this leads to the *evanescent* solution

$$u_r \sim K_1(\gamma r)\cos(k_z z)\sin\varpi t, \quad \gamma = k_z \sqrt{1 - (2\Omega)^2/\varpi^2},$$

where K_1 is the usual modified Bessel function (which decays exponentially without oscillation). Compare this with the evanescent gravity wave of Exercise 9.4. Note that $\gamma \to 0$ as ϖ approaches 2Ω from above, which represents a radially elongated disturbance.

10.8 *The source of eyewall vorticity in a tropical cyclone.* Consider the simple model of a 'dry' tropical cyclone described in §10.6.2. Near the eye and eyewall the Coriolis and buoyancy forces are negligible and so, outside the boundary layer, the governing equations (10.65a,b) reduce to

$$\frac{D\Gamma}{Dt} = 0, \quad \frac{D}{Dt}\left(\frac{\omega_\theta}{r}\right) = \frac{\partial}{\partial z}\left(\frac{\Gamma^2}{r^4}\right).$$

The first of these yields $\Gamma = \Gamma(\Psi)$, where Ψ is the Stokes streamfunction (3.36). As noted in the derivation of (3.71), the azimuthal vorticity equation can be rewritten as

$$\mathbf{u} \cdot \nabla \left(\omega_\theta/r - \Gamma\Gamma'(\Psi)/r^2\right) = 0,$$

so that $\omega_\theta/r - \Gamma\Gamma'(\Psi)/r^2$ is conserved along a streamline. Apply this conservation law to a streamline in Figure 10.24 which lies outside the boundary layer. Let the streamline enter the domain near the sea surface and exit at the *same radius* near the tropopause. Show that the net change in azimuthal vorticity is zero, so the negative azimuthal vorticity in the eyewall cannot arise from axial gradients in Γ. Rather, it comes from the boundary layer.

10.9 *Rayleigh's analogy between stratified and rotating fluids.* In Exercise 8.1 we described a swirling vortex bursting radially outward under the influence of the centrifugal force. Use Rayleigh's analogy between a stratified and rotating flow, as discussed in §10.1, to reinterpret this in terms of a blob of heavy fluid evolving under the influence of a radial gravitational field. Hence explain why the mushroom-like structure of the vortex in Figure 8.30 resembles that of a thermal, as shown in Figure 10.29.

10.10 *A simple model of the Ekman spiral in the atmospheric boundary layer.* Suppose that, in a rotating frame, there is an imposed uniform flow $\mathbf{u}_\infty = V\hat{\mathbf{e}}_x$ adjacent to a solid surface located at $z = 0$. The resulting motion is assumed to be steady, confined to

Figure 10.29 A buoyant blob evolves into a thermal.

the x–y plane, and a function of z only. Show that, since the Coriolis force far from the boundary is balanced by a pressure gradient in the y direction, the governing equations of motion are

$$\nu\frac{d^2 u_x}{dz^2} = -2\Omega u_y, \quad \nu\frac{d^2 u_y}{dz^2} = 2\Omega\left(u_x - V\right),$$

and that their solution is

$$u_x = V\left(1 - e^{-\eta}\cos\eta\right), \quad u_y = V e^{-\eta}\sin\eta,$$

where $\eta = z\big/\sqrt{\nu/\Omega}$. Evidently, this is a form of Ekman layer, and it might be taken as a simple model of the atmospheric boundary layer under neutral conditions, with ν interpreted as an eddy viscosity. Sketch the variation of $\mathbf{u}$ with z, i.e. the Ekman spiral. Note that the flow near the boundary is inclined at $45°$ to $\mathbf{u}_\infty$.

10.11 *Helicity transport within a transient Taylor column.* Consider the transient Taylor column induced by a disc slowly translating along the rotation axis, as shown in Figure 10.6. The fluid within the column is observed to rotate in a cyclonic sense ($\omega_z > 0$) below the disc, and in an anti-cyclonic sense ($\omega_z < 0$) above the disc. (See, for example, Figure 4.3 in Greenspan, 1968.) Explain this behaviour in terms of the helicity transported by inertial waves. Now suppose that the transient Taylor columns above and below the disc reflect off horizontal boundaries located some distance from the disc. Use the results of Exercise 10.3 to show that the axial velocity in the reflected waves is equal but opposite to that in the incident waves. Show also that the axial vorticity retains the same sign in both the incident and reflected waves. Hence confirm that a standing wave pattern is established in which $\omega_z < 0$ above the disc and $\omega_z > 0$ below the disc.

Show that the same conclusion can be reached by applying $\partial\omega_z/\partial t = 2\Omega\partial u_z/\partial z$ to the column of fluid trapped between the disc and the horizontal boundaries above and below the disc. If the disc were moved downwards, instead of upwards, what would be the sense of rotation above and below the disc? Consider both the bounded and unbounded cases.

....

REFERENCES

Atkinson, J.W., Davidson, P.A., Perry, J.E.D., 2019, Dynamics of a trapped vortex in rotating convection. *Phys. Rev. Fluids*, **4**, 074701.

Batchelor, G.K., 1967, *An introduction to fluid dynamics*, Cambridge University Press.

Bödewadt, U.T., 1940, Die Drehstromung uber festem Grunde, *Z. angew Math Mech.*, **20**, 241–53.

Bracewell, R.N., 1986, *The Fourier transform and its applications*, 2nd Ed., McGraw-Hill.

Davidson, P.A., 2013, *Turbulence in rotating, stratified and electrically conducting fluids*, Cambridge University Press.

Davidson, P.A., 2015, *Turbulence: an introduction for scientists and engineers*, 2nd Ed., Oxford University Press.

Davidson P.A. & Ranjan A., 2015, Planetary dynamos driven by helical waves: part 2. *Geophys. J. Int.*, **202**, 1646–62.

Davidson P.A. & Ranjan A., 2018, Are planetary dynamos driven by helical waves? *J. Plasma Phys.*, **84**.

Davidson P.A., Staplehurst P.J., & Dalziel S.B, 2006, On the evolution of eddies in a rapidly rotating system. *J. Fluid Mech.*, **557**, 135–45.

Goldstein, H., 1980, *Classical mechanics*, 2nd Ed., Addison-Wesley.

Greenspan, H.P., 1968, *The theory of rotating fluids*, Cambridge University Press.

Guyon, E., Hulin, J.-P., Petit, L., & Mitescu, C.D., 2015, *Physical hydrodynamics*, Oxford University Press.

Kármán, T., 1921, Uber laminare und turbulente Reibung, *Z. angew Math Mech.*, **1**, 233.

Kelvin, Lord, 1880, Vibrations of a columnar vortex, *Phil. Mag.*, **5**(61), 155–68.

Montgomery, M.T. & Smith, R.K., 2017, Recent developments in the fluid dynamics of tropical cyclones. *Ann. Rev. Fluid Mech.*, **49**, 541–74.

Oruba L., Davidson P.A., & Dormy E., 2017, Eye formation in rotating convection. *J. Fluid Mech.*, **812**, 890–904.

Pedlosky, J., 1987, *Geophysical fluid dynamics*, 2nd Ed., Springer-Verlag.

Ranjan A., Davidson P.A., Christensen U.R., & Wicht J., 2018. Internally driven inertial waves in dynamo simulations. *Geophys. J. Int.*, **213**(2).

Rayleigh, Lord, 1916, On the dynamics of revolving fluids. *Proc. Royal Soc., A*, **93**, 148.

Tritton, D.J., 1988, *Physical fluid dynamics*, Oxford University Press.

11

· · • · ·

Instability

Many well-known scientists and applied mathematicians are associated with the subject of hydrodynamic stability, but perhaps two of the most prominent names are the physicists Lord Rayleigh (left, 1842–1919) and Subrahmanyan Chandrasekhar (right, 1910–1995).

Steady solutions of the Navier–Stokes equations may be classified as either stable or unstable, depending on whether or not a disturbance introduced into the flow grows with time. Here, we restrict the discussion to infinitesimal perturbations so that the equations of motion describing the disturbance can be linearized about the base state. In such cases, the flow is deemed to be unstable if *any* disturbance can be found that grows continually.

11.1 The Centrifugal Instability

11.1.1 Rayleigh's Inviscid Criterion for Axisymmetric Disturbances

Let us start by returning to Rayleigh's instability criterion for the inviscid, rotating flow $\mathbf{u}_0 = (0, u_\theta(r), 0)$ described in (r, θ, z) coordinates. This is normally expressed in terms of Rayleigh's discriminant, defined as

$$\Phi(r) = \frac{1}{r^3} \frac{d\Gamma^2}{dr}, \qquad (11.1)$$

where $\Gamma = r u_\theta$. The criterion states that an inviscid swirling flow is stable to *axisymmetric* disturbances if and only if $\Phi(r) \geq 0$ at all points in the flow. In §10.1 we deduced this result using a simple energy argument, as did Rayleigh (1916(a)). Here, we shall derive the criterion using a more conventional perturbation analysis, in large part because a similar perturbation analysis is required for the equivalent viscous problem.

Let the velocity field of the perturbed flow be $\mathbf{u} = V(r)\hat{\mathbf{e}}_\theta + \mathbf{u}'(\mathbf{x}, t)$, where the perturbation $\mathbf{u}'$ is assumed small, $|\mathbf{u}'| \ll V$. If the disturbance is axisymmetric then the total flow is governed by (3.70) in the form

$$\frac{D\Gamma}{Dt} = 0, \qquad \frac{D}{Dt}\left(\frac{\omega_\theta}{r}\right) = \frac{\partial}{\partial z}\left(\frac{\Gamma^2}{r^4}\right), \qquad (11.2)$$

where the azimuthal vorticity is related to the Stokes streamfunction (3.35) by

$$\omega_\theta = \frac{\partial u_r}{\partial z} - \frac{\partial u_z}{\partial r} = -\frac{1}{r}\nabla_*^2 \Psi, \qquad (11.3)$$

∇_*^2 being the Stokes operator (3.31). Since $\mathbf{u}'$ is small, (11.2) may be linearized about the steady base flow to give

$$r\frac{\partial u_\theta'}{\partial t} + u_r' \frac{d}{dr}(rV) = 0, \qquad (11.4)$$

$$\frac{\partial}{\partial t}\left(\frac{\omega_\theta'}{r}\right) = \frac{2V}{r^2}\frac{\partial u_\theta'}{\partial z}. \qquad (11.5)$$

Next, we introduce the Stokes streamfunction for the perturbation in poloidal velocity

$$(u_r', 0, u_z') = \left(-\frac{1}{r}\frac{\partial \Psi'}{\partial z}, \, 0, \, \frac{1}{r}\frac{\partial \Psi'}{\partial r}\right),$$

and rewrite (11.4) and (11.5) in the form

$$\frac{\partial u_\theta'}{\partial t} = \frac{1}{r^2}\frac{d}{dr}(rV)\frac{\partial \Psi'}{\partial z}, \qquad (11.6)$$

$$\frac{\partial}{\partial t}\nabla_*^2 \Psi' = -2V\frac{\partial u_\theta'}{\partial z}. \qquad (11.7)$$

Finally, eliminating the azimuthal velocity perturbation from these equations yields the linearized governing equation for an inviscid, axisymmetric disturbance:

$$\frac{\partial^2}{\partial t^2}\nabla_*^2\Psi' + \Phi(r)\frac{\partial^2\Psi'}{\partial z^2} = 0.$$

(11.8)

If the fluid is bounded such that $R_1 < r < R_2$ then we have $\Psi'(R_1) = \Psi'(R_2) = 0$.

Notice that when the base flow is one of uniform rotation, say $V = \Omega r$, then we have $\Phi = (2\Omega)^2$, and (11.8) reverts to the governing equation for axisymmetric inertial waves, (10.31), as it must. Notice also that we have homogeneity in z and so we are free to look for normal modes of the form

$$\Psi' = \hat{\Psi}(r)\exp\left[j\left(k_z z - \varpi t\right)\right],$$

(11.9)

where we take k_z to be real but ϖ may be complex. If ϖ is real the flow is clearly stable, whereas a positive imaginary part of ϖ leads to instability. Combining (11.8) and (11.9) yields

$$r\frac{d}{dr}\frac{1}{r}\frac{d\hat{\Psi}}{dr} + \left(\frac{\Phi(r)}{\varpi^2} - 1\right)k_z^2\hat{\Psi} = 0,$$

(11.10)

which constitutes a classic Sturm–Liouville eigenvalue problem. Standard theory for such systems tell us that ϖ^2 is positive if Φ is everywhere positive, but ϖ^2 takes on negative values if Φ is anywhere negative in the range $R_1 < r < R_2$, the latter case representing instability. We have arrived back at Rayleigh's criterion.

It is instructive to multiply through (11.10) by $\hat{\Psi}/r$ and write the result in the form

$$\frac{d}{dr}\left(\frac{\hat{\Psi}}{r}\frac{d\hat{\Psi}}{dr}\right) - \frac{1}{r}\left(\frac{d\hat{\Psi}}{dr}\right)^2 + k_z^2\left(\frac{\Phi(r)}{\varpi^2} - 1\right)\frac{\hat{\Psi}^2}{r} = 0.$$

Integrating over the range $R_1 < r < R_2$ then gives

$$\varpi^2\int_{R_1}^{R_2}\left[\frac{1}{r}\left(\frac{d\hat{\Psi}}{dr}\right)^2 + k_z^2\frac{\hat{\Psi}^2}{r}\right]dr = k_z^2\int_{R_1}^{R_2}\left[\Phi(r)\frac{\hat{\Psi}^2}{r}\right]dr,$$

(11.11)

which confirms that ϖ^2 is positive (or negative) if Φ is everywhere positive (or negative).

11.1.2 Two-dimensional Inviscid Disturbances (Rayleigh again)

The two major weaknesses of Rayleigh's criterion are that it is restricted to axisymmetric perturbations and it ignores the stabilizing influence of viscous dissipation. We consider non-axisymmetric disturbances in this section and the role of viscous dissipation in §11.1.3.

The inviscid stability of a swirling flow to non-axisymmetric perturbations remains an active area of research (see, for example, Billant and Gallaire, 2005). Here, we restrict ourselves to two-dimensional disturbances, which provide a counter example to axisymmetric stability. The two-dimensional stability of a simple parallel shear flow of the form $\mathbf{u}_0 = u_x(y)\hat{\mathbf{e}}_x$ is discussed in §11.3. It is shown there that a *necessary* condition for inviscid instability is that the vorticity of the base flow, $\omega_z(y) = -\partial u_x/\partial y$, exhibits a local minimum or a local maximum, *i.e.* $\partial \omega_z/\partial y = -\partial^2 u_x/\partial y^2$ changes sign somewhere in the flow. This is known as *Rayleigh's inflection point theorem*. The equivalent result for the circular flow $\mathbf{u}_0 = (0, u_\theta(r), 0)$ was also obtained by Rayleigh (1880) and states that:

a necessary condition for the instability of the inviscid flow $\mathbf{u}_0 = (0, u_\theta(r), 0)$ to planar (r, θ) disturbances is that the radial gradient in the vorticity, $\partial \omega_z/\partial r$, or equivalently $\nabla_*^2(r u_\theta)$, changes sign somewhere in the flow.

It is not hard to identify velocity profiles $V(r)$ that are stable to axisymmetric disturbances yet unstable to planar perturbations. One example is an annular vortex sheet constructed from a sudden rise in $\Gamma = r u_\theta$, with radius, r, which is unstable to a planar roll-up of the sheet.

11.1.3 Viscous Instability and Taylor's Analysis

We now turn to the stability of the *viscous* flow generated in the annular gap between two concentric rotating cylinders of radius R_1 and R_2, say $\mathbf{u}_0 = (0, V(r), 0)$. This is called *Taylor–Couette flow* and in §3.3.3 we saw that the steady velocity profile is

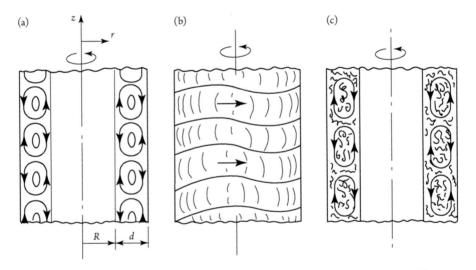

Figure 11.1 Flow between concentric cylinders where the inner cylinder rotates. (a) Taylor vortices. (b) Wavy Taylor vortices. (c) Turbulent Taylor vortices. (From Davidson, 2015.)

$$rV = \frac{\Omega_2 R_2^2 - \Omega_1 R_1^2}{R_2^2 - R_1^2} r^2 - \frac{(\Omega_2 - \Omega_1) R_1^2 R_2^2}{R_2^2 - R_1^2} = Ar^2 + B, \qquad (11.12)$$

where Ω_1 and Ω_2 are the rotation rates of the inner and outer cylinders (see Figure 11.2). We are interested particularly in the case where Ω_1 and Ω_2 are both positive (Ω_1 and Ω_2 having opposite signs leads to a more complex behaviour) and in which $\Omega_2 R_2^2 < \Omega_1 R_1^2$, as this is centrifugally unstable by Rayleigh's inviscid criterion of §11.1.1. It seems clear that a large viscosity should stabilize such a flow, and indeed it is found that this flow is stable provided that a suitably defined Reynolds number is small enough. However, it is observed experimentally that the same flow becomes unstable to an *axisymmetric* mode if the viscosity is decreased sufficiently, and we wish to characterize the onset of this instability. We shall restrict the discussion to axisymmetric disturbances, as these are observed to be the most dangerous in practice.

The governing equations for any axisymmetric disturbance in Taylor–Couette flow are the viscous counterparts of (11.2), which can be obtained from (3.63) and (3.69). These are

$$\frac{D\Gamma}{Dt} = \nu \nabla_*^2 \Gamma, \qquad (11.13)$$

$$\frac{D}{Dt}\left(\frac{\omega_\theta'}{r}\right) = \frac{\partial}{\partial z}\left(\frac{\Gamma^2}{r^4}\right) + \frac{\nu}{r^2}\nabla_*^2(r\omega_\theta'), \qquad (11.14)$$

where $\Gamma = rV(r) + \Gamma'$ and $r\omega_\theta' = -\nabla_*^2 \Psi'$ in accordance with (11.3). Linearizing about the base flow on the assumption that the disturbance is small gives us the viscous versions of (11.6) and (11.7), which are

$$\left(\frac{\partial}{\partial t} - \nu \nabla_*^2\right)\Gamma' = \frac{1}{r}\frac{d}{dr}(rV)\frac{\partial \Psi'}{\partial z}, \qquad (11.15)$$

$$\left(\frac{\partial}{\partial t} - \nu \nabla_*^2\right)\nabla_*^2 \Psi' = -\frac{2V}{r}\frac{\partial \Gamma'}{\partial z}. \qquad (11.16)$$

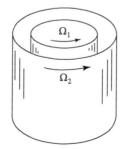

Figure 11.2 Rotating Taylor–Couette flow.

We now take advantage of the fact that

$$\frac{1}{r}\frac{d}{dr}(rV) = 2A,$$

and so is independent of r in Taylor–Couette flow. Applying the operator $(\partial/\partial t - \nu\nabla_*^2)\nabla_*^2$ to (11.15), and then substituting for Ψ' using (11.16), we obtain the governing equation for Γ':

$$\left(\frac{\partial}{\partial t} - \nu\nabla_*^2\right)^2 \nabla_*^2\Gamma' = -\frac{2(rV)}{r^3}\frac{d}{dr}(rV)\frac{\partial^2\Gamma'}{\partial z^2}. \tag{11.17}$$

This can be rewritten in terms of Rayleigh's discriminant as

$$\boxed{\left(\frac{\partial}{\partial t} - \nu\nabla_*^2\right)^2 \nabla_*^2\Gamma' + \Phi(r)\frac{\partial^2\Gamma'}{\partial z^2} = 0}, \tag{11.18}$$

where $\Phi = 4AV/r$. This is the viscous counterpart of (11.8).

So far, we have placed no restriction on the linear stability analysis, other than to assume axisymmetric motion. To focus our thoughts, we now take $\Omega_2 = 0$ and, following the example of Taylor (1923), we adopt the narrow-gap approximation which assumes that $d = R_2 - R_1 \ll R_1$. If we adopt the new radial coordinate $x = r - R_1$ then in the narrow-gap limit we have

$$V/r = \Omega_1(1 - x/d), \tag{11.19}$$

and hence

$$\Phi(x) = 4A\frac{V}{r} = 4A\Omega_1\left(1 - \frac{x}{d}\right) = -\frac{2\Omega_1^2 R_1}{d}\left(1 - \frac{x}{d}\right).$$

The narrow-gap approximation also allows us to replace ∇_*^2 by the conventional Laplacian operating on x and z. Our governing equation for viscous disturbances now becomes

$$\left(\frac{\partial}{\partial t} - \nu\nabla^2\right)^2 \nabla^2\Gamma' = \frac{2\Omega_1^2 R_1}{d}\left(1 - \frac{x}{d}\right)\frac{\partial^2\Gamma'}{\partial z^2}. \tag{11.20}$$

The final step is to once again follow Taylor (1923) and take advantage of the fact that the experiments show that the axisymmetric instability sets in as a non-oscillatory mode, so that at marginal stability we may suppress the time derivatives in (11.20). This gives

$$(\nabla^2)^3\Gamma' = \frac{2\Omega_1^2 R_1}{\nu^2 d}\left(1 - \frac{x}{d}\right)\frac{\partial^2\Gamma'}{\partial z^2} \tag{11.21}$$

at marginal stability. Since all spatial derivatives scale on d in the narrow-gap limit, this tells us that the key dimensionless control parameter for this instability is

$$\boxed{\text{Ta} = \frac{\Omega_1^2 R_1 d^3}{\nu^2}}, \tag{11.22}$$

which is known as the *Taylor number*, Ta. In effect, Ta is the square of a Reynolds number. (Readers should be aware that different authors use slightly different definitions of Ta.)

The determination of the critical value of Ta is now routine. One looks for solutions of the form $\Gamma' = \hat{\Gamma}'(x)\cos(k_z z)$ so that (11.21) becomes an eigenvalue problem in x. The solution of this eigenvalue problem is a little tricky, but a good approximation can be obtained by replacing the linear term in x by its average value (see Drazin and Reid, 1981), which amounts to replacing V/r in (11.19) by its mean, $\Omega_1/2$. This gives

$$\left(d^2/dx^2 - k_z^2\right)^3 \hat{\Gamma}' + \frac{\text{Ta}}{d^4} k_z^2 \hat{\Gamma}' = 0, \tag{11.23}$$

which is easier to solve. For a given k_z, there are non-trivial solutions of (11.23) only for certain discrete values of Ta and there will be a minimum eigenvalue $(\text{Ta})_{\min}$ for each k_z. The least of these values of $(\text{Ta})_{\min}$ yields the critical Taylor number $(\text{Ta})_c$ and associated critical wavenumber k_z. The end result is that (11.23) predicts that the most unstable axisymmetric mode corresponds to $k_z d = 3.117$ with $(\text{Ta})_c = 1708$, whereas a similar analysis of the *exact* equation, (11.21), yields $k_z d = 3.127$ and $(\text{Ta})_c = 1695$.

It is left as an exercise for the reader to show that, if we relax the assumption that $\Omega_2 = 0$, replacing it by $0 < \Omega_2 < \Omega_1$, but retain all the other approximations leading up to (11.23), then our approximate eigenvalue problem and associated Taylor number generalize to

$$\left(d^2/dx^2 - k_z^2\right)^3 \hat{\Gamma}' + \frac{\text{Ta}}{d^4} k_z^2 \hat{\Gamma}' = 0,$$
$$\text{Ta} = \frac{\left(\Omega_1 + \Omega_2\right)\left(\Omega_1 R_1^2 - \Omega_2 R_2^2\right) d^3}{\nu^2 R_1}. \tag{11.24}$$

Since the eigenvalue problem is unchanged, except for the new definition of Ta, we still have $(\text{Ta})_c = 1708$ at $k_z d = 3.117$, at least in the mean V/r approximation.

11.1.4 The Experimental Evidence

It is interesting to compare these predictions with laboratory experiments performed using long cylinders. Again, we restrict the discussion to a stationary outer cylinder, $\Omega_2 = 0$. At low Ta the flow is given by (11.12), except near the end-walls of the apparatus where a form of Ekman pumping sets up a secondary flow in the $r - z$ plane. However, when the cylinders are long, $L \gg d$, this Ekman pumping is a localized effect, limited to the ends of the apparatus. Suppose we now *slowly* increase Ta, say by increasing Ω_1. As Ta approaches $(\text{Ta})_c$ we find that a steady, axisymmetric, cellular motion appears in the $r - z$ plane, consisting

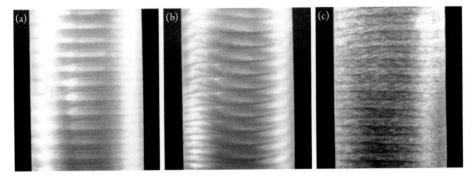

Figure 11.3 The flow between concentric cylinders where the inner cylinder rotates but the outer one is static. (a) Taylor vortices at Ta $\approx$ (Ta)$_c$. (b) Wavy Taylor vortices at Ta $=4$(Ta)$_c$ with an azimuthal wavenumber of $m=3$. (c) Turbulent Taylor vortices at Ta $=676$(Ta)$_c$. (Courtesy of Q. Xiao, T.T. Lim and Y.T. Chew, National University of Singapore.)

of vortices of alternating sense, as shown in Figures 11.1(a) and 11.3(a). This poloidal motion is superimposed on the azimuthal flow so that individual fluid particles now follow helical paths, confined to toroidal surfaces. The resulting steady, helical motion is said to be composed of *Taylor vortices*. Note that the flow is still laminar, if a little more complex than (11.12). Note also that the exponential growth predicted by linear stability theory at Ta $=$ (Ta)$_c$ does not produce a runaway situation. Rather, the instability quickly saturates through weakly non-linear interactions to yield steady Taylor vortices. Nevertheless, the axial wavelength of the observed steady vortices is close to that predicted by linear stability theory.

Actually, the theoretical analysis leading to (11.18), which assumes infinitely long cylinders, is a little misleading when it comes to laboratory experiments where the cylinders are long but finite. Rather than an abrupt appearance of Taylor vortices when Ta reaches a value of (Ta)$_c$, as predicted by the theory above, what happens in reality is more gradual. As Ta slowly approaches (Ta)$_c$ from below, the toroidal vortices established by Ekman pumping on the end-walls, which exist near the ends of the apparatus even at modest Ta, start to multiply and extend away from the end-walls and into the interior of the flow. This process accelerates rapidly as Ta $\rightarrow$ (Ta)$_c$, with vortices spreading inward from the two ends of the apparatus until they link up near the centre. So, by the time Ta reaches the value of Ta $=$ (Ta)$_c$, the flow shown in Figure 11.1 (a) is established throughout the annulus. This continuous, rather than abrupt, process occurs even for very long cylinders, where $L \gg d$.

Another simplification in the discussion above is that we have assumed that Ta is increased *slowly* from rest and we have ignored the possibility of hysteresis. In fact, it turns out that the size and sense of the Taylor vortices in any given experiment can be influenced by the history of how the steady state is approached, particularly if rapid changes in Ta are made. We shall, however, not pause to discuss this interesting topic, but rather refer the reader to Acheson (1990) for a succinct discussion and for more detailed references.

As Ta is increased beyond $(\mathrm{Ta})_c$, the flow pattern changes very little until suddenly, at a few multiples of $(\mathrm{Ta})_c$, the Taylor vortices themselves become unstable to non-axisymmetric disturbances. This heralds a second *bifurcation* (*i.e.* change in flow structure), this time to an unsteady, non-axisymmetric flow, as shown in Figure 11.1 (b). This new flow, which is referred to as the *wavy Taylor vortex* regime, consists of non-axisymmetric Taylor vortices that migrate around the inner cylinder with an azimuthal wavenumber m that depends on the value of Ta. Note that, although this unsteady flow is more complex than that comprised of steady Taylor vortices, it is non-chaotic and so still laminar. Note also that this is not some kind of second-order instability of the original Couette flow, but rather a primary instability of the Taylor vortices themselves.

Further increases in Ta cause yet more bifurcations to ever more complex flows until eventually, when Ta is large enough, the flow degenerates into a random motion which is chaotic in both time and space, *i.e.* it becomes turbulent. Interestingly, embedded within this chaotic motion there is a mean (time-averaged) component of the flow which resembles the laminar Taylor vortices of Figure 11.1 (a). This is shown in Figure 11.1 (c), and this mean motion is sometimes referred to as *turbulent Taylor vortices*. Eventually, however, when Ta becomes very large, say $\mathrm{Ta} \sim 10^5 (\mathrm{Ta})_c$, the turbulent Taylor vortices fade away, leaving only turbulence superimposed on a mean azimuthal motion.

11.2 The Stability of a Fluid Heated from Below

11.2.1 Rayleigh–Bénard Convection

Consider a layer of liquid bounded above and below by solid plates located at $z = 0$, d. These are held at fixed temperatures with an imposed temperature difference of ΔT, the lower plate being hotter than the top. It is intuitively obvious that such a configuration will exhibit natural convection in which hot, buoyant fluid rises up from the lower plate only to give up its heat on reaching the cooler upper plate, causing the fluid to sink back down. This often takes the form of two-dimensional convection rolls, as shown in Figure 11.4 (b). However, convection is not inevitable. This buoyant motion is opposed by viscous forces and a simple energy argument tells us that convection can occur only if the rate of working of the buoyancy forces exceeds the rate of viscous dissipation. When the heating is uniform across both plates, and the temperature difference between the plates low enough or the fluid viscosity particularly large, it turns out that no such motion occurs. Heat is then transferred by thermal conduction only, as suggested in Figure 11.4 (a).

The transition from a static equilibrium to steady convection is controlled by the *Rayleigh number*, Ra. This is defined as

$$\boxed{\mathrm{Ra} = \frac{g\beta\Delta T d^3}{\nu\alpha}},$$

(11.25)

where α is the thermal diffusivity of the fluid, $\beta = -(\partial\rho/\partial T)/\bar{\rho}$ the expansion coefficient, and $\bar{\rho}$ a mean fluid density. It is only when Ra exceeds a certain critical value, say $(\mathrm{Ra})_c$,

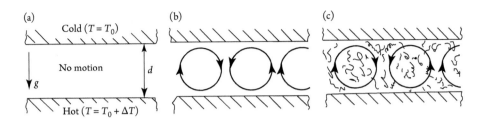

Figure 11.4 Rayleigh–Bénard convection in a fluid layer of depth d driven by a temperature difference of ΔT: (a) low ΔT, (b) Bénard cells, (c) high ΔT. (From Davidson, 2015.)

that a static equilibrium is destabilized and motion ensues. This is analogous to the sudden appearance of Taylor vortices in a rotating fluid at $\mathrm{Ta} = (\mathrm{Ta})_c$ and, as we shall see, there are close connections between these two flows. Note that large values of both ν and α are stabilizing, the latter because it allows hot rising fluid to shed its excess heat by diffusion.

This instability was first studied theoretically by Rayleigh (1916(b)), who was trying to understand the experiments of Bénard some sixteen years earlier. These experiments consisted of a shallow liquid pool sitting on a heated metallic plate and with a free upper surface. Consequently, the flow shown in Figure 11.4 (b), or the equivalent motion where the upper surface is free, is known as Rayleigh–Bénard convection. (Actually, it was James Thomson, Kelvin's brother, who first observed such convection cells, in 1882.) Rayleigh adopted the Boussinesq approximation, in which the variations in density are so small that they may be ignored, except to the extent that they introduce a buoyancy force, $\delta\rho\mathbf{g}$. In Rayleigh–Bénard convection it is convenient to rewrite this force as $-\bar{\rho}\beta(T - T_{\text{ref}})\mathbf{g}$, where T_{ref} is the temperature of one of the plates. Our governing equations are then

$$\frac{D\mathbf{u}}{Dt} = -\nabla(p/\bar{\rho}) + \nu\nabla^2\mathbf{u} - \beta(T - T_{\text{ref}})\mathbf{g}, \tag{11.26}$$

$$\frac{DT}{Dt} = \alpha\nabla^2 T, \tag{11.27}$$

where $\mathbf{g} = -g\hat{\mathbf{e}}_z$. Moreover, because the variations in density are assumed small, mass conservation allows us to take $\nabla \cdot \mathbf{u} = 0$.

11.2.2 Rayleigh's Stability Analysis

The equilibrium whose stability is in question is evidently $\mathbf{u}_0 = 0$ and $T = T_0(z)$, where T_0 is linear in z. Let us write the perturbation in temperature as $\vartheta = T - T_0(z)$. If we restrict ourselves to small-amplitude disturbances, we may linearize (11.26) and (11.27) about the equilibrium state to give

$$\frac{\partial\mathbf{u}}{\partial t} = -\nabla(\delta p/\bar{\rho}) + \nu\nabla^2\mathbf{u} - \beta\vartheta\mathbf{g}, \tag{11.28}$$

$$\frac{\partial\vartheta}{\partial t} + u_z\frac{dT_0}{dz} = \alpha\nabla^2\vartheta, \tag{11.29}$$

where δp is the perturbation in pressure. The vorticity equation corresponding to (11.28) is clearly

$$\left[\frac{\partial}{\partial t} - \nu\nabla^2\right]\boldsymbol{\omega} = -\beta(\nabla\vartheta) \times \mathbf{g},\tag{11.30}$$

and we rewrite (11.29) as

$$\left[\frac{\partial}{\partial t} - \alpha\nabla^2\right]\vartheta = u_z\frac{\Delta T}{d}.\tag{11.31}$$

The equations governing the instability are then (11.30) and (11.31).

We must now eliminate either ϑ or u_z from (11.31) using (11.30). To that end we take the curl of (11.30), which gives us

$$\left[\frac{\partial}{\partial t} - \nu\nabla^2\right]\nabla^2\mathbf{u} = g\beta\left[\hat{\mathbf{e}}_z\nabla^2\vartheta - \frac{\partial}{\partial z}\nabla\vartheta\right],$$

and whose z component is

$$\left[\frac{\partial}{\partial t} - \nu\nabla^2\right]\nabla^2 u_z = g\beta\nabla_\perp^2\vartheta,$$

where $\nabla_\perp^2 = \partial^2/\partial^2 x + \partial^2/\partial^2 y$. Next, we eliminate ϑ using (11.31), which yields the governing equation for u_z,

$$\boxed{\left[\frac{\partial}{\partial t} - \nu\nabla^2\right]\left[\frac{\partial}{\partial t} - \alpha\nabla^2\right]\nabla^2 u_z = \frac{g\beta\Delta T}{d}\nabla_\perp^2 u_z\,.}\tag{11.32}$$

It is readily confirmed that ϑ is governed by exactly the same equation.

We must also consider the boundary conditions, and these depend somewhat on whether the upper surface is a solid plate or a free surface, which we treat as a flat, stress-free boundary, i.e. we do not allow for surface waves. In either case we have $\vartheta = u_z = 0$ at $z = 0, d$. If the boundary at $z = 0$ or $z = d$ is a flat plate, then the no-slip condition $\mathbf{u} = 0$ plus continuity demands that $\partial u_z/\partial z = 0$. On the other hand, if the upper boundary is flat and stress free, then the vertical gradients in both u_x and u_y must be zero at $z = d$ and $\nabla \cdot \mathbf{u} = 0$ now requires $\partial^2 u_z/\partial z^2 = 0$. In summary, then, the boundary conditions are:

$$\vartheta = u_z = \partial u_z/\partial z = 0, \quad \text{(no-slip surface)}$$
$$\vartheta = u_z = \partial^2 u_z/\partial z^2 = 0. \quad \text{(stress-free surface)}$$

Now it is possible to show that the instability sets in as an exponential growth without oscillation (see Chandrasekhar, 1961) and consequently at *marginal stability* we have

$$(\nabla^2)^3 \vartheta = \frac{g\beta\Delta T}{\nu\alpha d}\nabla_\perp^2 \vartheta, \quad (\nabla^2)^3 u_z = \frac{g\beta\Delta T}{\nu\alpha d}\nabla_\perp^2 u_z, \tag{11.33}$$

where ϑ and u_z are related by (11.31), which requires

$$\alpha\nabla^2\vartheta = -(\Delta T/d)\,u_z. \tag{11.34}$$

Of course, the only geometric length scale here is d, and so (11.33) tells us that the stability threshold is determined simply by the value of $\mathrm{Ra} = g\beta\Delta T d^3/\nu\alpha$ and by the boundary conditions. In order to find the critical value of Ra, Rayleigh noted that (11.33) admits separable solutions of the form

$$\vartheta = \hat{\vartheta}(z)f(x,y), \quad u_z = \hat{u}_z(z)f(x,y), \tag{11.35}$$

provided that f satisfies Helmholtz's equation,

$$\nabla_\perp^2 f + k^2 f = 0. \tag{11.36}$$

Here, k is a horizontal wavenumber which is yet to be determined. Assuming solutions of this form, (11.33) yields

$$\left(\frac{d^2}{dz^2} - k^2\right)^3 \hat{u}_z + \frac{\mathrm{Ra}}{d^4}k^2\hat{u}_z = 0, \tag{11.37}$$

for marginally unstable modes. Of course, $\hat{\vartheta}$ is governed by the same equation. It is of interest to compare (11.37) with (11.23), which governs (approximately) the axisymmetric Taylor instability in the narrow-gap approximation,

$$\left(\frac{d^2}{dx^2} - k_z^2\right)^3 \hat{\Gamma}' + \frac{\mathrm{Ta}}{d^4}k_z^2\hat{\Gamma}' = 0. \tag{11.38}$$

It is the same equation, with Ta replacing Ra. This is the first hint that there are close links between these two problems. In any event, if (11.37) is combined with the boundary conditions above, we have a well-posed eigenvalue problem which can be solved for $(\mathrm{Ra})_c$.

11.2.3 Slip Boundaries Top and Bottom: an Artificial but Informative Case

In §11.2.4 we shall discuss the two most important configurations of no slip at $z = 0$ and no slip or stress free at $z = d$. First, however, it is informative to follow Rayleigh and consider the artificial case where the top and bottom surfaces are both stress free. This is readily solved since, by inspection, $\hat{\vartheta} \sim \hat{u}_z \sim \sin(n\pi z/d)$, $n = 1, 2, 3 \ldots$, satisfies the boundary conditions. Equation (11.37) then yields

$$\mathrm{Ra} = \left((n\pi)^2 + a^2\right)^3 \big/ a^2, \quad a = kd. \tag{11.39}$$

Clearly, $n = 1$ is the most unstable mode and so the critical value of a is the one that minimizes Ra in (11.39) for $n = 1$. This is $a = \pi/\sqrt{2}$, which yields $(\text{Ra})_c = 27\pi^4/4 = 657.5$.

There is an alternative way of getting the same result that is more intuitive and bypasses (or is equivalent to) the eigenvalue analysis. It is readily confirmed that (11.28) yields the energy equation

$$\frac{\partial}{\partial t}\left(\tfrac{1}{2}\mathbf{u}^2\right) = \nabla \cdot \left(\nu \mathbf{u} \times \boldsymbol{\omega} - \delta p \mathbf{u}/\bar{\rho}\right) - \nu \boldsymbol{\omega}^2 + g\beta \vartheta u_z. \tag{11.40}$$

On integrating over a single convection cell and invoking Gauss' theorem the divergence vanishes, leaving us with

$$\frac{d}{dt}\int \tfrac{1}{2}\mathbf{u}^2 dV = -\nu \int \boldsymbol{\omega}^2 dV + g\beta \int \vartheta u_z dV. \tag{11.41}$$

The integrals on the right are the rate of viscous dissipation and the rate of working of the buoyancy force. Clearly, the criterion for instability is that the rate of working of the buoyancy force exceeds the viscous dissipation,

$$g\beta \int \vartheta u_z dV \geq \nu \int \boldsymbol{\omega}^2 dV, \tag{11.42}$$

with the equality sign corresponding to marginal stability.

Let us now estimate these two integrals. Suppose that the instability takes the form of two-dimensional rolls in the x–z plane. The simplest guesses for u_z and θ that satisfy the boundary conditions are then

$$(u_z, \vartheta) = (u_0, \vartheta_0)\sin\left(\pi z/d\right)\sin(kx). \tag{11.43}$$

Since the instability sets in as an exponential growth without oscillation, at marginal stability ϑ is determined by (11.34), which fixes the relationship between u_0 and θ_0 as

$$\vartheta_0 = \frac{\Delta T}{\alpha d}\left[(\pi/d)^2 + k^2\right]^{-1} u_0.$$

Moreover, we can find u_x from u_z using continuity and the vorticity then follows. On integrating over a single cell, our dissipation and buoyancy integrals are then estimated as

$$\nu \int \boldsymbol{\omega}^2 dV = \frac{\nu}{4}\left[(\pi/d)^2 + k^2\right]^2 \frac{\pi d u_0^2}{k^3}, \tag{11.44}$$

$$g\beta \int \vartheta u_z dV = \frac{g\beta \Delta T}{4\alpha d}\left[(\pi/d)^2 + k^2\right]^{-1}\frac{\pi d u_0^2}{k}. \tag{11.45}$$

Equating these integrals and writing $a = kd$, we find that the *marginal stability curve* is

$$\frac{g\beta \Delta T d^3}{\alpha} \frac{a^2}{\pi^2 + a^2} = \nu \left[\pi^2 + a^2\right]^2,$$

(11.46)

or, in terms of the Rayleigh number,

$$(\text{Ra})_{\text{marginal}} = a^{-2} \left(\pi^2 + a^2\right)^3.$$

(11.47)

Of course, this is (11.39) with $n = 1$. Finally, suppose that $a = kd$ is chosen to maximize the rate of working of the buoyancy force relative to the rate of dissipation, *i.e.* a is chosen such that Ra is a minimum in (11.47). Then

$$a_c = \pi \big/ \sqrt{2}, \quad (\text{Ra})_c = 657.5,$$

(11.48)

which brings us back to the results of the eigenvalue problem (11.37).

11.2.4 No-slip Boundaries

For rigid–rigid and rigid–free boundaries, (11.37) needs to be solved numerically, although simple estimates of $(\text{Ra})_c$ accurate to better than 0.02% are readily found using (11.42) (see Exercise 11.3). The critical values for Ra and a are given in Table 11.1. For no-slip boundaries top and bottom we find $(\text{Ra})_c = 1708$ at $kd = 3.117$, as shown by the marginal stability curve in Figure 11.5. The value of $(\text{Ra})_c$ for a free upper surface is somewhat lower, $(\text{Ra})_c = 1101$, since there is now only one dissipative boundary layer, while that for two free surfaces is lower still, as there are no boundary layers at all and the dissipation now occurs in the bulk.

Perhaps some comments are in order. First, the predictions for $(\text{Ra})_c$ are close to those observed in experiments. Second, the case where the fluid is bounded above and below by rigid surfaces is particularly interesting. Recall that the critical threshold for the appearance of Taylor vortices in Couette flow is $(\text{Ta})_c = 1695$, corresponding to $kd = 3.127$. These are almost exactly the same as the values of $(\text{Ra})_c$ and $a = kd$ at the onset of Rayleigh–Bénard convection with no-slip boundaries. Of course this is no accident, and indeed we could have anticipated this by comparing (11.37) with (11.38). Third, our analysis has determined the critical value of Ra and kd, but not the form of $f(x, y)$ in (11.35). In fact, the only restrictions on f are that it must satisfy Helmholtz's equation and the lateral boundary conditions. Fourth, it is clear that the initial instability quickly saturates to give

Table 11.1 Critical values of Ra and $a = kd$ in Rayleigh–Bénard convection.

Boundary conditions	Free–free	Rigid–free	Rigid–rigid
$(\text{Ra})_c$	657.5	1101	1708
a_c	2.221	2.682	3.117

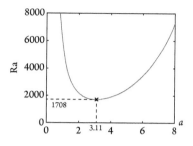

Figure 11.5 The marginal stability curve of Ra versus $a = kd$ for convection between two plates. Above the curve, the equilibrium is unstable to modes of the given wavenumber k.

way to steady convection, and this occurs through non-linear interactions. Consequently, linear stability theory has little to say about the pattern of steady convection cells as seen from above, other than to set the dominant length scale through the critical value of k. In practice, a variety of patterns can manifest themselves, with two-dimensional longitudinal rolls and hexagonal cells being two common examples. Fifth, suppose that Ra is slowly increased above the critical value in Table 11.1. Then there is a point at which the Bénard cells themselves become unstable, just as Taylor vortices eventually become unstable when Ta becomes large enough. More complex flows then emerge. Eventually, for large enough Ra, the convection becomes turbulent, as indicated in Figure 11.4 (c). However, embedded within this turbulence there is a time-averaged component of motion which is cellular and reminiscent of the laminar convection cells.

11.3 The Stability of Parallel Shear Flows

We now turn to one-dimensional shear flows of the form $\mathbf{u} = V(y)\hat{\mathbf{e}}_x$, where instability normally sets in as an oscillation, in marked contrast to the thermal and centrifugal instabilities considered so far. There is a theorem for shear flows, called *Squire's theorem*, which says that for every unstable mode in three dimensions there is a more unstable mode in two dimensions, and so it is sufficient to consider only two-dimensional stability. Let us start with inviscid flow and with Rayleigh's celebrated *inflection point theorem*.

11.3.1 Rayleigh's Inflection Point Theorem for Inviscid, Rectilinear Flow

Consider the inviscid flow $\mathbf{u}_0 = V(y)\hat{\mathbf{e}}_x$, which is confined to $0 < y < d$ as shown in Figure 11.6. The vorticity is

$$\boldsymbol{\omega}_0 = -dV/dy\hat{\mathbf{e}}_z = -\nabla^2\psi_0\hat{\mathbf{e}}_z, \tag{11.49}$$

where ψ_0 is the streamfunction (1.15) of the unperturbed flow. Now consider a two-dimensional disturbance of small amplitude, say $\boldsymbol{\omega}' = \boldsymbol{\omega} - \boldsymbol{\omega}_0$ and $\psi' = \psi - \psi_0$, whose linearized vorticity equation is

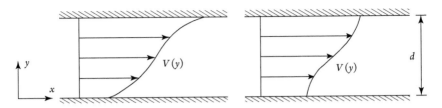

Figure 11.6 Both flows are potentially unstable by Rayleigh's inflection point theorem, but the flow on the left is stable by Fjørtoft's theorem.

$$\frac{\partial \omega'}{\partial t} + V \frac{\partial \omega'}{\partial x} = -u'_y \frac{\partial \omega_0}{\partial y} = -\frac{\partial \psi'}{\partial x} \frac{d^2 V}{dy^2},$$

or, equivalently,

$$\left(\frac{\partial}{\partial t} + V \frac{\partial}{\partial x} \right) \nabla^2 \psi' = \frac{d^2 V}{dy^2} \frac{\partial \psi'}{\partial x}. \tag{11.50}$$

Note that $u'_y = -\partial \psi'/\partial x = 0$ at the boundaries, and so we may take $\psi' = 0$ at $y = 0, d$.

Since we have homogeneity in x, it is natural to look for disturbances which are the real part of

$$\psi' = \hat{\psi}'(y) \exp\left[j \left(kx - \varpi t \right) \right], \tag{11.51}$$

where k is real but ϖ and $\hat{\psi}'$ may be real or complex. If ϖ is real the flow is neutrally stable, whereas a positive imaginary part of ϖ leads to instability. Combining (11.51) with (11.50) yields *Rayleigh's stability equation*,

$$\frac{d^2 \hat{\psi}'}{dy^2} - k^2 \hat{\psi}' = \frac{d^2 V/dy^2}{(V - \varpi/k)} \hat{\psi}'. \tag{11.52}$$

We now multiply through (11.52) by the complex conjugate of $\hat{\psi}'$, say $\tilde{\psi}'$, and integrate from $y = 0$ to $y = d$:

$$\int_0^d \tilde{\psi}' \frac{d^2 \hat{\psi}'}{dy^2} dy - k^2 \int_0^d \left| \hat{\psi}' \right|^2 dy = \int_0^d \frac{d^2 V/dy^2}{(V - \varpi/k)} \left| \hat{\psi}' \right|^2 dy. \tag{11.53}$$

The first integral on the left may be put into a more convenient form through integration by parts and using the boundary conditions on $\hat{\psi}'$. This yields

$$\int_0^d \left| \frac{d\hat{\psi}'}{dy} \right|^2 dy + k^2 \int_0^d \left| \hat{\psi}' \right|^2 dy = k \int_0^d \frac{d^2 V/dy^2}{\varpi - kV} \left| \hat{\psi}' \right|^2 dy$$

$$= k \int_0^d \frac{(\tilde{\varpi} - kV)}{|\varpi - kV|^2} \frac{d^2 V}{dy^2} \left| \hat{\psi}' \right|^2 dy, \tag{11.54}$$

where $\tilde{\varpi} = \varpi_r - j\varpi_i$ is the complex conjugate of ϖ. Crucially, the imaginary part of this integral equation is

$$\varpi_i \int_0^d \frac{\left|\hat{\psi}'\right|^2}{|\varpi - kV|^2} \frac{d^2V}{dy^2} dy = 0. \tag{11.55}$$

Now instability requires that ϖ has a positive imaginary part, $\varpi_i > 0$, yet (11.55) tells us that this is possible only if d^2V/dy^2 changes sign somewhere in the flow. Put another way, instability requires that the vorticity, $\omega_0 = -dV/dy$, exhibits a *local maximum* or a *local minimum*, and if no such maximum or minimum exists, this inviscid flow is stable. This is Rayleigh's inflection point theorem (Rayleigh, 1880), which states:

> a necessary condition for the linear instability of the inviscid shear flow $V(y)$ is that d^2V/dy^2 should change sign somewhere in the flow, *i.e.* there is an inflection point in $V(y)$.

Both of the flows shown in Figure 11.6 are potentially unstable by this criterion.

Note that this is a necessary but not a sufficient condition for instability, so the mere existence of an inflection point in $V(y)$ does not ensure instability. Indeed, this is clear from a stronger result obtained some 70 years later by the Norwegian meteorologist Fjørtoft. By examining the real part of (11.54), Fjørtoft showed that:

> a necessary condition for the linear instability of the inviscid shear flow $V(y)$ is that there is an inflection point, say at $y = y_*$, and also that $(V - V(y_*)) d^2V/dy^2 < 0$ at some point

(see Exercise 11.1). In terms of the background vorticity ω_0, Fjørtoft's theorem tells us that, for those cases where $V(y)$ increases monotonically, a necessary condition for the flow to be unstable is that the modulus of the vorticity should exhibit a *local maximum*, local minima being stable. Thus, for example, the flow on the left of Figure 11.6 is stable despite the inflection point, as discussed in Exercise 11.2. Although Fjørtoft's theorem provides only a necessary condition for instability, in practice it is often found to be both necessary and sufficient.

11.3.2 The Subtle Effects of Viscosity

Let us now consider the effects of a finite viscosity. When ν is finite, (11.52) generalizes to

$$j\nu \left(\frac{d^2}{dy^2} - k^2\right)^2 \hat{\psi}' + (Vk - \varpi)\left(\frac{d^2}{dy^2} - k^2\right)\hat{\psi}' - \frac{d^2V}{dy^2} k\hat{\psi}' = 0, \tag{11.56}$$

which is known as the *Orr–Sommerfeld equation*. The interpretation of solutions of this equation is notoriously subtle, and so we shall restrict ourselves to a few general observations, referring the reader to Drazin and Reid (1981) for more details. First, we note that, if the

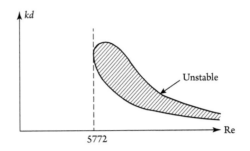

Figure 11.7 Schematic of the marginal stability curve for plane Poiseuille flow.

corresponding inviscid flow has an inflection point and is unstable, then the effect of a finite viscosity is usually stabilizing, giving rise to a critical value of Re below which the instability is suppressed. This is, of course, exactly what one might have expected from a finite viscous dissipation. However, contrary to intuition, a small but finite viscosity can *destabilize* a flow whose inviscid counterpart is stable. It would seem, therefore, that viscosity can be both stabilizing and destabilizing.

Consider, by way of an example, the plane Poiseuille flow

$$V(y) = V_0 \left(d/2 \right)^{-2} y \left(d - y \right), \quad 0 < y < d, \tag{11.57}$$

whose inviscid counterpart is stable according to Rayleigh's inflection point theorem. It turns out that this can be destabilized by viscosity and a schematic of the marginal stability curve in the kd–Re plane is shown in Figure 11.7. Evidently the flow is linearly unstable to a narrow band of stream-wise wavenumbers, k, provided that Re $= V_0 d/2\nu > 5772$. Note that this is *not* incompatible with the inviscid prediction, as the band of unstable wavenumbers tends to zero as Re $\rightarrow \infty$. Note also that most experiments yield a lower critical value of Re. This partly due to a non-parabolic velocity profile at the entrance to the duct and partly because finite-amplitude disturbances lead to non-linear effects.

11.4 The Kelvin–Helmholtz Instability

11.4.1 The Instability of an Inviscid Vortex Sheet

Fjørtoft's theorem tells us that, if $u_x(z)$ increases monotonically with z, a necessary condition for an inviscid flow to be unstable is that the modulus of the vorticity should exhibit a local maximum. The most extreme example of this is a vortex sheet created by a vertical jump in horizontal velocity, as shown in Figure 11.8. The instability of such a sheet was studied by both Kelvin and Helmholtz and is now named after them. It is convenient to take the unperturbed vortex sheet to be located at $z = 0$ and to adopt a frame of reference which moves with the mean velocity of the two layers. Hence, in the unperturbed state, we have $u_x = -V/2$ for $z < 0$ and $u_x = V/2$ for $z > 0$. We shall also allow for surface tension

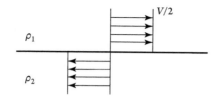

Figure 11.8 A vortex sheet.

between the two layers and take the density of the lower layer, ρ_2, to be larger than that of the upper layer, ρ_1, so that both surface tension and density have a stabilizing influence on the vortex sheet. Finally, we take the flow to be inviscid and the perturbation to be of small amplitude and irrotational.

Let the displacement of the vortex sheet be $\eta(x, y, t)$ and the velocity potentials for the disturbance in the top and bottom layers be ϕ_1 and ϕ_2, both of which obey Laplace's equation in accordance with (6.9). Now we have the kinematic condition that the fluid particles at the interface move with the interface. Consequently, immediately above the vortex sheet we have

$$\frac{\partial \phi_1}{\partial z} = \frac{D\eta}{Dt} = \frac{\partial \eta}{\partial t} + ((V/2)\hat{\mathbf{e}}_x + \mathbf{u}') \cdot \nabla \eta, \quad \text{at } z = \eta,$$

with a similar expression just below. These linearize to give the boundary conditions

$$\left(\frac{\partial \phi_1}{\partial z}\right)_{z=0} = \frac{\partial \eta}{\partial t} + \frac{V}{2}\frac{\partial \eta}{\partial x}, \quad \left(\frac{\partial \phi_2}{\partial z}\right)_{z=0} = \frac{\partial \eta}{\partial t} - \frac{V}{2}\frac{\partial \eta}{\partial x}. \tag{11.58}$$

Moreover, Bernoulli's equation in the form of (6.4) requires

$$\frac{\partial \phi_1}{\partial t} + \frac{p_1}{\rho_1} + g\eta + \frac{1}{2}\left(\nabla \phi_1 + (V/2)\hat{\mathbf{e}}_x\right)^2 = \frac{1}{2}(V/2)^2,$$

just above the interface, with a similar equation just below. Linearizing these equations and subtracting them yields

$$\left(\rho_1 \frac{\partial \phi_1}{\partial t} + p_1 + \frac{1}{2}\rho_1 V \frac{\partial \phi_1}{\partial x}\right)_{z=0} + \rho_1 g\eta = \left(\rho_2 \frac{\partial \phi_2}{\partial t} + p_2 - \frac{1}{2}\rho_2 V \frac{\partial \phi_2}{\partial x}\right)_{z=0} + \rho_2 g\eta,$$

where the jump in pressure across the interface is related to the surface tension coefficient γ by (7.49). Our dynamic boundary condition is therefore

$$\rho_1 \left(\frac{\partial \phi_1}{\partial t} + \frac{1}{2}V \frac{\partial \phi_1}{\partial x}\right)_{z=0} - \rho_2 \left(\frac{\partial \phi_2}{\partial t} - \frac{1}{2}V \frac{\partial \phi_2}{\partial x}\right)_{z=0} = (\rho_2 - \rho_1)g\eta - \gamma \nabla^2 \eta. \tag{11.59}$$

In anticipation that the most unstable mode is two-dimensional, we now look for solutions of the form

$$(\eta, \phi_1, \phi_2) = \left(\hat{\eta}, \hat{\phi}_1(z), \hat{\phi}_2(z)\right) \exp\left(j(kx - \varpi t)\right), \tag{11.60}$$

where k is real and ϖ may be real or complex. Since ϕ_1 and ϕ_2 are both harmonic, we have

$$\hat{\phi}_1 = A_1 e^{-|k|z}, \quad \hat{\phi}_2 = A_2 e^{|k|z}, \tag{11.61}$$

for two constants A_1 and A_2. Our boundary conditions (11.58) and (11.59) now yield

$$|k| A_1 = j\left(\varpi - kV/2\right)\hat{\eta}, \quad |k| A_2 = -j\left(\varpi + kV/2\right)\hat{\eta}, \tag{11.62}$$

and

$$\rho_2 j\left(\varpi + kV/2\right) A_2 - \rho_1 j\left(\varpi - kV/2\right) A_1 = \Delta\rho g\hat{\eta} + \gamma k^2 \hat{\eta}, \tag{11.63}$$

where $\Delta\rho = \rho_2 - \rho_1$. Finally, we eliminate A_1 and A_2 from (11.62) and (11.63) to give the dispersion relationship

$$(\rho_1 + \rho_2)\varpi = -\frac{1}{2}\Delta\rho V k \pm \sqrt{g|k|(\rho_2^2 - \rho_1^2) + (\rho_1 + \rho_2)\gamma|k|^3 - \rho_1\rho_2 V^2 k^2}. \tag{11.64}$$

Clearly, we have exponential growth whenever

$$\boxed{\frac{\rho_1\rho_2}{\rho_1 + \rho_2} V^2 > \frac{g\Delta\rho}{|k|} + \gamma|k|.} \tag{11.65}$$

As expected, shear is destabilizing while stratification and surface tension both help to stabilize the vortex sheet.

Perhaps some comments are in order. First, if we put $V = \rho_1 = 0$ in (11.64) we recover the dispersion relationship (7.52) for capillary–gravity waves. Second, in the absence of surface tension the vortex sheet is always unstable to short-wavelength perturbations, irrespective of the values of V and $\Delta\rho$. That is to say, the sheet is unstable to all wavenumbers which satisfy

$$|k| > \frac{\rho_1 + \rho_2}{\rho_1\rho_2}\frac{g\Delta\rho}{V^2} = \frac{2g\Delta\rho}{\rho_0 V^2}, \tag{11.66}$$

where ρ_0 is a representative density. So a finite surface tension is critical to stabilizing the short-wavelength modes. Third, the non-linear development of the instability causes the vortex sheet to roll up into spiralled vortex tubes, as shown in Figure 11.9. Fourth, for a finite

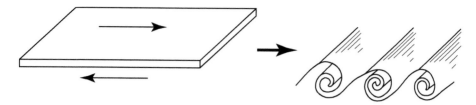

Figure 11.9 Development of the Kelvin–Helmholtz billows. (From Davidson, 2015.)

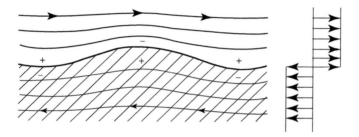

Figure 11.10 Prandtl's explanation of the K–H instability. (Adapted from Prandtl, 1952.)

surface tension, the right-hand side of (11.65) has a minimum at $k^2 = g\Delta\rho/\gamma$ and so, as Kelvin noted, a necessary and sufficient condition for instability is

$$\boxed{\frac{\rho_1\rho_2}{\rho_1+\rho_2}V^2 > 2\sqrt{\gamma g\Delta\rho}}. \tag{11.67}$$

Again, we see that stratification and surface tension both help to stabilize the interface, in the sense that increasing either γ or $\Delta\rho$ causes a rise in the minimum value of V required for instability. Expression (11.67) is well supported by laboratory experiments. For air blowing over sea-water, (11.67) gives a minimum wind speed for instability of $V \sim 6.5$ m/s, with a corresponding wavelength of $\lambda \sim 1.7$ cm.

The physical nature of this instability is most readily understood following an argument given by Prandtl (1952). Consider a slowly growing mode close to marginal stability and move into a frame of reference in which the perturbed interface does not translate in the x direction. The motion may then be viewed as quasi-steady. As the interface slowly deforms, the fluid speeds up at those points where it passes over a crest or below a trough, and it slows down where it passes over a trough or below a crest. From Bernoulli's equation there is a corresponding fall in pressure above the crests and below the troughs, these changes being marked by '−' in Figure 11.10. Similarly, there is a rise in pressure above the troughs and below the crests, as marked by '+' in Figure 11.10. Crucially, these changes in pressure are such as to reinforce the initial deformation of the interface, and hence drive an instability.

An alternative explanation of the instability is offered by Batchelor (1967) in terms of vortex dynamics. In effect, flow either side of a rippled vortex sheet sweeps vorticity towards certain accumulation points, locally thickening the sheet, while thinning the sheet at other points. This redistribution of vorticity creates a perturbed velocity field via the Biot–Savart law and this accentuates the initial distortion of the sheet.

11.4.2 The Inviscid Instability of a Layer of Vorticity of Finite Thickness

It is striking that, in the absence of surface tension, a vortex sheet of infinitesimal thickness is unstable to *all* wavenumbers greater than $2g\Delta\rho/\rho_0 V^2$, i.e. there is no lower cut-off in wavelength. It is natural to wonder if this is linked to the assumption that the sheet has zero thickness, something that cannot occur in a real, viscous fluid. Some of the consequences of allowing for a *finite* sheet thickness are provided by a model problem analysed by Taylor (1931). We shall follow the description given in Chandrasekhar (1961), in which there is no surface tension. Consider the following distributions of ρ, u_x, and ω_y:

$$
\begin{array}{llll}
z > d: & \rho = \rho_0 (1 - \varepsilon), & u_x = V/2, & \omega_y = 0, \\
-d < z < d: & \rho = \rho_0, & u_x = Vz/2d, & \omega_y = V/2d, \\
z < -d: & \rho = \rho_0 (1 + \varepsilon), & u_x = -V/2, & \omega_y = 0,
\end{array}
$$

as shown in Figure 11.11. We now have a layer of vorticity of finite thickness, although there are still discontinuities in the density field, which will turn out to be important.

Chandrasekhar (1961) gives a detailed derivation of the linear stability criterion for this configuration. Because the analysis is somewhat involved, we shall simply summarize the end result. Here the key control parameter is the *global Richardson number*, Ri, defined as

$$
\text{Ri} = \frac{g(\Delta\rho/2d)}{\rho_0 (du_x/dz)^2} = \frac{2dg\Delta\rho}{\rho_0 V^2}, \tag{11.68}
$$

where $\Delta\rho = 2\varepsilon\rho_0$ and $du_x/dz = V/2d$. For a Boussinesq fluid, in which density changes are small, instability occurs when

$$
\frac{\kappa}{1 + e^{-\kappa}} < \text{Ri} + 1 < \frac{\kappa}{1 - e^{-\kappa}}, \tag{11.69}
$$

where $\kappa = 2|k|d$. Outside this range of wavenumbers, the motion is stable.

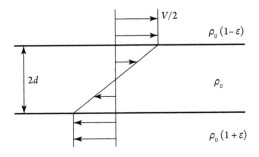

Figure 11.11 Taylor's model problem for a vortex sheet of finite thickness.

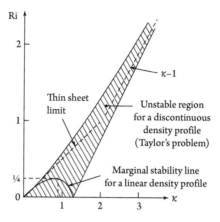

Figure 11.12 The upper two curves are the marginal stability boundaries for Taylor's problem. The lowest curve represents marginal stability for a *linear* variation in density.

This criterion is illustrated in Figure 11.12. It might be compared with expression (11.66) for a vortex sheet of *zero* thickness which for a Boussinesq fluid can be written as $\kappa > 2\mathrm{Ri}$. For $d \to 0$, and hence $\kappa \to 0$, the right of (11.69) does indeed simplify to $\mathrm{Ri} < \kappa/2$, as it must. More generally, the upper boundary in Figure 11.12 is a continuation of the thin-sheet criterion, while the adjacent curve to the right has arisen from the introduction of a second interface. For any given Ri, there is always a narrow band of wavenumbers to which the flow is unstable, and this band gets narrower as Ri increases. The fact that the bottom-right of Figure 11.12 is stable, in contrast to the prediction of (11.66) which would have it unstable, tells us that the introduction of a finite sheet thickness has stabilized the higher wavenumbers and introduced a lower cut-off in wavelength. For example, if $\mathrm{Ri} = 0$ (*i.e.* zero density difference) the finite vortex sheet is stable for $\kappa > 1.28$, which corresponds to wavelengths less than $9.83d$, whereas (11.66) would have instability for all values of k.

It is important to recall that there are discontinuities in the density field in Taylor's problem, and it turns out that these play a key role in the instability. Crucially, it may be shown that if the density varies linearly across the vortex sheet, rather than discontinuously, then the flow is linearly stable to *all wavenumbers* whenever $\mathrm{Ri} > 1/4$. The corresponding marginal stability curve is shown in Figure 11.12. We have our first glimpse of what turns out to be a powerful stability criterion for continuously stratified shear flows.

11.5 The Stability of Continuously Stratified Shear Flow

We now turn to the stability of stratified shear flows of the form $\mathbf{u} = V(z)\hat{\mathbf{e}}_x$. We shall assume that the fluid is inviscid, the stratification $\rho = \rho_0(z)$ is stable, density variations are sufficiently small for the Boussinesq approximation to apply, and $V(z)$ and $\rho_0(z)$ are smooth functions of z, as shown in Figure 11.13. As noted by Richardson, the key control parameter for such a flow is

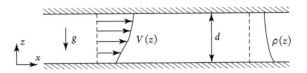

Figure 11.13 A continuously stratified shear flow.

$$\boxed{\text{Ri} = -\frac{g\,(d\rho_0/dz)}{\bar{\rho}\,(du_x/dz)^2} = \frac{g\,|d\rho_0/dz|}{\bar{\rho}\,(du_x/dz)^2}},$$

(11.70)

where $\bar{\rho}$ is a mean density. This dimensionless parameter is known as the *local* (or *gradient*) *Richardson number*. In 1931 Taylor conjectured that stratified shear flows are linearly stable provided that Ri > 1/4 at all points, although it took a further 30 years for a formal proof to be established. Our primary aim here is to provide both a proof and a simple physical interpretation of this criterion. We start by deriving the governing equation for small-amplitude perturbations to the flow.

11.5.1 The Taylor–Goldstein Equation for Fluctuations in a Stratified Shear Flow

Consider an inviscid flow $\mathbf{u}_0 = V(z)\hat{\mathbf{e}}_x$ which is confined to $0 < z < d$ and has a stable stratification, $\rho_0(z)$. The vorticity is evidently

$$\boldsymbol{\omega}_0 = dV/dz\,\hat{\mathbf{e}}_y = -\nabla^2\psi_0\hat{\mathbf{e}}_y,$$

(11.71)

where ψ is the streamfunction defined by $\mathbf{u} = \nabla \times (\psi\hat{\mathbf{e}}_y)$. Suppose that this is subject to a disturbance which is two-dimensional and of small amplitude, say

$$\rho'(x,z,t) = \rho - \rho_0(z), \quad \boldsymbol{\omega}'(x,z,t) = \boldsymbol{\omega} - \boldsymbol{\omega}_0(z).$$

The density and vorticity of the perturbed flow are then governed by (9.3) and (9.5) in the form

$$\frac{D\rho'}{Dt} = \frac{\bar{\rho}N^2}{g}u_z', \quad \frac{D\omega_y}{Dt} - \frac{g}{\bar{\rho}}\frac{\partial\rho'}{\partial x} = 0,$$

(11.72)

where $N(z)$ is the Väisälä–Brunt frequency,

$$N^2 = \frac{g}{\bar{\rho}}\left|\frac{d\rho_0}{dz}\right|,$$

(11.73)

and $\bar{\rho}$ is a mean density. The linearized density and vorticity equations are therefore

$$\left(\frac{\partial}{\partial t} + V\frac{\partial}{\partial x}\right)\rho' = \frac{\bar{\rho}N^2}{g}u_z' = \frac{\bar{\rho}N^2}{g}\frac{\partial\psi'}{\partial x},$$

(11.74)

and

$$\left(\frac{\partial}{\partial t} + V\frac{\partial}{\partial x}\right)\omega_y' - \frac{g}{\bar{\rho}}\frac{\partial\rho'}{\partial x} = -u_z'\frac{d\omega_0}{dz} = -\frac{\partial\psi'}{\partial x}\frac{d^2V}{dz^2}, \tag{11.75}$$

from which we can eliminate ρ' to give

$$\left(\frac{\partial}{\partial t} + V\frac{\partial}{\partial x}\right)^2\omega_y' - N^2\frac{\partial^2\psi'}{\partial x^2} = -\frac{d^2V}{dz^2}\left(\frac{\partial}{\partial t} + V\frac{\partial}{\partial x}\right)\frac{\partial\psi'}{\partial x}. \tag{11.76}$$

Finally, noting that $\omega_y = -\nabla^2\psi$, this yields our governing equation for the perturbed streamfunction:

$$\left(\frac{\partial}{\partial t} + V\frac{\partial}{\partial x}\right)^2\nabla^2\psi' + N^2\frac{\partial^2\psi'}{\partial x^2} = \frac{d^2V}{dz^2}\left(\frac{\partial}{\partial t} + V\frac{\partial}{\partial x}\right)\frac{\partial\psi'}{\partial x}. \tag{11.77}$$

For $V = 0$ this reverts to the (9.48) for internal gravity waves, while $N = 0$ returns us to (11.50) for an inviscid shear flow without stratification. Note that $u_z' = \partial\psi'/\partial x = 0$ at the boundaries, and so we may take $\psi' = 0$ at $z = 0$, d.

Since we have homogeneity in x, it is natural to look for disturbances of the form

$$\psi' = \hat{\psi}'(z)\exp\left[j\left(kx - \varpi t\right)\right], \tag{11.78}$$

where k is real but ϖ may be real or complex. If ϖ is real the flow is neutrally stable, whereas a positive imaginary part of ϖ indicates instability. Substituting (11.78) into (11.77) yields

$$\frac{d^2\hat{\psi}'}{dz^2} - k^2\hat{\psi}' + \frac{N^2}{(V - \varpi/k)^2}\hat{\psi}' = \frac{d^2V/dz^2}{(V - \varpi/k)}\hat{\psi}', \tag{11.79}$$

which is known as the *Taylor–Goldstein equation*. Note that, for $N = 0$, (11.79) reverts to Rayleigh's stability equation, (11.52).

11.5.2 The Richardson Number Criterion for the Stability of a Stratified Shear Flow

Our task now is to use (11.79) to formulate a stability criterion. To that end, it is convenient to replace ϖ by the phase velocity $c = \varpi/k$ and introduce an auxiliary function defined by

$$\phi = \hat{\psi}'\big/\sqrt{V - c}.$$

It is routine, if tedious, to rewrite (11.79) in terms of ϕ and the end result is

$$\frac{d}{dz}\left[(V - c)\frac{d\phi}{dz}\right] - k^2(V - c)\phi + \frac{4N^2 - (dV/dz)^2}{4(V - c)}\phi = \frac{1}{2}\frac{d^2V}{dz^2}\phi. \tag{11.80}$$

We now multiply through (11.80) by the complex conjugate of ϕ, say $\tilde{\phi}$, and integrate from $z = 0$ to $z = d$. This gives

$$\int_0^d \left\{ \tilde{\phi} \frac{d}{dz} \left[(V - c) \frac{d\phi}{dz} \right] - k^2 (V - c) |\phi|^2 \right\} dz + \int_0^d \frac{4N^2 - (dV/dz)^2}{4(V - c)} |\phi|^2 \, dz$$
$$= \int_0^d \frac{d^2 V}{dz^2} \frac{|\phi|^2}{2} \, dz.$$

The first integral on the left may be put into a more convenient form through integration by parts and using the boundary conditions on $\hat{\psi}'$, which yields

$$\int_0^d (c - V) \left(\left| \frac{d\phi}{dz} \right|^2 + k^2 |\phi|^2 \right) dz - \int_0^d \frac{4N^2 - (dV/dz)^2}{4(c - V)} |\phi|^2 \, dz = \int_0^d \frac{d^2 V}{dz^2} \frac{|\phi|^2}{2} \, dz.$$

$$(11.81)$$

Crucially, since

$$\frac{1}{c - V} = \frac{(c_r - V) - jc_i}{|c - V|^2},$$

where $c = c_r + jc_i$, the imaginary part of this integral equation is

$$c_i \int_0^d \left(\left| \frac{d\phi}{dz} \right|^2 + k^2 |\phi|^2 + \frac{4N^2 - (dV/dz)^2}{4|c - V|^2} |\phi|^2 \right) dz = 0. \qquad (11.82)$$

Now instability requires that c has a positive imaginary part, $c_i > 0$, yet (11.82) tells us that this is impossible if $4N^2 > (dV/dz)^2$ for all z, as the integral is then non-zero. Moreover, from definition (11.70), we have $\mathrm{Ri} = N^2 \big/ (dV/dz)^2$. It follows that

linear stability is guaranteed if $\qquad \boxed{\mathrm{Ri} = \dfrac{g|d\rho_0/dz|}{\bar{\rho}(du_x/dz)^2} > \dfrac{1}{4}}, \qquad (11.83)$

at all points in the flow. This is precisely what we observe in Figure 11.12 for the marginal stability curve relating to the case of a linear variation in density and velocity. We conclude, therefore, that a *necessary* condition for instability is that the local Richardson number drops below $1/4$ at some point in the flow. Note, however, that this is not a *sufficient* criterion for instability. In practice, many (but not all) shear layers do indeed have a stability threshold of $(\mathrm{Ri})_{min} = 1/4$, with a wavelength of around seven times the layer thickness.

11.5.3 An Interpretation of the Stability Criterion in terms of Energy

It remains to understand the physical origin of this criterion. Clearly, shear is destabilizing while stratification stabilizes the flow, with Ri representing a balance between the two effects. Moreover, *locally*, the stability of the flow is characterized by just N and $\partial u_x / \partial z$, with Ri the key local dimensionless control parameter. So, it might be argued that any local stability threshold should be of the form $\text{Ri} = (\text{Ri})_{\text{crit}}$. However, this does not explain why it takes the value $(\text{Ri})_{\text{crit}} = 1/4$, at least as far as a sufficient condition for stability is concerned. This also assumes that stability can be assessed purely on the basis of local events, with no regard to the boundaries, yet boundary conditions can and do affect stability thresholds.

It is common to invoke an energy argument to explain (11.83), with the precise details varying from author to author. We shall present an argument similar to that given in Chandrasekhar (1961) and Drazin and Reid (1981). Move into a frame of reference in which the instability does not translate in the x-direction and consider two adjacent parcels of fluid of equal volume, A and B, which are located at z and $z + dz$ respectively, as shown in Figure 11.14. Now suppose that these slowly exchange position as part of a *marginally unstable* mode. Then the *increase* in potential energy (per unit volume) resulting from this exchange is

$$\Delta(\text{P.E.}) = \rho_A g(dz) - \rho_B g(dz) = -\frac{d\rho_0}{dz} g(dz)^2 = \left| \frac{d\rho_0}{dz} \right| g(dz)^2. \tag{11.84}$$

This rise in potential energy must be compensated for by a corresponding drop in kinetic energy. If it is not, then the exchange is not permissible and there is no instability.

We cannot readily calculate the change in kinetic energy because we do not know what velocity the two parcels will have when they arrive at their new locations. Certainly, it is unlikely that they will retain their original velocities as they exchange position, because pressure forces will act on the two parcels during the exchange to change their momentum. It seems plausible, however, that on arrival at their new positions the fluid parcels have a velocity which is somewhere in between their original speed and the ambient speed at the new location. So let us write

$$u_x^A(z+dz) = u_x^A(z) + k_A \frac{du_x}{dz} dz, \quad u_x^B(z) = u_x^B(z+dz) - k_B \frac{du_x}{dz} dz$$

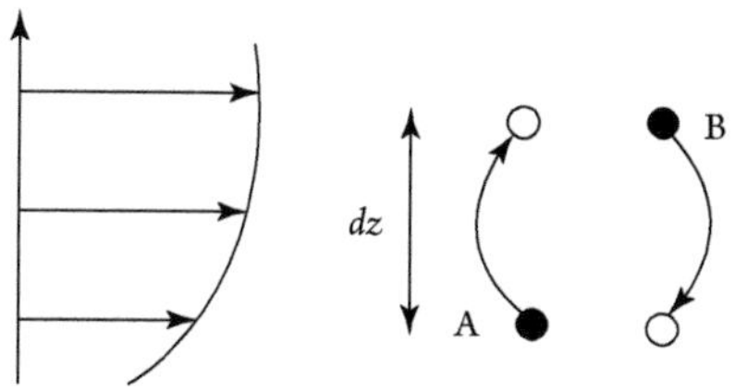

Figure 11.14 Two adjacent fluid parcels exchange height as part of a marginal instability.

for two constants k_A and k_B which are expected to take values somewhere in the range $0 \to 1$. However, in order to retain the same volumetric flow rate, we require $k_A = k_B$, and so

$$u_x^A(z+dz) = u_x^A(z) + k\frac{du_x}{dz}dz, \quad u_x^B(z) = u_x^B(z+dz) - k\frac{du_x}{dz}dz.$$

The change in kinetic energy (per unit volume) of parcel A is then

$$\Delta T_A = \frac{1}{2}\bar{\rho}\left[2ku_x(z)\frac{du_x}{dz}dz + k^2\left(\frac{du_x}{dz}\right)^2(dz)^2\right], \tag{11.85}$$

while that of parcel B is

$$\Delta T_B = \frac{1}{2}\bar{\rho}\left[-2ku_x(z+dz)\frac{du_x}{dz}dz + k^2\left(\frac{du_x}{dz}\right)^2(dz)^2\right]. \tag{11.86}$$

(We use T for kinetic energy on the grounds that it cannot be confused with temperature in this discussion.) The total change in kinetic energy is therefore

$$\Delta T = -k(1-k)\bar{\rho}\left(\frac{du_x}{dz}\right)^2(dz)^2. \tag{11.87}$$

Since $0 < k < 1$, this represents a *reduction* in kinetic energy. Of course, we do not know what value k will take. Nevertheless, we can at least place an upper bound on $|\Delta T|$ as a function of k. According to (11.87), this maximum corresponds to $k = 1/2$ and is

$$|\Delta T|_{\max} = \frac{1}{4}\bar{\rho}\left(\frac{du_x}{dz}\right)^2(dz)^2. \tag{11.88}$$

Now, marginal instability is energetically permissible if the fall in kinetic energy compensates for the rise in potential energy, whereas it is not realizable if $|\Delta T|_{max}$ is less than $\Delta(\text{P.E.})$. Comparing (11.84) with (11.88), we conclude that stability is guaranteed when

$$|\Delta T|_{\max} = \frac{1}{4}\bar{\rho}\left(\frac{du_x}{dz}\right)^2(dz)^2 < \Delta(\text{P.E.}) = \left|\frac{d\rho_0}{dz}\right|g(dz)^2. \tag{11.89}$$

So a sufficient condition for stability is

$$\frac{g\,|d\rho_0/dz|}{\bar{\rho}(du_x/dz)^2} > \frac{1}{4}. \tag{11.90}$$

We have arrived back at (11.83). Although such an energy argument may seem plausible, it is ultimately heuristic, the weakness being the assumed form for the change in velocity. Still, it does at least capture the physical content of the stability criterion.

11.6 The Kelvin–Arnold Variational Principle for Inviscid Flows

11.6.1 A Statement of the Theorem

We now turn to an important theorem in the stability of inviscid flows. It was first stated by Kelvin (1887) and subsequently rediscovered and popularized by the Russian mathematician Vladimir Arnold some 80 years later (Arnold, 1966). It takes the form of a variational principle, and like many theorems in the calculus of variations it has the allure of generality. Indeed, as we shall see, Rayleigh's inflection point theorem, as well as his centrifugal stability criterion, are both special cases of Kelvin's variational principle. Kelvin (1887) stated the principle as follows:

> The condition for steady motion of an incompressible inviscid fluid filling a finite fixed portion of space … is that, with given vorticity, the energy is a thorough maximum, or a thorough minimum, or a minimax. The further condition of stability is secured, by consideration of energy alone, for any case of steady motion for which the energy is a thorough maximum or a thorough minimum; because when the boundary is held fixed the energy is of necessity constant. But the mere consideration of energy does not decide the question of stability for any case of steady motion in which the energy is a minimax.

In effect, Kelvin noted that the kinetic energy of a steady Euler flow is stationary under a small, volume-preserving disturbance in which the vortex lines are frozen into the fluid. That is to say, if the vortex lines of a steady Euler flow are slightly displaced, then there is no change in the kinetic energy of the fluid, at least to *leading order* in the disturbance. He also noted that such flows are stable provided that their kinetic energy is a maximum or a minimum under such a perturbation, *i.e.* stability is ensured if the *second variation* in kinetic energy is strictly positive or strictly negative. Finally, Kelvin observed that nothing can be concluded about the stability of a flow when its equilibrium represents a saddle point in energy, such flows being potentially unstable. In short, a sufficient condition for the stability of an Euler flow is that the kinetic energy of the flow is an extremum under a perturbation of the vortex lines. This is known as the *Kelvin–Arnold variational principle*, and many specific stability criteria are readily derived from this theorem. Although the Kelvin–Arnold theorem is valid for three-dimensional disturbances, in practice it is usually used to determine sufficient conditions for two-dimensional or axisymmetric stability.

11.6.2 A Derivation of the Theorem

Let us try to see where the Kelvin–Arnold variational principle comes from. We are dealing with three-dimensional, inviscid flows governed by

$$\frac{\partial \boldsymbol{\omega}}{\partial t} = \nabla \times (\mathbf{u} \times \boldsymbol{\omega}), \quad \mathbf{u} \cdot d\mathbf{S} = 0 \text{ on the boundary,}$$

whose steady solutions obey

$$\mathbf{u}_0 \times \boldsymbol{\omega}_0 = \nabla H_0.$$

In order to establish the theorem, we need to examine the vorticity fields (and associated kinetic energy) that can be obtained from a steady vorticity distribution, $\boldsymbol{\omega}_0$, by a small, smooth *virtual displacement* of the vortex lines (a so-called *isovortical perturbation*). Following Moffatt (1986), we can find these vorticity fields as follows. Suppose that we freeze the flow and, for a short time, τ, we apply an imaginary, steady velocity field, $\mathbf{v}(\mathbf{x})$, to the fluid (and hence to the vortex lines) which shifts the vortex lines from their equilibrium position, $\mathbf{x}$, to $\mathbf{x} + \boldsymbol{\xi}$. This is shown schematically in Figure 11.15. Since the fluid is incompressible, our imaginary velocity field must be solenoidal, $\nabla \cdot \mathbf{v} = 0$. It must also satisfy $\mathbf{v} \cdot d\mathbf{S} = 0$ on the boundary. We shall refer to $\mathbf{v}(\mathbf{x})$ as a *virtual velocity field*. The application of this virtual velocity field creates a new vorticity distribution, $\boldsymbol{\omega} = \boldsymbol{\omega}_0 + \delta\boldsymbol{\omega}$, which can be uncurled to provide an associated velocity field, $\mathbf{u} = \mathbf{u}_0 + \delta\mathbf{u}$.

Since the vortex lines are frozen into the fluid during the application of this virtual velocity field, we have

$$\frac{\partial \boldsymbol{\omega}}{\partial t_{\mathrm{v}}} = \nabla \times (\mathbf{v} \times \boldsymbol{\omega}), \quad 0 < t_{\mathrm{v}} < \tau,$$

where t_{v} is virtual time. If we write $\delta^1 \boldsymbol{\omega} = \nabla \times (\mathbf{v} \times \boldsymbol{\omega}_0) \tau$, this integrates to give

$$\boldsymbol{\omega}(\mathbf{x}, \tau) = \boldsymbol{\omega}_0(\mathbf{x}) + \delta^1 \boldsymbol{\omega} + \tfrac{1}{2} \nabla \times (\mathbf{v} \times \delta^1 \boldsymbol{\omega}) \tau + O(\tau^3).$$

At this point it turns out to be convenient to introduce a second displacement field, $\boldsymbol{\eta}(\mathbf{x})$, which is distinct from, though closely related to, the true displacement $\boldsymbol{\xi}(\mathbf{x})$. Following Moffatt (1986), we define the *virtual displacement field* to be $\boldsymbol{\eta} = \mathbf{v}(\mathbf{x})\tau$, where $\nabla \cdot \boldsymbol{\eta} = 0$ and $\boldsymbol{\eta} \cdot d\mathbf{S} = 0$ on the boundary. In terms of $\boldsymbol{\eta}(\mathbf{x})$, the changes to $\boldsymbol{\omega}_0$ resulting from an application of $\mathbf{v}(\mathbf{x})$ are

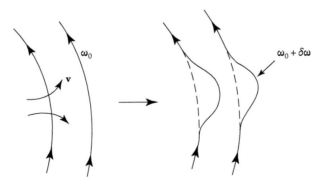

Figure 11.15 The perturbation of vortex lines by a virtual velocity field $\mathbf{v}$.

$$\delta^1 \boldsymbol{\omega} = \nabla \times (\boldsymbol{\eta} \times \boldsymbol{\omega}_0), \quad \delta^2 \boldsymbol{\omega} = \frac{1}{2} \nabla \times (\boldsymbol{\eta} \times \delta^1 \boldsymbol{\omega}). \tag{11.91}$$

The corresponding changes in velocity, $\delta^1 \mathbf{u}$ and $\delta^2 \mathbf{u}$, can be obtained by uncurling (11.91) to give

$$\delta^1 \mathbf{u} = \boldsymbol{\eta} \times \boldsymbol{\omega}_0 + \nabla \varphi_1, \quad \delta^2 \mathbf{u} = \frac{1}{2} \boldsymbol{\eta} \times \delta^1 \boldsymbol{\omega} + \nabla \varphi_2, \tag{11.92}$$

for some φ_1 and φ_2 that are chosen to ensure that $\nabla \cdot (\delta^1 \mathbf{u}) = 0$ and $\nabla \cdot (\delta^2 \mathbf{u}) = 0$.

Note that, during the application of our imaginary velocity field, we have

$$\frac{\partial \boldsymbol{\xi}}{\partial t_v} = \mathbf{v}(\mathbf{x} + \boldsymbol{\xi}) = \mathbf{v}(\mathbf{x}) + (\boldsymbol{\xi} \cdot \nabla)\mathbf{v} + \cdots,$$

and it follows that

$$\boldsymbol{\xi} = \boldsymbol{\eta} + \frac{1}{2}(\boldsymbol{\eta} \cdot \nabla)\boldsymbol{\eta} + \cdots.$$

Hence the true displacement of the vortex lines, $\boldsymbol{\xi}(\mathbf{x})$, and the virtual displacement field, $\boldsymbol{\eta}(\mathbf{x})$, are equal only at leading order. It turns out to be $\boldsymbol{\eta}(\mathbf{x})$, rather that $\boldsymbol{\xi}(\mathbf{x})$, which provides the more convenient description of the disturbance, largely because $\nabla \cdot \boldsymbol{\eta} = 0$.

Let us now calculate the change in the kinetic energy E resulting from changes in the vorticity field. We write $E - E_0 = \delta^1 E + \delta^2 E + \cdots$, where $\delta^1 E$ and $\delta^2 E$ are the first- and second-order changes in kinetic energy. Noting that $\mathbf{u}_0 \cdot \nabla \varphi_1 = \nabla \cdot (\varphi_1 \mathbf{u}_0)$ integrates to zero, the first-order change in energy is

$$\delta^1 E = \int (\mathbf{u}_0 \cdot \delta^1 \mathbf{u}) \, dV = \int (\mathbf{u}_0 \cdot (\boldsymbol{\eta} \times \boldsymbol{\omega}_0)) \, dV = -\int (\boldsymbol{\eta} \cdot (\mathbf{u}_0 \times \boldsymbol{\omega}_0)) \, dV,$$

and given that $\mathbf{u}_0 \times \boldsymbol{\omega}_0 = \nabla H_0$, we find

$$\boxed{\delta^1 E = -\int (\boldsymbol{\eta} \cdot \nabla H_0) \, dV = -\oint H_0 \boldsymbol{\eta} \cdot d\mathbf{S} = 0}. \tag{11.93}$$

We conclude that the kinetic energy of an Euler flow is stationary at equilibrium with respect to an isovortical perturbation, just as Kelvin predicted. The second-order change is

$$\delta^2 E(\boldsymbol{\eta}) = \frac{1}{2} \int \left[(\delta^1 \mathbf{u})^2 + 2\mathbf{u}_0 \cdot \delta^2 \mathbf{u} \right] dV.$$

On substituting for $\delta^2 \mathbf{u}$, noting that $\mathbf{u}_0 \cdot \nabla \varphi_2 = \nabla \cdot (\varphi_2 \mathbf{u}_0)$ integrates to zero, and rearranging the vector triple product $\mathbf{u}_0 \cdot (\boldsymbol{\eta} \times \delta^1 \boldsymbol{\omega})$, we find

$$\boxed{\delta^2 E(\boldsymbol{\eta}) = \frac{1}{2} \int \left[(\delta^1 \mathbf{u})^2 - \delta^1 \boldsymbol{\omega} \cdot (\boldsymbol{\eta} \times \mathbf{u}_0) \right] dV}. \tag{11.94}$$

Crucially, we can use (11.94) to establish sufficient conditions for the stability of a steady Euler flow. We proceed as follows. Suppose we perform an isovortical perturbation on a steady Euler flow to provide an initial condition for a second, unsteady flow. Then, because an isovortical perturbation will be maintained under Euler dynamics, we can relate the perturbed and base flows through an *evolving* virtual displacement field, $\boldsymbol{\eta}(\mathbf{x}, t)$. Now the kinetic energy of the perturbed flow is conserved, and, since $\delta^1 E = 0$ at all times, we conclude that, in the linear approximation, $\delta^2 E(\boldsymbol{\eta})$ remains constant as the perturbed flow evolves. We have therefore created a conserved, quadratic measure of the disturbance. Now suppose that $\delta^2 E(\boldsymbol{\eta})$ is strictly positive, or else strictly negative, for all admissible $\boldsymbol{\eta}$, so that the equilibrium flow represents a minimum or a maximum in kinetic energy with respect to isovortical perturbations. Then $\delta^2 E$ can be used to bound the growth of any disturbance. For example, suppose that $\|\delta \mathbf{u}\|^2$ is some suitable measure of the disturbance, say $\|\delta \mathbf{u}\|^2 = \int (\mathbf{u} - \mathbf{u}_0)^2 dV$. Then the flow will be unstable if $\|\delta \mathbf{u}\|^2$ grows continually despite the conservation of $\delta^2 E(\boldsymbol{\eta})$, and so a prerequisite for instability is that $\left| \delta^2 E \right| \big/ \|\delta \mathbf{u}\|^2 \to 0$. Consequently, if there exists a bound of the form $\left| \delta^2 E \right| \geq \lambda \|\delta \mathbf{u}\|^2$ for all possible $\delta \mathbf{u}$, λ being a positive constant, then the flow cannot become unstable. When $\delta^2 E(\boldsymbol{\eta})$ is strictly positive, or else strictly negative, for all admissible $\boldsymbol{\eta}$, such a bound can usually be found and the flow is then said to be stable *in the sense Liapounov*.

The Kelvin–Arnold theorem is often represented symbolically as shown in Figure 11.16. This is a cartoon of an infinite-dimensional function space (reduced to three dimensions for ease of representation) of all possible solenoidal velocity fields which satisfy $\mathbf{u} \cdot d\mathbf{S} = 0$ on the boundary. The various axes represent the values of the velocity components at each point in the flow. Consequently, an Euler flow traces out a curve within such a function space while preserving its vortex-line topology, as well as its energy. It is natural, therefore, to divide up this function space into lower dimensional sub-domains in which the vorticity fields are all linked one to another by a smooth, volume-preserving displacement field. These

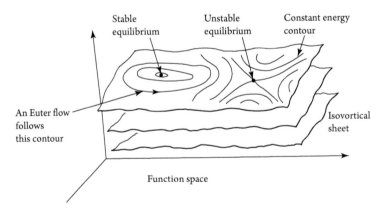

Figure 11.16 Euler flows follow constant energy contours on isovortical sheets in function space. Steady flows are stationary points on such sheets. Extrema in energy are stable flows.

sub-domains are referred to as *isovortical sheets* and the function space is said to be foliated by such sheets. An Euler flow is constrained to follow a constant energy contour on such a sheet, and so steady Euler flows are stationary points with respect to energy on isovortical sheets. These stationary points can be either extrema or saddle points.

Kelvin's original vision is, perhaps, most readily understood in terms of this cartoon. Suppose a steady flow is perturbed isovortically onto an adjacent energy contour. Then the perturbed flow must follow that contour for all time. It follows that energy extrema represent stable Euler flows, since the perturbed flow always stays close to the local stationary point. Saddle points, on the other hand, represent potentially unstable flows, since energy conservation places no restriction on the migration of the flow away from the stationary point.

11.6.3 Some Simple Applications of the Theorem

The Kelvin–Arnold theorem is, perhaps, most productive when used for two-dimensional stability. Consider a steady, inviscid, two-dimensional flow governed by $\mathbf{u}_0 \cdot \nabla \omega_0 = 0$. Since the vorticity is constant along a streamline we may write $\omega_0 = \omega_0(\psi_0)$, where ψ_0 is the two-dimensional streamfunction defined by (1.15). Now suppose that the spatial gradients in ω_0 are everywhere smooth and non-zero, *i.e.* we exclude vortex sheets and vortex patches, and let $\phi(x, y) = \delta^1 \psi$ be the first-order change in ψ_0 under an isovortical perturbation. Then

$$\delta^1 \omega = -\boldsymbol{\eta} \cdot \nabla \omega_0 = -(d\omega_0/d\psi_0)\boldsymbol{\eta} \cdot \nabla \psi_0, \quad \text{and} \quad (\boldsymbol{\eta} \times \mathbf{u}_0)_z = -\boldsymbol{\eta} \cdot \nabla \psi_0.$$

It follows that

$$\delta^1 \boldsymbol{\omega} \cdot (\boldsymbol{\eta} \times \mathbf{u}_0) = -(\delta^1 \omega)\boldsymbol{\eta} \cdot \nabla \psi_0 = (\delta^1 \omega)^2 / (d\omega_0/d\psi_0),$$

and so (11.94) simplifies to

$$\delta^2 E = \frac{1}{2} \int \left[(\nabla \phi)^2 - (\nabla^2 \phi)^2 / (d\omega_0/d\psi_0) \right] dV. \tag{11.95}$$

Clearly, such flows are stable by virtue of the Kelvin–Arnold theorem provided that either:

(i) $d\omega_0/d\psi_0 < 0$ at all points in the flow, in which case $\delta^2 E > 0$ for all $\boldsymbol{\eta}$, or else;

(ii) $d\omega_0/d\psi_0 > 0$ at all points in the flow and the second term in integral (11.95) is dominant for all admissible ϕ, in which case $\delta^2 E < 0$ for all $\boldsymbol{\eta}$.

It is possible to use the calculus of variations to show that, in case (ii), we have $\delta^2 E(\boldsymbol{\eta}) < 0$ for all $\boldsymbol{\eta}$ (and so stability is ensured) provided that the minimum eigenvalue, $\lambda_{\min}$, of

$$\nabla^2 \phi + \lambda (d\omega_0/d\psi_0)\phi = 0, \quad \phi = 0 \text{ on the boundary},$$

is greater than unity. The simplest way to prove this is to normalize $\delta^2 E$ by $\frac{1}{2} \int (\nabla \phi)^2 \, dV$ and look for the maximum of that ratio. The resulting eigenvalue problem then gives

$$\delta^2 E_{\max} = (1 - \lambda_{\min}) \frac{1}{2} \int (\nabla \phi)^2 \, dV.$$

(We shall not pause to work through the details, but rather refer the reader to Davidson, 1994, and references therein.) The equivalent stability criteria for axisymmetric poloidal flows, $\mathbf{u}(r, z) = (u_r, 0, u_z)$, is discussed in Exercise 11.4.

The stability of many two-dimensional flows follow from (11.95). For example, consider the flow $\mathbf{u}_0 = V(y)\hat{\mathbf{e}}_x$ for $0 < y < d$, as shown in Figure 11.17. Here $\omega_0 = -dV/dy$ while $V = d\psi_0/dy$, and so

$$V^2 \frac{d\omega_0}{d\psi_0} = V \frac{d\omega_0}{dy} = -V \frac{d^2 V}{dy^2}.$$

We conclude that $d\omega_0/d\psi_0 < 0$, and hence stability is ensured, provided that

$$V \frac{d^2 V}{dy^2} > 0, \quad \text{for } 0 < y < d.$$

However, since this is an inviscid flow with slip boundary conditions, we are free to change the frame of reference and this does not alter the stability of the flow. Hence stability is ensured whenever

$$(V - U) \frac{d^2 V}{dy^2} > 0, \quad \text{for } 0 < y < d, \tag{11.96}$$

for some constant U. It is always possible to find a U that satisfies (11.96) provided that $d^2 V/dy^2$ does not change sign anywhere in the flow, and so a sufficient condition for stability is that $d^2 V/dy^2$ does not change sign. We have arrived back at Rayleigh's inflection point theorem.

We can also derive Rayleigh's centrifugal stability criterion from the Kelvin–Arnold theorem. Consider the flow

$$\mathbf{u}_0 = \frac{\Gamma(r)}{r} \hat{\mathbf{e}}_\theta, \quad \boldsymbol{\omega}_0 = \frac{1}{r} \frac{d\Gamma}{dr} \hat{\mathbf{e}}_z, \tag{11.97}$$

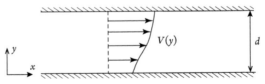

Figure 11.17 An inviscid shear flow.

written in (r, θ, z) coordinates. We can investigate the axisymmetric stability of this flow using the virtual displacement field $\boldsymbol{\eta}(r, z) = (\eta_r, \eta_\theta, \eta_z)$. For such a displacement field we have

$$\delta^1 \mathbf{u} = \delta^1 \mathbf{u}_p - \frac{\eta_r}{r}\frac{d\Gamma}{dr}\hat{\mathbf{e}}_\theta, \qquad \boldsymbol{\eta} \times \mathbf{u}_0 = \frac{\Gamma}{r}(-\eta_z, 0, \eta_r),$$

and

$$\delta^1 \boldsymbol{\omega}_p = \nabla \times (\boldsymbol{\eta}_p \times \boldsymbol{\omega}_0) = \frac{1}{r}\left(\frac{d\Gamma}{dr}\frac{\partial\eta_r}{\partial z}, 0, -\frac{\partial}{\partial r}\left(\eta_r\frac{d\Gamma}{dr}\right)\right),$$

where the subscript p stands for poloidal. Some careful algebra then yields

$$\left(\delta^1 \mathbf{u}\right)^2 - \delta^1 \boldsymbol{\omega} \cdot (\boldsymbol{\eta} \times \mathbf{u}_0) = \left(\delta^1 \mathbf{u}_p\right)^2 + \Phi(r)\eta_r^2 + \nabla \cdot \left(\frac{\eta_r}{2r^2}\frac{d\Gamma^2}{dr}\boldsymbol{\eta}\right), \qquad (11.98)$$

where

$$\Phi(r) = \frac{1}{r^3}\frac{d\Gamma^2}{dr}$$

is Rayleigh's discriminant. Thus

$$\delta^2 E(\boldsymbol{\eta}) = \frac{1}{2}\int\left[\left(\delta^1 \mathbf{u}_p\right)^2 + \Phi(r)\eta_r^2\right]dV, \qquad (11.99)$$

and the Kelvin–Arnold theorem tells us that this inviscid flow is stable to axisymmetric disturbances provided that $\Phi(r) > 0$ for all r. This is, of course, Rayleigh's centrifugal stability criterion. Note that $\delta^1 \mathbf{u}_p = 0$ when $\eta_\theta = 0$, and in such cases the change in kinetic energy arising from a rearrangement of the underlying angular momentum distribution, $\frac{1}{2}\int \Phi(r)\eta_r^2 dV$, is exactly that predicted by (10.7).

11.7 A Variational Principle for Inviscid Flows based on the Lagrangian

There is a second variational principle that is directly related to the Kelvin–Arnold theorem, but is also closely aligned with conventional Hamiltonian mechanics. Before introducing the principle, we must first revisit the idea of a *Lagrangian displacement*, and also introduce a second class of variation. This second variation is *not* isovortical, but rather of the type used in Hamilton's principle. Let us start with the Lagrangian displacement.

As before, suppose that $\mathbf{u}_0(\mathbf{x})$ and $\boldsymbol{\omega}_0(\mathbf{x})$ are perturbed to give $\mathbf{u}(\mathbf{x}, t) = \mathbf{u}_0 + \delta\mathbf{u}$ and $\boldsymbol{\omega}(\mathbf{x}, t) = \boldsymbol{\omega}_0 + \delta\boldsymbol{\omega}$. Moreover, let $\boldsymbol{\zeta}(\mathbf{x}, t)$ be the Lagrangian displacement of a particle p from its position in the unperturbed flow, as shown in Figure 11.18. If $\mathbf{x}_{p0}(t)$ is the position

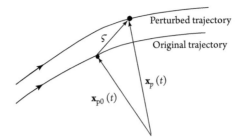

Figure 11.18 The perturbed, $\mathbf{x}_p(t)$, and unperturbed, $\mathbf{x}_{p0}(t)$, trajectories of a fluid particle.

vector of the particle in the original flow, and $\mathbf{x}_p(t)$ its trajectory in the perturbed flow, then $\zeta(\mathbf{x}, t)$ is defined by

$$\zeta = \mathbf{x}_p(t) - \mathbf{x}_{p0}(t). \tag{11.100}$$

The rate of change of ζ as seen by an observer at $\mathbf{x}_{p0}(t)$ is then

$$\frac{D\zeta}{Dt} = \frac{\partial \zeta}{\partial t} + (\mathbf{u}_0 \cdot \nabla)\zeta = \mathbf{u}(\mathbf{x}_{p0} + \zeta) - \mathbf{u}_0(\mathbf{x}_{p0}), \tag{11.101}$$

which can be rewritten as

$$\frac{\partial \zeta}{\partial t} + (\mathbf{u}_0 \cdot \nabla)\zeta = \delta \mathbf{u}(\mathbf{x} + \zeta) + [\mathbf{u}_0(\mathbf{x} + \zeta) - \mathbf{u}_0(\mathbf{x})], \tag{11.102}$$

where we have dropped the subscripts on $\mathbf{x}$. This may be linearized for small $|\zeta|$ using the approximations $\delta \mathbf{u}(\mathbf{x} + \zeta) \approx \delta \mathbf{u}(\mathbf{x})$ and $\mathbf{u}_0(\mathbf{x} + \zeta) - \mathbf{u}_0(\mathbf{x}) \approx (\zeta \cdot \nabla)\mathbf{u}_0(\mathbf{x})$, which gives us

$$\frac{\partial \zeta}{\partial t} + (\mathbf{u}_0 \cdot \nabla)\zeta = \delta^1 \mathbf{u}(\mathbf{x}, t) + (\zeta \cdot \nabla)\mathbf{u}_0(\mathbf{x}), \tag{11.103}$$

to leading order in $|\zeta|$. As in §11.6.2, it is more convenient to work with a *solenoidal* virtual displacement field $\boldsymbol{\eta}(\mathbf{x}, t)$ which, with the flow frozen in time, achieves the same instantaneous particle displacement ζ through the application of a virtual velocity field. Since $\boldsymbol{\eta}$ and ζ are equal to leading order, (11.103) gives us a convenient relationship between $\boldsymbol{\eta}$ and the Eulerian perturbation in velocity:

$$\boxed{\delta^1 \mathbf{u}(\mathbf{x}, t) = \frac{\partial \boldsymbol{\eta}}{\partial t} + \nabla \times (\boldsymbol{\eta} \times \mathbf{u}_0)}. \tag{11.104}$$

We now introduce a second class of perturbation. Though not isovortical and hence *not conserved under Euler dynamics*, this is still a legitimate virtual displacement field which identifies adjacent flows in function space. Using d rather than δ for this perturbation, it is defined by

$$d^1\mathbf{u} = \nabla \times (\boldsymbol{\eta} \times \mathbf{u}_0), \qquad d^2\mathbf{u}(\mathbf{x},t) = \tfrac{1}{2}\nabla \times (\boldsymbol{\eta} \times d^1\mathbf{u}). \qquad (11.105)$$

When compared with (11.91), we see that such a 'd-variation' represents a virtual displacement of the fluid in which the *streamlines*, rather than the vortex lines, are frozen into the fluid. The first-order change in kinetic energy under such a virtual displacement is zero, since

$$d^1 E(\boldsymbol{\eta}) = \int \left[\mathbf{u}_0 \cdot d^1\mathbf{u}\right] dV = \int \left[\mathbf{u}_0 \cdot \nabla \times (\boldsymbol{\eta} \times \mathbf{u}_0)\right] dV = \int \left[\boldsymbol{\omega}_0 \cdot (\boldsymbol{\eta} \times \mathbf{u}_0)\right] dV,$$

from which

$$d^1 E(\boldsymbol{\eta}) = \int \left[\boldsymbol{\eta} \cdot (\mathbf{u}_0 \times \boldsymbol{\omega}_0)\right] dV = \int \left[\boldsymbol{\eta} \cdot \nabla H_0\right] dV = \oint H_0 \boldsymbol{\eta} \cdot d\mathbf{S} = 0. \qquad (11.106)$$

Moreover, the second-order change in energy is

$$d^2 E = \frac{1}{2} \int \left[(d^1\mathbf{u})^2 + 2\mathbf{u}_0 \cdot d^2\mathbf{u}\right] dV = \frac{1}{2} \int \left[(d^1\mathbf{u})^2 + \mathbf{u}_0 \cdot \nabla \times (\boldsymbol{\eta} \times d^1\mathbf{u})\right] dV.$$

$$(11.107)$$

Introducing the notation $\hat{\mathbf{u}} = d^1\mathbf{u}$, this may be conveniently rewritten using the identity

$$\nabla \cdot (\mathbf{u}_0 \times (\boldsymbol{\eta} \times \hat{\mathbf{u}})) = \boldsymbol{\omega}_0 \cdot (\boldsymbol{\eta} \times \hat{\mathbf{u}}) - \mathbf{u}_0 \cdot \nabla \times (\boldsymbol{\eta} \times \hat{\mathbf{u}})$$

as

$$d^2 E(\boldsymbol{\eta}) = \frac{1}{2} \int \left[\hat{\mathbf{u}}^2 + \boldsymbol{\omega}_0 \cdot (\boldsymbol{\eta} \times \hat{\mathbf{u}})\right] dV = \frac{1}{2} \int \left[\hat{\mathbf{u}}^2 - \hat{\mathbf{u}} \cdot (\boldsymbol{\eta} \times \boldsymbol{\omega}_0)\right] dV, \qquad (11.108)$$

where we have used the fact that $\boldsymbol{\eta} \cdot d\mathbf{S} = \hat{\mathbf{u}} \cdot d\mathbf{S} = 0$ on the boundary.

The significance of the 'freezing-in' of the streamlines during a virtual displacement of the 'd-type' turns out to be rather classical. The point is that such a variation creates a new set of particle trajectories, which have the special property that their time of flight is the same as for the original trajectories. To see that this is so, think of a stream-tube which is frozen into the fluid during such a d—perturbation and consider a short element of that tube with length ℓ and cross-sectional area A. During a 'd-variation' both the volume of that element, ℓA, and the flux of $\mathbf{u}$ along the tube, uA, are conserved. It follows that ℓ/u is conserved for each short element of the stream-tube, which in turn is the time taken for fluid to pass through that element. We conclude that the time of flight of all fluid trajectories is conserved during a d-variation. This is exactly the kind of virtual displacement associated with Hamilton's principle. Indeed, the fact that $d^1 E(\boldsymbol{\eta}) = 0$ for a steady Euler flow can be shown to follow directly from the principle of least action.

Armed with (11.104) and (11.108), we now return to §11.6.2 and to Euler flows which have been subject to an *isovortical* perturbation. We use the identity

$$\nabla \cdot (\delta^1 \mathbf{u} \times (\eta \times \mathbf{u}_0)) - \delta^1 \boldsymbol{\omega} \cdot (\eta \times \mathbf{u}_0) = -\delta^1 \mathbf{u} \cdot \nabla \times (\eta \times \mathbf{u}_0) = -\delta^1 \mathbf{u} \cdot \hat{\mathbf{u}},$$

and the fact that $\eta \cdot d\mathbf{S} = \mathbf{u}_0 \cdot d\mathbf{S} = 0$ on the boundary, to rewrite (11.94) as

$$\delta^2 E = \frac{1}{2} \int \left[(\delta^1 \mathbf{u})^2 - \delta^1 \mathbf{u} \cdot \hat{\mathbf{u}} \right] dV = \frac{1}{2} \int \delta^1 \mathbf{u} \cdot \dot{\eta} dV = \frac{1}{2} \int (\dot{\eta} + \hat{\mathbf{u}}) \cdot \dot{\eta} dV,$$

(11.109)

where we have used (11.104) in the form $\delta^1 \mathbf{u} = \dot{\eta} + \hat{\mathbf{u}}$. Next, we use the fact that $\delta^1 \mathbf{u} = \eta \times \boldsymbol{\omega}_0 + \nabla \varphi_1$ for an isovortical perturbation to rewrite this as

$$\delta^2 E = \frac{1}{2} \int \dot{\eta}^2 dV - \frac{1}{2} \int \left[\hat{\mathbf{u}} \cdot (\hat{\mathbf{u}} - \delta^1 \mathbf{u}) \right] dV$$

$$= \frac{1}{2} \int \dot{\eta}^2 dV - \frac{1}{2} \int \left[\hat{\mathbf{u}}^2 - \hat{\mathbf{u}} \cdot (\eta \times \boldsymbol{\omega}_0) \right] dV.$$

Finally, a comparison with (11.108) yields

$$\boxed{\delta^2 E = \frac{1}{2} \int \dot{\eta}^2 dV - d^2 E(\eta) = \text{constant}, \quad \delta^1 E = d^1 E = 0}.$$

(11.110)

Evidently, this relates the second variation in energy under d-type and δ-type perturbations.

Crucially, (11.110) may be generalized to ideal flows that are subject to a conservative body force, such as the Lorentz force of ideal magnetohydrodynamics, or gravitational forces. It then reads

$$\boxed{\frac{1}{2} \int \dot{\eta}^2 dV + d^2 V(\eta) - d^2 E(\eta) = \text{constant} = e_0, \quad d^1(E - V) = 0},$$

(11.111)

where V is the potential energy of the body force (Davidson, 2000). This is the key result. If $\frac{1}{2} \int \dot{\eta}^2 dV$ is adopted as a measure of the disturbance, a sufficient condition for stability is that the Lagrangian, $L = E - V$, is a maximum at equilibrium, since $\frac{1}{2} \int \dot{\eta}^2 dV$ is then bounded from above by e_0. This, in turn, yields a number of useful stability criteria. (Note that *static* equilibria are then stable when V is a minimum.) However, we shall not pause to discuss such cases, but rather consider the consequences of (11.110) for two-dimensional Euler flows.

Consider a steady, two-dimensional Euler flow governed by $\mathbf{u}_0 \cdot \nabla \omega_0 = 0$, or equivalently $\omega_0 = \omega_0(\psi_0)$, where ψ is the streamfunction. We restrict ourselves to cases where $d\omega_0/d\psi_0$ is single signed, smooth and non-zero, i.e. we exclude vortex patches and vortex sheets. Let $\phi(x, y) = d^1 \psi$ be the first-order change in ψ *under a d-variation*. Then

$(\boldsymbol{\eta} \times \mathbf{u}_0)_z = -\boldsymbol{\eta} \cdot \nabla \psi_0$, so that $d^1\mathbf{u} = \nabla \times (\boldsymbol{\eta} \times \mathbf{u}_0)$ uncurls to give $\phi = -\boldsymbol{\eta} \cdot \nabla \psi_0$. Moreover, $(\boldsymbol{\eta} \times d^1\mathbf{u})_z = -\boldsymbol{\eta} \cdot \nabla \phi$, and so the second contribution to d^2E in (11.108) can be rewritten as

$$\int \boldsymbol{\omega}_0 \cdot (\boldsymbol{\eta} \times d^1\mathbf{u}) dV = \int \omega_0(-\boldsymbol{\eta} \cdot \nabla \phi) dV = \int \phi \boldsymbol{\eta} \cdot \nabla \omega_0 dV = -\int \frac{d\omega_0}{d\psi_0} \phi^2 dV,$$

from which

$$d^2E(\phi) = \frac{1}{2} \int \left[(\nabla \phi)^2 - \frac{d\omega_0}{d\psi_0} \phi^2 \right] dV. \tag{11.112}$$

This should be compared with (11.95) for $\delta^2 E$. When $d\omega_0/d\psi_0 < 0$, both d^2E and $\delta^2 E$ are positive for all ϕ. Moreover, when $d\omega_0/d\psi_0 > 0$, (11.110) ensures that $d^2E > 0$ whenever $\delta^2 E < 0$. So stable flows for which $\delta^2 E$ is of definite sign all have $d^2E > 0$. In fact, using the calculus of variations it may be shown that, provided $d\omega_0/d\psi_0$ is single signed, the conditions under which d^2E is strictly positive correspond to those under which $\delta^2 E$ is of definite sign (Davidson, 1994). We conclude that, subject to the caveats above, the stability of a two-dimensional Euler flow is assured whenever $d^2E(\phi) > 0$ for all admissible ϕ; that is:

a two-dimensional Euler flow is stable to two-dimensional disturbances if its kinetic energy is a minimum at equilibrium under a 'd-variation', in which the streamlines are 'frozen in'.

11.8 The Stability of Pipe Flow: a Qualitative Discussion

So far, we have restricted the discussion to small-amplitude disturbances and to linear stability. However, as noted in §2.6.2, finite-amplitude perturbations can play an important role in destabilizing an otherwise linearly stable flow, such as in the transition to turbulence of fully developed pipe flow. So we shall close this chapter with a few comments about the stability of flow in a pipe and the possible role of finite-amplitude disturbances.

Let us return to Reynolds' famous paper where he identified the key role played by $\mathrm{Re} = ud/\nu$ in the transition to turbulence of pipe flow, d being the pipe diameter and u the mean speed of the fluid. Crucially, Reynolds noted that the value of Re at which turbulence first appears is extremely sensitive to the level of the disturbances at the entrance to the pipe, and to the shape of the pipe inlet. This led him to suggest that the instability which initiates turbulence might require a perturbation of finite magnitude to take root, that magnitude being a function of Re. For example, Reynolds found that turbulence typically appears at $\mathrm{Re} \approx 2200$ if no particular effort is made to minimize inlet disturbances, yet when those disturbances were minimized, Reynolds was able to maintain laminar flow up to $\mathrm{Re} \sim 13{,}000$. In fact, it turns out that *fully developed* laminar pipe flow is *stable* to infinitesimal disturbances for all values of Re! Indeed, recent experiments with special inlet conditions have achieved laminar flow for values of Re up to $\sim 90{,}000$. So, as Reynolds suggested, the

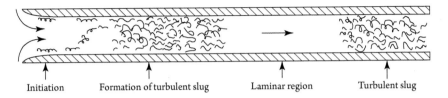

Figure 11.19 A schematic of turbulent slugs near the inlet of a pipe. (From Davidson, 2015.)

Figure 11.20 Simulation of a turbulent puff in a pipe at Re ≈ 2100. (Courtesy of A. P. Willis.)

transition to turbulence is controlled by the size of the disturbances, and by the conditions near the inlet prior to the formation of a parabolic velocity profile. Reynolds also showed that, when turbulence is artificially created in a pipe, there is a value of Re below which the turbulence invariably dies away, that value being around Re ∼ 2000.

In brief, the current view is the following. For a straight inlet, such as that shown in Figure 11.19, and when inlet disturbances are carefully minimized, turbulence tends to appear first at around Re ∼ 10^4, initiated in the annular boundary layer near the inlet. This turbulence first takes the form of small, localized patches, called turbulent spots. These then spread and merge until a 'slug' of turbulence fills a finite portion of the pipe. For Re < 10^4, on the other hand, the inlet boundary layer appears to be stable to small-amplitude disturbances. The implication is that, for transition in the range $2{,}200 \rightarrow 10^4$, either the perturbations are large enough to trigger some kind of finite-amplitude instability of the inlet boundary layer, or else pre-existing turbulence is swept into the entrance of the pipe.

Either way, whatever the origin of the turbulence, transition starts with a series of intermittent turbulent slugs passing down the pipe. Provided Re exceeds ∼2200, these slugs progressively grow in length, eventually merging to form fully developed turbulence. However, when Re is less than ∼2000, any turbulence that enters the pipe, or is locally generated at the inlet, simply decays, exactly as suggested by Reynolds. For the intermediate range of 2000 < Re < 2200, weak patches of turbulence (sometimes called puffs) can persist, but they usually do not lead to fully developed turbulence.

There has been renewed interest in recent years into the nature of these turbulent puffs and slugs, and in particular their internal structure. One example of a numerical simulation of a turbulent puff at Re ≈ 2100 is shown in Figure 11.20. A curious feature of these turbulent patches is that they move, more or less, with the mean speed of the fluid, which for a parabolic profile is one half of the centreline velocity. This may seem unremarkable until one notices that it implies that laminar fluid on the centre-line of the pipe approaches the turbulent puff from behind, becomes turbulent on entering the puff, and then emerges from the front of the turbulent patch as a quiescent, laminar flow!

Perhaps the most important lesson we learn from transitional flow in a pipe is that linear stability theory is only a first step. We shall pursue this idea in the next chapter.

○ ○ ○

This concludes our introduction to stability theory. Two classical books on the subject are Chandrasekhar (1961) and Drazin and Reid (1981). Both are comprehensive. The Kelvin–Arnold theorem is discussed in Moffatt (1986) and Davidson (1994), while Song et al. (2017) provide a recent review of turbulent slugs in transitional pipe flow.

....

EXERCISES

11.1 *A proof of Fjørtoft's stability theorem for inviscid shear flows.* Consider an inviscid shear flow $V(y)$ such as that shown in Figure 11.17. The real part of (11.54) is evidently

$$\int_0^d \left| \frac{d\hat{\psi}'}{dy} \right|^2 dy + k^2 \int_0^d \left| \hat{\psi}' \right|^2 dy = k \int_0^d \frac{(\varpi_r - kV)}{|\varpi - kV|^2} \frac{d^2V}{dy^2} \left| \hat{\psi}' \right|^2 dy,$$

where $\varpi = \varpi_r + j\varpi_i$. Now suppose the flow is unstable with an inflection point at $y = y_*$. Then (11.55) tells us that

$$(\varpi_r - kV(y_*)) \int_0^d \frac{\left| \hat{\psi}' \right|^2}{|\varpi - kV|^2} \frac{d^2V}{dy^2} dy = 0.$$

Hence show that

$$k^2 \int_0^d \frac{(V(y_*) - V)}{|\varpi - kV|^2} \frac{d^2V}{dy^2} \left| \hat{\psi}' \right|^2 dy = \int_0^d \left| \frac{d\hat{\psi}'}{dy} \right|^2 dy + k^2 \int_0^d \left| \hat{\psi}' \right|^2 dy > 0$$

and deduce Fjørtoft's theorem.

11.2 *An application of Fjørtoft's theorem.* Consider the two inviscid shear flows $V(y)$ shown in Figure 11.6. Both contain an inflection point, say at $y = y_*$, and so both are potentially unstable by Rayleigh's inflection point theorem. Show that, for the flow on the left, $(V - V(y_*)) d^2V/dy^2 > 0$, while on the right $(V - V(y_*)) d^2V/dy^2 < 0$.

Hence use Fjørtoft's theorem to show that the flow on the left of Figure 11.6 is in fact stable, whilst that on the right is potentially unstable. Notice that, for the stable flow on the left, the modulus of the vorticity exhibits a local *minimum*, while the potentially unstable flow on the right exhibits a local *maximum* in the modulus of the vorticity.

11.3 *An estimate of the critical Rayleigh number for the transition to Bénard convection of a fluid held between two parallel plates.* Consider the linear stability of a fluid heated from below and confined by two plates at $z = 0, d$, as shown in Figure 11.4. The boundary conditions for the vertical velocity and the perturbation in temperature, ϑ, are

$$\vartheta = u_z = \partial u_z / \partial z = 0, \quad \text{at } z = 0, d.$$

Suppose that the instability takes the form of two-dimensional rolls in the x–z plane. Then a vertical velocity field which is admissible, in the sense that it satisfies the boundary conditions, is

$$u_z = \frac{u_0}{d^4} z^2 (d-z)^2 \sin(kx),$$

where u_0 is a constant and k is a horizontal wavenumber. Show that, for marginal stability, combining this estimate of u_z with (11.34) requires

$$\vartheta = \frac{\Gamma \sin kx}{a^4} \left[a^4 \eta^2 (1-\eta)^2 + 2a^2 (1 - 6\eta + 6\eta^2) + 24 \right]$$
$$- \frac{2(a^2 + 12)\Gamma \sin kx}{a^4 \cosh(a/2)} \cosh(a\eta - a/2),$$

where

$$a = kd, \quad \eta = z/d, \quad \Gamma = \Delta T u_0 d / \alpha a^2.$$

Here ΔT is the temperature difference between the two plates and α the thermal diffusivity of the fluid. Next, find u_x from u_z using continuity and show that the associated vorticity is

$$\omega_y = \frac{u_0 \cos kx}{da} \left[2 \left(1 - 6\eta + 6\eta^2 \right) - a^2 \eta^2 (1-\eta)^2 \right].$$

Now integrate over a single convection cell to show that the corresponding dissipation and buoyancy integrals are

$$\nu \int \omega^2 dV = \frac{2\pi \nu u_0^2}{a^3} \left[\frac{1}{5} + \frac{a^2}{105} + \frac{a^4/4}{630} \right],$$

and

$$g\beta \int \vartheta u_z dV$$
$$= \frac{\pi g \beta \Delta T d^3 u_0^2}{2\alpha a^{11}} \left[\frac{a^8}{630} - \frac{2a^6}{105} + \frac{4a^4}{5} + 48a^2 + 576 - 4 \left(a^2 + 12 \right)^2 \frac{\tanh(a/2)}{a/2} \right].$$

You may find the definite integral

$$\int_0^1 (1 - p^2)^2 \cosh \left(\frac{ap}{2} \right) dp = \frac{8 \left[(a/2)^2 + 3 \right]}{(a/2)^5} \sinh(a/2) - \frac{24}{(a/2)^4} \cosh(a/2)$$

of some help. Finally, use (11.42) to show that, at marginal stability,

$$\left[\frac{4}{5} + \frac{4a^2}{105} + \frac{a^4}{630} \right]$$
$$= \frac{\mathrm{Ra}}{a^8} \left[\frac{a^8}{630} - \frac{2a^6}{105} + \frac{4a^4}{5} + 48a^2 + 576 - 4\left(a^2 + 12\right)^2 \frac{\tanh(a/2)}{a/2} \right],$$

where Ra is the Rayleigh number defined by (11.25). The corresponding marginal stability curve in the a–Ra plane is shown in Figure 11.5. Of course, this is only an estimate, as the entire calculation is based on the guess

$$u_z \sim z^2 (d - z)^2 \sin(kx).$$

However, the end result turns out to be surprisingly accurate, predicting a critical value of $(\mathrm{Ra})_c = 1708.1$ corresponding to $a = 3.113$. The true values are $(\mathrm{Ra})_c = 1707.8$ and $a = 3.117$, as shown in Table 11.1, and so the error in $(\mathrm{Ra})_c$ is less than 0.02%!

11.4 *Application of the Kelvin–Arnold stability theorem to axisymmetric, poloidal flows.* Consider a steady, inviscid, axisymmetric flow of the form $\mathbf{u}_0 = (u_r, 0, u_z)$, $\boldsymbol{\omega}_0 = (0, \omega_\theta, 0)$, which is governed by $\mathbf{u}_0 \cdot \nabla(\omega_0/r) = 0$ in accordance with (3.70). Since ω_0/r is constant along a streamline, we may write $\omega_0/r = \Omega(\Psi_0)$, where Ψ_0 is the Stokes streamfunction defined by (1.17). Now suppose that the spatial gradients in ω_0 are everywhere smooth and non-zero, and let $\phi = \delta^1\Psi$ be the first-order change in Ψ_0 under an axisymmetric perturbation of the vortex lines. Show that, if $\boldsymbol{\eta}(r, z)$ is the virtual displacement field, then

$$r\left(\boldsymbol{\eta} \times \mathbf{u}_0\right)_\theta = -\boldsymbol{\eta} \cdot \nabla\Psi_0,$$

and

$$\delta^1\left(\omega_\theta/r\right) = -\boldsymbol{\eta} \cdot \nabla\left(\omega_0/r\right) = -\left(d\Omega/d\Psi_0\right)\boldsymbol{\eta} \cdot \nabla\Psi_0,$$

from which

$$\delta^1\boldsymbol{\omega} \cdot \left(\boldsymbol{\eta} \times \mathbf{u}_0\right) = -\left(\delta^1(\omega_\theta/r)\right)\boldsymbol{\eta} \cdot \nabla\Psi_0 = \left(\delta^1(\omega_\theta/r)\right)^2 \Big/ \left(d\Omega/d\Psi_0\right).$$

Hence confirm that (11.94) gives the second variation in energy as

$$\delta^2 E = \frac{1}{2} \int \left[(\nabla\phi)^2 - \left(\nabla_*^2\phi\right)^2 \Big/ r^2(d\Omega/d\Psi_0) \right] r^{-2} dV,$$

where ∇_*^2 is the Stokes operator (3.31). Such flows are stable by virtue of the Kelvin–Arnold theorem provided that either (i)$d\Omega/d\Psi_0 < 0$ at all points, or (ii) $d\Omega/d\Psi_0 > 0$ and the second term in the integral is dominant for all admissible ϕ. Finally, use the calculus of variations to show that, in case (ii),

$$\delta^2 E_{\max} = (1 - \lambda_{\min}) \frac{1}{2} \int (\nabla \phi)^2 \, r^{-2} \, dV,$$

where $\lambda_{\min}$ is the least eigenvalue of

$$\nabla_*^2 \phi + \lambda (r^2 d\Omega/d\Psi_0) \phi = 0, \quad \phi = 0 \text{ on the boundary}$$

Hence show that stability is ensured for case (ii) provided that $\lambda_{\min}$ is greater than unity.

11.5 *The capillary instability of a liquid jet.* Consider a liquid jet in air. It has a uniform speed and a radius of $r = R$ when viewed in (r, z) coordinates. Surface tension acts on the jet in such a way as to pressurize the liquid and we wish to determine the stability of the jet. For simplicity, we ignore viscosity and consider only axisymmetric disturbances (Figure 11.21), which turn out to be the most dangerous.

Moving into a frame of reference in which the unperturbed jet is stationary, we let $\eta(z, t)$ be the radial displacement of the surface of the jet, and $\mathbf{u}'$ and p' be the perturbations in velocity and pressure arising from the deformation of the jet surface. The equations of motion and linearized kinematic boundary condition are then

$$\rho \partial \mathbf{u}'/\partial t = -\nabla p', \quad \nabla \cdot \mathbf{u}' = 0, \quad u_r' = \partial \eta/\partial t \text{ at } r = R,$$

from which we deduce that

$$\nabla^2 p' = 0, \quad \rho \frac{\partial u_r'}{\partial t} = -\frac{\partial p'}{\partial r} = \rho \frac{\partial^2 \eta}{\partial t^2}, \quad \text{at } r = R. \tag{11.113}$$

We also need a dynamic boundary condition. Show that (7.48) gives

$$p(r = R + \eta) = p_0 + p'(r = R + \eta) = \gamma \left(\frac{1}{R + \eta} - \frac{\partial^2 \eta}{\partial z^2} \right),$$

where p_0 is the unperturbed jet pressure and γ the surface tension coefficient. Confirm this linearizes to give the boundary condition

$$p'(r = R) = -\gamma \left(\frac{\eta}{R^2} + \frac{\partial^2 \eta}{\partial z^2} \right).$$

Now consider normal modes of the form

$$(\eta, p', u_r') = (\hat{\eta}, \hat{p}(r), \hat{u}_r(r)) \exp(j(kz - \varpi t)), \quad a = kR,$$

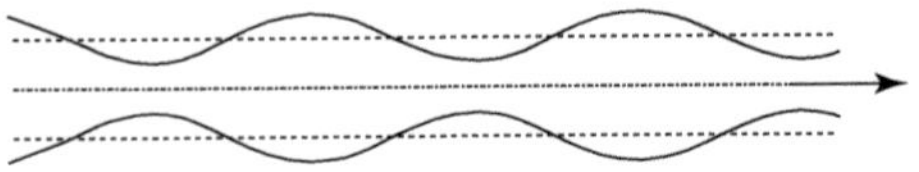

Figure 11.21 Capillary-driven instability of a jet.

and confirm that $\hat{p}(r)$ satisfies

$$\frac{d^2\hat{p}}{dr^2} + \frac{1}{r}\frac{d\hat{p}}{dr} - k^2\hat{p} = 0, \ \hat{p}(r=R) = (a^2-1)\frac{\gamma\hat{\eta}}{R^2}. \tag{11.114}$$

Solve (11.114) for $\hat{p}(r)$ and show that

$$\hat{p}(r) = (a^2-1)\frac{\gamma\hat{\eta}}{R^2}\frac{I_0(kr)}{I_0(a)}, \quad \frac{\partial\hat{p}}{\partial r} = (a^2-1)\frac{\gamma\hat{\eta}}{R^3}\frac{aI_1(kr)}{I_0(a)},$$

where I_0 and I_1 are the usual modified Bessel functions. Finally, combine this with (11.113) to give *Rayleigh's dispersion relationship*:

$$\varpi^2 = (a^2-1)\frac{\gamma}{\rho R^3}\frac{aI_1(a)}{I_0(a)}.$$

Show that the jet is unstable for $kR < 1$, with $kR = 0.697$ being the fastest growing mode.

11.6 *The centrifugal instability of a boundary layer on the surface of a rotating cylinder.* A long cylinder of radius R is suddenly set into rotation at a speed Ω and a thin, laminar boundary layer, of thickness $\delta(t)$, diffuses out from its surface. Within the boundary

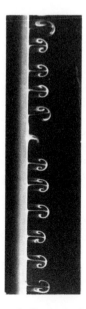

Figure 11.22 (a) Simulation of a blob of swirling fluid bursting radially outward under the action of the centrifugal force, as discussed in Exercise 8.1. The image is a cross-section in the r–z plane coloured by the magnitude of $\Gamma = ru_\theta$. (b) Similar structures are generated by the centrifugal instability of a boundary layer on a rotating cylinder. (From Davidson, 2015.)

layer the angular momentum $\Gamma = ru_\theta$ drops from $\Gamma = \Omega r^2$ at $r = R$ to $\Gamma = 0$ at $r = R + \delta$. Use the analysis of §11.1.3 to show that this boundary layer becomes centrifugally unstable at a critical value of $Ta = \Omega^2 R \delta^3 / \nu^2$, say $Ta \sim 1700$. Just such a centrifugal instability is shown in Figure 11.22 (b). Now use the diffusion estimate $\delta \sim \sqrt{2\nu t}$ to *suggest* that this instability cuts in at around $\Omega t_{\text{crit}} \sim 70 \left(\Omega R^2 / \nu \right)^{-1/3}$. In practice, it is observed at $\Omega t_{\text{crit}} \approx 100 \left(\Omega R^2 / \nu \right)^{-1/3}$.

....

REFERENCES

Acheson, D.J., 1990, *Elementary fluid dynamics*, Oxford University Press.

Arnold, V.I., 1966, Sur un principe variationnel pour les écoulements stationaires des liquides parfaits et ses applications aux problèmes de stabilité non-linéaires. *J. Méc.*, **5**, 29–43.

Batchelor, G.K., 1967, *Introduction to fluid dynamics*, Cambridge University Press.

Billant, P. & Gallaire, F., 2005, Generalized Rayleigh criterion for non-axisymmetric centrifugal instabilities. *J. Fluid Mech.*, **542**, 365–79.

Chandrasekhar, S., 1961, *Hydrodynamic and hydromagnetic stability*, Dover.

Davidson, P.A., 1994, Global stability of two-dimensional and axisymmetric Euler flows. *J. Fluid Mech.*, **276**, 273–305.

Davidson, P.A., 2000, Energy criteria for the linear stability of conservative flows. *J. Fluid Mech.*, **402**, 329–48.

Davidson, P.A., 2015, *Turbulence: an introduction for scientists and engineers*, 2nd Ed., Oxford University Press.

Drazin P.G, Reid, W.H., 1981, *Hydrodynamic stability*, Cambridge University Press.

Kelvin, Lord, 1887, On the stability of steady and of periodic fluid motion.—Maximum and minimum energy in vortex motion. *Phil. Mag.*, **23**, 529.

Moffatt, H.K., 1986, Magnetostatic equilibria and analogous Euler flows of arbitrary complex topology. Part 2. Stability considerations, *J. Fluid Mech.*, **166**, 359–79.

Prandtl, L., 1952, *Essentials of fluid dynamics*, Blackie & Son.

Rayleigh, Lord, 1880, On the stability, or instability, of certain fluid motions. *Proc. Lond. Math. Soc.*, **11**, 57–70.

Rayleigh, Lord, 1916(a), On the dynamics of revolving fluids. *Proc. Royal Soc.*, A, **93**, 148–54.

Rayleigh, Lord, 1916(b), On the convection currents in a horizontal layer of fluid, when the higher temperature is on the underside. *Phil. Mag.*, **32**(6), 529–46.

Song, B., Barkley, D., Hof, B., & Avila, M., 2017, Speed and structure of turbulent fronts in pipe flow. *J. Fluid Mech.*, **813**, 1045–59.

Taylor, G.I., 1923, Stability of a viscous liquid contained between two rotating cylinders. *Phil. Trans. Roy. Soc. Lond.* A, **223**.

Taylor, G.I., 1931, Effect of variation in density on the stability of superposed streams of fluid. *Proc. Roy. Soc.* A, **132**.

12

· · • · ·

The Transition to Turbulence and the Nature of Chaos

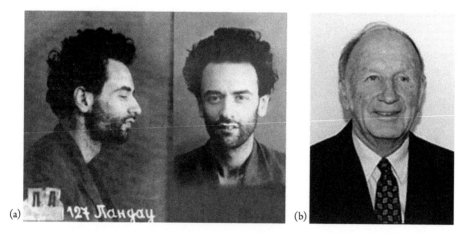

(a) The Soviet physicist Lev Landau (1908–1968) was probably the first to suggest that a sequence of bifurcations leads to turbulence. (b) The US mathematician and meteorologist Edward Lorenz (1917–2008) was a pioneer in the field of deterministic chaos.

12.1 Some Common Themes in the Transition to Turbulence

The transition from laminar to turbulent flow is commonplace. Consider the buoyant plume from a cigarette, as shown in Figure 12.1. As the plume rises, the fluid accelerates and Re becomes larger. Eventually an instability sets in and the laminar plume starts to exhibit a complex, three-dimensional structure. Shortly thereafter, the plume becomes fully turbulent.

From the discussion in Chapter 11, it would appear that there are at least two (and in practice many more) routes to turbulence. There are some flows, such as Taylor–Couette flow and Rayleigh–Bénard convection, in which an increasing Taylor or Rayleigh number

Figure 12.1 Sketch of a cigarette plume. (By F.C. Davidson, after photograph by Stanley Corrsin.)

initiates a sequence of bifurcations (abrupt changes in flow structure), which eventually lead to chaos. This type of transition typically starts out as a simple instability of the mean flow, that quickly saturates, giving rise to a more complicated laminar motion. This, in turn, becomes unstable at a slightly higher value of Ta or Ra, eventually giving way to an even more complicated flow. A sequence of such bifurcations leads to ever more complex flows, and quite quickly to a motion that is everywhere chaotic in both space and time, *i.e.* fully developed turbulence (see Figures 11.3 and 11.4).

Then there are flows, such as inlet pipe flow, in which turbulent motion appears first in small patches as a result of a *local* breakdown of the laminar flow. Provided Re is large enough, these turbulent patches grow and merge until fully developed turbulence is established (see Figures 11.19 and 11.20). This is also typical of transition to turbulence in a flat-plate boundary layer. If the plate is smooth and the level of free-stream turbulence is low, a two-dimensional instability, known as a Tollmien–Schlichting wave, develops some distance downstream of the leading edge. However, these waves rapidly succumb to a three-dimensional instability, which leads to so-called *hairpin vortices* (vortex loops that arch up from the boundary). This, in turn, gives rise to intermittent patches of turbulence (called *turbulent spots*) in regions of particularly high local shear (Figure 12.2). In such cases the turbulence is initially intermittent, being interspersed by laminar flow, and fully developed turbulence is established only once the turbulent spots have merged.

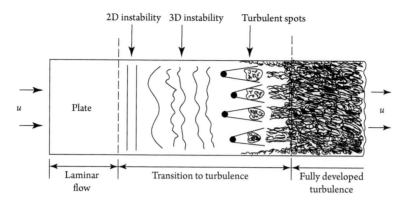

Figure 12.2 Transition to turbulence in a boundary layer in which the plate is flat and smooth and the level of free-stream turbulence is low.

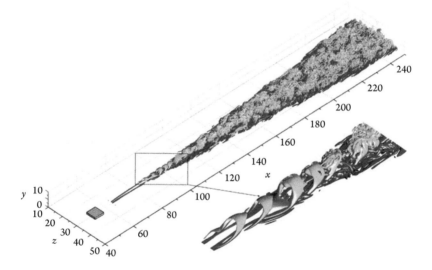

Figure 12.3 Numerical simulation of turbulence which has been triggered by a small rectangular object mounted on a plate. (Figure courtesy of Neil Sandham.)

Note that the transition to turbulence at pipe inlets and in boundary layers tends to be very sensitive to the amplitude of the disturbances, so inlet turbulence in a pipe, or free-stream turbulence above a boundary layer, can trigger transition in an otherwise stable flow. Similarly, small protrusions on a pipe wall or on the surface of a plate can trigger turbulence. This is illustrated in Figure 12.3, which shows a numerical simulation of turbulence which has been triggered in a flat-plate boundary layer by a small rectangular object whose height is somewhat less than the boundary-layer thickness. Laminar hairpin vortices are seen to develop just downstream of the object, and these quickly break down into patches of turbulence, which eventually merge to give an extensive wedge of fully developed turbulence. It is striking that such a small object can create an extensive and persistent stretch of turbulence.

Perhaps the most important point to emphasize is that almost all flows are inherently unstable, and potential instabilities are suppressed only if the viscous dissipation is high enough. Moreover, the viscosity of most common fluids is very small. This is clearly the case for air, but is also true of water, blood, and even the molten metal in the core of the Earth. Thus turbulence is the natural state of things, with laminar flow being the exception and not the rule.

12.2 A Definition of Turbulence

Note that, so far, we have avoided giving a precise definition of turbulence, and indeed many authors avoid doing so on the grounds that turbulence is self-evident. Rather, these authors note that, when Re is sufficiently large, all flows develop a random component of motion and they group these chaotic flows together and call them turbulent. They then note some of their common characteristics, such as:

(i) the velocity fluctuates randomly in both space and time;

(ii) the flow exhibits a wide range of length and time scales;

(iii) the velocity field is unpredictable in the sense that a minute change to the initial conditions will produce a large change to the subsequent motion.

While accurate, this is an incomplete critique of turbulence. A more formal definition might be:

Turbulence is a spatially complex distribution of *vorticity* which advects itself in a chaotic manner. The vorticity field is random in both space and time, and exhibits a wide and continuous distribution of length and time scales.

Note that we have defined turbulence in terms of a chaotic *vorticity* field, and not a chaotic velocity field. That is because, as we shall see, chaos is the hallmark of a non-linear system, and it is the presence of vorticity in the form of $\mathbf{u} \times \boldsymbol{\omega}$ that makes the Navier–Stokes equation, (2.43), non-linear. In short, without vorticity, there can be no turbulence. Thus, for example, we define a turbulent boundary layer to be that region of space adjacent to a surface where there exists a chaotic vorticity field (Figure 12.4). Further away from the surface one can still detect random fluctuations in velocity, fluctuations which are generated by the adjacent vorticity field in accordance with the Biot–Savart law. However, these fluctuations are irrotational and so we choose *not* to call this turbulence. Rather, we think of these fluctuations in $\mathbf{u}$ as a passive response to the adjacent turbulent vorticity field.

It is instructive to return to point (iii) above. Consider an experiment in which a cylinder is placed in a wind-tunnel and the speed in the tunnel is sufficiently large to generate

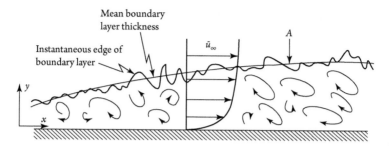

Figure 12.4 A turbulent boundary layer is a region of space filled with chaotic vorticity.

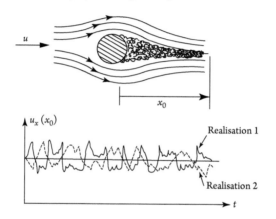

Figure 12.5 The speed u_x is measured at location x_0 behind a cylinder. Plots of $u_x(t)$ in two nominally identical realizations of the experiment are different. (From Davidson, 2015.)

a turbulent wake, as shown in Figure 12.5. We repeat the experiment many times and on each occasion we measure the stream-wise velocity component $u_x(t)$ at a point x_0 downstream of the cylinder. Despite the nominally identical conditions, we find that the function $u_x(x_0, t)$ is radically different on each occasion. This is because there are always minute differences in the way an experiment is carried out and it is in the nature of turbulence to amplify those differences. It should be noted that this is not because we are poor experimentalists and that if only we could control conditions better this randomness would disappear. Almost *any* change in the initial conditions, no matter how small, eventually leads to a large change in $u_x(x_0, t)$. It is striking that, although the governing equations of this incompressible flow are simple and deterministic, the exact details of $u_x(x_0, t)$ appear to be, for all practical purposes, random and unpredictable. This extreme sensitivity to initial conditions, long known to engineers and meteorologists, is now recognized as the hallmark of chaos, and it is exhibited by many non-linear systems.

Now suppose that we measure, not $u_x(x_0, t)$, but rather some of its statistical properties, such as the time average of u_x, or perhaps the time average of u_x^2. This time we will obtain identical results in all experiments. Evidently, although the velocity field appears to be random and unpredictable, its statistical properties are not. This is also the case for the buoyant plume in Figure 12.1: despite the chaotic particle paths, the time-averaged velocity and temperature fields will be the same from one realization of the experiment to another. This is why most theories of turbulence are statistical theories.

12.3 The Nature of Chaos: the Logistic Map as an Example

Of course, chaos is not unique to turbulence, but rather a feature of many non-linear systems. Perhaps the simplest, yet surprisingly rich, system exhibiting chaos is the *logistic equation* (a rather obscure name), which we now discuss. Consider the difference equation

$$x_{n+1} = F(x_n) = ax_n(1 - x_n), \quad 1 < a \leq 4, \, 0 \leq x_n \leq 1. \tag{12.1}$$

This was originally introduced to model population growth, with x_n the normalized population of the nth generation. Steady solutions of a difference equation of the form $X_{n+1} = X_n$, or $X = F(X)$, are called *fixed points* and in the case of the logistic equation the fixed points are evidently $X = 0$ and $X = (a - 1)/a$. As with the steady solutions of a differential equation, a fixed point may be stable or unstable, and we say that a fixed point is linearly unstable if $x_0 = X + \delta x$ leads to a sequence $x_0, x_1, \ldots, x_n$ that diverges from X.

The solutions of (12.1) are discussed in detail in, for example, Drazin (1992). The fixed point, $X = 0$, is linearly unstable for all a, while $X = (a - 1)/a$ is linearly stable for $1 < a \leq 3$, but unstable for $a > 3$ (see Exercise 12.1). Interestingly, just as we lose stability of the fixed point $X = (a - 1)/a$, a new periodic solution of (12.1) emerges. This takes the form $X_2 = F(X_1)$ and $X_1 = F(X_2)$, and so x_n bounces back and forth between X_1 and X_2 in a so-called *two-cycle*. For the logistic equation that two-cycle is

$$X_1, X_2 = \left[(a + 1) \pm \sqrt{(a + 1)(a - 3)}\right] / 2a \tag{12.2}$$

(see Exercise 12.2), and we say that the fixed-point has *bifurcated* to a two-cycle through a *flip bifurcation*.

This two-cycle can be shown to be linearly stable for $3 < a \leq 1 + \sqrt{6}$. In fact, these various linear stability characteristics extend to non-linear stability. That is to say, it may be shown that, for $0 < x_0 < 1$, the iterates converge to the fixed point $X = (a - 1)/a$ when $1 < a \leq 3$, and to the two-cycle if $3 < a \leq 1 + \sqrt{6}$. This is shown in Figure 12.6 (a). The fixed point is said to have a *domain of attraction* of $x_0 = 0 \rightarrow 1$ for $1 < a \leq 3$, and the two-cycle the same domain of attraction for $3 < a \leq 1 + \sqrt{6}$.

At $a = 1 + \sqrt{6} = 3.449$ there is second bifurcation to a more complicated periodic state, called a *four-cycle*, in which x_n bounces back and forth between four curves in the x–a plane. This bifurcation from a two-cycle to a four-cycle is an example of *period doubling*. A third bifurcation then occurs at $a = 3.544$, this time to an eight-cycle, and so it goes on, with an infinite sequence of period-doubling bifurcations to ever more complex states. Such a hierarchy of transitions is known as a *Feigenbaum sequence*. Rather remarkably, all of these bifurcations occur within the finite range $3 < a < 3.5700$. Crucially, for $a > 3.5700$ the

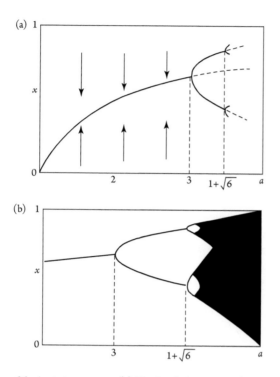

Figure 12.6 Iterates of the logistic equation. (a) The first bifurcation, where stable fixed points and two-cycles are represented by a continuous curve and unstable fixed points and two-cycles by a dashed curve. (b) The first two bifurcations. (From Davidson, 2015.)

solutions of (12.1) become aperiodic and the sequence x_n becomes chaotic in the sense that it may be regarded as a sample of a random variable.

The special case of $a = 4$ has an exact solution that yields some insight into the chaotic behaviour for $a > 3.5700$. This solution is $x_n = \sin^2(\pi\theta_n)$, where $0 \leq \theta_n < 1$ and θ_{n+1} is defined as the *fractional part* of $2\theta_n$. Equivalently, $x_n = \sin^2(2^n \pi\theta_0)$ where $0 \leq \theta_0 < 1$, as shown in Exercise 12.3. This has several interesting properties, which are discussed in detail in Drazin (1992). First, it may be shown that, if θ_0 is irrational, then the sequence θ_n is aperiodic. Second, as n gets large, each aperiodic sequence, θ_n, eventually gets arbitrarily close to any given point in the range $0 < x_n < 1$. In fact, provided θ_0 is irrational, θ_n may be regarded as the sample of a random variable that visits all points in the range with equal likelihood. Third, consider two sequences, θ_n and θ_n^*, which are generated from irrational initial conditions, and let $\gamma_n = \theta_n - \theta_n^*$ where $|\gamma_0| = \varepsilon \ll 1$. Then it may be shown that γ_n grows exponentially, so two initially close sequences diverge rapidly. Specifically, two irrational initial conditions, θ_0 and $\theta_0 + \varepsilon$, typically lead to sequences that diverge as $\gamma_n \sim 2^n \varepsilon$. This extreme sensitivity to initial conditions is the hallmark of mathematical chaos.

Let us summarize these results for $a = 4$. Although the sequence x_n is fixed by the deterministic equation $x_{n+1} = 4x_n(1 - x_n)$, the variable x_n jumps around in a chaotic fashion as if it were a sample of a random variable. This type of chaos is known as *deterministic chaos*. Moreover, the sequence x_n is very sensitive to initial conditions in that two initially close states usually diverge exponentially fast. Thus we cannot track a given sequence for long since a minute amount of rounding error quickly swamps the calculation. Despite this, the statistical properties of x_n are relatively simple. These characteristics are also observed for most values of a in the range $3.5700 < a < 4$, although there are some narrow windows of a for which periodic behaviour is recovered. The same properties are also observed in other non-linear systems. For example, in many non-linear systems a small perturbation ε in the initial condition results in the trajectories of the original and the perturbed solutions diverging as $\varepsilon \exp(\lambda n)$, the constant λ being called a *Liapounov exponent*.

Much of this behaviour is also reminiscent of turbulence, especially the extreme sensitivity to initial conditions and the contrast between the complexity of individual trajectories and the simplicity of the statistical behaviour of the system. Moreover, in the experiments of Taylor and Bénard the flow passes through a sequence of bifurcations before turbulence is established, and period doubling is sometimes observed in such a sequence. This qualitative similarity between the logistic equation and turbulence is striking, and so it is natural to ask if a formal link can be established between bifurcation theory and the transition to turbulence.

12.4 Landau's Inspired (but Incomplete) Vision of the Transition to Turbulence

The link between bifurcation theory and hydrodynamic stability was explored by Hopf in 1942. However, it was probably Landau who, in 1944, first suggested a link between bifurcations and the transition to turbulence. It is impressive that Landau proposed this link some thirty years before the formal development of deterministic chaos.

Let R be a control parameter in a hydrodynamic experiment, say Re, Ra, or Ta, and let R_c be the critical value of R at which infinitesimal disturbances start to grow. For example, R_c might correspond to the critical value of Ta at which Taylor cells first appear. Moreover, let $A(t)$ and σ be the amplitude and growth rate of the unstable mode that is triggered at $R = R_c$. Linear theory then gives

$$A(t) = A_0 \exp\left[(\sigma + j\varpi)\, t\right], \tag{12.3}$$

for some frequency ϖ, and if we average over many cycles we obtain

$$\frac{d|A|^2}{dt} = 2\sigma |A|^2. \tag{12.4}$$

Now suppose $R - R_c$ is small so that all other modes are stable. We have $\sigma < 0$ for $R < R_c$ and $\sigma > 0$ when $R > R_c$, and so one usually finds

$$\sigma = c^2 \left(R - R_c\right) + O\left[(R - R_c)^2\right], \quad |R - R_c| \ll R_c, \tag{12.5}$$

for some real constant, c. As the unstable mode grows it soon becomes large enough to distort the base flow and linear theory is then no longer appropriate. In such cases, Landau suggested that the magnitude of $|A|^2$, averaged over many cycles, is governed by a generalization of (12.4):

$$\frac{d|A|^2}{dt} = 2\sigma|A|^2 - \alpha|A|^4 - \beta|A|^6 + \dots \tag{12.6}$$

where α and β are constants. For small-to-moderate $|A|$ we might neglect terms larger than $|A|^4$ in (12.6) and we arrive at the *Landau equation*,

$$\frac{d|A|^2}{dt} = 2\sigma|A|^2 - \alpha|A|^4, \tag{12.7}$$

where the coefficient α is called *Landau's constant*. The term $\alpha|A|^4$ in (12.7) represents the non-linear distortion of the base flow by the disturbance and this may either accentuate (for $\alpha < 0$) or moderate (for $\alpha > 0$) the growth rate of the disturbance.

On rewriting (12.7) as an evolution equation for $|A|^{-2}$, it is readily confirmed that it has the exact solution,

$$\frac{|A|^2}{A_0^2} = \frac{e^{2\sigma t}}{1 + \lambda\left(e^{2\sigma t} - 1\right)} \quad , \quad \lambda = \frac{\alpha A_0^2}{2\sigma}. \tag{12.8}$$

The behaviour of this solution depends critically on the sign of α. If $\alpha > 0$ then we find that, at large times, $|A| \to 0$ when $R < R_c$ and $|A| \to (2\sigma/\alpha)^{1/2} = A_\infty$ for $R > R_c$. Moreover, since (12.5) requires $\sigma = c^2(R - R_c)$, we have $A_\infty \sim \sqrt{R - R_c}$. The key point here is that, even though the flow is linearly unstable for $R > R_c$, the instability soon saturates and the flow settles down to a new laminar motion. Moreover, the amplitude of the new motion grows with R as $\sqrt{R - R_c}$. This kind of behaviour is called a *supercritical bifurcation* and it is illustrated in Figure 12.7 (a). It is not unlike of the emergence of Taylor cells in Taylor–Couette flow, or convection rolls in the Rayleigh–Bénard problem.

The situation is very different for $\alpha < 0$. Consider first the case of $R > R_c$. Here, the disturbance grows rapidly and $|A|$ is predicted to become infinite within a finite time, at $t^* = (2\sigma)^{-1}\ln\left[1 + |\lambda|^{-1}\right]$. However, long before t reaches t^*, the Landau equation, (12.7), ceases to be valid and higher-order terms need to be considered, such as $-\beta|A|^6$ in (12.6). These higher-order terms can lead to a restabilization of the disturbance at larger amplitude, as indicated by the solid line in Figure 12.7 (b) and as discussed in Exercise 12.4. For $R < R_c$, on the other hand, we have $|A| \to 0$ as $t \to \infty$ provided that the initial disturbance is small enough, and in particular when $A_0 < \sqrt{2|\sigma|/|\alpha|}$. However, if the disturbance is sufficiently large, $i.e. A_0 > \sqrt{2|\sigma|/|\alpha|}$, then (12.8) predicts that $|A| \to \infty$ within a finite time, although long before this occurs the higher-order terms in (12.6) become important, typically moderating the growth. This is called a *subcritical bifurcation* and it is reminiscent of transition in a boundary layer or at the inlet to a pipe. In conclusion, then, when $\alpha < 0$ and $R > R_c$, the flow is both linearly and non-linearly unstable. However, when $\alpha < 0$ and $R < R_c$, it is linearly stable but non-linearly unstable for large enough A_0.

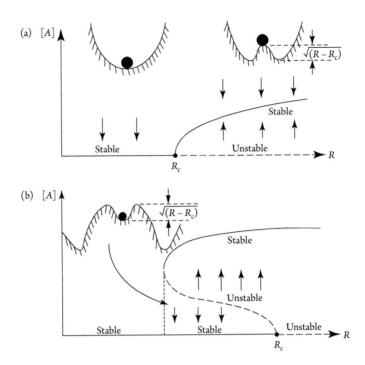

Figure 12.7 Bifurcations of Landau's equation. (a) A supercritical bifurcation, $\alpha > 0$. (b) A subcritical bifurcation, $\alpha < 0$. (From Davidson, 2015.)

All of this assumes that $R - R_c$ is small. For finite $R - R_c$, Landau proposed the following sequence of events for a supercritical bifurcation. He suggested that, as $R - R_c$ increases, the flow resulting from the first bifurcation will itself become unstable and this second instability will also saturate, yielding yet another laminar flow. He then proposed that, as R rises, more and more complex flows will appear through a succession of such supercritical bifurcations. He also suggested that the difference in the value of R between successive bifurcations will decrease rapidly, so very quickly the flow becomes extremely complex. The suggestion, then, is that turbulence is the result of an infinite sequence of bifurcations, which is an appealing idea in the light of the behaviour of the logistic equation. (Note the similarity between (12.1) and (12.7).)

Landau's theory is important as it provided the first qualitative insight into the influence of non-linearity. However, this picture has required substantial modification over the years. For example, while Landau foresaw the emergence of increasingly complex behaviour through a sequence of bifurcations, consistent with the behaviour of the logistic equation for $3 < a < 3.57$, he did not foresee the sudden transition to chaotic behaviour that occurs in the logistic equation at the *accumulation point*, $a = 3.57$. Moreover, the infinite sequence of bifurcations exhibited by the logistic equation turns out to be only one of many routes to chaos. It now seems likely that there are several routes to chaos and to turbulence, including period doubling, intermittent turbulence, and three or four Hopf-like bifurcations (see Exercise 12.5) leading directly to chaotic behaviour.

○ ○ ○

There is a substantial literature devoted to the relationship between chaos theory and turbulence, especially the transition to turbulence. A measured introduction is provided by Tritton (1988), while Drazin and Reid (1981) give a detailed account of Landau's (inspirational, if overly simplified) theory of transition. There are, of course, many specialist texts on chaos theory. Drazin (1992), for example, provides a detailed discussion of the logistic equation.

EXERCISES

12.1 *The stability of the fixed points of the logistic equation.* Show that a linear stability analysis of the fixed point $X = (a - 1)/a$ of the logistic equation, (12.1), leads to

$$(\delta x)_{n+1} = (2 - a)\delta x_n,$$

and hence that the fixed point is stable for $1 < a \leq 3$, but unstable for $a > 3$.

12.2 *The two-cycle of the logistic equation.* Use the relationships $X_2 = F(X_1)$ and $X_1 = F(X_2)$ to confirm that the two-cycle of the logistic equation, (12.1), is given by

$$X_1, X_2 = \left[(a + 1) \pm \sqrt{(a + 1)(a - 3)}\right]\Big/2a.$$

Now show that this two-cycle is stable for $3 < a \leq 1 + \sqrt{6}$ by considering the stability of the so-called second-generation map, $x_{n+2} = F(F(x_n))$.

12.3 *Solution of the logistic equation for $a = 4$.* Consider $x_n = \sin^2(\pi\theta_n)$ where $0 \leq \theta_n < 1$ and θ_{n+1} is defined to be the *fractional part* of $2\theta_n$, i.e. $0 \leq \theta_{n+1} < 1$. Confirm that

$$x_{n+1} = \sin^2(\pi\theta_{n+1}) = \sin^2(2\pi\theta_n) = 4\sin^2(\pi\theta_n)\cos^2(\pi\theta_n) = 4x_n(1 - x_n),$$

and deduce that $x_n = \sin^2(2^n\pi\theta_0)$ is the solution of (12.1) for $a = 4$. The properties of this solution are best explored by expressing θ_n as a *binary number*. That is, since $0 \leq \theta_0 < 1$ and $1/2 + 1/4 + 1/8 + \ldots = 1$, we can write θ_0 as

$$\theta_0 = b_1/2 + b_2/2^2 + b_3/2^3 + \ldots., b_i = 0, 1.$$

Show that

$$\theta_1 = b_2/2 + b_3/2^2 + b_4/2^3 + \ldots..$$
$$\theta_2 = b_3/2 + b_4/2^2 + b_5/2^3 + \ldots..$$
$$\theta_3 = b_4/2 + b_5/2^2 + b_6/2^3 + \ldots..$$

and so on. This property of θ_n is known as the *Bernoulli shift*. It can be used to help establish the various of properties of (12.1) listed in §12.3 for the case of $a = 4$. For example, consider two initially adjacent sequences, θ_n and θ_n^*, which are generated from irrational initial conditions. In particular, suppose that we perturb θ_0 by changing the coefficient b_{20} from 1 to 0. In general, this represents only a minute change to θ_0. However, considering the binary expansion for θ_n given above, after only 19 iterations this small change to θ_0 has become a first-order change to θ_n^*. In general, two irrational initial conditions, say θ_0 and $\theta_0 + \varepsilon$, will diverge at the rate $\varepsilon 2^n$.

12.4 *The effect of higher-order terms in Landau's equation.* Consider the extended Landau equation, (12.6), truncated at $|A|^6$,

$$\frac{d|A|^2}{dt} = 2\sigma|A|^2 - \alpha|A|^4 - \beta|A|^6,$$

where $\beta > 0$ and σ is given by (12.5). Find the shape of the bifurcation diagram in the vicinity of the critical point $R = R_c$ for the case of $\alpha < 0$.

12.5 *An example of a Hopf bifurcation.* Consider the two-dimensional system,

$$\frac{d\mathbf{x}}{dt} = \mathbf{\Omega} \times \mathbf{x} + \left(a - |\mathbf{x}|^2\right)\mathbf{x}, \quad \frac{d}{dt}|\mathbf{x}|^2 = 2\left(a - |\mathbf{x}|^2\right)|\mathbf{x}|^2,$$

where $\mathbf{x} = (x, y)$, $\mathbf{\Omega} = \Omega\hat{\mathbf{e}}_z$, and Ω and a are constants. Evidently, this is a combination of rigid-body rotation plus a radial movement which is inward for $a - |\mathbf{x}|^2 < 0$ and outward for $a - |\mathbf{x}|^2 > 0$. It also has the steady solution $\mathbf{x} = (0,0)$. Show that this steady solution is linearly stable for $a < 0$ but unstable if $a > 0$. Confirm that $r^2 = |\mathbf{x}|^2$ satisfies the supercritical Landau equation and find its solution. Hence show that, for $a < 0$, solutions in the x–y plane spiral radially inward towards the origin, whereas solutions for $a > 0$ spiral radially outwards from the origin to converge on the circle $r = a^{1/2}$, which is called a *limit cycle*. Also show that, if $a > 0$ and $r_0 > a^{1/2}$, the solutions spiral inwards towards $r = a^{1/2}$. This kind of supercritical bifurcation to an oscillatory state is called a *Hopf bifurcation*.

....

REFERENCES

Davidson, P.A., 2015, *Turbulence: an introduction for scientists and engineers.* 2nd Ed., Oxford University Press.

Drazin, P.G., 1992, *Nonlinear systems.* Cambridge University Press.

Drazin, P.G. & Reid, W.H., 1981, *Hydrodynamic stability.* Cambridge University Press.

Tritton, D.J.,1988, *Physical Fluid dynamics*, 2nd Ed., Oxford University Press.

13

. . • . .

An Introduction to Turbulence and to Kolmogorov's Theory

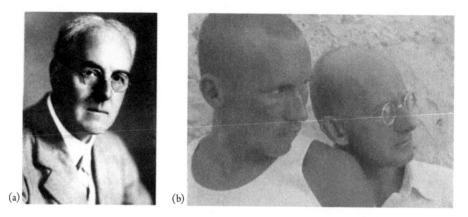

(a) The British physicist and meteorologist Lewis Fry Richardson (1881–1953) was a pioneer in turbulence, being the first to suggest that the transfer of energy from large to small scales takes the form of a multi-step cascade. (b) The Russian mathematician Andrey Kolmogorov (1903–1987) is on the left. Influenced by Richardson, he transformed our understanding of turbulence by providing a quantitative picture of the energy cascade.

13.1 Elementary Properties of Turbulence: a Qualitative Overview

We now provide a brief introduction to turbulence. This chapter, and the next, is based, at least in part, on the more extensive discussion given in Davidson (2015).

13.1.1 The Need for a Statistical Approach and the Problem of Closure

In Chapter 12 we noted the need for a statistical approach to turbulence. Consider Figure 12.1. As the smoke twists and turns the velocity field fluctuates chaotically, constantly evolving and never repeating itself. The same is true of the turbulent wake behind a cylinder, as shown in Figure 12.5. However, despite this randomness, the statistical properties of the velocity field shown in Figure 12.5, such as the time average of $\mathbf{u}(\mathbf{x}, t)$, or perhaps the time average of $\mathbf{u}^2(\mathbf{x}, t)$, are perfectly reproducible from one realization of an experiment to another. The same is true of the temperature field in Figure 12.1. It is for this reason that 'theories of turbulence' seek to make predictions about the statistical properties of a flow, rather than the temporal and spatial variations of the full velocity field.

There are different ways of forming averages in turbulence. Consider a turbulent boundary layer which is *steady-on-average*, but has statistical properties which depend on the distance from the wall (Figure 13.1). Such a flow is said to be statistically steady but inhomogeneous. In such cases one usually forms averages by time-averaging, denoted here by an overbar. Thus we write $\mathbf{u}(\mathbf{x}, t) = \bar{\mathbf{u}}(\mathbf{x}) + \mathbf{u}'(\mathbf{x}, t)$ where $\overline{\mathbf{u}'} = 0$, and refer to $\bar{\mathbf{u}}(\mathbf{x})$ as the mean velocity field and to $\mathbf{u}'(\mathbf{x}, t)$ as the turbulent fluctuations. We might note in passing that the time-average of the continuity equation yields $\nabla \cdot \bar{\mathbf{u}} = 0$, and hence $\nabla \cdot \mathbf{u}' = 0$, so that $\bar{\mathbf{u}}$ and $\mathbf{u}'$ are individually solenoidal. The kinetic energy density of the turbulence is then $\frac{1}{2}\rho\overline{(\mathbf{u}')^2}$, while the quantity

$$\tau_{ij}^R = -\rho\overline{u_i' u_j'}$$

(13.1)

is known as the *Reynolds stress*, for reasons that will become apparent. For a statistically steady flow, time averages can be shown to be equivalent to averaging over many realizations of an experiment. This is known as an *ensemble average* and denoted by angled brackets. Thus, for example, we can rewrite the Reynolds stress as $\tau_{ij}^R = -\rho\langle u_i' u_j' \rangle$.

In Figure 13.1 the turbulence is kept alive, despite the viscous dissipation, by the action of the mean shear, $\partial\bar{u}_x/\partial y$, which transfers kinetic energy from the mean flow to the turbulence. (We shall discuss how this works in §14.2.) However, often one encounters turbulence in which there is no mean shear and so the turbulent kinetic energy must decay. If the statistical properties of this decaying turbulence are only a weak function of position,

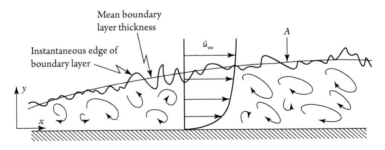

Mean boundary
layer thickness

$\bar{u}_\infty$

A

Instantaneous edge of
boundary layer

y

x

Figure 13.1 A steady-on-average turbulent boundary layer.

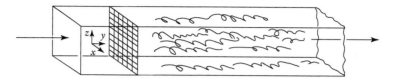

Figure 13.2 Turbulence induced by a grid in a wind tunnel is (almost) locally homogeneous.

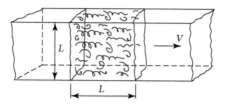

Figure 13.3 A patch of turbulence moves along the wind tunnel.

the turbulence is said to be approximately *statistically homogeneous*. For example, suppose we create turbulence in a wind tunnel using a grid of bars, as shown in Figure 13.2. If we adopt a laboratory frame of reference, then this is a statistically steady flow. Moreover, outside the side-wall boundary layers there is no mean shear and $\bar{\mathbf{u}}$ is uniform, and so viscous dissipation ensures that the turbulent kinetic energy generated by the grid decays with distance along the tunnel. However, the mean flow, V, is always much larger than $|\mathbf{u}'|$ and so the turbulence is rapidly swept through the tunnel. Hence, the decay of turbulent kinetic energy causes only weak spatial gradients in the statistical properties of the turbulence. At any one location, then, we have a good approximation to *locally homogeneous turbulence*.

Now suppose we adopt a frame of reference that moves with the mean flow V, so that we follow a patch of turbulence as it is swept along the tunnel (see Figure 13.3). In this moving frame we have $\bar{\mathbf{u}} = 0$ and the statistical properties of the turbulence now evolve with *time*, rather than with position. Since the turbulence is more or less spatially uniform, this is referred to as *freely decaying, homogeneous turbulence*. In such a time-dependent situation one usually defines an average in terms of a *local volume average*, and if the turbulence is homogeneous, that volume average is equivalent to an ensemble average.

Given the importance of the statistical properties of $\mathbf{u}$ in turbulence theory, it is natural to ask if we can use the Navier–Stokes equation to write down the governing equations for these various statistical quantities. Consider, for example, the case of a steady-on-average shear flow, such as that shown in Figure 13.1. Then the Navier–Stokes equation, or equivalently Cauchy's equation, (2.14), can be written as

$$\rho \frac{D\mathbf{u}}{Dt} = \rho\,\frac{\partial u_i'}{\partial t} + \rho(\bar{\mathbf{u}} + \mathbf{u}')\cdot\nabla(\bar{u}_i + u_i') = \frac{\partial \tau_{ij}}{\partial x_j}, \tag{13.2}$$

where

$$\tau_{ij} = \bar{\tau}_{ij} + \tau_{ij}' = -p\delta_{ij} + 2\rho\nu S_{ij} \tag{13.3}$$

is the usual stress tensor. Since $\overline{\bar{\mathbf{u}} \cdot \nabla \mathbf{u}'} = \overline{\mathbf{u}' \cdot \nabla \bar{\mathbf{u}}} = 0$, while $\nabla \cdot \mathbf{u}' = 0$ ensures that $\overline{\rho \mathbf{u}' \cdot \nabla u_i'} = -\partial \tau_{ij}^R / \partial x_j$, the time average of Cauchy's equation yields

$$\rho \left(\bar{\mathbf{u}} \cdot \nabla \right) \bar{u}_i = \frac{\partial}{\partial x_j} \left[\bar{\tau}_{ij} + \tau_{ij}^R \right], \tag{13.4}$$

which is known as the *Reynolds-averaged equation of motion*. Evidently, the effect of this averaging is to supplement the usual stress tensor with the Reynolds stress (hence the name). The key point to note, however, is that, because of the non-linearity of (13.2), the mean flow is a function of the Reynolds stress, $\tau_{ij}^R = -\rho \overline{u_i' u_j'}$, and so to determine $\bar{\mathbf{u}}(\mathbf{x})$ we need an evolution equation for τ_{ij}^R. We shall see in Chapter 14 that this takes the form

$$\bar{\mathbf{u}} \cdot \nabla \tau_{ij}^R = \bar{\mathbf{u}} \cdot \nabla \left(-\rho \overline{u_i' u_j'} \right) = \frac{\partial}{\partial x_k} \left[\rho \overline{u_i' u_j' u_k'} \right] + \cdots \tag{13.5}$$

and so our evolution equation for τ_{ij}^R involves the triple-velocity average, $\overline{u_i' u_j' u_k'}$. We now need an evolution equation for this triple-velocity average, but unfortunately it is of the form

$$\bar{\mathbf{u}} \cdot \nabla \left(\overline{u_i' u_j' u_k'} \right) = \frac{\partial}{\partial x_m} \left[-\overline{u_i' u_j' u_k' u_m'} \right] + \cdots, \tag{13.6}$$

so we have yet another set of unknowns, $\overline{u_i' u_j' u_k' u_m'}$. Of course, the governing equation for these involves fifth-order terms, and so it goes on. In short, because of non-linearity, we always have more unknowns than equations. This is the *closure problem of turbulence*.

An equivalent problem arises in the study of freely decaying homogeneous turbulence where, in a suitable frame of reference, there is no mean flow. Here one is interested in the structure of the turbulence itself and, as we shall show, a manipulation of the Navier–Stokes equation leads, once again, to an *unclosed* hierarchy of statistical equations.

There is a striking irony here. It seems that the velocity field, **u**, is controlled by a simple deterministic equation (the Navier–Stokes equation), yet to all intents and purposes **u** is random and unpredictable. Conversely, the various statistical properties of **u** are all well behaved, in the sense that they are reproducible in any experiment, yet we know of no closed set of equations that govern them! The closure problem of turbulence has very deep consequences. In effect, it means that there is no rigorous theory of turbulence. If we wish to make statistical predictions about a turbulent flow, we almost always need to supplement our rigorous equations with some additional information, usually in the form of an empirical equation or a statistical assumption. This additional information constitutes a *closure model*.

13.1.2 *The Various Stages of Development of Freely Decaying Turbulence*

Let us consider freely decaying turbulence in a wind tunnel in a little more detail. In Chapter 12 we defined turbulence to be a vorticity field which advects itself in a chaotic manner in accordance with the vorticity evolution equation, with the instantaneous velocity

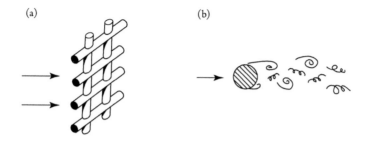

Figure 13.4 (a) Part of a grid. (b) Vortices shed from one of the bars in a grid.

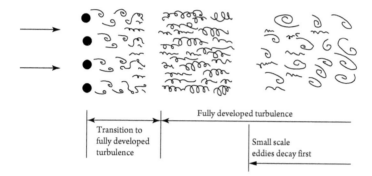

Figure 13.5 Schematic of the various stages of development of grid turbulence.

field determined from the vorticity in accordance with the Biot–Savart law. It follows that, to produce turbulence, we first need to create vorticity. In a wind tunnel, this is achieved by blowing air past a grid of bars which then shed vortices, as shown in Figure 13.4.

A schematic of the evolution of grid turbulence is shown in Figure 13.5, which highlights a number of different stages. Initially the motion consists of an array of vortices which are shed from the bars of the grid. These soon interact and the vorticity field then begins to shred itself, teasing out increasingly fine filaments of vorticity through the action of vortex stretching. This filamentation of the vorticity field is driven by inertia and it continues to stretch and twist the vortex tubes until such time as they are thin enough for the viscous diffusion of vorticity to counter the thinning by stretching (see §8.5.3), at least in some statistical sense. Figure 13.6 shows an example of these very fine vortex filaments.

The thinnest vortex filaments are reminiscent of a Burgers' vortex (see §8.5), and they are subject to order-one viscous stresses. Once these Burgers-like vortices have been established, which is some distance downstream of the grid, we talk of having *fully developed, freely decaying turbulence*, and this contains a wide range of length scales.

In fully developed grid turbulence, the size of the large eddies (*i.e.* the largest blobs of vorticity) is roughly the size of the bars in the grid. This is called the *integral length scale* of the turbulence, denoted here by ℓ. The characteristic radius of the thinnest vortex filaments, on the other hand, is called the *Kolmogorov microscale*, and is usually given the symbol η. In grid turbulence we might have, say, $\ell \sim 2$ cm and $\eta \sim 0.1$ mm, so there may be two or

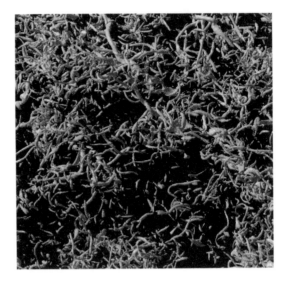

Figure 13.6 A numerical simulation of turbulence visualized by vorticity contours. The vorticity threshold is high, so the vortices shown are very thin Kolmogorov-scale vortex filaments. (Courtesy of Y. Kaneda, T. Ishihara, A. Uno, K. Itakura, and M. Yokokawa.)

three decades of scale. We can also associate characteristic velocities with these large and small vortices. Let u be a characteristic value of the velocity fluctuations associated with the largest eddies, called the *integral velocity scale*, and υ be a typical velocity associated with the thinnest vortex filaments, called the *Kolmogorov velocity*. Then $Re = u\ell/\nu$ is invariably large in grid turbulence, perhaps around 10^3, whereas by definition we have $\upsilon\eta/\nu \sim 1$, because the viscous stresses are of order unity at the Kolmogorov scales. The characteristic strain rate of the eddies, S_{ij}, and the typical vorticity magnitude, $|\boldsymbol{\omega}|$, varies from u/ℓ at the large scales to υ/η at the Kolmogorov scale, being largest in the thin, intense vortex filaments. It follows that most of the dissipation per unit mass, $\varepsilon = 2\nu \langle S_{ij}S_{ij}\rangle = \nu \langle \boldsymbol{\omega}^2\rangle$, and enstrophy, $\langle \frac{1}{2}\boldsymbol{\omega}^2\rangle$, is concentrated at the Kolmogorov scale. (See Exercise 13.2 for the proof that $\varepsilon = \nu \langle \boldsymbol{\omega}^2\rangle$ in homogeneous turbulence.)

Let us now try to refine this picture. It is conventional in turbulence to use the Fourier transform to distinguish between different scales in a flow. Suppose, for example, that we measure the *instantaneous* span-wise velocity fluctuations across the tunnel at several stream-wise locations, giving a series of signals of the form $u'_x(x)$. If we Fourier transform each of these signals, then there will appear to be many wavenumbers contributing to u'_x, and of course this is because the velocity at any one location is the result of a multitude of vortical structures in the vicinity of that point, each contributing to u'_x via the Biot–Savart law (Figure 13.7). Moreover, each vortex has associated with it a range of characteristic length scales, say s, and hence a range of characteristic wavenumbers, $k \sim \pi/s$.

Suppose we now go further and estimate the relative contributions to $\langle \frac{1}{2}\mathbf{u}'^2\rangle$ that come from each of the various eddy sizes (or wavenumber ranges) at a given instant. This results in

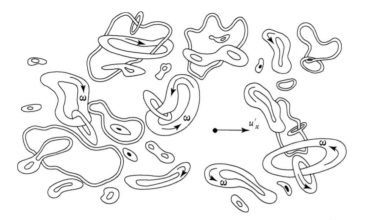

Figure 13.7 The velocity u'_x is a result of the surrounding vortices and the Biot–Savart law.

a function $E(k)$ called the *energy spectrum*, which has the property $\langle \frac{1}{2}\mathbf{u}'^2 \rangle = \int E(k)dk$. (The way in which this is done using Fourier analysis is discussed in §13.2.3.) When we plot the energy spectrum against wavenumber, we get something that looks like Figure 13.8, where we have adopted a frame of reference that moves with the mean flow. Close to the grid, much of the energy is centred around a wavenumber of order $k \sim (\text{bar size})^{-1}$, corresponding to the size of the vortices shed from the bars of the grid. This energy then spreads over a broader range of wavenumbers as the vorticity field develops an increasingly fine-scale structure. As some of the vortex filaments approach the Kolmogorov scale, dissipation starts to become significant and eventually we obtain fully developed turbulence in which the Kolmogorov scales are fully excited and the viscous dissipation of energy is a maximum. We now have energy distributed over a wide range of scales, although most of the energy is contained in the large *energy-containing eddies*, as indicated in Figure 13.8 (b). So, if u is a typical fluctuating velocity of the large eddies, then $\langle \mathbf{u}'^2 \rangle \sim 3u^2$. We now enter the regime of freely decaying, fully developed turbulence, as shown in Figure 13.8 (c). The energy per unit mass of the turbulence now declines as a result of viscous dissipation, and it turns out that the smallest eddies decay fastest.

As already noted, in fully developed turbulence the kinetic energy is concentrated at the integral scale, whilst the enstrophy is predominantly associated with eddies of size η, as shown in Figure 13.9. The fact that the energy and enstrophy are dominated by vortices of very different scale can be understood as follows. The vorticity that underpins the large eddies (via the Biot–Savart law) is weak and dispersed, and so makes little contribution to the net enstrophy, but nevertheless dominates the kinetic energy because it is coherent over large length scales. The small vortices, on the other hand, are composed of particularly intense vortex tubes and so they dominate the enstrophy, though they make little contribution to the net kinetic energy (via the Biot–Savart law) because they are thin and randomly orientated, and hence there is a great deal of cancellation when their velocity fields are superimposed.

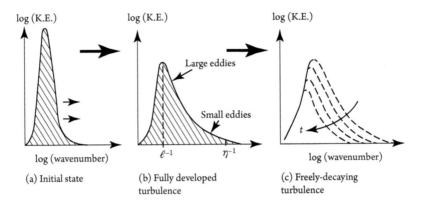

Figure 13.8 The energy spectrum, $E(k)$, at various stages in the decay of grid turbulence.

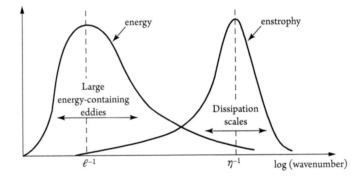

Figure 13.9 The distribution of energy and enstrophy in fully developed turbulence.

13.1.3 Richardson's Energy Cascade

Although we focussed on grid turbulence in the previous section, many of our conclusions extend to turbulence as a whole. In particular, at high Reynolds number there is always a wide range of length scales, with the kinetic energy dominated by the large eddies, and the dissipation by Burgers-like vortex filaments of scale η. This wide range of scales is nicely captured by Leonardo's sketch of water falling into a pool (Figure 13.10).

Another common feature of turbulence is that the kinetic energy enters the turbulence at the large scales, which is the bar size in the case of grid turbulence, or the width of the plunging jet in the case of Leonardo's sketch. Given that the dissipation occurs at the Kolmogorov microscale, η, this raises the question of how the kinetic energy gets from the integral scale, where it enters the turbulence, down to the Kolmogorov scale, where it is dissipated. It was Richardson who first proposed an answer to this question.

Richardson suggested that the flux of kinetic energy down through *scale space* (*i.e.* the transfer of energy from scale to scale) takes the form of a *multi-step cascade*. That is to say, the

Figure 13.10 Copy of Leonardo's sketch of water entering a pool. Note the different scales of motion, suggestive of an energy cascade. (Drawing by F. C. Davidson, with permission.)

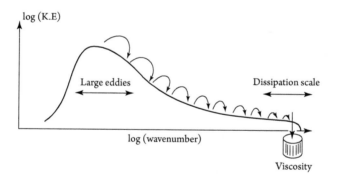

Figure 13.11 The energy cascade in terms of the energy spectrum, $E(k)$, versus wavenumber.

largest eddies mostly pass their energy onto slightly smaller vortices, perhaps a factor three times smaller, and these in turn pass their energy onto somewhat smaller vortices, and so it goes on, all the way down to the Kolmogorov scale. The energy transfer is said to be *localized in scale space*, as illustrated in Figure 13.11. Numerical simulations suggest that, statistically, this picture is indeed correct (see, for example, Leung et al., 2012).

The mechanism of energy transfer is vortex stretching. That is, large energetic vortices tend to stretch smaller, weaker ones and this increases their kinetic energy. Since the whole process is effectively inviscid, provided that we are away from the Kolmogorov scales, the energy gained by the smaller vortices must have come from the eddies doing the stretching, and so we have an energy transfer from large to small scales. (Actually, there is almost as much compression as there is stretching, and this pushes energy to *larger* scales. However, statistically, the stretching wins out.) The reason for the localization of energy transfer in scale space is that the stretching process is highly inefficient if the eddies involved have very different sizes. This is because the velocity field of a large vortex is almost uniform on the scale of a much smaller one, and so when a small vortex falls prey to a much larger eddy, the primary effect is simply to sweep the smaller vortex through space with very little stretching. It turns out that, statistically, the maximum stretching, and hence maximum energy transfer, occurs when there is a difference in scale of a factor of three or four between the eddies doing the stretching and those being stretched (Leung et al., 2012).

Notice that, in Richardson's top–down picture, the rate of transfer of energy to small scales is controlled by the largest eddies, with the small dissipative vortices playing a passive role, mopping up whatever energy cascades down from above. This suggests that the flux of energy down through scale space will be of the order of the kinetic energy of the large eddies times their characteristic strain rate, u/ℓ, where u and ℓ are the integral scales of the turbulence. If we use the symbol Π to represent the energy transfer (or flux) down through scale space, then $\Pi \sim u^3/\ell$. Moreover, since all of this energy is eventually dissipated at the small Kolmogorov scales, we conclude that the viscous dissipation rate per unit mass in a turbulent flow is

$$\boxed{\varepsilon \sim \Pi \sim u^3/\ell}. \tag{13.7}$$

There is a second important consequence of Richardson's multistep cascade. It seems likely that each time energy is passed from one generation of vortices to the next, some of the information about the larger scale vortices is lost. After many such transfers we would expect that the small vortices know very little about their great-great grandparents, other than the fact that the rate of transfer of energy (or energy flux) is $\Pi \sim u^3/\ell$. This raises the possibility that, for Re $\gg 1$, the small scales in turbulence may have some universal characteristics which are independent of the structure of the large-scale eddies, and hence the same in boundary layers, jets, and so on. We shall see that this is indeed the case, and in fact this provides the starting point for *Kolmogorov's universal theory of the small scales.*

13.1.4 The Rate of Destruction of Energy and an Estimate of Kolmogorov's Microscales

Expression (13.7) is intriguing, as is suggests that the viscous dissipation rate in high-Re turbulence is independent of viscosity. This is indeed observed in all conventional turbulent flows, provided that Re is large enough, and it is known as the *zeroth law of turbulence*. For example, in grid turbulence we find that, on tracking the turbulence down the tunnel,

$$\frac{du^2}{dt} = -\frac{Au^3}{\ell}, \quad A \sim 1, \tag{13.8}$$

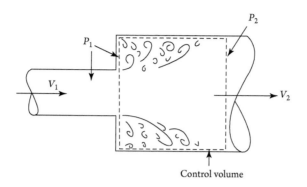

Figure 13.12 Energy loss due to turbulence in a sudden expansion in a pipe.

where A is more or less constant. (We note in passing that, as discussed in Exercise 13.4, this leads to the decay law, $u^2 \sim t^{-6/5}$.) Similarly, in Chapter 1 we saw that the net rate of loss of energy due to turbulence in a sudden pipe expansion is

$$\text{rate of loss of energy} = \frac{1}{2}\dot{m}\,(V_1 - V_2)^2$$

(see equation 1.57), where V_1 and V_2 are the mean velocities at sections 1 and 2, as shown in Figure 13.12. Now, $\dot{m} = \rho V_2 A_2$, while the net rate of loss of energy in the expansion is $\rho \int \varepsilon dV = \rho \bar{\varepsilon} A_2 L$, where $\bar{\varepsilon}$ is the mean dissipation rate in a control volume of axial length L. It follows that

$$\bar{\varepsilon} = V_2\,(V_1 - V_2)^2 \big/ 2L, \tag{13.9}$$

which is consistent with (13.7).

Let us now determine the smallest scales in a turbulent flow: the Kolmogorov scales η and υ. The rate of dissipation of energy at the smallest scales is $\varepsilon = 2\nu S_{ij}S_{ij}$, where S_{ij} is the strain rate associated with the small eddies, $S_{ij} \sim \upsilon/\eta$. This yields $\varepsilon \sim \nu\,(\upsilon^2/\eta^2)$. Since the dissipation, ε, must match the rate at which energy enters the cascade, Π, we have

$$\varepsilon \sim u^3/\ell \sim \nu\,(\upsilon/\eta)^2. \tag{13.10}$$

We also know that the Reynolds number based on υ and η is of order unity, $\upsilon\eta/\nu \sim 1$, and combining these we find

$$\boxed{\eta \sim \ell \mathrm{Re}^{-3/4}, \quad \upsilon \sim u\mathrm{Re}^{-1/4}}, \tag{13.11}$$

or

$$\boxed{\eta = \left(\nu^3/\varepsilon\right)^{1/4}, \quad \upsilon = (\nu\varepsilon)^{1/4}}, \tag{13.12}$$

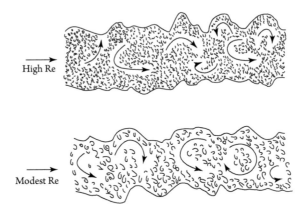

High Re

Modest Re

Figure 13.13 The effect of Re on the small scales in a wake. (After Tennekes and Lumley, 1972.)

where Re is the Reynolds number based on the integral scales: $\text{Re} = u\ell/\nu$. Note that it is conventional to use the estimates for η and υ in terms of ε and ν to *define* η and υ, hence the use of equality signs in (13.12), but not in (13.11). Note also that the higher the Reynolds number, the finer the small-scale turbulence. This is illustrated in Figure 13.13, which shows two nominally similar wake flows at different values of Re.

13.2 A Digression into the Kinematics of Homogeneous Turbulence

13.2.1 Two Useful Diagnostic Tools: Correlation Functions and Structure Functions

The discussion so far has been entirely qualitative. To pave the way for a more quantitative description of turbulence, we now introduce some statistical functions that will help us quantify the state of a field of turbulence. There are three related quantities commonly used for that purpose: the velocity *correlation function*, the *structure function*, and the *energy spectrum*, the third of which we have already mentioned. We focus here on the first two.

One quantity of central interest in turbulence is the *two-point, second-order, velocity correlation tensor*,

$$Q_{ij}(\mathbf{x}, \mathbf{r}, t) = \langle u_i'(\mathbf{x})u_j'(\mathbf{x}+\mathbf{r})\rangle,$$
(13.13)

where the angled brackets represent an ensemble average. Q_{ij} tells us about the degree to which the velocities at $\mathbf{x}$ and $\mathbf{x}' = \mathbf{x} + \mathbf{r}$ are correlated. If Q_{ij} does not depend on time the turbulence is statistically steady, whereas when Q_{ij} does not depend on $\mathbf{x}$ the turbulence is

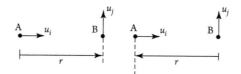

Figure 13.14 The geometrical property $Q_{ij}(\mathbf{r}) = Q_{ji}(-\mathbf{r})$.

statistically homogeneous. For simplicity, we shall restrict ourselves here to homogeneous turbulence, in which case this correlation tensor has the geometric property

$$Q_{ij}(\mathbf{r},t) = Q_{ji}(-\mathbf{r},t) \tag{13.14}$$

(see Figure 13.14), since reversing $\mathbf{r}$ simply amounts to switching $\mathbf{x}$ and $\mathbf{x}'$. Moreover, the continuity equation, $\nabla \cdot \mathbf{u}' = 0$, requires

$$\partial Q_{ij}/\partial r_i = \partial Q_{ij}/\partial r_j = 0, \tag{13.15}$$

since $u_i'(\mathbf{x})$ is independent of $\mathbf{x}'$ and $u_j'(\mathbf{x}')$ independent of $\mathbf{x}$, $\partial/\partial r_i$ and $\partial/\partial r_j$ operating on averages may be replaced by $-\partial/\partial x_i$ and $\partial/\partial x_j'$, and the operations $\langle \sim \rangle$ and $\partial/\partial x_i$ commute.

Two important components of Q_{ij} are

$$Q_{xx}(r\hat{\mathbf{e}}_x) = \langle u_x'(\mathbf{x})u_x'(\mathbf{x}+r\hat{\mathbf{e}}_x)\rangle = \langle u_x'^2 \rangle f(r), \tag{13.16}$$

$$Q_{yy}(r\hat{\mathbf{e}}_x) = \langle u_y'(\mathbf{x})u_y'(\mathbf{x}+r\hat{\mathbf{e}}_x)\rangle = \langle u_y'^2 \rangle g(r). \tag{13.17}$$

The functions f and g are called the *longitudinal* and *lateral* velocity correlation functions. They are dimensionless, satisfy $f(0) = g(0) = 1$, and have the shape shown in Figure 13.15 (a). Consider, for example, $f(r)$. This tells us whether or not u_x' at one point, say A, is correlated to u_x' at an adjacent point, B (Figure 13.15 (b)). If the fluctuations at A and B are statistically independent, because no eddies span the gap, then $Q_{xx} = \langle u_x'(\mathbf{x})\rangle\langle u_x'(\mathbf{x}')\rangle = 0$. This will be the case if r is much greater than the largest eddies. The integral scale ℓ is often defined in terms of f, as

$$\ell = \int_0^\infty f(r)dr. \tag{13.18}$$

This provides a convenient measure of the extent of the region over which velocities are appreciably correlated, *i.e.* the size of the large eddies.

For dynamic reasons, it turns out to be useful to introduce the *two-point, third-order velocity correlation tensor*, defined as

$$\boxed{S_{ijk}(\mathbf{r}) = \langle u_i'(\mathbf{x})\, u_j'(\mathbf{x})\, u_k'(\mathbf{x}+\mathbf{r})\rangle}, \tag{13.19}$$

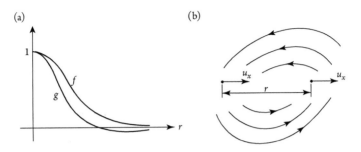

Figure 13.15 (a) The shape of the longitudinal and lateral velocity correlation functions. (b) The longitudinal correlation function $f(r)$ is non-zero if eddies span the gap r.

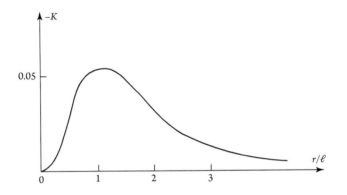

Figure 13.16 Schematic shape of $-K(r)$.

a special case of which is

$$\langle u_x'^2 \rangle^{3/2} K(r) = \langle u_x'^2(\mathbf{x})\, u_x'(\mathbf{x}+r\hat{\mathbf{e}}_x)\rangle. \tag{13.20}$$

$K(r)$ is known as the *longitudinal triple correlation function*. It turns out to be odd in r and scales as $K(r) \sim -r^3$ for small r. The shape of K is shown schematically in Figure 13.16.

Now, we have already noted that, in fully developed turbulence, eddies vary in size from ℓ down to η, and it is natural to ask how the energy of the turbulence is distributed across the various eddy sizes. The question then arises as to how best to reconstruct the energy distribution in scale space from a turbulent signal like $u_x'(x, t)$. It turns out that this is a non-trivial task, made all the more difficult by the fact that the characteristic shape of the eddies varies with scale, being tube-like at small scales (Figure 13.6) and more blob-like at large scales. Nevertheless, there are ways (albeit imperfect ways) of extracting information about the distribution of energy with scale from velocity measurements.

Consider, for example, the *second-order longitudinal structure function*, defined in terms of the *longitudinal velocity increment*, $\Delta v = u_x'(\mathbf{x}+r\hat{\mathbf{e}}_x) - u_x'(\mathbf{x})$. This is

$$\boxed{\left\langle [\Delta v]^2 \right\rangle (r) = \left\langle [u'_x (\mathbf{x} + r\hat{\mathbf{e}}_x) - u'_x (\mathbf{x})]^2 \right\rangle},$$ (13.21)

whose generalization, $\langle [\Delta v]^p \rangle$, is the *longitudinal structure function of order p*. It seems plausible that only eddies of size $\sim r$, or less, make a significant contribution to $\Delta v(r)$, since eddies much larger than r will induce similar velocities at $\mathbf{x}$ and $\mathbf{x} + r\hat{\mathbf{e}}_x$. Consequently, $< (\Delta v)^2 > (r)$ is often taken as indicative of the *cumulative kinetic energy per unit mass* held in eddies of size r or less. This interpretation is given tentative support by the following argument. Suppose that $\langle u'^2_x \rangle \approx \langle u'^2_y \rangle \approx \langle u'^2_z \rangle$. Then expanding $\left\langle [\Delta v]^2 \right\rangle$ gives

$$\left\langle [\Delta v]^2 \right\rangle (r) = 2 \left\langle u'^2_x \right\rangle (1 - f(r)) \approx \frac{2}{3} \left\langle \mathbf{u}'^2 \right\rangle (1 - f(r)).$$ (13.22)

Since $f = 0$ for $r \gg \ell$, this gives us

$$\frac{3}{4} \left\langle [\Delta v]^2 \right\rangle_{r > \ell} \approx \frac{1}{2} \left\langle \mathbf{u}'^2 \right\rangle.$$ (13.23)

Since these results are essentially kinematic, we may apply (13.23) to $\mathbf{u}'$ after it has been (somehow) low-pass filtered at an intermediate scale, say s, effectively replacing ℓ by s. Then

$$\frac{3}{4} \left\langle [\Delta v]^2 \right\rangle (r) \approx \text{[all energy held in eddies of size } s < r \text{]},$$ (13.24)

consistent with the estimate above. This, then, is the usual interpretation of $< (\Delta v)^2 > (r)$.

Unfortunately, this is a little too simplistic. Eddies larger than r (*i.e.* the scales $s > r$) *do* make a contribution to $\Delta v(r)$. For example, for $r < \eta$ we have $\Delta v \approx (\partial u'_x / \partial x) r$ and hence

$$\frac{3}{4} \left\langle (\Delta v)^2 \right\rangle_{r < \eta} \approx \frac{3}{4} \left\langle (\partial u'_x / \partial x)^2 \right\rangle r^2 = \left\langle \frac{1}{2} \boldsymbol{\omega}^2 \right\rangle \frac{r^2}{10} \approx \left\langle \frac{1}{2} \boldsymbol{\omega}^2 \right\rangle \frac{r^2}{\pi^2},$$ (13.25)

since, as we shall see in §13.2.2, $\langle \boldsymbol{\omega}^2 \rangle = 15 \langle (\partial u'_x / \partial x)^2 \rangle$. If we apply (13.25) to $\mathbf{u}'$ after it has been *high-pass filtered* at some intermediate scale, s, effectively replacing η by s, we get an additional contribution to $< (\Delta v)^2 > (r)$, of $(r^2/\pi^2) \times$ (enstrophy of eddies of size $s > r$). This then leads to a refined interpretation of $< (\Delta v)^2 >$:

$$\boxed{\begin{aligned} \frac{3}{4} \left\langle [\Delta v]^2 \right\rangle (r) &\approx \text{[all energy held in eddies of size } s < r \text{]} \\ &+ \frac{r^2}{\pi^2} \text{[all enstrophy held in eddies of size } s > r \text{]} \end{aligned}}$$ (13.26)

(see Figure 13.17). In those cases where the second contribution to (13.26) is important, we lose the usual interpretation of the structure function. We shall return to (13.26) in §13.2.4.

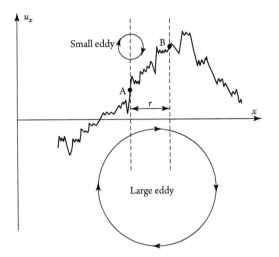

Figure 13.17 Eddies of size s $(s < r)$ make a contribution to $< (\Delta v)^2 > (r)$ of the order of their energy, while larger eddies $(s > r)$ make a contribution proportional to their enstrophy.

13.2.2 The Simplifications of Isotropy and the Taylor Scale

Sometimes it turns out that homogeneous turbulence has statistical properties that are not only independent of position, but also independent of direction. That is, all averages have reflectional symmetry and are invariant under rotations of the frame of reference. This is referred to as *isotropic turbulence*. Fully developed turbulence in a wind tunnel is approximately homogeneous and isotropic, as are the *small scales* in both homogeneous and inhomogeneous turbulent flows, provided the Reynolds number is large enough. (We shall discuss the reasons for this small-scale isotropy in §13.3.1.) The idealization of isotropic turbulence is sufficiently important for us to devote all of this subsection to it.

Let us start by introducing some notation. We define the integral scales, ℓ and u, through the expressions

$$\ell = \int_0^\infty f(r)dr, \quad u^2 = \langle u_x'^2 \rangle = \langle u_y'^2 \rangle = \langle u_z'^2 \rangle. \tag{13.27}$$

Since the turbulence is homogeneous, the mean flow must be uniform and so we may eliminate it through a suitable change in frame of reference. (The only exception to this is homogeneous turbulence subject to a uniform mean shear, which we shall not discuss.) There is then no reason to place a prime on **u** to indicate a fluctuating quantity. So, when discussing homogeneous isotropic turbulence, we shall use a prime for a completely different purpose: to indicate a quantity measured at $\mathbf{x}' = \mathbf{x} + \mathbf{r}$. Thus we now write

$$Q_{ij}(\mathbf{r}, t) = \langle u_i(\mathbf{x}) u_j(\mathbf{x} + \mathbf{r}) \rangle = \langle u_i u_j' \rangle, \tag{13.28}$$

$$S_{ijk}(\mathbf{r}, t) = \langle u_i(\mathbf{x}) u_j(\mathbf{x}) u_k(\mathbf{x} + \mathbf{r}) \rangle = \langle u_i u_j u_k' \rangle, \tag{13.29}$$

with special cases

$$Q_{xx}(r\hat{\mathbf{e}}_x) = u^2 f(r), \qquad\qquad S_{xxx}(r\hat{\mathbf{e}}_x) = u^3 K(r), \qquad (13.30)$$

$$\left\langle [\Delta v]^2 \right\rangle = 2u^2 \left(1 - f(r)\right), \qquad\qquad \left\langle [\Delta v]^3 \right\rangle (r) = 6u^3 K(r). \qquad (13.31)$$

The restrictions imposed by isotropy, as well as by continuity in the form of (13.15), enforce connections between the various components of Q_{ij}. In fact, it turns out that Q_{ij} can be written in terms of f alone,

$$Q_{ij} = \frac{u^2}{2r} \left[\frac{\partial}{\partial r} \left(r^2 f\right) \delta_{ij} - \frac{\partial f}{\partial r} r_i r_j \right], \qquad (13.32)$$

(see Batchelor, 1953, or Davidson, 2015). Similarly, S_{ijk} can be written as a function of $K(r)$ only,

$$S_{ijk} = u^3 \left[\frac{K - rK'}{2r^3} r_i r_j r_k + \frac{2K + rK'}{4r} \left(r_i \delta_{jk} + r_j \delta_{ik}\right) - \frac{K}{2r} r_k \delta_{ij} \right], \qquad (13.33)$$

where $K' = \partial K / \partial r$. Expressions (13.32) and (13.33) represent an enormous simplification and are very useful. For example, it will prove convenient to introduce the notation $R = \frac{1}{2} \langle \mathbf{u} \cdot \mathbf{u}' \rangle$, and (13.32) tells us that

$$R(r) = \frac{1}{2} \langle \mathbf{u} \cdot \mathbf{u}' \rangle = \frac{u^2}{2r^2} \frac{\partial}{\partial r} \left(r^3 f\right). \qquad (13.34)$$

Let us now introduce the *vorticity correlation tensor*, $\langle \omega_i \omega_j' \rangle$. With a little patience, it may be confirmed that $\boldsymbol{\omega} = \nabla \times \mathbf{u}$ yields

$$\langle \omega_i \omega_j' \rangle = \nabla^2 Q_{ij} + \frac{\partial Q_{kk}}{\partial r_i \partial r_j} - \left(\nabla^2 Q_{kk}\right) \delta_{ij}, \qquad (13.35)$$

(see Batchelor, 1953), of which an important the special case is

$$\langle \boldsymbol{\omega} \cdot \boldsymbol{\omega}' \rangle = -\nabla^2 \langle \mathbf{u} \cdot \mathbf{u}' \rangle. \qquad (13.36)$$

This combines with (13.34) to give

$$\langle \boldsymbol{\omega} \cdot \boldsymbol{\omega}' \rangle = -\frac{1}{r^2} \frac{\partial}{\partial r} \left[r^2 \frac{\partial}{\partial r} \left(\frac{u^2}{r^2} \frac{\partial}{\partial r} \left(r^3 f\right) \right) \right]. \qquad (13.37)$$

The significance of (13.37) is that the shape of f near $r = 0$ is determined by $\langle \boldsymbol{\omega}^2 \rangle$, and hence by the dissipation per unit mass, $\varepsilon = \nu \langle \boldsymbol{\omega}^2 \rangle$, as we now discuss.

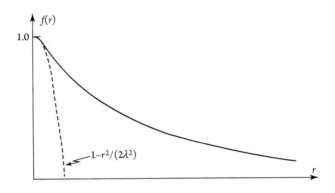

Figure 13.18 Shape of $f(r)$ near $r = 0$.

Since f is even in r, $f(0) = 1$, and $f \leq 1$, we may introduce a new length scale, λ, defined by the expression

$$f(r) = 1 - \frac{r^2}{2\lambda^2} + O(r^4).$$

(13.38)

Substituting this into (13.37) now gives us

$$\lambda^2 = \frac{15u^2}{\langle \omega^2 \rangle} = \frac{15\nu u^2}{\varepsilon},$$

(13.39)

and

$$u^2 f(r) = u^2 - \frac{\varepsilon r^2}{30\nu} + \cdots.$$

(13.40)

However, the zeroth law gives us $\varepsilon \sim u^3/\ell$ and it follows that, when $\mathrm{Re} = u\ell/\nu \gg 1$, the curvature of $f(r)$ near $r = 0$ is very large, as shown in Figure 13.18. The length scale λ is known as the *Taylor scale*. From (13.39) we have $\lambda^2/\ell^2 \sim 15\mathrm{Re}^{-1}$, and so the Taylor scale lies somewhere between the integral scale and the Kolmogorov microscale, $\eta \sim \ell(\mathrm{Re})^{-3/4}$, with $\lambda^2/\eta^2 \sim 15\mathrm{Re}^{1/2}$.

Note that (13.34) and (13.36) combine to give us

$$\langle (\nabla \times \boldsymbol{\omega}) \cdot (\nabla \times \boldsymbol{\omega})' \rangle = -\nabla^2 \langle \boldsymbol{\omega} \cdot \boldsymbol{\omega}' \rangle = \nabla^4 \langle \mathbf{u} \cdot \mathbf{u}' \rangle = \nabla^4 \left[\frac{u^2}{r^2} \frac{\partial}{\partial r} (r^3 f) \right],$$

(13.41)

which allows us to determine the next term in expansion (13.40). A little algebra then gives

$$u^2 f(r) = u^2 - \frac{\langle \omega^2 \rangle}{30} r^2 + \frac{\langle (\nabla \times \boldsymbol{\omega})^2 \rangle}{840} r^4 - \frac{\langle (\nabla^2 \boldsymbol{\omega})^2 \rangle}{45360} r^6 + \cdots,$$

(13.42)

from which

$$\left\langle [\Delta v]^2 \right\rangle (r) = 2u^2 (1 - f) = \frac{\left\langle \omega^2 \right\rangle}{15} r^2 - \frac{\left\langle (\nabla \times \omega)^2 \right\rangle}{420} r^4 + \cdots . \qquad (13.43)$$

It follows that

$$\left\langle (\partial u_x / \partial x)^2 \right\rangle = \lim_{r \to 0} \frac{< (\Delta v)^2 >}{r^2} = \left\langle \omega^2 \right\rangle / 15, \qquad (13.44)$$

as noted below (13.25).

13.2.3 Scale-by-scale Energy Distributions in Fourier Space: the Energy Spectrum

Given that the second-order structure function mixes information about energy and enstrophy, and about large and small scales, we should look for an alternative means of describing the scale-by-scale variation of kinetic energy in a turbulent flow. We do so now, while continuing to restrict the discussion to isotropic turbulence. In Appendix 4 we show how the Fourier transform can be used to distinguish between different scales in a random signal, and so it is natural to seek a spectral description of the turbulent kinetic energy per unit mass. Let us start with the three-dimensional transform of the velocity correlation tensor, $Q_{ij}(\mathbf{r})$, which is called the *spectral tensor* and denoted $\Phi_{ij}(\mathbf{k})$.

The three-dimensional transform pair (A4.13) and (A4.14) tell us that Φ_{ij} and Q_{ij} are related by

$$\Phi_{ij}(\mathbf{k}) = \frac{1}{(2\pi)^3} \int Q_{ij}(\mathbf{r}) e^{-j\mathbf{k} \cdot \mathbf{r}} d\mathbf{r}, \qquad (13.45)$$

$$Q_{ij}(\mathbf{r}) = \int \Phi_{ij}(\mathbf{k}) e^{j\mathbf{k} \cdot \mathbf{r}} d\mathbf{k}. \qquad (13.46)$$

Note that, since Q_{xx}, Q_{yy}, and Q_{zz} are three-dimensional autocorrelation functions for u_x, u_y, and u_z, the three-dimensional version of the autocorrelation theorem (A4.8) tells us that the diagonal components of Φ_{ij} are non-negative, and in particular, $\Phi_{ii} \geq 0$. Let us focus, therefore, on Φ_{ii}. Since $Q_{ii} = \langle \mathbf{u} \cdot \mathbf{u}' \rangle$ is a spherically symmetric function in $\mathbf{r}$-space, $\Phi_{ii}(\mathbf{k})$ must be spherically symmetric in $\mathbf{k}$-space. Setting $\mathbf{r} = 0$ and $i = j$ in (13.46) then gives us

$$\frac{1}{2} \langle \mathbf{u}^2 \rangle = \frac{1}{2} \int \Phi_{ii} d\mathbf{k} = \int_0^\infty 2\pi k^2 \Phi_{ii} dk, \qquad (13.47)$$

and so integrating $\Phi_{ii}(k)$ over all $\mathbf{k}$-space gives the turbulent kinetic energy per unit mass.

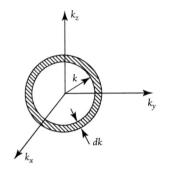

Figure 13.19 $E(k)\,dk$ represents the contribution to $\frac{1}{2}\langle \mathbf{u}^2 \rangle$ from Φ_{ii} which lies within a spherical annulus in $\mathbf{k}$-space.

We now introduce the *three-dimensional energy spectrum* of the velocity field, $E(k)$, defined by

$$E(k) = 2\pi k^2 \Phi_{ii}, \tag{13.48}$$

which evidently has the properties

$$E(k) \geq 0, \quad \frac{1}{2}\langle \mathbf{u}^2 \rangle = \int_0^\infty E(k)dk. \tag{13.49}$$

Clearly, $E(k)\,dk$ represents the contribution of Φ_{ii} to $\frac{1}{2}\langle \mathbf{u}^2 \rangle$ that is contained in a spherical annulus in $\mathbf{k}$-space of thickness dk, as shown in Figure 13.19. We say that $E(k)$ represents the *distribution of energy in spectral space*. Some authors go further and suggest that $E(k)$ represents the energy distribution in *scale space*. However, as we shall see, this can be misleading as scale space and spectral space need not be the same thing.

Next, we note that Φ_{ii} and Q_{ii} are three-dimensional transform pairs which have spherical symmetry. So (A4.17) and (A4.18) tell us that integrating over the polar and azimuthal angles in (13.45) and (13.46) yields

$$\Phi_{ii} = \frac{1}{2\pi^2}\int_0^\infty r^2 Q_{ii}\frac{\sin(kr)}{kr}dr\,, \quad Q_{ii} = 4\pi\int_0^\infty k^2\Phi_{ii}\frac{\sin(kr)}{kr}dk. \tag{13.50}$$

Finally, we substitute for Φ_{ii} and Q_{ii} in terms of $E(k)$ and $R(r) = \frac{1}{2}\langle \mathbf{u}\cdot\mathbf{u}'\rangle$ to give:

$$E(k) = \frac{2}{\pi}\int_0^\infty R(r)\,kr\sin(kr)dr\,, \tag{13.51}$$

$$R(r) = \int_0^\infty E(k)\frac{\sin(kr)}{kr}dk. \tag{13.52}$$

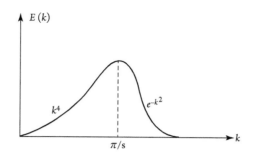

Figure 13.20 The shape of the energy spectrum for a random array of simple Gaussian eddies of fixed size s.

Now we have already seen that $E(k)$ defined by this transform pair has the important properties

$$E(k) \geq 0, \quad \frac{1}{2} \langle \mathbf{u}^2 \rangle = \int_0^\infty E(k)dk. \tag{13.53}$$

However, this is not enough to ensure that $E(k)$ represents the distribution of kinetic energy in *scale space*. To show that, we need one additional result. Consider, the Gaussian eddy

$$\mathbf{u} = \Omega r \, \exp\left(-2\mathbf{x}^2/s^2\right) \hat{\mathbf{e}}_\theta \tag{13.54}$$

in (r, θ, z) coordinates. This represents a blob of swirling fluid of characteristic size s. An ensemble of such eddies whose centres are randomly but uniformly distributed in space, and whose axes of rotation are randomly orientated, is statistically isotropic. Moreover, the ensemble has a longitudinal correlation function and energy spectrum of

$$f(r) = \exp\left[-r^2/s^2\right], \tag{13.55}$$

$$E(k) = \frac{\langle \mathbf{u}^2 \rangle s}{24\sqrt{\pi}} (ks)^4 \exp\left[-s^2k^2/4\right] \tag{13.56}$$

(see Davidson, 2015). Crucially, $kE(k)$ is sharply peaked at $k = \sqrt{10}/s \approx \pi/s$, as shown in Figure 13.20.

Now suppose we have a random array of Gaussian eddies of size $s = L_1$, plus a second, independent array of eddies of size $s = L_2$, and so on, up to $s = L_N$. Then the energy spectrum $E(k)$ will look a bit like that shown in Figure 13.21, with each eddy size contributing to $E(k)$ primarily around $k \sim \pi/L_i$. In view of this third property of $E(k)$, it is customary to interpret $E(k)dk$ as the contribution to $\frac{1}{2}\langle \mathbf{u}^2 \rangle$ from all eddies with wavenumbers in the range $k \to k + dk$, where $k \sim \pi/s$. This then provides a convenient measure of how energy is distributed across the various eddy sizes in scale space.

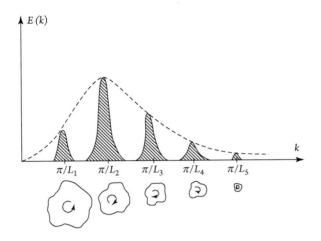

Figure 13.21 The shape of $E(k)$ for a hierarchy of well-separated eddy sizes.

It should be emphasized, however, that this is a flawed view of $E(k)$. It is true that simple eddies of scale, s, contribute to $E(k)$ primarily in the vicinity of $k \sim \pi/s$. However, they also contribute to $E(k)$ across the full range of wavenumbers, albeit peaked around π/s. Nevertheless, the conventional interpretation of $E(k)$, that it represents the energy in eddies of size π/k, is convenient and works well for many purposes. Consequently, we shall adopt this as a kind of shorthand. However, we shall see in §13.2.5 that this is a false interpretation of $E(k)$ if one considers the wavenumber ranges $k < \pi/\ell$ or $k > \pi/\eta$. It is also a flawed interpretation if the eddies are characterized by two or more distinct length scales.

The function $E(k)$ has one further useful property. It follows from (13.36) and (13.46) that

$$\langle \boldsymbol{\omega} \cdot \boldsymbol{\omega}' \rangle = -\nabla^2 \langle \mathbf{u} \cdot \mathbf{u}' \rangle = \int k^2 \Phi_{ii}(\mathbf{k}) e^{j\mathbf{k} \cdot \mathbf{r}} d\mathbf{k},$$

from which

$$\boxed{\frac{1}{2} \langle \boldsymbol{\omega}^2 \rangle = \int_0^\infty k^2 E(k) dk}. \tag{13.57}$$

So we interpret $k^2 E(k) dk$ as the contribution to the enstrophy, $\frac{1}{2} \langle \boldsymbol{\omega}^2 \rangle$, from the range of wavenumbers $k \to k + dk$.

We close this section with second note of caution. We argue in Appendix 4 that Fourier analysis can distinguish between different scales in a random signal because, when Fourier-transformed, small-scale features contribute to high-wavenumber components of the transform, while large-scale features contribute to low wavenumbers. Following that logic, one might well wonder why we went to all the trouble of taking a *three-dimensional* transform

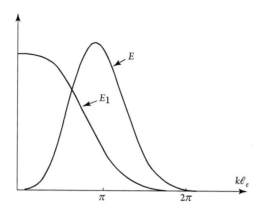

Figure 13.22 Spectra $E(k)$ and $E_{1D}(k)$ for a random field of Gaussian eddies of size ℓ_e.

of $\langle \mathbf{u} \cdot \mathbf{u}' \rangle$. Why not simply define the energy spectrum in terms of a one-dimensional transform? For example, we might introduce the transform pair

$$E_{1D}(k) = \frac{1}{\pi} \int_0^\infty \langle \mathbf{u} \cdot \mathbf{u}' \rangle \cos(kr)\, dr, \tag{13.58}$$

$$\langle \mathbf{u} \cdot \mathbf{u}' \rangle = 2 \int_0^\infty E_{1D}(k) \cos(kr)\, dk. \tag{13.59}$$

Certainly, $E_{1D}(k)$ has the properties

$$E_{1D}(k) \geq 0, \qquad \frac{1}{2}\langle \mathbf{u}^2 \rangle = \int_0^\infty E_{1D}(k)\, dk, \tag{13.60}$$

expected of an energy spectrum. However, $E(k)$ and $E_{1D}(k)$ have very different shapes. This can be seen by differentiating (13.58) with respect to k to give

$$\frac{dE_{1D}}{dk} = -\frac{E}{k}, \quad \text{or} \quad E_{1D}(k) = \int_k^\infty \frac{E(p)}{p}\, dp. \tag{13.61}$$

Evidently, $E_{1D}(k)$ is a cumulative measure of $E(p)/p$ for $p > k$. Figure 13.22 shows the relative shapes of $E(k)$ and $E_{1D}(k)$ for a random field of Gaussian eddies of fixed size, ℓ_e, and it is clear that, in the case of $E_{1D}(k)$, we have lost the crucial property that the spectrum peaks at around $\pi/(\text{eddy size})$. So, if we work with a one-dimensional transform, the distribution of energy in Fourier space does *not* reflect the distribution of energy in scale space. The reasons for this unexpected behaviour are discussed in Tennekes and Lumley (1972).

13.2.4 Relating Real-space and Spectral-space Estimates of the Energy Distribution

We now explore the relationship between $E(k)$ and $\langle[\Delta v]^2\rangle(r)$, both of which make some claim to measure the scale-by-scale energy distribution. Recall from §13.2.1 that it is common to associate $\langle[\Delta v]^2\rangle(r)$ with the cumulative energy of eddies of size r or less. However, we also suggested that this can be overly simplistic, in that eddies whose size is greater than r can also make a contribution to $<(\Delta v)^2>$ of $(r/\pi)^2\times$(enstrophy of eddies). We shall now confirm that this is the case, at least for isotropic turbulence. We start by combining (13.34) and (13.52) to give

$$R(r) = \frac{u^2}{2r^2}\frac{\partial}{\partial r}\left(r^3 f\right) = \int_0^\infty E(k)\frac{\sin kr}{kr}\,dk.$$

Integrating to find f and noting that $\langle[\Delta v]^2\rangle(r) = 2u^2(1 - f(r))$, we obtain

$$\boxed{\frac{3}{4}\left\langle[\Delta v]^2\right\rangle = \int_0^\infty E(k)H(kr)\,dk}, \tag{13.62}$$

where

$$H(x) = 1 + \frac{3\cos x}{x^2} - \frac{3\sin x}{x^3}. \tag{13.63}$$

The function $H(x)$ is shown in Figure 13.23. For small x it grows as $H(x) \sim x^2/10$, while for large x it approaches the asymptote $H = 1$ in an oscillatory manner.

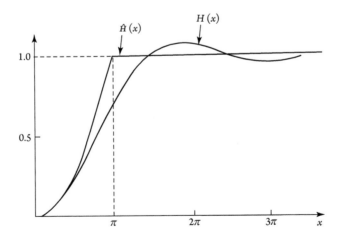

Figure 13.23 The shape of $H(x)$ in (13.62) as well as $\hat{H}(x)$.

Also shown in Figure 13.23 is an approximation to $H(x)$, given by

$$\hat{H}(x) = \begin{cases} (x/\pi)^2, & x < \pi \\ 1, & x > \pi \end{cases}$$

If we think of $H(kr)$ as a weighting function acting on $E(k)$ in (13.62), then we should get a reasonable estimate of $<(\Delta v)^2>$ by replacing H by $\hat{H}$. Making this substitution in (13.62) gives

$$\boxed{\frac{3}{4}\left\langle [\Delta v]^2 \right\rangle \approx \int_{\pi/r}^{\infty} E(k)\,dk + \frac{r^2}{\pi^2}\int_0^{\pi/r} k^2 E\,dk}. \qquad (13.64)$$

Given our interpretation of $E(k)$, we can rewrite this as

$$\frac{3}{4}\left\langle [\Delta v]^2 \right\rangle \approx [\text{energy in eddies of size } s < r] + \frac{r^2}{\pi^2}[\text{enstrophy in eddies of size } s > r],$$

which brings us back to (13.26). Evidently $E(k)$ is a cleaner diagnostic than $\langle[\Delta v]^2\rangle(r)$.

13.2.5 A Common Error in the Interpretation of Energy Spectra

We have already suggested that it is necessary to interpret energy spectra with care. We close our discussion of kinematics by highlighting a common misinterpretation of $E(k)$. Let us return to Figure 13.21. Suppose we have a *continuous* range of eddy sizes, s, from ℓ down to η, but no eddies outside this range. Then at each scale, s, eddies will contribute to $E(k)$ in a manner similar to that shown in Figure 13.20. For each scale there will be a low-k tail in the form of a power law, a peak in the contribution to $E(k)$ at $k \sim \pi/s$, and then a Gaussian tail, $E \sim \exp\left(-s^2k^2/4\right)$, for $ks \gg 1$. Moreover, we shall see in §13.3.2 that the peaks of the individual contributions, which are centred at $k \sim \pi/s$, vary with k according to $E_{\text{peak}} \sim k^{-5/3}$, at least in high-Re turbulence. The overall situation is then roughly the following. The spectrum to the left of $k \sim \pi/\ell$ consists of the sum of lots of terms of the form $E \sim k^4$, say $E = Ik^4$, while we have $E \sim k^{-5/3}$ for $\pi/\ell < k < \pi/\eta$. Finally, to the right of $k \sim \pi/\eta$, $E(k)$ will be an integral over $\eta < s < \ell$ of contributions of the form $E \sim \exp\left(-s^2k^2/4\right)$. Such an integral takes the form of $E \sim \exp\left(-k\eta\right)$, as discussed in Davidson (2015). Overall, then, the situation is as shown in Figure 13.24.

The key point, however, is that the shaded regions in Figure 13.24 do *not* represent the energy of eddies of size $s \sim \pi/k$. Rather, these regions represent the tails of spectra generated by eddies that sit in the range $\eta < s < \ell$. In fact, the shaded region on the left, $E = Ik^4$, has a quite different physical interpretation. I is a measure of how much *angular impulse* the eddies possess, and it is an *invariant* of fully developed, freely decaying turbulence (see Exercise 13.4). Often one encounters turbulence where the low-k spectrum takes the alternative form $E = Lk^2$. In such cases, L is a measure of how much *linear impulse* the eddies possess, and it too is an *invariant* of freely decaying turbulence.

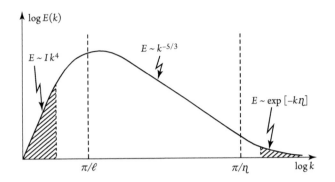

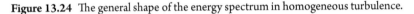

Figure 13.24 The general shape of the energy spectrum in homogeneous turbulence.

13.3 Kolmogorov's Universal Equilibrium Theory of the Small Scales (K41)

13.3.1 Does Small-scale Turbulence have a Universal, Isotropic Structure at Large Re?

We are finally in a position to present Kolmogorov's celebrated 1941 theory of the small scales of turbulence. Our starting point, as was Kolmogorov's, is Richardson's notion of a multistep energy cascade. The key point which Kolmogorov appreciated is that such a cascade should be *information losing*. This is important because the large scales in most turbulent flows are strongly inhomogeneous and anisotropic, and vary in character from one class of flow to another. (Jets are different to boundary layers.) However, the energy of the large eddies needs to transition all the way from ℓ down to η through a hierarchy of intermediate vortical structures, with each generation of vortices passing its energy onto somewhat smaller structures through vortex stretching. As we pass down this information-losing cascade the large-scale inhomogeneity and anisotropy is progressively lost. That is, the anisotropy gets progressively weaker because the relative orientation of the nth generation vortices, and hence the vortex stretching associated with those vortices, becomes increasingly decoupled from any orientation bias at the large scales, there being almost no *direct* interaction between the large and intermediate-sized vortices. Moreover, any large-scale *inhomogeneity* becomes increasingly irrelevant as we focus on smaller and smaller scales, and so become more localized in space. So Kolmogorov suggested that, at some point, we should reach a scale at which the large-scale anisotropy and inhomogeneity are more or less undetectable.

This led Kolmogorov to the conclusion that, provided Re is very large so that there are many steps to this information-losing cascade, we eventually reach a scale which is *statistically homogeneous and isotropic*. This is referred to as *local isotropy*. Below this scale the turbulence is oblivious to the details of the largest eddies, except to the extent that they dictate the flux of kinetic energy to small scales, Π, in accordance with (13.7). Kolmogorov's final step was to conjecture that all of the turbulence below this scale is *statistically universal*, being the same for *all* turbulent flows, such as boundary layers, jets and wakes. This hypothesis of universality has since found considerable experimental support, though it is often noted that a painfully large value of Re is required to eradicate the influence of the large-scale anisotropy. (Note, also, the caveat discussed in §13.4.1.)

Kolmogorov then divided scale space into three asymptotically distinct regimes. On the one hand we have the large, anisotropic, non-universal eddies whose dynamics are independent of viscosity, because Re is assumed to be very large. At the other extreme, we have scales of order η which are directly influenced by viscosity, but are isotropic and universal. This is called the *dissipation range*. Finally, provided that Re $\gg 1$, there are the intermediate scales, $\eta \ll s \ll \ell$, which are large enough to be inviscid yet small enough to be universal and locally homogeneous and isotropic. This is called the *inertial subrange*.

13.3.2 Kolmogorov's Universal Equilibrium Theory: the Two-thirds and Five-thirds Laws

The union of the inertial subrange and the dissipation range, which are the scales for which $s \ll \ell$, is called the *universal equilibrium range*. Kolmogorov's *universal equilibrium theory* of 1941 applies to this range. The use of the term *equilibrium* here requires some explanation. From (13.11) the characteristic timescale of the small eddies is $\eta/\upsilon \sim \mathrm{Re}^{-1/2}\ell/u$, which is extremely fast by comparison with the large scales. Now, in decaying turbulence, the large scales lose their energy at the rate

$$\frac{du^2}{dt} \sim -\Pi \sim -\frac{u^3}{\ell}, \tag{13.65}$$

i.e. on a timescale of ℓ/u. For the scales $s \ll \ell$, this is extremely slow by comparison with the characteristic timescale, or adjustment time, of the small eddies. Hence, to the eddies in the universal equilibrium range, the large scales evolve very slowly and so the turbulence appears to be statistically steady, or perhaps we should say quasi-steady. So, the fact that the large scales may not be steady-on-average is important to the small scales only to the extent that the energy flux varies slowly in time, $\Pi = \Pi(t)$. The small scales are therefore said to be in *statistical equilibrium* with the *instantaneous* energy flux.

Kolmogorov's theory can be developed either in terms of the energy spectrum, $E(k)$, or the second-order structure function, the latter being favoured by Kolmogorov but the former technically more sound (as we shall see). We start by ignoring the large-scale enstrophy contribution to the second-order structure function and adopt the simplistic view

$$\frac{3}{4}\left\langle [\Delta \upsilon]^2 \right\rangle (r) \approx [\text{all energy held in eddies of size } s < r]. \tag{13.66}$$

(We shall check retrospectively to see if this is self-consistent.) Kolmogorov now made the crucial assumption that the turbulence in the universal equilibrium range cares about the large scales only to the extent that it sets the instantaneous flux of energy down through scale space. Since we have excluded any large-scale information in (13.66), we then have

$$\left\langle [\Delta v]^2 \right\rangle = F(\Pi, \nu, r), \quad r \ll \ell. \tag{13.67a}$$

Note that, since these scales are assumed to be in statistical equilibrium with the instantaneous energy flux, there is no local accumulation or depletion of energy at any one scale in the equilibrium range. The flux Π must therefore be independent of r and equal to ε. So, according to Kolmogorov's 1941 theory, often denoted K41, we can write

$$\left\langle [\Delta v]^2 \right\rangle = F(\varepsilon, \nu, r), \quad r \ll \ell, \tag{13.67b}$$

where ε acts as a *proxy* for Π. Equation (13.67a) should not be passed over lightly as it is a remarkable statement. Consider Figure 13.6, which shows the small-scale eddies in a computer simulation. According to (13.67a), the statistical evolution of this complex tangle of vortex tubes is controlled exclusively by Π, r, and ν. If we now introduce the Kolmogorov microscales, $\eta = \left(\nu^3 / \varepsilon \right)^{1/4}$ and $\upsilon = (\nu \varepsilon)^{1/4}$, we can rewrite (13.67b) in the dimensionless form

$$\boxed{\left\langle [\Delta v]^2 \right\rangle (r) = \upsilon^2 F(r/\eta), \quad r \ll \ell}. \tag{13.68}$$

Moreover, according to Kolmogorov, F is a *universal* function, that is, it is the same for all types of turbulence (jets, wakes, boundary layers, etc.).

Now suppose that Re is very large. Then it is possible that there exists a range of eddies that satisfy (13.68), yet they are still large enough for ν to have no influence, *i.e.* there exists an inertial subrange in which ν is irrelevant. The only dimensionally consistent possibility is then $F(x) \sim x^{2/3}$, or

$$\boxed{\left\langle [\Delta v]^2 \right\rangle = \beta \varepsilon^{2/3} r^{2/3}, \quad \eta \ll r \ll \ell}, \tag{13.69}$$

where β (Kolmogorov's constant) is a universal constant, observed to have a value of $\beta \sim 2$. This is known as *Kolmogorov's two-thirds law* and its validity rests entirely on the hypothesis that the intermediate eddies are controlled exclusively by Π (or equivalently ε).

When formulated in terms of the energy spectrum the equivalent result is

$$\boxed{E(k) = \alpha \varepsilon^{2/3} k^{-5/3}, \quad \ell^{-1} \ll k \ll \eta^{-1}}, \tag{13.70}$$

where (13.62) gives $\alpha \approx 0.76\beta$ for Re $\to \infty$. Equation (13.70) is known as Kolmogorov's *five-thirds law*, and it can be derived by dimensional arguments in exactly the same way as (13.69), starting from

$$E(k) = \upsilon^2 \eta F(k\eta), \quad k\ell \gg 1, \tag{13.71}$$

for the universal equilibrium range, and then noting that ν is irrelevant in the inertial range.

Perhaps some comments are in order at this point. First, the experimental data supports (13.68)–(13.71), at least to within experimental error, and indeed there is early German data (Gödecke, 1935) that shows a clear $r^{2/3}$ dependence for the second-order structure function. (It is likely that Kolmogorov was aware of this data in 1941.) Second, Kolmogorov's student, Obukhov, obtained the two-thirds law in a different way. He hypothesized that at each scale in the inertial subrange the flux of energy to smaller scales is of the order of the energy at that scale multiplied by the strain rate of those eddies. If v_s is a typical velocity at scale s, then Obukhov's assumption can be written $\Pi \sim v_s^3/s$. The two-thirds law then follows if one assumes that $v_s^2 \sim \langle [\Delta v]^2 \rangle (s)$. Third, the enstrophy spectrum *grows* with wavenumber according to $k^2 E(k) \sim \varepsilon^{2/3} k^{1/3}$, which is consistent with Figure 13.9. Fourth, we neglected the contribution of the large-scale enstrophy in (13.66), and hence in (13.67). To see if this makes a difference, let us consider the case of $\mathrm{Re} \gg 1$, so that (13.70) can be taken to apply throughout most of the spectral range of $E(k)$. Then (13.64) gives us

$$\frac{3}{4} \left\langle [\Delta v]^2 \right\rangle \approx \int_{\pi/r}^{\pi/\eta} E(k)\,dk + \frac{r^2}{\pi^2} \int_{\pi/\ell}^{\pi/r} k^2 E\,dk,$$

from which,

$$\frac{3}{4} \left\langle [\Delta v]^2 \right\rangle \approx \frac{3\alpha\varepsilon^{2/3}}{2\pi^{2/3}} \left[\left(r^{2/3} - \eta^{2/3} \right) + \frac{1}{2} \left(r^{2/3} - \frac{r^2}{\ell^{4/3}} \right) \right]. \tag{13.72}$$

For $\eta \ll r \ll \ell$ this takes us back to (13.69), with $\beta \approx 3\alpha/\pi^{2/3}$. Evidently, the large-scale enstrophy does indeed make a first-order contribution to the structure function. We have hit the problem that Δv is a rather leaky filter in scale space. This is a potential issue because our assumption that the turbulence cares about the large scales only to the extent that it determines Π applies *only* in the universal equilibrium range. Given that Δv admits large-scale information, perhaps the safest way to derive the two-thirds law is to first deduce the five-thirds law on dimensional grounds (recall that $E(k)$ is a less leaky filter than Δv) and then derive the two-thirds law from (13.62), assuming $\mathrm{Re} \to \infty$.

There is one final comment that is is important to make. If we *tentatively* neglect the large-scale enstrophy contribution to the longitudinal velocity increment, dimensional analysis allows us to generalize (13.69) to higher-order structure functions. That is to say, *if the statistical properties of Δv in the inertial subrange depend on only Π and r, then*

$$\boxed{\left\langle [\Delta v]^p \right\rangle = \beta_p \varepsilon^{p/3} r^{p/3}, \quad \eta \ll r \ll \ell}, \tag{13.73}$$

where the β_p are universal constants. In fact, as we shall see, the case of $p = 3$ is *exact*,

$$\boxed{\left\langle [\Delta v]^3 \right\rangle = -\frac{4}{5} \varepsilon r, \quad \eta \ll r \ll \ell}. \tag{13.74}$$

This is Kolmogorov's *four-fifths law*. One weakness of (13.73), however, is that the large scales *do* contribute to Δv, as $\Delta v \sim (\partial u_x / \partial x)_s r$, where $(\partial u_x / \partial x)_s$ is a typical velocity gradient at scale s. So the leaky nature of Δv as a filter in scale space opens up the possibility that the statistical moments of Δv know more about the large scales than simply Π. For example, the presence of ℓ in (13.72) may corrupt the two-thirds law when Re is modest and the inertial range short.

13.3.3 The Kármán–Howarth Equation

Rather remarkably, we have yet to deploy the Navier–Stokes equation in our analysis. So let us see if we can get an evolution equation for the two-point correlation $\langle \mathbf{u} \cdot \mathbf{u}' \rangle$ directly from the Navier–Stokes equation. Since we have in mind Kolmogorov's theory, and this applies only to the small scales, we shall limit the discussion to homogeneous, isotropic turbulence. As before, we adopt a frame of reference in which there is no mean flow and use the notation $\mathbf{x}' = \mathbf{x} + \mathbf{r}$ and $\mathbf{u}(\mathbf{x}') = \mathbf{u}'$. The Navier–Stokes equation applied at $\mathbf{x}$ and $\mathbf{x}'$ then gives us

$$\frac{\partial u_i}{\partial t} = -\frac{\partial (u_i u_k)}{\partial x_k} - \frac{\partial (p/\rho)}{\partial x_i} + \nu \nabla_x^2 u_i,$$

$$\frac{\partial u_j'}{\partial t} = -\frac{\partial (u_j' u_k')}{\partial x_k'} - \frac{\partial (p'/\rho)}{\partial x_j'} + \nu \nabla_{x'}^2 u_j'.$$

Multiplying the first of these by u_j', and the second by u_i, then adding and averaging, yields

$$\frac{\partial}{\partial t} \langle u_i u_j' \rangle = -\left\langle u_i \frac{\partial u_j' u_k'}{\partial x_k'} + u_j' \frac{\partial u_i u_k}{\partial x_k} \right\rangle - \frac{1}{\rho} \left\langle u_i \frac{\partial p'}{\partial x_j'} + u_j' \frac{\partial p}{\partial x_i} \right\rangle$$
$$+ \nu \left\langle u_i \nabla_{x'}^2 u_j' + u_j' \nabla_x^2 u_i \right\rangle.$$

This complicated expression may be simplified if we note that:

(i) u_i is independent of $\mathbf{x}'$ while u_j' is independent of $\mathbf{x}$;

(ii) the operations of differentiation and taking averages, $\langle \sim \rangle$, commute;

(iii) $\partial/\partial x_i$ and $\partial/\partial x_j'$ operating on averages may be replaced by $-\partial/\partial r_i$ and $\partial/\partial r_j$;

(iv) in isotropic turbulence, $\langle u_i u_j' u_k' \rangle (\mathbf{r}) = \langle u_j u_k u_i' \rangle (-\mathbf{r}) = -\langle u_j u_k u_i' \rangle (\mathbf{r})$;

(v) in isotropic turbulence, conservation of mass demands $\langle \mathbf{u} p' \rangle = 0$.

We have already used (i), (ii), and (iii) in the derivation of (13.15), while (iv) follows from Figure 13.14 combined with the observation that S_{ijk} is odd in $\mathbf{r}$. The fact that $\langle \mathbf{u} p' \rangle = 0$ in isotropic turbulence is far from obvious and involves a rather detailed calculation. We shall not pause to prove the result here but rather refer interested readers to Batchelor (1953) or Davidson (2015). In any event, after a little algebra, our evolution equation simplifies to

$$\frac{\partial Q_{ij}}{\partial t} = \frac{\partial}{\partial r_k} \left[S_{ikj} + S_{jki} \right] + 2\nu \nabla^2 Q_{ij}. \tag{13.75}$$

We now set $i = j$ and substitute for S_{ijk} using (13.33). This yields, after some algebra,

$$\frac{\partial}{\partial t} \langle \mathbf{u} \cdot \mathbf{u}' \rangle = 2\Gamma(r,t) + 2\nu\nabla^2 \langle \mathbf{u} \cdot \mathbf{u}' \rangle, \quad \Gamma = \frac{1}{2r^2} \frac{\partial}{\partial r} \left[\frac{1}{r} \frac{\partial}{\partial r} \left(r^4 u^3 K \right) \right]. \tag{13.76}$$

This is known as the *Kármán–Howarth equation*. If we substitute for $\langle \mathbf{u} \cdot \mathbf{u}' \rangle$ using (13.34), and integrate, (13.76) may also be written in terms of $f(r,t)$:

$$\frac{\partial}{\partial t} \left[u^2 r^4 f(r,t) \right] = \frac{\partial}{\partial r} \left[r^4 u^3 K(r,t) \right] + 2\nu \frac{\partial}{\partial r} \left[r^4 u^2 \frac{\partial f}{\partial r} \right]. \tag{13.77}$$

One problem with (13.76) is that we cannot predict the evolution of $\langle \mathbf{u} \cdot \mathbf{u}' \rangle$ without a knowledge of $K(r, t)$. Unfortunately, if we use the Navier–Stokes equation to find $\partial S_{ijk}/\partial t$ we obtain

$$\rho \frac{\partial S_{ijk}}{\partial t} = \rho \langle uuuu \rangle - \frac{\partial}{\partial r_k} \langle u_i u_j p' \rangle - \left\langle u_k' \left(u_i \frac{\partial p}{\partial x_j} + u_j \frac{\partial p}{\partial x_i} \right) \right\rangle + \rho \nu (\sim), \tag{13.78}$$

where $\langle uuuu \rangle$ is a symbolic representation of terms involving fourth-order, two-point correlations, and $\nu(\sim)$ represents a viscous term. Evidently, since (13.78) involves $\langle uuuu \rangle$, we do not have a closed set of equations. Of course, the evolution equation for $\langle uuuu \rangle$ involves fifth-order correlations, and so on. We have hit the closure problem of turbulence. Still, we can extract a great deal of useful information from (13.76), as we shall see shortly.

Finally, we note in passing that Γ can be expressed in terms of the third-order structure function written in terms of the *vector* velocity increment $\Delta\mathbf{v} = \mathbf{u}(\mathbf{x}') - \mathbf{u}(\mathbf{x})$, i.e.

$$\left\langle (\Delta\mathbf{v})^2 \Delta\mathbf{v} \right\rangle = \left\langle (\mathbf{u}' - \mathbf{u})^2 (\mathbf{u}' - \mathbf{u}) \right\rangle.$$

The details are spelt out in Davidson (2015), and we merely note here that the end result is

$$\frac{\partial}{\partial t} \langle \mathbf{u} \cdot \mathbf{u}' \rangle = \frac{1}{2} \nabla \cdot \left[\left\langle (\Delta\mathbf{v})^2 \Delta\mathbf{v} \right\rangle \right] + 2\nu\nabla^2 \langle \mathbf{u} \cdot \mathbf{u}' \rangle. \tag{13.79}$$

This version of the Kármán–Howarth equation turns out to equally valid in anisotropic, though still homogeneous, turbulence, as discussed in Monin and Yaglom (1975).

13.3.4 Kolmogorov's Four-fifths Law

We shall now derive Kolmogorov's four-fifths law (13.74) directly from the Kármán–Howarth equation. Consider the Kármán–Howarth equation in the form of (13.77). This may be rewritten in terms of the second- and third-order structure functions using (13.31):

$$-\frac{2}{3}r^4\varepsilon - \frac{r^4}{2}\frac{\partial}{\partial t}\left\langle[\Delta v]^2\right\rangle = \frac{\partial}{\partial r}\left[\frac{r^4}{6}\left\langle[\Delta v]^3\right\rangle\right] - \nu\frac{\partial}{\partial r}\left[r^4\frac{\partial}{\partial r}\left\langle[\Delta v]^2\right\rangle\right]. \quad (13.80)$$

Moreover, in the universal equilibrium range we have $\left\langle[\Delta v]^2\right\rangle \sim \varepsilon^{2/3}r^{2/3}$, or less for $r < \eta$. Hence, for $r \ll \ell$, the second term on the left of (13.80) is, at most, of order

$$r^4\frac{\partial}{\partial t}\left(\varepsilon^{2/3}r^{2/3}\right) \sim r^4\left(u/\ell\right)\varepsilon^{2/3}r^{2/3} \sim r^4\varepsilon\left(r/\ell\right)^{2/3}. \quad (13.81)$$

This is negligible by comparison with the first term and so, whenever $r \ll \ell$, we may neglect $\partial\langle[\Delta v]^2\rangle/\partial t$ in (13.80). Physically, this is a consequence of the fact that we are in the *equilibrium* range of scales. On integrating the remaining terms in (13.80) we obtain,

$$\left\langle[\Delta v]^3\right\rangle = -\frac{4}{5}\varepsilon r + 6\nu\frac{\partial}{\partial r}\left\langle[\Delta v]^2\right\rangle, r \ll \ell. \quad (13.82)$$

Equation (13.82) is important for the dynamics of the universal equilibrium range and was first obtained by Kolmogorov in 1941. It is sometimes convenient to write it in an alternative way by introducing the so-called *skewness factor*, defined by

$$S(r) = \left\langle[\Delta v]^3\right\rangle \Big/ \left\langle[\Delta v]^2\right\rangle^{3/2}. \quad (13.83)$$

In terms of $S(r)$ we have

$$6\nu\frac{\partial}{\partial r}\left\langle[\Delta v]^2\right\rangle - S\left\langle[\Delta v]^2\right\rangle^{3/2} = \frac{4}{5}\varepsilon r, \quad r \ll \ell. \quad (13.84)$$

We shall revisit (13.84) in the next section. In the meantime, we note that in the inertial subrange, viscous effects are unimportant and so (13.82) reduces to

$$\boxed{\left\langle[\Delta v]^3\right\rangle = -\frac{4}{5}\varepsilon r}, \quad \eta \ll r \ll \ell, \quad (13.85)$$

which is Kolmogorov's four-fifths law. This is one of the few exact, non-trivial results in turbulence theory and is a special case of K41, and in particular of (13.73).

13.3.5 Obukhov's Constant Skewness Closure Model

Because of the closure problem, there is no rigorous means of using Kolmogorov's equation, (13.82), to evaluate the second-order structure function across the entire equilibrium range. However, there is a simple *closure model* that yields estimates that are very close to the observed values. Our starting point is to note that, according to Kolmogorov's two-thirds and four-fifths laws, we have a skewness factor that is constant and equal to $S = -\frac{4}{5}\beta^{-3/2}$ across the inertial subrange. Given that $\beta \sim 2$, this yields $S \sim -0.3$, while measured values of S in the dissipation range are around -0.4. Since the measured values of S are almost constant for $r \ll \ell$, Obukhov suggested the simple expedient of taking $S = -\frac{4}{5}\beta^{-3/2}$ across the entire universal equilibrium range. Kolmogorov's equation, (13.84), then becomes

$$\frac{15\nu}{2}\frac{\partial}{\partial r}\left\langle[\Delta v]^2\right\rangle + \frac{1}{\beta^{3/2}}\left\langle[\Delta v]^2\right\rangle^{3/2} = \varepsilon r, \quad r \ll \ell, \tag{13.86}$$

which is readily integrated to give $\langle[\Delta v]^2\rangle$ and hence $\langle[\Delta v]^3\rangle$. Note that, for small r, (13.43) requires $\langle[\Delta v]^2\rangle = \varepsilon r^2/15\nu$, which is consistent with ignoring the inertial term on the left of (13.86). In the inertial subrange, on the other hand, we may ignore the viscous term in (13.86), which then yields the expected $\langle[\Delta v]^2\rangle = \beta\varepsilon^{2/3}r^{2/3}$.

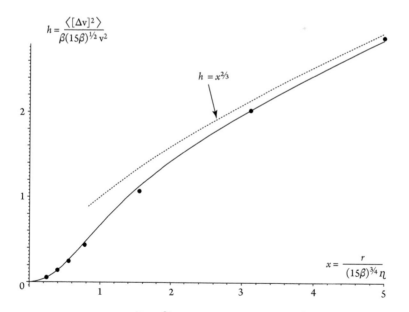

Figure 13.25 The solid line is $\left\langle[\Delta v]^2\right\rangle$, normalized by $\beta\sqrt{15\beta}\,v^2$, obtained from (13.88). • indicates the results of direct numerical simulations of turbulence at $u\lambda/\nu = 460$.

It is convenient to rewrite (13.86) using the Kolmogorov microscales $\eta = \left(\nu^3/\varepsilon\right)^{1/4}$ and $\upsilon = (\nu\varepsilon)^{1/4}$. In particular, if we introduce normalized versions of $\langle[\Delta\upsilon]^2\rangle$ and r, defined as

$$h = \frac{\langle[\Delta\upsilon]^2\rangle}{\beta\sqrt{15\beta}\,\upsilon^2}, \qquad x = \frac{r}{(15\beta)^{3/4}\,\eta}, \tag{13.87}$$

then $h(x)$ is governed by

$$\frac{1}{2}\frac{dh}{dx} + h^{3/2} = x, \qquad r \ll \ell. \tag{13.88}$$

The boundary conditions on h are then $h \to x^2$ for $x \to 0$ and $h \to x^{2/3}$ for $x \to \infty$. Figure 13.25 shows h obtained by integrating (13.88), as well as the results of direct numerical simulations (DNS) of turbulence at $u\lambda/\nu = 460$. The predictions of the closure model and the DNS are remarkably close.

A good approximation to the solution of (13.88), which is continuous in $h(x)$ and its first and second derivatives, is given by

$$h(x) = x^2 - \frac{1}{2}x^4 + \frac{1}{4}x^6 + \frac{1}{918}x^8 - \frac{167}{3240}x^{10}, \qquad x \le 1\,,$$

$$h(x) = x^{2/3} - \frac{2}{9}x^{-2/3} - \frac{5}{81}x^{-2} - \frac{101}{6120}x^{-10/3}, \qquad x \ge 1\,.$$

13.4 Subsequent Refinements to K41

13.4.1 Landau's Objection to K41 Based on Large-scale Intermittency of the Dissipation

Kolmogorov's claim that all the β_p in (13.73) are *universal* constants, the same for turbulent jets and boundary layers, say, has come under attack. This goes back to the first edition of Landau and Lifshitz's *Fluid Mechanics* (English translation, 1959, p.126), where there appears the following footnote:

> It might be thought that a possibility exists in principle of obtaining a universal formula, applicable to any turbulent flow, which should give $< (\Delta \upsilon)^2 >$ for all distances r that are small compared with ℓ. In fact, however, there can be no such formula, as we see from the following argument. The instantaneous value of $(\Delta \upsilon)^2$ might in principle be expressed as a universal function of the energy dissipation ε at the instant considered. When we average these expressions, however, an important part will be played by the law of variation of ε over times of the order of the periods of the large eddies (of size $\sim \ell$), and this law is different for different flows. The result of the averaging therefore cannot be universal.

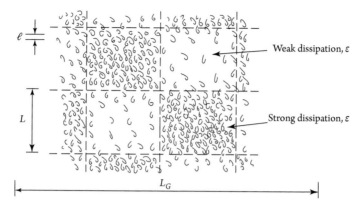

ℓ

L

Weak dissipation, ε

Strong dissipation, ε

L_G

Figure 13.26 Turbulence which is weakly inhomogeneous on the scale L.

Following Monin and Yaglom (1975), it has become conventional to reinterpret this objection in terms of spatial, rather than temporal, averaging. Indeed, Monin and Yaglom offer the following model problem to illustrate Landau's objection. Let ε be the spatially averaged dissipation, averaged over a volume of the order of the integral scale, $V \sim \ell^3$. Now consider a turbulent flow in which ε varies slowly in space on a length scale much greater than ℓ, say L. This is shown schematically in Figure 13.26.

Next, suppose that the nature of this inhomogeneity is such that when considered on a global scale, L_g, where $L_g \gg L \gg \ell$, the turbulence is statistically homogeneous. Then, according to Kolmogorov's theory, the structure function of order p, averaged over a volume of order ℓ^3, satisfies

$$\langle [\Delta v]^p \rangle = \beta_p \left(\varepsilon r \right)^{p/3}, \qquad \eta \ll r \ll \ell, \tag{13.89}$$

where the β_p in (13.89) are *universal*. Now suppose that we evaluate ε and $\langle [\Delta v]^p \rangle$ at N well-separated points in the flow so that ε is different at each point. We can then form a super-average of ε and $\langle [\Delta v]^p \rangle$ as follows:

$$\bar{\varepsilon} = \frac{1}{N} \sum_{i=1}^{N} \varepsilon_i, \quad \overline{\langle [\Delta v]^p \rangle} = \frac{1}{N} \sum_{i=1}^{N} \langle [\Delta v]^p \rangle_i. \tag{13.90}$$

However, according to Kolmogorov's theory, $\bar{\varepsilon}$ and $\overline{\langle [\Delta v]^p \rangle}$ are also related by

$$\overline{\langle [\Delta v]^p \rangle} = \beta_p \left(\bar{\varepsilon} r \right)^{p/3},$$

from which

$$\frac{1}{N} \sum \beta_p \varepsilon_i^{p/3} r^{p/3} = \beta_p r^{p/3} \left[\frac{1}{N} \sum \varepsilon_i \right]^{p/3}. \tag{13.91}$$

This, in turn, requires that ε_i satisfies

$$\frac{1}{N}\sum_i \varepsilon_i^{p/3} = \left[\frac{1}{N}\sum \varepsilon_i\right]^{p/3}. \tag{13.92}$$

Equation (13.92) is satisfied if $p = 3$, as it must be since the four-fifths law is exact. However, it is not satisfied if ε is non-uniform and $p \neq 3$, the violation of the equality being guaranteed by Hölder's inequality for $p > 3$. Evidently there is a problem.

Let us see what happens if we relax the assumption that each β_p in the super-average are the same as those in a local average, leaving all other aspects of K41 intact. To focus thoughts, consider the case of $p = 2$, where there is reasonably convincing experimental evidence in favour of $\left\langle [\Delta v]^2 \right\rangle \sim r^{2/3}$ (though see §13.4.2 below). At each point we may apply Kolmogorov's law locally to give

$$\left\langle [\Delta v]^2 \right\rangle_i = (\beta_2)_L (\varepsilon_i r)^{2/3},$$

where $(\beta_2)_L$ is the local value of β_2. Now, suppose that $\varepsilon_i = (1 - \gamma)\,\bar{\varepsilon}$ in half of the measurements and $\varepsilon_i = (1 + \gamma)\,\bar{\varepsilon}$ in the other half. The super-average then gives us

$$\overline{\left\langle [\Delta v]^2 \right\rangle} = \tfrac{1}{2}\left[(1 - \gamma)^{2/3} + (1 + \gamma)^{2/3}\right](\beta_2)_L\,\bar{\varepsilon}^{2/3} r^{2/3}, \tag{13.93}$$

and Kolmogorov's constant appearing in this super-average is evidently

$$\overline{\beta_2} = \tfrac{1}{2}\left[(1 - \gamma)^{2/3} + (1 + \gamma)^{2/3}\right](\beta_2)_L.$$

A similar calculation may be done for the higher-order structure functions and we conclude that the value of β_p depends on the method of averaging. Moreover, since $\overline{\beta_p}$ is a function of γ, Kolmogorov's constants evaluated in this way cannot be truly universal.

In summary, any inhomogeneity in ε at the large scales can lead to variations in the Kolmogorov coefficients, $\overline{\beta_p}$. Moreover, the nature of the large-scale fluctuations in ε varies from one class of turbulent flow to another, being different in, say, turbulent jets and boundary layers. It follows that the values of $\overline{\beta_p}$ can vary from flow to flow. Note, however, that a particularly strong inhomogeneity of $\gamma = 1/2$ in (13.93) results in only a 3% difference between $(\beta_2)_L$ and $\overline{\beta_2}$, which is probably within the experimental uncertainty in the value of β_2. Perhaps because this correction is so small, this deficiency in K41 has been largely ignored and attention has shifted to a second potential weakness of the theory.

13.4.2 Kolmogorov's 1961 Refinement of K41 based on Inertial-range Intermittency

The second criticism of K41 came from Kolmogorov himself, who offered a refined version of his original theory in 1961. Inspired by Landau's concerns over the consequences of *large-scale* fluctuations of the energy flux, Π, or equivalently the locally averaged dissipation, ε,

Kolmogorov considered the consequences of Π being intermittent in the *inertial subrange*. Crucially, he showed that this raises doubts over the power-law exponent $p/3$ in $\langle [\Delta v]^p \rangle = \beta_p \varepsilon^{p/3} r^{p/3}$.

The issue is the following. The rate of dissipation of energy, $2\nu S_{ij} S_{ij}$, is highly intermittent in space. There are regions of large dissipation and regions of weak dissipation. In a particular region of size r, r being assumed much less than ℓ, the instantaneous flux of energy to the small scales, $\Pi(r, t)$, should be equal to the spatially averaged dissipation in that region. Moreover, according to the original 1941 theory, the dynamics of the eddies of size r are controlled by the instantaneous local energy flux $\Pi(r, t)$. In K41 this energy flux is taken to be equal to the *globally* averaged dissipation,

$$\bar{\varepsilon} = 2\nu \langle S_{ij} S_{ij} \rangle = \nu \langle \omega^2 \rangle.$$

However, a more careful formulation of Kolmogorov's theory should work with a *local* average of $2\nu S_{ij} S_{ij}$ for each point $\mathbf{x}$ in the flow, that average being taken over a volume of size r, centred on $\mathbf{x}$. So let us introduce the stochastic variable

$$\varepsilon_{AV}(r, \mathbf{x}, t) = \frac{1}{V_r} \int_{V_r} (2\nu S_{ij} S_{ij}) dV,$$

where V_r is a spherical volume of radius r, centred on $\mathbf{x}$. A more careful statement of Kolmogorov's original theory is then

$$\langle [\Delta v]^p (r) \rangle = \beta_P \left\langle \varepsilon_{AV}^{p/3}(r, \mathbf{x}, t) \right\rangle r^{p/3}, \quad \eta \ll r \ll \ell, \tag{13.94}$$

which is known as *Kolmogorov's refined hypothesis*. As in his original theory, Kolmogorov took all the β_p in (13.94) to be universal (the same for all types of turbulent flow). Note that $\langle \varepsilon_{AV}(r) \rangle = \bar{\varepsilon}$, so the four-fifths law (*i.e.* $p = 3$) is unchanged by this refined theory. However, for $p \neq 3$ we have the possibility that the form of $\langle [\Delta v]^p \rangle$ departs from that predicted by the original theory.

In order to determine $\langle [\Delta v]^p \rangle$ we need to examine the statistics of $\varepsilon_{AV}(r, \mathbf{x}, t)$ and hence estimate $\left\langle \varepsilon_{AV}^{p/3}(r, \mathbf{x}, t) \right\rangle$. To that end, Kolmogorov proposed a simple statistical model for $\varepsilon_{AV}(r, \mathbf{x}, t)$, known as the log-normal model. This leads to a correction to (13.73) of the form

$$\langle [\Delta v]^p \rangle = C_p \bar{\varepsilon}^{p/3} r^{p/3} (r/\ell)^{\mu p(3-p)/18}, \tag{13.95}$$

(see Frisch, 1995) where μ is a free parameter known as the *intermittency exponent*. The parameter μ is taken to be a *universal* constant in this refined theory and is usually assigned a value in the range $0.2 < \mu < 0.3$ in order to be consistent with the experimental measurements. By way of contrast, Kolmogorov's modelling of $\varepsilon_{AV}(r, \mathbf{x}, t)$ results in C_p being *non-universal*.

The precise form of Kolmogorov's correction to K41 is often criticized, although the essential idea, that the intermittency of the energy flux in the inertial range necessitates a

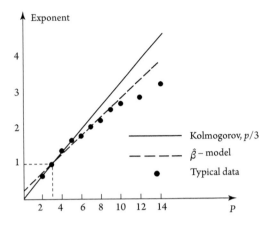

Figure 13.27 A comparison of K41 with the β model and with typical data.

correction to Kolmogorov's original theory, is now widely accepted. A popular alternative to the log-normal model is the so-called β-model. This provides an alternative means of estimating the statistical behaviour of $\varepsilon_{AV}(r, \mathbf{x}, t)$ and predicts

$$\langle [\Delta v]^p \rangle \sim \bar{\varepsilon}^{p/3} r^{p/3} \left(r/\ell \right)^{(3-p)(3-D)/3}, \tag{13.96}$$

(again, see Frisch, 1995) where the free parameter is now D, usually given a value of $D = 2.8$. Figure 13.27 shows the power-law exponent ς_p in $\langle [\Delta v]^p \rangle \sim r^{\varsigma_p}$ for the β-model (with $D = 2.8$) alongside typical experimental data. Note that, for $p = 2$, any supposed correction to the original two-thirds law is small and probably lies within the range of experimental uncertainty.

More generally, a comparison between the various intermittency models of the inertial range and the experimental measurements is fraught with difficulty, in part because reliable measurements of ς_p for large p are rare, while departures from K41 for low p are small and hard to detect. Moreover, the various intermittency models all assume that Re is asymptotically large, whereas the experimental measurements are usually made at modest values of Re, which creates its own systematic departures from K41, as illustrated by (13.72). It is probably fair to say that the entire subject is still a little uncertain.

13.5 The Probability Distribution of the Velocity Field

13.5.1 The Skewness and Flatness Factors

We close this chapter with a discussion of the probability distribution of $\mathbf{u}$. There are two points in particular we wish to make. First, we shall provide a statistical measure of intermittency, that is important for refined versions of K41. Second, we shall show that turbulence is fundamentally a *non-Gaussian* process.

Let us start with some simple definitions in statistics. The probability distribution of a random variable, say X, is represented by a *probability density function* (or pdf), which is defined as follows. The probability that X lies in the range $a \to b$, denoted $P(a < X < b)$, is

$$P(a < X < b) = \int_a^b f(x)dx,$$

where $f(x)$ is the pdf of the distribution. Thus $f(x)dx$ represents the *relative likelihood* that X lies in the range $x \to x + dx$. Evidently,

$$\int_{-\infty}^{\infty} f(x)dx = 1,$$

as the sum of the relative likelihoods must come to 1. The *mean* of a distribution is given by

$$\mu = \int_{-\infty}^{\infty} xf(x)dx = \langle X \rangle.$$

The *variance*, σ^2, measures the spread of the distribution about this mean, and is defined as

$$\sigma^2 = \int_{-\infty}^{\infty} (x - \mu)^2 f(x)dx.$$

In turbulence, we are primarily concerned with distributions with zero mean, in which case

$$\langle X \rangle = 0, \quad \sigma^2 = \int_{-\infty}^{\infty} x^2 f(x)dx = \langle X^2 \rangle.$$

The *skewness factor* for a distribution of zero mean is a normalized version of the third moment of f,

$$S = \int_{-\infty}^{\infty} x^3 f(x)dx \Big/ \sigma^3 = \langle X^3 \rangle \Big/ \langle X^2 \rangle^{3/2},$$

while the *flatness factor* is a normalized version of the fourth moment,

$$\delta = \int_{-\infty}^{\infty} x^4 f(x)dx \Big/ \sigma^4 = \langle X^4 \rangle \Big/ \langle X^2 \rangle^2.$$

A common distribution is the *Gaussian* or *normal* distribution which, when f has zero mean, takes the form

$$f(x) = \frac{1}{\sigma\sqrt{2\pi}} \exp\left[-x^2 \big/ (2\sigma^2)\right].$$

This has zero skewness, because of symmetry, and a flatness factor of $\delta = 3$. The Gaussian distribution is important since the *central limit theorem* tells us that any random variable that is the sum of many other independent random variables is approximately Gaussian.

Let us now consider grid turbulence. Suppose that, well downstream of a grid, we make many measurements of the transverse component of **u**, say u_x. We now plot the relative number of times that u_x attains a particular value, *i.e.* we plot the density function $f(x)$ for u_x. Provided the turbulence is fully developed, the probability density has a flatness factor of 2.9–3.0 and can be closely fitted by a Gaussian. One interpretation of this normal distribution for u_x is that the velocity at any one point is the consequence of a large number of randomly orientated vortical structures, the relationship between u_x and the surrounding vorticity being fixed by the Biot–Savart law (Figure 13.7). If these vortices are randomly distributed and there are many of them, the central limit theorem suggests that u_x should be almost Gaussian, and indeed it is.

The situation is different, however, if we examine the probability distribution of the velocity increment $\Delta v = u_x (\mathbf{x} + r\hat{\mathbf{e}}_x) - u_x(\mathbf{x})$, or the velocity gradient $\partial u_x / \partial x$. Here, we find that the probability density function is not at all Gaussian, and indeed this non-Gaussian behaviour is essential to the dynamics of turbulence. Consider, for example, the fourth-order structure function $\left\langle [\Delta v]^4 \right\rangle$ normalized by $\left\langle [\Delta v]^2 \right\rangle^2$, which is the flatness factor

$$\delta(r) = \left\langle [\Delta v]^4 \right\rangle \Big/ \left\langle [\Delta v]^2 \right\rangle^2. \tag{13.97}$$

This has the shape shown in Figure 13.28. It approaches the Gaussian value of 3.0 for large r, although crucially it is greater than 3 for small r. The fact that $\delta(r \to \infty) = 3$ indicates that remote points are (almost) statistically independent. That is, if u_x at $\mathbf{x}$ and $\mathbf{x} + r\hat{\mathbf{e}}_x$ are statistically independent, then expanding $\langle [\Delta v]^4 \rangle$ gives

$$\delta (r \to \infty) = \left\langle [\Delta v]^4 \right\rangle_\infty \Big/ \left\langle [\Delta v]^2 \right\rangle_\infty^2 \to \frac{3}{2} + \frac{1}{2} \left\langle u_x^4 \right\rangle \Big/ \left\langle u_x^2 \right\rangle^2 \tag{13.98}$$

(see Exercise 13.1) and we have already noted that the flatness factor for u_x is ~ 3.0. For $r \to 0$, on the other hand, we find that δ is a function of the Reynolds number $u \ell / \nu$. For the modest values of Re found in a wind tunnel we have $\delta(0) = \delta_0 \sim 4 \to 6$. However, it is observed that higher values of Re favour larger flatness factors, with $\delta_0 \sim 3 + \frac{1}{2} (u\ell/\nu)^{0.25}$ for $10^3 < \text{Re} < 10^7$.

The fact that δ_0 is large, relative to a normal distribution, tells us that the probability density function for Δv has a high central peak and relatively broad skirts, so that both near zero and unexpectedly large excursions from zero are common. As we shall see, this is consistent with a signal which is dormant much of the time and occasionally bursts into life, *i.e.* an intermittent signal.

The situation is similar for the third-order structure function. The skewness of Δv is

$$S(r) = \left\langle [\Delta v]^3 \right\rangle \Big/ \left\langle [\Delta v]^2 \right\rangle^{3/2} \tag{13.99}$$

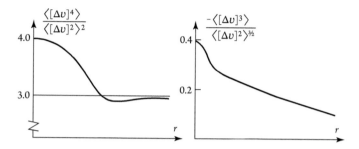

Figure 13.28 Schematic of the flatness and skewness factors for Δv in grid turbulence.

and, unlike $\langle u_x^3 \rangle / \langle u_x^2 \rangle^{3/2}$, this is non-zero. It has a value of around -0.4 for small r and decays slowly with r. The shape of $\langle [\Delta v]^3 \rangle / \langle [\Delta v]^2 \rangle^{3/2}$ is also shown in Figure 13.28. Note that, according to Kolmogorov's two-thirds and four-fifths laws, we have

$$S = \frac{\langle [\Delta v]^3 \rangle}{\langle [\Delta v]^2 \rangle^{3/2}} = -\frac{4}{5}\beta^{-3/2} \tag{13.100}$$

in the inertial range, where β is Kolmogorov's constant. Given that $\beta \approx 2$, this predicts that $S \approx -0.3$, in line with the measurements. Evidently Δv is necessarily *non-Gaussian*.

For small r, the probability distributions for Δv and $\partial u_x / \partial x$ become identical, and so

$$S(r \to 0) = S_0 = \frac{\langle [\Delta v]^3 \rangle_{r \to 0}}{\langle [\Delta v]^2 \rangle_{r \to 0}^{3/2}} = \frac{\langle (\partial u_x / \partial x)^3 \rangle}{\langle (\partial u_x / \partial x)^2 \rangle^{3/2}}. \tag{13.101}$$

It follows that $\partial u_x / \partial x$ is non-Gaussian, with a skewness factor of $S_0 \approx -0.4$ and a flatness factor of $\delta_0 \sim 4 \to 30$, depending on Re. The probability density function for $\partial u_x / \partial x$ is shown schematically in Figure 13.29. Small positive values of $\partial u_x / \partial x$ are more likely than small negative ones. However, large negative values of $\partial u_x / \partial x$ are more likely than large positive ones, and this dominates the skewness, giving a negative value of S.

13.5.2 The Flatness Factor as a Measure of Intermittency

Consider a probability distribution $f(x)$ which consists of a δ-function of area $1-\gamma$ at the origin surrounded by a Gaussian distribution of area γ:

$$f(x) = \frac{\gamma}{\sigma_* \sqrt{2\pi}} \exp\left[-x^2 / \left(2\sigma_*^2\right)\right], x \neq 0.$$

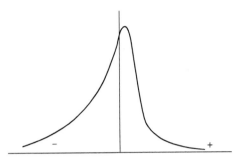

Figure 13.29 Schematic of the probability density function for $\partial u_x / \partial x$.

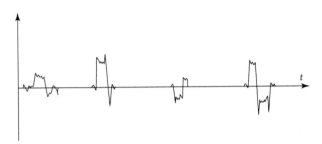

Figure 13.30 An example of an intermittent signal.

This represents an intermittent signal that is dormant for $(1—\gamma)$ percent of its time (see Figure 13.30). The variance of $f(x)$ is readily shown to be $\sigma^2 = \gamma\,\sigma_*^2$ while the flatness factor is $3/\gamma$, which is larger than that of a Gaussian. (Readers may wish to confirm this for themselves.) We therefore have

$$ f(x) = \frac{\gamma^{3/2}}{\sigma\sqrt{2\pi}} \exp\left[-\gamma x^2 / \left(2\sigma^2\right)\right], \quad x \neq 0. $$

Let us now compare this with a Gaussian distribution of the *same variance*. The three main differences are:

(i) $f(x)$ has a high central peak;

(ii) for large x, $f(x)$ is greater than the equivalent Gaussian;

(iii) $f(x)$ has a relatively large flatness factor.

This is typical of an intermittent signal in which both near-zero and unexpectedly high values are common. Their pdfs have a large flatness factor resulting from broad skirts and a high central peak.

The fact that the flatness factor for $\partial u_x / \partial x$ is not Gaussian, adopting increasingly large values as Re increases, tells us something interesting about the spatial structure of turbulence. The large values of δ_0 suggest that the spatial distribution of the velocity gradients (*e.g.* the vorticity) is somewhat spotty, with the regions of intense enstrophy localized in thin filaments, the filaments themselves being sparsely distributed throughout space. Moreover, it seems that the higher the value of Re, the sparser the vorticity distribution becomes.

We should not be surprised by the observed intermittency in vorticity. Consider the evolution of the vorticity field in grid turbulence. All of the vorticity originates from the boundary layers on the grid bars. This spills out into the flow in the form of Kármán vortices which then intermingle, leading to turbulence. The resulting chaotic velocity is a manifestation of the evolving vorticity field, with the instantaneous distribution of **u** determined, via the Biot–Savart law, by the instantaneous distribution of **ω**. So, as the vorticity is swept down the tunnel, it twists, turns, and stretches as a result of its self-induced velocity field. Moreover, this occurs at large Re, so that the vortex lines are more or less frozen into the fluid. A glance at a turbulent cloud of smoke rising from a lit cigarette (see Figure 12.1) is enough to suggest that a chaotic velocity field continually teases out any marker in the fluid (see Exercise 13.5), and this is also true of the vorticity. Thus filaments of vorticity are teased out into finer and finer structures, being continually stretched and folded. This process continues until the tubes are thin enough for diffusion to set in, *i.e.* we have reached the Kolmogorov microscale, η, leaving a sparse network of thin filaments (see Figure 13.6). Moreover, we would expect this intermittency to become more pronounced as Re is increased since η shrinks as Re grows, and so more stretching and folding is required to reach the Kolmogorov scale.

13.5.3 The Skewness Factor as a Measure of Enstrophy Production

We shall now show that the rate of generation of enstrophy in isotropic turbulence is determined by the skewness factor $S_0 = S(r \to 0)$. Our starting point is (13.42),

$$
u^2 f(r) = u^2 - \frac{\langle \omega^2 \rangle}{30} r^2 + \frac{\langle (\nabla \times \omega)^2 \rangle}{840} r^4 - \frac{\langle (\nabla^2 \omega)^2 \rangle}{45360} r^6 + \cdots,
$$

which we substitute into the Kármán–Howarth equation, (13.77). Since $K(r \to 0) \sim r^3$, we find

$$
10 \frac{d}{dt} \frac{1}{2} \langle \mathbf{u}^2 \rangle - \frac{d}{dt} \frac{1}{2} \langle \omega^2 \rangle r^2 + \cdots = 105 \left[u^3 K/r \right]_{r \to 0} - 10 \nu \langle \omega^2 \rangle
$$
$$
+ \nu \left\langle (\nabla \times \omega)^2 \right\rangle r^2 + \cdots,
$$

where, according to (13.31), $\left\langle [\Delta v]^3 \right\rangle = 6u^3 K(r)$. The terms of zero order now yield the energy equation

$$
\frac{d}{dt} \frac{1}{2} \langle \mathbf{u}^2 \rangle = -\nu \langle \omega^2 \rangle,
$$

whilst the terms of order r^2 give us

$$\frac{d}{dt}\frac{1}{2}\langle \boldsymbol{\omega}^2 \rangle = -\frac{35}{2}\left[\langle (\Delta v)^3 \rangle \big/ r^3\right]_{r\to 0} - \nu\left\langle (\nabla \times \boldsymbol{\omega})^2 \right\rangle. \qquad (13.102)$$

Compare this with the enstrophy equation obtained by averaging (2.52),

$$\frac{d}{dt}\frac{1}{2}\langle \boldsymbol{\omega}^2 \rangle = \langle \omega_i \omega_j S_{ij} \rangle - \nu\left\langle (\nabla \times \boldsymbol{\omega})^2 \right\rangle. \qquad (13.103)$$

(Note that, as shown in Exercise 13.2, the average of a divergence is zero in homogeneous turbulence, so the divergences in (2.52) vanish on averaging.) Evidently, the generation of enstrophy by vortex-line stretching is related to the third-order structure function at small r by

$$\left\langle [\Delta v]^3 \right\rangle_{r\to 0} = -\frac{2}{35}\langle \omega_i \omega_j S_{ij} \rangle\, r^3. \qquad (13.104)$$

However, since $\left\langle (\Delta v)^3 \right\rangle = S(r)\left\langle (\Delta v)^2 \right\rangle^{3/2}$, we have

$$\left\langle (\Delta v)^3 \right\rangle_{r\to 0} = S_0\left\langle (\Delta v)^2 \right\rangle_{r\to 0}^{3/2} = S_0\left\langle (\partial u_x/\partial x)^2 \right\rangle^{3/2} r^3. \qquad (13.105)$$

We may therefore rewrite (13.104) in terms of the skewness factor $S_0 = S(0)$. Replacing $\left\langle (\Delta v)^3 \right\rangle_{r\to 0}$ by (13.105), and noting that $\left\langle (\partial u_x/\partial x)^2 \right\rangle = \langle \boldsymbol{\omega}^2 \rangle/15$, we obtain,

$$\boxed{\langle \omega_i \omega_j S_{ij} \rangle = -\frac{7}{6\sqrt{15}}S_0 \langle \boldsymbol{\omega}^2 \rangle^{3/2}}. \qquad (13.106)$$

We conclude that the rate of generation of enstrophy is determined by S_0. Evidently, a negative skewness is essential to the production of enstrophy and in this sense turbulence is fundamentally a *non-Gaussian process*.

∘ ∘ ∘

This concludes our introduction to homogeneous turbulence. Readers may want to consult Tennekes and Lumley (1972) or Davidson (2015) for a more extensive introduction. For an advanced treatment of homogeneous turbulence see Batchelor (1953) or Frisch (1995). English translations of Kolmogorov's 1941 papers may be found in Kolmogorov (1991).

....

EXERCISES

13.1 *The flatness factor for the velocity increment at large r.* If a and b are statistically independent then $\langle ab \rangle = \langle a \rangle \langle b \rangle$. Show that, for $r \to \infty$, $\langle [\Delta v]^4 \rangle = 2\langle u_x^4 \rangle + 6\langle u_x^2 \rangle^2$ and so

$$\delta\,(r \to \infty) = \left\langle [\Delta v]^4 \right\rangle_\infty \Big/ \left\langle [\Delta v]^2 \right\rangle_\infty^2 = \tfrac{3}{2} + \tfrac{1}{2} \left\langle u_x^4 \right\rangle \Big/ \left\langle u_x^2 \right\rangle^2.$$

13.2 *The mean dissipation rate per unit mass in homogeneous turbulence.* Show that, since the operations of differentiation and taking averages commute, the average of a divergence is zero in homogeneous turbulence, $\langle \nabla \cdot (\sim) \rangle = 0$. Hence use (2.30) to show that the mean viscous dissipation rate in homogeneous turbulence can be written as $\varepsilon = \nu \left\langle \omega^2 \right\rangle$.

13.3 *Scale-by-scale energy distributions measured in real (rather than Fourier) space.* A real-space analogue of the energy spectrum, called the *signature function*, is proposed in Davidson (2015). It is defined as

$$V\,(r) = -\frac{r^2}{2} \frac{\partial}{\partial r} \frac{1}{r} \frac{\partial}{\partial r} \left[\frac{3}{4} \left\langle [\Delta v]^2 \right\rangle \right].$$

Show, through integration by parts, and with the help of (13.43), that

$$\frac{1}{2} \left\langle \mathbf{u}^2 \right\rangle = \int_0^\infty V\,(s)\,ds, \quad \frac{1}{2} \left\langle \omega^2 \right\rangle = \int_0^\infty \frac{10V\,(s)}{s^2}\,ds \approx \int_0^\infty (\pi/s)^2\,V\,(s)\,ds.$$

Also show that $V\,(r)$ satisfies

$$\frac{3}{4} \left\langle [\Delta v]^2 \right\rangle (r) = \int_0^r V\,(s)\,ds + \frac{r^2}{\pi^2} \int_r^\infty (\pi/s)^2\,V\,(s)\,ds,$$

and note the similarity to (13.64):

$$\frac{3}{4} \left\langle [\Delta v]^2 \right\rangle (r) \approx \int_{\pi/r}^\infty E\,(k)\,dk + \frac{r^2}{\pi^2} \int_0^{\pi/r} k^2 E\,(k)\,dk.$$

This suggests that $V\,(r)$ is indeed the real-space analogue of $E(k)$, with $k \approx \pi/r$. Next, use (13.55) to show that, for a random sea of Gaussian eddies of scale s,

$$V\,(r) = \left\langle \mathbf{u}^2 \right\rangle s^{-1}\,(r/s)^3 \exp\left[-(r/s)^2 \right].$$

This is sharply peaked at $r \approx 1.22s$ and might be compared with (13.56) for $E(k)$. It would seem that simple eddies of scale s contribute to $V\,(r)$ predominantly at $r \sim s$. Finally, use (13.62), together with the definition of $V\,(r)$, to show that $E(k)$ and $V\,(r)$ are related by the *Hankel transform*,

$$rV\,(r) = \frac{3\sqrt{\pi}}{2\sqrt{2}} \int_0^\infty E\,(k)\,\sqrt{rk}\,\ \mathrm{J}_{7/2}\,(rk)\,dk,$$

where $J_{7/2}$ is the usual Bessel function. Integrating this yields

$$\int_0^r V(r)dr = \int_0^\infty E(k)G(rk)dk,$$

where the shape of $G(x)$ is shown in Fig. 13.31(a). Since $G(x) > 0$, we have $\int_0^r V(r)dr > 0$. Moreover, it may be shown that, in *fully developed* turbulence, the stronger condition of $V(r) > 0$ holds (Davidson, 2015). Thus $V(r)$ has the three essential properties of an energy density:

(i) $V(r) > 0$;

(ii) $V(r)$ integrates to give the kinetic energy $\frac{1}{2}\langle \mathbf{u}^2 \rangle$;

(iii) $V(r)$ is sharply peaked at $r \sim s$ for simple eddies of scale s.

An approximation to $G(x)$ is $G(x) = 0$ for $x < \pi$ and $G(x) = 1$ for $x > \pi$. It follows that

$$\int_0^r V(r)dr \approx \int_{\pi/r}^\infty E(k)dk.$$

Show that this integral equation is satisfied provided $rV(r) = [kE(k)]_{k=\pi/r}$, a property of $V(r)$ that is illustrated in Figure 13.31 (b). This provides independent confirmation that $V(r)$ is an energy density, at least for the range of scales $\eta < r < \ell$. Note that $V(r)$ is a sharper diagnostic than the second-order structure function, as it does not mix information about small-scale energy and large-scale enstrophy.

13.4 *The integral invariants of Loitsyansky and Saffman in freely decaying, homogeneous turbulence.* Show that a low-k expansion of (13.51) yields, for isotropic turbulence,

$$E(k) = \frac{Lk^2}{4\pi^2} + \frac{Ik^4}{24\pi^2} + \cdots\cdots,$$

where

$$L = \int \langle \mathbf{u} \cdot \mathbf{u}' \rangle d\mathbf{r} = 4\pi u^2 \left[r^3 f \right]_{r\to\infty}, \qquad I = -\int r^2 \langle \mathbf{u} \cdot \mathbf{u}' \rangle \, d\mathbf{r}.$$

The integrals L and I are known as the *Saffman* and *Loitsyansky integrals*, respectively. For turbulence in which the eddies (vortices) possess very little *linear impulse*, the long-range vortex–vortex interactions are relatively weak, as shown by expansion (8.60). (See §8.7 for the definition of linear and angular impulse.) The long-range velocity correlations are then weak and it turns out that $[r^3 f]_\infty = 0$ (see Davidson 2015). In such cases we have $L = 0$ and hence $E(k \to 0) \sim Ik^4$. Show that the Kármán–Howarth equation then integrates to give

$$\frac{dI}{dt} = 8\pi \left[u^3 r^4 K \right]_{r\to\infty}.$$

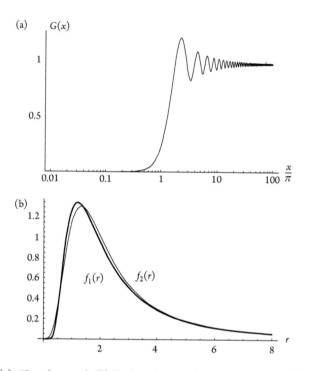

Figure 13.31 (a) The shape of $G(x)$ plotted as a function of x/π. (b) The functions $f_1(r) = [kE(k)]_{k=\pi/r}$ and $f_2(r) = rV$ corresponding to the energy spectrum $E = k^2 \exp(-k)$.

In such turbulence we find $K_\infty \leq O(r^{-4})$ and so $\left(r^4 K\right)_\infty$ can be finite, and indeed $\left(r^4 K\right)_\infty$ is often significant in an initial transient immediately following the creation of the turbulence. However, it is found to be negligible in *fully developed*, freely decaying turbulence, so I is an invariant in such cases. The turbulence then decays as indicated in Figure 13.32. Landau suggested that I is a statistical measure of how much *angular momentum*, or *angular impulse*, the eddies possess within some large control volume embedded within the turbulence:

$$I = \lim_{V \to \infty} \langle \mathbf{H}^2 \rangle / V, \quad \text{where} \quad H = \int_V (\mathbf{x} \times \mathbf{u}) dV.$$

He also suggested that conservation of angular momentum underlies the invariance of I in fully developed turbulence (see Landau and Lifshitz, 1959). However, the story is somewhat more complicated than that, as discussed in Davidson (2015).

When the eddies possess a significant amount of linear impulse, the far-field expansion (8.60) ensures that the long-range interactions are stronger, and for isotropic turbulence we find that $f_\infty = O(r^{-3})$ and $K_\infty \leq O(r^{-4})$. In such cases we have $E(k \to 0) \sim Lk^2$. Show that the Kármán–Howarth equation now integrates to give

$dL/dt = 0$, so Saffman's integral is an invariant of freely decaying turbulence. Saffman showed that L is a measure of how much *linear impulse* the eddies possess:

$$L = \lim_{V \to \infty} \left\langle \left[\int \mathbf{u} dV \right]^2 \right\rangle \Big/ V,$$

where V is the volume of some large control volume embedded within the turbulence. At first sight it may be thought that, through the appropriate choice of the frame of reference, we could set the linear momentum in V, and hence L, to zero. However, the choice of frame of reference is set by the requirement that $\langle \mathbf{u} \rangle = 0$, or

$$\lim_{V \to \infty} \frac{1}{V} \int_V \mathbf{u} dV = 0,$$

and the central limit theorem then tells us that, in that frame, L is in general non-zero. (Again, see the discussion in Davidson, 2015). Once again, the turbulence decays as shown in Figure 13.32, with the left-hand side of the spectrum fixed throughout the decay.

Both L and I are dominated by the integral scales and so are of the order of $L \sim u^2 \ell^3$ and $I \sim u^2 \ell^5$. Combine $u^2 \ell^3 = $ constant with (13.8) to show that $u^2 \sim t^{-6/5}$ in $E(k \to 0) \sim Lk^2$ turbulence. This is known as *Saffman's decay law* and it seems that grid turbulence is of this type (see Krogstad and Davidson, 2010). Now combine $u^2 \ell^5 = $ constant with (13.8) to show that $u^2 \sim t^{-10/7}$ in $E(k \to 0) \sim Ik^4$ turbulence. This is called *Kolmogorov's decay law*, following Kolmogorov's prediction of such a decay law in 1941.

13.5 *Richardson's law for two-particle dispersion.* In 1926 Richardson asked the question: 'how rapidly, on average, will a pair of initially adjacent particles separate in isotropic turbulence?'. This is tantamount to asking how rapidly a small cloud (or puff) of contaminant grows in size, as indicated in Figure 13.33.

Let $R(t)$ be the mean radius of the cloud at time t. We are interested in clouds for which $\eta \ll R \ll \ell$, where ℓ is the integral scale. Richardson suggested that eddies of

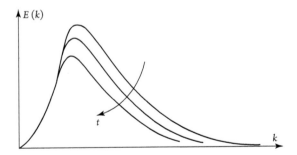

Figure 13.32 The decay of $E(k)$ in fully developed, isotropic turbulence.

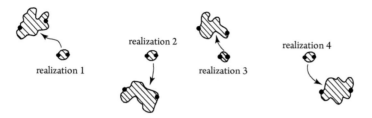

Figure 13.33 Spreading of a small cloud of contaminant in several realizations.

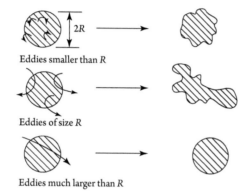

Eddies smaller than R

Eddies of size R

Eddies much larger than R

Figure 13.34 The effect of eddy size on the spreading of a small cloud of contaminant.

size $\sim R$ are the most important for the growth of the puff, since much smaller eddies merely ripple the surface of the cloud, while larger ones advect it without a significant change in shape (see Figure 13.34). Now, we have already suggested in our derivation of Kolmogorov's law that eddies of size s tend to pass their energy onto smaller scales on a timescale of their turn-over time, and so $\Pi \sim v_s^3/s$, or, equivalently, $v_s \sim (\varepsilon s)^{1/3}$. Consider a cloud that has grown sufficiently to have forgotten the conditions of its initial release, yet whose mean radius is still significantly smaller than the integral scale. Then, according to Richardson, we expect that the average rate of change of R should be independent of v, u, and ℓ and depend only on v_R and t: $dR/dt = f(v_R, t)$. Show that the only dimensionally consistent possibility is

$$\frac{dR}{dt} \sim v_R \sim (\varepsilon R)^{1/3},$$

and hence

$$\frac{dR^2}{dt} \sim \varepsilon^{1/3} R^{4/3}, \quad \eta \ll R \ll \ell.$$

This is known as *Richardson's four-thirds law*, and it was first established empirically.

....

REFERENCES

Batchelor, G.K., 1953, *The theory of homogeneous turbulence,* Cambridge University Press.

Davidson, P.A., 2015, *Turbulence: an introduction for scientists and engineers,* 2nd Ed., Oxford University Press.

Frisch, U., 1995, *Turbulence.* Cambridge University Press.

Gödecke, K., 1935, Messungen der atmospharischen Turbulenz in Bodennähe mit einer Hitzdraht-methode, *Ann. Hydrogr.,* **10**, 400–10.

Kolmogorov, A.N., 1991, The local structure of turbulence in incompressible viscous fluid for very large Reynolds numbers. *and* Dissipation of energy in the locally isotropic turbulence. *Proc. Roy. Soc.* A, **434**, pp 9–13, 15–17. (These are English translations of Kolmogorov's 1941 papers.)

Krogstad, P-A,, & Davidson P.A., 2010, Is grid turbulence Saffman turbulence? *J. Fluid Mech.,* **642**, 373–94.

Landau, L.D. & Lifshitz, E.M., 1959, *Fluid mechanics,* 1st Ed., Pergamon Press.

Leung T., Swaminathan N., & Davidson, P.A., 2012, Geometry and interaction of structures in homogeneous isotropic turbulence. *J. Fluid Mech.,* **710**, 453–81.

Monin, A.S. & Yaglom, A.M., 1975, *Statistical fluid mechanics II.* MIT Press.

Tennekes, H. & Lumley, J.L., 1972, *A first course in turbulence.* MIT Press.

14

· · **·** · ·

Turbulent Shear Flows and Simple Closure Models

(a)   (b)

(a) The British physicist G.I. Taylor (1886–1975) is probably best known in the field of turbulence for introducing a statistical approach to the subject, in which two-point correlations are used to characterize the turbulence. (b) The Chinese physicist Pei-Yuan Zhou (1902–1993), pictured here with his wife, led the development of turbulence theory in China.

We now turn to turbulent shear flows, by which we mean flows in which the time-averaged velocity is dominated by a single component, such as in a boundary layer, pipe flow, or a submerged jet (Figure 14.1). The discussion is based, at least in part, on that found in Davidson (2015). We shall restrict ourselves to *statistically steady* flows and use time averages, $\overline{(\sim)}$, rather than ensemble averages. As in §13.1, we take $\bar{\mathbf{u}}$ and $\mathbf{u}'$ to represent the

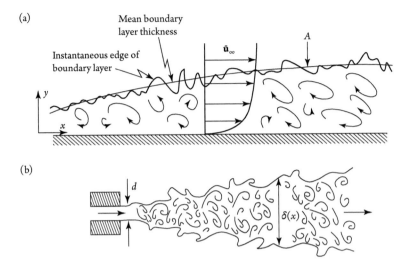

Figure 14.1 Two examples of turbulent shear flows: (a) boundary layer, (b) submerged jet.

mean and fluctuating components of **u**, where the time-average of the continuity equation tells us that $\bar{\mathbf{u}}$ and $\mathbf{u}'$ are individually solenoidal, *i.e.* $\nabla \cdot \bar{\mathbf{u}} = 0, \nabla \cdot \mathbf{u}' = 0$.

One of the striking features of *external* shear flows, such as a boundary layers or jets, is the sharp interface between the turbulent and non-turbulent motion; that is, between the chaotic vorticity field that constitutes the turbulence and the surrounding irrotational flow. This vorticity is generated at a solid surface, such as the inside surface of the nozzle in the case of a jet, and is then swept downstream. The sharp interface between the turbulent and non-turbulent motion reflects the fact that, at high Re, vorticity is (almost) frozen into the fluid. Thus, at location A in Figure 14.1 (a), the turbulence is intermittent.

14.1 Reynolds Stresses, Energy Budgets, and the Concept of Eddy Viscosity

14.1.1 Reynolds Stresses and the Closure Problem (Reprise)

We introduced the Reynolds stress $\tau_{ij}^R = -\rho \overline{u_i' u_j'}$ in §13.1.1. The significance of this quantity becomes clear when we time-average the Navier–Stokes equation, *i.e.* we average

$$\rho \frac{\partial u_i'}{\partial t} + \rho(\bar{\mathbf{u}} + \mathbf{u}') \cdot \nabla (\bar{u}_i + u_i') = \frac{\partial \tau_{ij}}{\partial x_j}, \tag{14.1}$$

where $\tau_{ij} = -p\delta_{ij} + 2\rho\nu S_{ij}$. This yields the *Reynolds-averaged equation of motion,*

$$\rho(\bar{\mathbf{u}} \cdot \nabla) \bar{u}_i = \frac{\partial}{\partial x_j} \left[\bar{\tau}_{ij} + \tau_{ij}^R \right], \tag{14.2}$$

or equivalently,

$$\rho\left(\bar{\mathbf{u}}\cdot\nabla\right)\bar{u}_i = -\frac{\partial\bar{p}}{\partial x_i} + \frac{\partial}{\partial x_j}\left[2\rho\nu\bar{S}_{ij} - \rho\overline{u_i'u_j'}\right].$$

(14.3)

Crucially, the non-linear term in (14.1) has given rise to a Reynolds stress in (14.3), which supplements the expected terms involving $\bar{\mathbf{u}}$ to $\bar{p}$. This gradient in Reynolds stress couples the mean flow to the turbulence, so that the turbulence helps shape the mean flow.

Note that the corresponding energy equation, obtained from the product of (14.3) with $\bar{\mathbf{u}}$, is

$$\bar{\mathbf{u}}\cdot\nabla\left(\tfrac{1}{2}\rho\bar{\mathbf{u}}^2\right) = -\frac{\partial}{\partial x_i}\left[\bar{p}\bar{u}_i\right] + \bar{u}_i\frac{\partial}{\partial x_j}\left[2\rho\nu\bar{S}_{ij} + \tau_{ij}^R\right].$$

Since $\bar{S}_{ij}$ and τ_{ij}^R are symmetric, this can be conveniently rewritten as

$$\bar{\mathbf{u}}\cdot\nabla\left(\tfrac{1}{2}\rho\bar{\mathbf{u}}^2\right) = -\tau_{ij}^R\bar{S}_{ij} - 2\rho\nu\bar{S}_{ij}\bar{S}_{ij} + \frac{\partial}{\partial x_j}\left[\bar{u}_i\tau_{ij}^R - \bar{p}\bar{u}_j + 2\rho\nu\bar{u}_i\bar{S}_{ij}\right].$$

(14.4)

(mean flow energy equation)

We recognize the second term on the right of (14.4) as the mean flow dissipation, while the divergence on the far right transports mean flow kinetic energy from place to place, but integrates to zero in a closed domain. Conservation of energy then tells us that the term $\tau_{ij}^R\bar{S}_{ij}$ must represent an exchange of energy between the mean flow and the turbulence.

The physical origin of the Reynolds stresses becomes clearer if we rewrite (14.1) in integral form

$$\frac{d}{dt}\int_V\left(\rho u_i\right)dV + \oint_S\left(\rho u_i\right)\mathbf{u}\cdot d\mathbf{S} = \oint_S\tau_{ij}dS_j,$$

whose time average is

$$\oint_S\left(\rho\bar{u}_i\right)\bar{\mathbf{u}}\cdot d\mathbf{S} = \oint_S\left(2\rho\nu\bar{S}_{ij} - \rho\overline{u_i'u_j'}\right)dS_j - \oint_S\bar{p}dS_i.$$

(14.5)

The new terms, $-\rho\overline{u_i'u_j'}$, which appear to act like stresses, really represent the flux of momentum in or out of the volume, V, caused by the turbulent fluctuations in velocity.

Returning now to (14.3), it is clear that if we are to predict the behaviour of the mean flow we need to know the Reynolds stress distribution. We can get an evolution equation for τ_{ij}^R as follows. First, we subtract (14.2) from (14.1) to give

$$\frac{\partial u_i'}{\partial t} + \bar{u}_k\frac{\partial u_i'}{\partial x_k} + u_k'\frac{\partial\bar{u}_i}{\partial x_k} + u_k'\frac{\partial u_i'}{\partial x_k} = \frac{1}{\rho}\frac{\partial}{\partial x_k}\left(\tau_{ik}' - \tau_{ik}^R\right),$$

(14.6)

with a similar equation for u'_j,

$$\frac{\partial u'_j}{\partial t} + \bar{u}_k \frac{\partial u'_j}{\partial x_k} + u'_k \frac{\partial \bar{u}_j}{\partial x_k} + u'_k \frac{\partial u'_j}{\partial x_k} = \frac{1}{\rho} \frac{\partial}{\partial x_k} \left(\tau'_{jk} - \tau^R_{jk} \right). \tag{14.7}$$

Next, we multiply (14.6) by u'_j and (14.7) by u'_i, add the resulting equations and average. This yields

$$\bar{u}_k \frac{\partial}{\partial x_k} \left(\overline{u'_i u'_j} \right) + \overline{u'_i u'_k} \frac{\partial \bar{u}_j}{\partial x_k} + \overline{u'_j u'_k} \frac{\partial \bar{u}_i}{\partial x_k} + \frac{\partial}{\partial x_k} \left(\overline{u'_i u'_j u'_k} \right) = \frac{\overline{u'_i \frac{\partial \tau'_{jk}}{\partial x_k}}}{\rho} + \frac{\overline{u'_j \frac{\partial \tau'_{ik}}{\partial x_k}}}{\rho}. \tag{14.8}$$

Finally, we substitute for τ'_{ij} using $\tau'_{ij} = -p'\delta_{ij} + 2\rho\nu S'_{ij}$, which gives us

$$u'_i \frac{\partial \tau'_{jk}}{\partial x_k} + u'_j \frac{\partial \tau'_{ik}}{\partial x_k} = -u'_i \frac{\partial p'}{\partial x_j} - u'_j \frac{\partial p'}{\partial x_i} + 2\rho\nu \left(u'_i \frac{\partial S'_{jk}}{\partial x_k} + u'_j \frac{\partial S'_{ik}}{\partial x_k} \right), \tag{14.9}$$

where the terms involving S'_{ij} can be rewritten as

$$u'_i \frac{\partial S'_{jk}}{\partial x_k} + u'_j \frac{\partial S'_{ik}}{\partial x_k} = \frac{\partial}{\partial x_k} \left(u'_i S'_{jk} + u'_j S'_{ik} \right) - S'_{jk} \frac{\partial u'_i}{\partial x_k} - S'_{ik} \frac{\partial u'_j}{\partial x_k}.$$

Combining (14.8) with (14.9) yields an evolution equation for the Reynolds stresses:

$$\boxed{\begin{aligned} \bar{\mathbf{u}} \cdot \nabla \left(\rho \overline{u'_i u'_j} \right) &= \tau^R_{ik} \frac{\partial \bar{u}_j}{\partial x_k} + \tau^R_{jk} \frac{\partial \bar{u}_i}{\partial x_k} - \frac{\partial}{\partial x_k} \left[\rho \overline{u'_i u'_j u'_k} \right] - \overline{u'_i \frac{\partial p'}{\partial x_j}} - \overline{u'_j \frac{\partial p'}{\partial x_i}} \\ &+ 2\rho\nu \left(\frac{\partial}{\partial x_k} \left(\overline{u'_i S'_{jk} + u'_j S'_{ik}} \right) - \overline{S'_{jk} \frac{\partial u'_i}{\partial x_k}} - \overline{S'_{ik} \frac{\partial u'_j}{\partial x_k}} \right) \end{aligned}} \tag{14.10}$$

Perhaps the main point to note about (14.10) is that our evolution equation for τ^R_{ij} involves new quantities of the form $\overline{u'_i u'_j u'_k}$. Moreover, the evolution equation for these triple correlations introduces yet another set of unknowns, $\overline{u'_i u'_j u'_k u'_m}$, and so it goes on. We have hit the closure problem of turbulence. It seems that, if we are to make definite predictions about turbulent shear flows, we are obliged to introduce additional information in the form of empirical hypotheses, which is known as a *turbulence closure model*.

Note that, by setting $i = j$ in (14.10), we obtain an energy equation for the turbulence

$$\boxed{\bar{\mathbf{u}} \cdot \nabla \left[\frac{1}{2} \overline{\rho(\mathbf{u}')^2} \right] = \tau^R_{ik} \bar{S}_{ik} - 2\rho\nu \overline{S'_{ik} S'_{ik}} + \nabla \cdot \left[-\frac{1}{2} \overline{\rho(\mathbf{u}')^2 \mathbf{u}'} - \overline{p'\mathbf{u}'} + 2\rho\nu \overline{u'_i S'_{ik}} \right].} \tag{14.11}$$

(energy equation for the turbulence)

As in (14.4), the divergence on the far right transports kinetic energy from place to place, but integrates to zero in a closed domain, while we recognize the second term on the right as the turbulent dissipation. Crucially, $\tau_{ik}^R \bar{S}_{ik}$ appears with opposite signs in (14.4) and (14.11), confirming that it represents an exchange of energy between the mean flow and the turbulence. We shall return to these energy equations shortly.

14.1.2 The Eddy Viscosity Model of Boussinesq, Taylor, and Prandtl

The first turbulence closure model dates back to Boussinesq, who proposed that, for simple shear flows of the type shown in Figure 14.1,

$$\bar{\tau}_{xy} + \tau_{xy}^R = \rho\left(\nu + \nu_t\right)\frac{\partial \bar{u}_x}{\partial y}, \tag{14.12}$$

where ν_t is called an *eddy viscosity*. The idea is that the effect of the turbulent mixing of momentum, which gives rise to the Reynolds stress, is analogous to the molecular mixing of momentum, which lies behind the laminar shear stress. This suggests that the role of turbulence is simply to increase the effective viscosity from ν to $\nu + \nu_t$. This notion of an eddy viscosity is sometimes used for flows that are more complex than a simple shear flow. In such cases, the three-dimensional generalization of (14.12) is

$$\tau_{ij}^R = -\rho\overline{u_i' u_j'} = \rho\nu_t \left[\frac{\partial \bar{u}_i}{\partial x_j} + \frac{\partial \bar{u}_j}{\partial x_i}\right] - \frac{\rho}{3}\overline{u_k' u_k'}\delta_{ij}. \tag{14.13}$$

Many authors refer to (14.13) as *Boussinesq's equation*. Note that the additional term on the right is necessary to ensure that the sum of the normal stresses adds up to $-\rho\overline{(\mathbf{u}')}^2$.

To make progress, we must now determine ν_t. Prandtl was probably the first to set out a systematic means of estimating ν_t, based in part on G.I. Taylor's notion of a *mixing length*. Both Taylor and Prandtl noted the success of the kinetic theory of gases in predicting the property of viscosity. This theory predicts that $\nu \sim lV$, where l is the mean-free-path length of the molecules and V their characteristic speed. Moreover, there is an analogy between Newton's law of viscosity and the Reynolds stress. In a laminar flow, the relative sliding of layers of fluid gives rise to a shear stress because molecules bounce around between the layers, exchanging momentum. This is illustrated in Figure 14.2 (a). A similar sort of thing happens in a turbulent shear flow, in which fluid elements are jostled by the turbulence. Thus we can equally interpret Figure 14.2 (a) as fluid elements being thrown from one layer to another, carrying some of their momentum with them. For example, a fluid element at A may move up to B, slowing down the faster moving fluid above. Conversely, an element moving from C to D will speed up the slower fluid below. This momentum exchange results in a Reynolds stress. The (tentative) analogy between these macroscopic and molecular processes led Prandtl to suggest

$$\nu_t = \ell_m V_t, \tag{14.14}$$

where ℓ_m is called the mixing length and V_t is a suitable measure of u'.

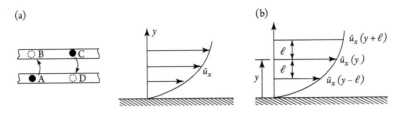

Figure 14.2 (a) Molecules or fluid elements exchanging momentum in a simple shear flow creates a time-averaged shear stress between layers. (b) Prandtl's mixing length.

In a sense, (14.14) just transfers the problem from one of finding ν_t to one of determining ℓ_m and V_t. However, Prandtl's *mixing-length theory* provides a means of estimating these quantities. Consider the mean flow $\bar{u}_x(y)$ shown in Figure 14.2 (b). The fluid at y will, on average, have come from levels $y \pm \ell$, where ℓ is some measure of the size of the large eddies. Suppose that a fluid element moves from $y + \ell$ to y and that it retains a significant fraction of its linear momentum in the process. Then it arrives at y with a velocity of order $\bar{u}_x(y + \ell)$. If ℓ is small by comparison with the characteristic transverse scale of the mean flow, then the spread of velocities at y will be of the order of $\bar{u}_x \pm \ell \partial \bar{u}_x / \partial y$, and so we have $|u'_x| \sim \ell |\partial \bar{u}_x / \partial y|$. Next, we note that, when $\partial \bar{u}_x / \partial y > 0$, there is a negative correlation between u'_x and u'_y. That is, a positive u'_x is consistent with fluid coming from $y + \ell$, which requires a negative u'_y. Similarly, u'_x and u'_y are likely to have a positive correlation when $\partial \bar{u}_x / \partial y < 0$. So, provided $|u'_x| \sim |u'_y|$, we have

$$\overline{u'_x u'_y} \sim -\ell^2 \left| \frac{\partial \bar{u}_x}{\partial y} \right| \frac{\partial \bar{u}_x}{\partial y}. \tag{14.15}$$

We now absorb the unknown constant in (14.15) into the definition of ℓ_m and we find that, for this one-dimensional shear flow,

$$\tau_{xy}^R = -\rho \overline{u'_x u'_y} = \rho \ell_m^2 \left| \frac{\partial \bar{u}_x}{\partial y} \right| \frac{\partial \bar{u}_x}{\partial y}. \tag{14.16}$$

Comparing this with Boussinesq's equation, (14.12), we conclude that,

$$\boxed{\nu_t = \ell_m^2 \left| \frac{\partial \bar{u}_x}{\partial y} \right|.} \tag{14.17}$$

This is Prandtl's mixing-length model, which relies on ℓ_m being determined by experiment.

Although a useful first approximation, this model is often heavily criticized. For example, the use of a Taylor expansion is not justified in estimating $\bar{u}_x(y + \ell) - \bar{u}_x(y)$, as the size of the large eddies is comparable with the transverse scale of the mean velocity. Moreover, there is no justification for assuming that the fluid elements retain their linear momentum as they move between layers. Nevertheless, Prandtl's model can provide reasonable estimates

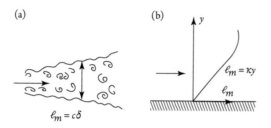

Figure 14.3 Mixing length. (a) A jet, where $\ell_m = c\delta$, δ being the jet width and c a constant of order unity. (b) A boundary layer, where $\ell_m = \kappa y$, y being the distance from the wall.

for *simple one-dimensional shear flows,* provided that ℓ_m is chosen appropriately. For jets and wakes it is found that ℓ_m is of the order of δ, the *local* thickness of the jet or wake (Figure 14.3). In boundary layers, on the other hand, one finds $\ell_m = \kappa y$ near a wall, where $\kappa \approx 0.4$ is known as Kármán's constant. Perhaps the main defect of (14.17) for simple shear flows is that it gives the anomalous result that $\nu_t = 0$ on the centreline of a jet. However, as we shall see, this is can be remedied by returning to (14.14).

It should be emphasized, however, that if we try to use (14.13) and (14.14) to extend Prandtl's ideas to more complex flows, significant problems arise, as discussed in §14.6.

14.2 The Transfer of Energy from the Mean Flow to the Turbulence

Let us return to the mean-flow energy equation, (14.4), integrated over an open control volume V fixed in space and with bounding surface S:

$$\oint_S \left(\tfrac{1}{2}\rho\bar{\mathbf{u}}^2\right)\bar{\mathbf{u}} \cdot d\mathbf{S} =$$

$$(1)$$

$$\oint_S \left(2\rho\nu\bar{S}_{ij} + \tau_{ij}^R\right)\bar{u}_i dS_j - \oint_S \bar{p}\bar{\mathbf{u}} \cdot d\mathbf{S} - \int_V 2\rho\nu\bar{S}_{ij}\bar{S}_{ij}dV - \int_V \tau_{ij}^R\bar{S}_{ij}dV \qquad (14.18)$$

$$(2) \qquad\qquad\qquad (3) \qquad\qquad\qquad (4)$$

We interpret the various terms in (14.18) as:

(1) the convection of mean flow kinetic energy across the bounding surface S;

(2) the rate of working of the mean viscous and Reynolds stresses on the boundary S;

(3) the rate of working of the mean pressure forces on the boundary S;

(4) the rate of dissipation of energy associated with the mean flow.

In a steady-on-average turbulent flow, the mean flow kinetic energy within V does not change with time. Consequently, as noted earlier, conservation of energy suggests that the final term on the right must represent the rate of loss of kinetic energy to the turbulence. Similarly, integrating (14.11) over V yields

$$
\underbrace{\oint_S \left(\tfrac{1}{2}\rho\overline{(\mathbf{u}')^2} \right) \bar{\mathbf{u}} \cdot d\mathbf{S}}_{(1)} = \underbrace{\oint_S \left(2\rho\nu\overline{u_i' S_{ij}'} - \tfrac{1}{2}\rho\overline{(\mathbf{u}')^2 u_j'} \right) dS_j}_{(2)} - \underbrace{\oint_S \overline{p'\mathbf{u}'} \cdot d\mathbf{S}}_{(3)}
$$

$$
\underbrace{- \int_V 2\rho\nu\overline{S_{ij}' S_{ij}'}\, dV}_{(4)} + \int_V \tau_{ij}^R \bar{S}_{ij}\, dV
$$

(14.19)

Again, we interpret the various terms in (14.19) as:

(1) the mean convection of turbulent kinetic energy across the bounding surface S;

(2) the rate of working of the fluctuating viscous stresses acting on S and the turbulent flux of turbulent kinetic energy across S;

(3) the rate of working of the turbulent pressure forces on the boundary S;

(4) the rate of dissipation of turbulent kinetic energy.

As before, conservation of energy suggests that the final term on the right must represent the rate of transfer of kinetic energy from the mean flow to the turbulence. This tells us that, in order to keep the turbulence alive, we require a finite mean shear, $\bar{S}_{ij}$. In the absence of such a shear (or of a body force, such as buoyancy), the turbulence simply dies. Moreover, in order to counter the viscous dissipation of turbulent energy, the turbulent fluctuations must be such that, on average, we have $\tau_{ij}^R \bar{S}_{ij} > 0$.

However, perhaps it is important to bear in mind that this is a kind of artificial book-keeping, brought about by our insistence on dividing the flow into its mean and fluctuating components. In reality, there is only one fluid and one flow. Nevertheless, it is true that, in the absence of body forces, the turbulence will die unless there is a mean shear, $\bar{S}_{ij}$. It is also true that the turbulence arranges itself such that, on average, $\tau_{ij}^R \bar{S}_{ij} > 0$. The need for a finite mean strain, and the mechanism of energy transfer from the mean flow to the turbulence, can be pictured as shown in Figure 14.4. We might visualize the turbulent vorticity field as a seething tangle of vortex tubes, rather like a pot of boiling spaghetti. In the presence of a mean shear these tubes will be systematically elongated along the direction of maximum positive strain, which is $45°$ to the horizontal for a uniform mean shear. As the vortex tubes are stretched, their kinetic energy rises and this represents an exchange of energy from the mean flow to the turbulence.

It is natural to ask how the flow ensures that $\tau_{ij}^R \bar{S}_{ij} > 0$. Consider a simple shear flow of the form $\bar{u}_x(y)$, say part of a boundary layer, with $d\bar{u}_x/dy > 0$. Then we have

$$
\tau_{ij}^R \bar{S}_{ij} = -\rho\overline{u_x' u_y'} \frac{\partial \bar{u}_x}{\partial y}.
$$

(14.20)

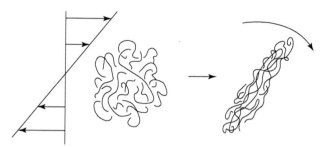

Figure 14.4 A mean shear teases out the vortex tubes in the turbulence. The kinetic energy of the turbulence rises as the tubes are stretched.

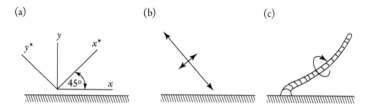

Figure 14.5 (a) The coordinate system x^*, y^*. (b) The velocity fluctuations needed for a positive Reynolds stress. (c) Side view of a vortex tube stretched by the mean shear.

Now consider a coordinate system $(x*, y*)$ which is inclined at $45°$ to (x, y), as shown in Figure 14.5 (a). Our Reynolds stress can be expressed in terms of these new coordinates as

$$\tau_{xy}^R = \frac{\rho}{2} \left\{ \overline{(u'_{y*})^2} - \overline{(u'_{x*})^2} \right\}.$$
(14.21)

So, a positive Reynolds stress is associated with large fluctuations in the y^* direction and weak fluctuations in the x^* direction (Figure 14.5 (b)). However, this is precisely what is achieved by stretching a vortex tube along the direction of maximum positive mean strain, as shown in Figure 14.5 (c).

So we conclude that $\tau_{ij}^R \bar{S}_{ij}/\rho$ represents the rate at which energy enters the turbulence (per unit mass) and passes down the energy cascade. We shall denote this by G, for *generation*. Also, we recognize the fourth term in (14.19) as the rate of dissipation of turbulent energy by the fluctuating viscous stresses. As usual, we denote this by ε (per unit mass), and so

$$G = -\overline{u'_i u'_j} \bar{S}_{ij}, \quad \varepsilon = 2\nu \overline{S'_{ij} S'_{ij}}.$$
(14.22)

If we introduce one last label,

$$T_i = \frac{1}{2} \overline{(\mathbf{u}')^2 u'_i} + \frac{1}{\rho} \overline{p' u'_i} - 2\nu \overline{u'_j S'_{ij}},$$
(14.23)

where **T** stands for *transport*, our turbulent kinetic energy equation, (14.11), becomes

$$\bar{\mathbf{u}} \cdot \nabla \left[\tfrac{1}{2} \overline{\mathbf{u}'^2} \right] = G - \varepsilon - \nabla \cdot \mathbf{T}. \tag{14.24}$$

When the turbulence is statistically homogeneous, which is admittedly rare for a turbulent shear flow, the divergence of any statistically averaged quantity vanishes and (14.24) reduces to the statement that $G = \varepsilon$. That is to say, the local rate of generation of turbulent energy is equal to the rate of viscous dissipation. Moreover, in Chapter 13 we introduced the symbol Π to represent the flux of energy down the turbulent energy cascade, from the large to the small scales. We also argued that $\Pi \sim u^3/\ell$, where u and ℓ are the integral scales. In statistically steady turbulence Π is the same at all points in the cascade and so $\varepsilon = \Pi$. It follows that, for the rare case of statistically steady, homogeneous turbulence we have

$$G = \varepsilon = \Pi \sim u^3/\ell. \tag{14.25}$$

(homogeneous, steady-on-average shear flow)

However, even when a shear flow is inhomogeneous, which is nearly always the case, G and ε are usually of the same order of magnitude, in which case

$$\boxed{G \sim \varepsilon = \Pi \sim u^3/\ell}. \tag{14.26}$$

(inhomogeneous, steady-on-average shear flow)

14.3 Turbulent Jets

14.3.1 The Plane Jet

Perhaps one of the simpler turbulent shear flows to characterize is a submerged jet, so let us consider such jets before moving onto the more difficult topic of boundary layers. We start with a *planar jet*, which can be created by pushing fluid out through a slot, as shown in Figure 14.6. The mean velocity is then two-dimensional, and we take x and y to represent the stream-wise and transverse coordinates respectively, with $\bar{u}_0(x)$ and $\delta(x)$ being the local mean centreline velocity and local mean jet width. The mean flow is then characterized by $\bar{u}_x \gg \bar{u}_y$ and $\partial/\partial x \ll \partial/\partial y$. It follows that:

(i) axial gradients in the Reynolds stresses are much smaller than transverse gradients;

(ii) the transverse component of the mean inertia $(\bar{\mathbf{u}} \cdot \nabla) \bar{u}_y$, which is of the order of $\bar{u}^2 \times$ (curvature of the mean streamlines), is very small.

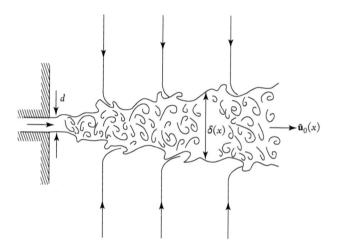

Figure 14.6 Schematic of planar jet.

Given these simplifications, and neglecting the laminar stresses in (14.2), the axial and transverse equations of motion become

$$\rho\left(\bar{\mathbf{u}} \cdot \nabla\right)\bar{u}_x = \frac{\partial \tau_{xy}^R}{\partial y} - \frac{\partial \bar{p}}{\partial x}, \tag{14.27}$$

$$0 = \frac{\partial \tau_{yy}^R}{\partial y} - \frac{\partial \bar{p}}{\partial y}, \tag{14.28}$$

the second of which integrates to give

$$\bar{p} + \rho\overline{\left(u_y'\right)^2} = \bar{p}_\infty = \text{constant},$$

where $\bar{p}_\infty$ is the pressure in the far field. Evidently $\bar{p}$ is a function of x only to the extent that $\overline{\left(u_y'\right)^2}$ depends on x. However, the longitudinal gradients in the Reynolds stresses may be neglected, and so (14.27) simplifies to

$$\rho\left(\bar{\mathbf{u}} \cdot \nabla\right)\bar{u}_x = \frac{\partial \tau_{xy}^R}{\partial y}, \tag{14.29}$$

where $\nabla \cdot \bar{\mathbf{u}} = 0$. It is convenient to rewrite this as,

$$\frac{\partial}{\partial x}\left[\rho\bar{u}_x^2\right] + \frac{\partial}{\partial y}\left[\rho\bar{u}_y\bar{u}_x\right] = \frac{\partial \tau_{xy}^R}{\partial y}. \tag{14.30}$$

Now both $\bar{u}_x$ and τ_{xy}^R tend to zero for large $|y|$. It follows that, on integrating (14.30) from $y = -\infty$ to $y = +\infty$, we find,

$$M = \int_{-\infty}^{\infty} \rho \bar{u}_x^2 \, dy = \text{constant}, \tag{14.31}$$

which tells us that the momentum flux is independent of x. Note that, although momentum is conserved, the mass flux is not. Rather, a turbulent jet, like its laminar counterpart, entrains ambient fluid, causing its mass flux to increase. In a laminar jet, entrainment is a result of viscous drag, while in the turbulent case, it is a consequence of the convoluted outer edge of the jet, which continually engulfs external fluid. In either case, there is a weak inflow towards the jet whose role is to maintain its growing mass flux (see Figure 14.6).

Now suppose the initial jet width is d. Then it is observed that, after a distance of around $30d$, the structure of the jet is independent of the precise details of its initiation and its *local* structure is controlled primarily by the local centre-line velocity, $\bar{u}_0$, and the jet width, $\delta(x)$: i.e. $\bar{u}_x(x, y) = f(y, \bar{u}_0(x), \delta(x))$. Dimensional analysis now tells us that the jet must adopt the self-similar form

$$\bar{u}_x(x, y) = \bar{u}_0(x) f(y/\delta(x)) = \bar{u}_0(x) f(\eta), \tag{14.32}$$

which constitutes a considerable simplification. Two immediate consequences of (14.32) are that, for large enough x,

$$M = \int_{-\infty}^{\infty} \rho \bar{u}_x^2 \, dy = \rho \bar{u}_0^2 \delta \int_{-\infty}^{\infty} f^2 \, d\eta = \text{constant},$$

$$\dot{m} = \int_{-\infty}^{\infty} \rho \bar{u}_x \, dy = \rho \bar{u}_0 \delta \int_{-\infty}^{\infty} f \, d\eta \sim \rho \bar{u}_0 \delta,$$

which demands that $\bar{u}_0^2 \delta = \text{constant}$ while $\bar{u}_0 \delta$ is an increasing function of x.

There are two common approaches to modelling turbulent jets, and they yield essentially the same result. One method rests on characterizing the rate of entrainment into the jet, while the other invokes the idea of an eddy viscosity. We start with the entrainment argument. Let us assume that the rate of entrainment of mass at any one stream-wise location is proportional to the local fluctuations in velocity, $\mathbf{u}'$. These, in turn, are proportional to the local mean velocity in the jet, and so on dimensional grounds we expect

$$\frac{d}{dx}(\rho \bar{u}_0 \delta) = \frac{1}{2} \alpha \rho \bar{u}_0,$$

which defines a dimensionless constant α, called the *entrainment coefficient*. It follows that, sufficiently far downstream,

$$\bar{u}_0^2 \delta = \text{constant}, \qquad \frac{d}{dx}(\bar{u}_0 \delta) = \frac{1}{2} \alpha \bar{u}_0. \tag{14.33}$$

These integrate to give

$$\frac{\delta}{\delta_0} = 1 + \frac{\alpha x}{\delta_0}, \quad \frac{\bar{u}_0}{V_0} = \left[1 + \frac{\alpha x}{\delta_0}\right]^{-1/2}, \tag{14.34}$$

where the origin for x is taken to be the start of the self-similar region and δ_0 and V_0 are the values of δ and $\bar{u}_0$ at $x = 0$. The first half of (14.34) may be rewritten in the form

$$\boxed{\frac{d\delta}{dx} = \alpha = \text{constant}}, \tag{14.35}$$

which is a good fit to the experimental data for $\alpha \approx 0.42$. The decay law $\bar{u}_0 \sim x^{-1/2}$ is also well supported by the data.

Let us now consider the eddy viscosity hypothesis (14.14) applied to a plane jet. It is observed that, for jets, ν_t is more or less independent of y, with a mixing length of order δ. Equation (14.14) then suggests $\nu_t \sim \delta \bar{u}_0$, and so we shall take $\nu_t = b\delta(x)\bar{u}_0(x)$ for some constant b. (This estimate of ν_t for a jet has an advantage over (14.17) in that it does not yield the unphysical result that $\nu_t = 0$ on the jet centreline.) Equation (14.30) now becomes

$$\bar{u}_x \frac{\partial \bar{u}_x}{\partial x} + \bar{u}_y \frac{\partial \bar{u}_x}{\partial y} = b\delta(x)\bar{u}_0(x) \frac{\partial^2 \bar{u}_x}{\partial y^2}. \tag{14.36}$$

This admits the self-similar solution (14.32), with $\bar{u}_0^2\delta = \text{constant}$, provided that $d\delta/dx$ is also a constant. If we make this additional assumption, then δ and $\bar{u}_0$ are once again given by (14.34), where α is now *defined* by (14.35). Moreover, after some careful algebra, (14.36) reduces to

$$F'^2 + FF'' + (2b/\alpha) F''' = 0, \quad F(\eta) = \int_0^\eta f\, d\eta,$$

from which

$$FF' + (2b/\alpha) F'' = 0. \tag{14.37}$$

This may be integrated again to give

$$\bar{u}_x/\bar{u}_0 = f(\eta) = \text{sech}^2 (y/\lambda\delta), \quad \lambda^2 = 4b/\alpha, \tag{14.38}$$

which is the same as (5.65) for a laminar jet.

So far we have not given a precise definition of δ and it is this that determines the value of λ, and hence the value of b. Given that any definition is somewhat arbitrary, since the time-averaged velocity declines exponentially with y, we adopt the simple definition that the jet velocity has dropped to 10% of $\bar{u}_0$ when $y = \pm\delta/2$. This requires that $\lambda = 0.275$, and given that $\alpha \approx 0.42$, we find $b \approx 8.0 \times 10^{-3}$. The velocity profile (14.38) is a good match to the experimental data (see Schlichting, 1979).

14.3.2 The Round Jet

We now consider a statistically axisymmetric jet described in polar coordinates (r, θ, z). This is characterized by $\bar{u}_z \gg \bar{u}_r$ and $\partial/\partial z \ll \partial/\partial r$, and so we may assume that:

(i) axial gradients in the Reynolds stresses are much weaker than radial gradients;

(ii) the radial component of the mean inertial force is negligible.

Moreover, as with planar jets, for distances well downstream of the source, $\bar{u}_z(r, z)$ adopts a self-similar form,

$$\frac{\bar{u}_z(r, z)}{\bar{u}_0(z)} = f(r/\delta) = f(\eta), \tag{14.39}$$

where $\bar{u}_0 = \bar{u}_z(0, z)$ and δ is the local mean diameter of the jet. The edge of the jet is, of course, highly convoluted. That is to say, the turbulence is a manifestation of the vorticity that is stripped off the inside surface of the nozzle, vorticity which turns chaotic as it is swept downstream (Figure 14.7). There is, therefore, entrainment of the ambient fluid into the jet as the convoluted outer edge of the jet engulfs irrotational fluid. Hence the mass flux of a round jet increases with distance from the source, just like that of a plane jet.

Adopting the simplifying assumptions above, and neglecting laminar stresses, the time-averaged Navier–Stokes equation (see Appendix 3) yields

$$\nabla \cdot (\rho \bar{u}_z \bar{\mathbf{u}}) = -\frac{\partial \bar{p}}{\partial z} + \frac{1}{r}\frac{\partial}{\partial r}\left[r\tau_{rz}^R\right], \tag{14.40}$$

$$0 = -\frac{\partial \bar{p}}{\partial r} + \frac{1}{r}\frac{\partial}{\partial r}\left[r\tau_{rr}^R\right] - \frac{\tau_{\theta\theta}^R}{r}. \tag{14.41}$$

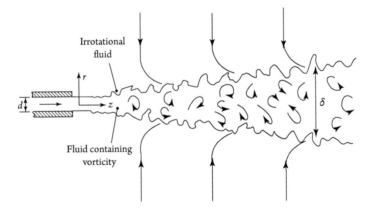

Figure 14.7 The entrainment of ambient fluid into a round jet of local mean diameter δ.

Since axial gradients in the Reynolds stresses are negligible, the radial equation, (14.41), tells us that we may neglect $\partial \bar{p}/\partial z$ in (14.40). The net result is

$$\frac{1}{r}\frac{\partial}{\partial r}\left[r\rho\bar{u}_r\bar{u}_z\right] + \frac{\partial}{\partial z}\left[\rho\bar{u}_z^2\right] = \frac{1}{r}\frac{\partial}{\partial r}\left[r\tau_{rz}^R\right], \tag{14.42}$$

from which we see that the momentum flux is independent of z,

$$\boxed{M = \int_0^\infty \left[\rho\bar{u}_z^2\right]2\pi r\,dr = \text{constant}}. \tag{14.43}$$

If we now invoke the self-similar approximation $\bar{u}_z = \bar{u}_0(z)f(\eta)$, the momentum flux becomes

$$M = \rho\bar{u}_0^2\delta^2\int_0^\infty 2\pi\eta\,f^2(\eta)d\eta = \text{constant},$$

which demands $\rho\bar{u}_0^2\delta^2 = $ constant. The mass flux, on the other hand, scales as $\dot{m} \sim \rho\bar{u}_0\delta^2$.

As with the planar jet, we may determine the rate of change of $\bar{u}_0(z)$ and $\delta(z)$ using an entrainment argument, or else by invoking an eddy viscosity. Let us start with the entrainment approach. The rate of entrainment of mass per unit length of the jet is proportional to the perimeter of the jet and the local intensity of the turbulent fluctuations. These fluctuations are, in turn, proportional to $\bar{u}_0(z)$, and so we have

$$\frac{d\dot{m}}{dz} \sim \rho\bar{u}_0\pi\delta,$$

which we can rewrite as

$$\frac{d}{dz}\left[\rho\bar{u}_0\delta^2\right] = \alpha\rho\bar{u}_0\delta. \tag{14.44}$$

This defines the *entrainment coefficient*, α. Combined with $\bar{u}_0^2\delta^2 = $ constant, this yields

$$\frac{\delta}{\delta_0} = 1 + \frac{\alpha z}{\delta_0}, \quad \frac{\bar{u}_0}{V_0} = \left[1 + \frac{\alpha z}{\delta_0}\right]^{-1}, \tag{14.45}$$

where z is now measured from the start of the self-similar region of the jet, and δ_0 and V_0 are the values of δ and $\bar{u}_0$ at $z = 0$. As for the planar jet, we predict

$$\boxed{\frac{d\delta}{dz} = \alpha = \text{constant}}, \tag{14.46}$$

and indeed just such a linear growth of δ is observed. If we define δ via the requirement that $\bar{u}_x/\bar{u}_0 = 0.1$ at $r = \delta/2$, then we find that $\alpha \approx 0.43$, which is very close to $\alpha \approx 0.42$ for a plane jet.

Let us now consider the eddy viscosity approach, which is similar to that for a plane jet. We take the eddy viscosity to be of the form $\nu_t = b\delta(z)\bar{u}_0(z)$, for some constant b, and seek a self-similar solution of

$$\bar{\mathbf{u}} \cdot \nabla \bar{u}_z = (b\delta\bar{u}_0)\frac{1}{r}\frac{\partial}{\partial r}\left(r\frac{\partial \bar{u}_z}{\partial r}\right). \tag{14.47}$$

Equation (14.47) does indeed admit a self-similar solution, in which $\bar{u}_0^2\delta^2 = $ constant, but only if $d\delta/dz$ is also made a constant. If we make this additional assumption, then δ and $\bar{u}_0$ are once again given by (14.45), with α now *defined* by (14.46). Moreover, with a little effort, it may be shown that (14.47) reduces to

$$\eta f^2 + f'\int_0^\eta \eta f d\eta + \frac{b}{\alpha}\frac{d}{d\eta}(\eta f') = 0, \tag{14.48}$$

which may be integrated to give

$$\frac{\bar{u}_z(r,z)}{\bar{u}_0(z)} = f(\eta) = \frac{1}{[1 + (\eta/\lambda)^2]^2}, \quad \lambda^2 = 8b/\alpha. \tag{14.49}$$

This velocity profile agrees well with the experimental data (see Schlichting, 1979).

It remains to determine the coefficient b. To that end we now recall that δ is defined so that $\bar{u}_z/\bar{u}_0 = 0.1$ at $r = \delta/2$. This requires $\lambda = 0.340$, and since observations give us $\alpha \approx 0.43$, we find that $b \approx 6.2 \times 10^{-3}$, which is close to the value of b for a planar jet.

Note that, since $\nu_t = b\delta\bar{u}_0$, $b = \alpha(\lambda^2/8)$ and $\bar{u}_0\delta \sim \sqrt{M/\rho}$, we have the estimate $\nu_t \sim \alpha\sqrt{M/\rho}$. It follows that, for large z, we can rewrite (14.45) and (14.49) in the form

$$\delta(z) \sim \frac{\nu_t z}{\sqrt{M/\rho}}, \quad \bar{u}_0(z) \sim \frac{M/\rho}{\nu_t z}, \quad \frac{\bar{u}_z(r,z)}{\bar{u}_0(z)} = \left[1 + (r/\lambda\delta)^2\right]^{-2}. \tag{14.50}$$

A comparison with (5.75) shows that the structure of a turbulent round jet is identical to that of a laminar jet, but with ν_t replaced by ν. This is because ν_t is *uniform* for a round jet.

14.4 Turbulent Flow near a Smooth Boundary: the Log-law of the Wall

14.4.1 The Log-law of the Wall in Channel Flow

The presence of a boundary fundamentally alters the structure of a turbulent shear flow, in large part because the velocity fluctuations must fall to zero at the wall. Perhaps the simplest case to consider is a unidirectional flow $\bar{\mathbf{u}} = (\bar{u}_x(y), 0, 0)$ between two smooth, parallel plates, as shown in Figure 14.8. If the flow is fully developed, all statistical properties except $\bar{p}$ are independent of x, and so the x and y components of (14.3) yield:

$$\frac{d}{dy}\left[\rho\nu\frac{d\bar{u}_x}{dy} - \rho\overline{u'_x u'_y}\right] = \frac{\partial\bar{p}}{\partial x}, \tag{14.51}$$

$$\frac{d}{dy}\left[-\rho\overline{u'_y u'_y}\right] = \frac{\partial\bar{p}}{\partial y}. \tag{14.52}$$

Evidently, $\bar{p} + \rho\overline{u'_y u'_y}$, which we shall label $\bar{p}_w$, is independent of y, and since $\mathbf{u}' = 0$ at $y = 0$, we have, $\bar{p}_w = \bar{p}(y=0)$. Clearly, $\bar{p}_w$ is the time-averaged pressure at the wall; hence the subscript w. Moreover, $\rho\overline{u'_y u'_y}$ is independent of x and so (14.51) becomes

$$\frac{d}{dy}\left[\rho\nu\frac{d\bar{u}_x}{dy} - \rho\overline{u'_x u'_y}\right] = \frac{d\bar{p}_w}{dx}. \tag{14.53}$$

Now the left-hand side of (14.53) is independent of x while the right is independent of y, and so $d\bar{p}_w/dx$ is a negative constant. We may therefore integrate (14.53) to give

$$\bar{\tau}_{xy} + \tau_{xy}^R = \rho\nu\frac{d\bar{u}_x}{dy} - \rho\overline{u'_x u'_y} = \left|\frac{d\bar{p}_w}{dx}\right|(W - y) = \tau_w\left(1 - \frac{y}{W}\right), \tag{14.54}$$

where $\tau_w = W|d\bar{p}_w/dx|$ is the wall shear stress and the constant of integration is fixed by the fact that the mean flow is symmetric about $y = W$. It is conventional at this point to introduce the notation $\tau_w = \rho V_*^2$, where V_* is known as the *friction velocity*. Equation (14.54) then becomes

$$\boxed{\frac{\tau}{\rho} = \nu\frac{d\bar{u}_x}{dy} - \overline{u'_x u'_y} = V_*^2\left(1 - \frac{y}{W}\right)}, \tag{14.55}$$

where τ is the net shear stress, $\tau = \bar{\tau}_{xy} + \tau_{xy}^R$.

Perhaps the key question now is how to get from a specified shear stress distribution to the mean velocity profile. To that end, we invoke the rather general tools of dimensional

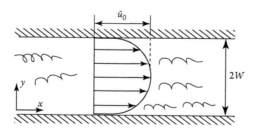

Figure 14.8 Flow between parallel plates.

analysis and *asymptotic matching*. First, we note that, given the variables available, there are only two ways of making y dimensionless, and these are $\eta = y/W$ and $y^+ = V_* y/\nu$. It is instructive, therefore, to consider separately a region close to the wall, defined by $y \ll W$, and a region that is somewhat removed from the wall, in the sense that $V_* y/\nu \gg 1$. These regions are shown in Figure 14.9.

Close to the wall, where $y \ll W$, the variation of shear stress in (14.55) is negligible and we may assume that τ is constant and equal to τ_w. This is known as the *inner region* and it is governed by

$$\frac{\tau}{\rho} = \nu \frac{d\bar{u}_x}{dy} - \overline{u_x' u_y'} = V_*^2, \quad y/W \ll 1. \tag{14.56}$$

Note that we must retain the viscous term in (14.56) because, adjacent to the wall, the Reynolds stress vanishes and the entire shear stress is laminar. The inner region is characterized by rapid variations in $\bar{\tau}_{xy}$ and τ_{xy}^R, as shown in Figure 14.10, although the sum of the two stresses is (almost) constant. Specifically, we transition from a purely viscous stress at $y = 0$ to negligible viscous stresses a short distance from the wall, characterized by $V_* y/\nu \gg 1$. This suggests that we introduce a second region, the *outer region*, where $V_* y/\nu \gg 1$ and the laminar stress is negligible:

$$\tau/\rho = -\overline{u_x' u_y'} = V_*^2 (1 - y/W), \quad V_* y/\nu \gg 1. \tag{14.57}$$

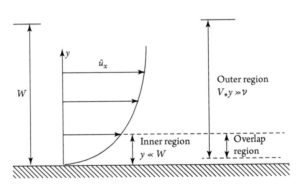

Figure 14.9 Different regions in a turbulent duct flow.

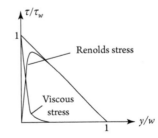

Figure 14.10 The variation of Reynolds stress and viscous stress with y.

Now the eddies centred a distance y from the wall tend to have a size of order $2y$. So within the inner region the important eddies are typically much smaller than W. This tentatively suggests that W is not a relevant parameter for the near-wall turbulence because the turbulence does not know about the second boundary a distance $2W$ away. We might speculate, therefore, that in the inner region the only parameters on which $\bar{u}_x$ can depend are V_*, y, and ν. By way of contrast, in the outer region we would expect mean velocity gradients to scale with W since the largest eddies, which are most effective at transporting momentum, are of the order of W. Moreover, ν is not a relevant parameter because the viscous stresses are negligible. So we might anticipate that departures from the centre-line velocity, $\Delta\bar{u}_x = \bar{u}_0 - \bar{u}_x$, will be independent of ν but a function of W. Provided this is all true (and we shall see in §14.4.3 that it is almost, though not quite, correct), then we have

$$\text{inner region}: \quad \bar{u}_x = \bar{u}_x(y, \nu, V_*), \quad y/W \ll 1,$$
$$\text{outer region}: \quad \bar{u}_0 - \bar{u}_x = \Delta\bar{u}_x(y, W, V_*), \quad V_*y/\nu \gg 1.$$

Recalling that $\eta = y/W$ and $y^+ = V_*y/\nu$, this may be put into dimensionless form:

$$\boxed{\bar{u}_x/V_* = f(y^+), \quad \eta \ll 1}, \tag{14.58}$$

$$\boxed{(\bar{u}_0 - \bar{u}_x)/V_* = g(\eta), \quad y^+ \gg 1}. \tag{14.59}$$

The first of these is called the *law of the wall*, while the second is the *velocity defect law*.

Now, in a turbulent duct flow we have $\text{Re} = WV_*/\nu \gg 1$, and so there exists an *overlap region* in which y is small when normalized by W, but large when normalized by ν/V_*. This region has the property that τ is (almost) constant, because $\eta \ll 1$, and the laminar stress is negligible, since $y^+ \gg 1$. Both (14.58) and (14.59) apply and so we have

$$\frac{y}{V_*}\frac{d\bar{u}_x}{dy} = y^+ f'(y^+) = -\eta g'(\eta). \tag{14.60}$$

However, y^+ and η are independent variables in the sense that we can change y^+, but not η, by varying ν or V_*, and we can change η, but not y^+, by varying W. It follows that

$$y^+ f'(y^+) = -\eta g'(\eta) = \text{constant} = 1/\kappa, \tag{14.61}$$

where the constant κ is called *Kármán's constant*. This integrates to give

$$\boxed{\frac{\bar{u}_x}{V_*} = \frac{1}{\kappa}\ln y^+ + A}, \tag{14.62}$$

$$\boxed{\frac{\bar{u}_0 - \bar{u}_x}{V_*} = -\frac{1}{\kappa}\ln\eta + B}. \tag{14.63}$$

(overlap region, $y^+ \gg 1, \eta \ll 1$)

Equation (14.62) is the celebrated *log-law of the wall*, and, along with Kolmogorov's 2/3rds and 4/5ths laws, it is one of the cornerstones of turbulence theory. Most experimental estimates of κ lie in the range 0.38–0.43 and we shall take $\kappa = 0.4$. The log-law is a good fit to the experimental data in the range $y^+ > 60$ and $\eta < 0.2$, with $A \approx 5.5$ and $B \approx 1.0$. Also, combining (14.62) and (14.63) gives

$$\frac{\bar{u}_0}{V_*} = \frac{1}{\kappa} \ln\left(\frac{V_* W}{\nu}\right) + A + B, \tag{14.64}$$

which relates the centreline velocity to $\mathrm{Re} = WV_*/\nu$ and V_*, and hence to the pressure gradient, $|d\bar{p}_w/dx| = \rho V_*^2/W$. This is a good fit to the experimental data for $\mathrm{Re} > 4{,}000$.

For $y^+ < 5$ the flow is largely, though not entirely, laminar and so $\nu d\bar{u}_x/\partial y \approx V_*^2$ integrates to give $\bar{u}_x \approx V_*^2 y/\nu$. This region is called the *viscous sublayer* and the adjacent region, $5 < y^+ < 60$, is called the *buffer layer*. These are shown in Figure 14.11.

Perhaps two footnotes are in order. First, the log-law of the wall may also be derived using Prandtl's mixing-length theory (14.17), by taking $\ell_m = \kappa y$. (This is discussed in Exercise 14.1.) The rationale for $\ell_m \sim y$ is that the average eddy size grows as we move away from the wall. Second, in the outer region, it is conventional to write the velocity defect law (14.59) as

$$\frac{\bar{u}_0 - \bar{u}_x}{V_*} = -\frac{1}{\kappa} \ln \eta + B - \Pi_w(\eta), \tag{14.65}$$

where $\Pi_w(\eta)$, which is called the *wake function*, is the difference between the log-law (14.63) and the defect law.

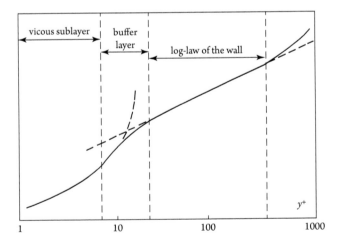

Figure 14.11 $\bar{u}_x/V_*$ versus y^+ showing the viscous sublayer and the log-law of the wall.

14.4.2 *The Log-law and Viscous Sublayer for Other Smooth-walled Flows*

The derivation of the log-law in §14.4.1 hinges critically on the idea that, when $\mathrm{Re} \gg 1$, there is a region close to the wall where the flow does not know about the details of the remote, outer flow, or about the existence of a second boundary. This suggests that the *statistical properties of the inner region should be universal* (*i.e.* the same for ducts, pipes, and external boundary layers) and depend only on V_*, y, and ν. We might expect, therefore, to find a log-law adjacent to *any* smooth, solid surface over which there is a mean turbulent flow. Moreover, the values of κ and A in (14.62) should also be universal. The only requirement, over and above $\mathrm{Re} \gg 1$, is that the stream-wise variations of $\bar{u}_x$, $\overline{u'_x u'_y}$, etc. are small. Hence the log-law is often awarded a status akin to that of Kolmogorov's theory.

However, just as there are weaknesses in K41, and its claim for universality has been attacked, so there are potential problems with the log-law, again relating to universality. For example, we shall see in §14.4.3 that, for rather subtle reasons, the various contributions to the turbulent kinetic energy near the wall *do* depend on the outer flow, and so are different for ducts, pipes, and boundary layers. However, theoretical considerations (again see §14.4.3) suggest that this lack of universality is largely restricted to the turbulent kinetic energy and does not alter the arguments leading to the log-law.

Certainly, this is consistent with the experimental evidence, where a log-law is found to exist whenever $\mathrm{Re} \gg 1$, and κ does indeed appear to be uniform across a range of flows, at least to within the accuracy of the experimental data. For example, expressions (14.58), (14.59), and (14.62) apply equally to pipe flow, provided we replace W by the pipe radius, R. Similarly, in a turbulent boundary layer on a flat plate, we find

$$\text{inner region}: \quad \bar{u}_x = V_* \, f(y^+), \quad y \ll \delta,$$

$$\text{outer region}: \quad \bar{u}_\infty - \bar{u}_x = V_* \, g(y/\delta), \quad y^+ \gg 1,$$

$$\text{overlap region}: \quad \bar{u}_x = V_* \left[\frac{1}{\kappa} \ln y^+ + A \right], \quad y^+ \gg 1, y \ll \delta,$$

where δ is the local thickness of the boundary layer and $\bar{u}_\infty$ is the free-stream velocity, as indicated in Figure 14.12.

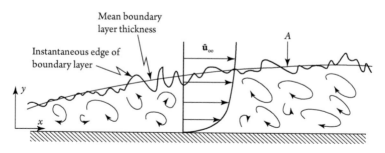

Figure 14.12 Schematic of a turbulent boundary layer.

Note that, even though the form of $\bar{u}_x$ in the inner region is (probably) universal, provided that Re $\gg 1$, the details of the turbulence in the outer region depend crucially on the global characteristics of the flow, being different in ducts, pipes, and external boundary layers. Moreover, in the case of boundary layers, any free-stream pressure gradient can have a substantial influence on the statistical properties of the outer layer.

The viscous sublayer plays a crucial role in all wall-bounded flows. This is because velocity gradients are particularly large there, and so we may regard this layer as a major reservoir of vorticity. Moreover, as suggested earlier, the viscous sublayer is not entirely quiescent. Rather, it is subjected to so-called *bursts*, in which fluid is suddenly ejected away from the wall, carrying its intense vorticity with it. Such bursts are important because the chaotic vorticity field, which constitutes the turbulence, has its origins at the wall. That is, while vortex stretching within the boundary layer can and does intensify the vorticity, ultimately the only source of vorticity is the wall itself. This vorticity may spread upward by diffusion or else by advection, and within a quiescent viscous sublayer the dominant mechanism is slow diffusion. However, when a turbulent burst occurs, the local value of Re increases, allowing intense vorticity to be propelled up into the upper regions of the boundary layer.

14.4.3 Inactive Motion: a Problem for the Universality of the Log-law?

The weakest aspect of our derivation of the log-law in §14.4.1 is the assumption that the turbulence near the wall is independent of W. Indeed, the Biot–Savart law tells us that this cannot be strictly correct. Recall that, when we talk of eddies, we really mean blobs of vorticity. The fluctuating velocity field is then an auxiliary field, dictated by the instantaneous vorticity distribution through the Biot–Savart law. It follows that an eddy in the core of a channel or boundary layer induces a velocity field that extends throughout all of the fluid, including the near-wall region. Moreover, the core eddies are influenced by the channel width, W, or boundary-layer thickness, δ. Hence there *are* velocity fluctuations near the wall which depend on W or δ.

However, all is not lost for the log-law. It turns out that the near-wall velocity fluctuations associated with the remote core eddies contribute very little to τ_{xy}^R, and hence do not influence the mean velocity profile in the inner region. To see why, we note that kinematics gives us

$$\frac{\partial}{\partial y}\left[-\overline{u_x' u_y'}\right] = \overline{u_y' \omega_z'} - \overline{u_z' \omega_y'} + \frac{1}{2}\frac{\partial}{\partial x}\left[\overline{(u_x')^2} - \overline{(u_y')^2} - \overline{(u_z')^2}\right],$$

and since there is no x-dependence of the statistical variables in fully developed duct flow, or in a slowly developing boundary layer, this simplifies to

$$\frac{\partial}{\partial y}\left[\frac{\tau_{xy}^R}{\rho}\right] = \left[\overline{\mathbf{u}' \times \boldsymbol{\omega}'}\right]_x. \tag{14.66}$$

We now divide the near-wall velocity fluctuations into two parts. There are those fluctuations associated with the remote, core eddies, which are more or less irrotational, and those which are associated with the small-scale vortices near the wall, which are largely (but not completely) rotational. So let us write $\mathbf{u}' = \mathbf{u}'_{rot} + \mathbf{u}'_{irrot}$ to represent these two sets of fluctuations, from which we have

$$\frac{\partial}{\partial y}\left[\frac{\tau^R_{xy}}{\rho}\right] = \left[\overline{\mathbf{u}'_{rot} \times \boldsymbol{\omega}'}\right]_x + \left[\overline{\mathbf{u}'_{irrot} \times \boldsymbol{\omega}'}\right]_x. \tag{14.67}$$

Now the near-wall velocity fluctuations induced by the remote, core vortices consist of a large, planar, sweeping motion parallel to the surface. This irrotational motion operates over length and time scales much greater than those of the near-wall vorticity fluctuations. Thus we might expect $\mathbf{u}'_{irrot}$ and $\boldsymbol{\omega}'$ to be only weakly correlated and indeed this is what is observed in practice. So the second term on the right of (14.67) makes almost no contribution to the near-wall Reynolds stress and we have

$$\frac{\partial}{\partial y}\left[\frac{\tau^R_{xy}}{\rho}\right] \approx \left[\overline{\mathbf{u}'_{rot} \times \boldsymbol{\omega}'}\right]_x. \tag{14.68}$$

The near-wall dynamics are now once again independent of W or δ, and this accounts for the success of the log-law. For this reason, $\mathbf{u}'_{irrot}$ is known as the *inactive motion* (Townsend, 1976). Indeed, as far as the near-wall turbulence is concerned, the planar sweeping motion induced by the core vortices looks a bit like a slow, random modulation of the mean flow far from the boundary, rather like the prevailing wind changing direction. Note, however, that $\mathbf{u}'_{irrot}$ *will* influence the turbulent kinetic energy distribution near the wall, so the variation of $\overline{(u'_x)^2}$ and $\overline{(u'_z)^2}$ with y in the log-layer will be non-universal and depend on the channel width or boundary-layer thickness.

It would seem, then, that the log-law has reasonably sound experimental and theoretical support. However, this is not quite the end of the story. As Townsend noticed, inactive motion near the wall, which looks like a slow, *non-universal* meandering of the external flow, brings into question the universality of Kármán's constant. The underlying issue is essentially the same as Landau's objection to the assumed universality of Kolmogorov's constant β; *i.e.* that a lack of universality arises from averaging over slow, non-universal fluctuations (see §13.4.1). In short, it is reasonable to suppose that, when averaging over the fast timescale of the near-wall eddies, one might indeed observe a universal log-law based on the local, *instantaneous* value of V_*. The problem is that the inactive motion causes slow, non-universal fluctuations in V_* at any given location, and in a laboratory experiment we do not measure the instantaneous value of V_*, but rather τ_w averaged over long times. Unfortunately, when the (universal) log-law is averaged over the slow, non-universal fluctuations in τ_w, the results of that averaging are not universal.

The consequences of non-universal averaging for Kolmogorov's 2/3rds law are outlined in §13.4.1, while its implications for the log-law are discussed in Davidson (2015). In the latter case, the log-law survives intact, though κ is possibly non-universal. However, as with Landau's criticism of Kolmogorov's 2/3rds law, the suggested non-universality of the relevant pre-factor (κ in the case of the log-law, β in the case of the 2/3rds law) is so small that it probably lies within the bounds of experimental uncertainty. The conclusion, then, is that κ may indeed be non-universal, but such variations are so slight that, for all practical purposes, we probably should not care too much.

14.4.4 Energy Balances and Structure Functions in the Log-law Layer

In the log-law layer the rate of generation of turbulent kinetic energy falls off with y as

$$G = \frac{\tau_{xy}^R}{\rho} \frac{d\bar{u}_x}{dy} = \frac{V_*^3}{\kappa y}, \tag{14.69}$$

whereas τ_{xy}^R, and hence G, falls to zero at the wall. It follows that G must be a maximum somewhere in the buffer region, and in fact it peaks at around $y^+ \approx 12$. The kinetic energy, $k = \frac{1}{2}\overline{(\mathbf{u}')^2}$, and the dimensionless ratio G/ε, also peak in the lower buffer region, with the maximum value of G/ε being around $G/\varepsilon \approx 1.8$. The fact that both G and k are maximal in this region means that it is a location of violent turbulent activity. Moreover, since G significantly exceeds ε, there must be a flux of turbulent kinetic energy away from the buffer region and into the log-law layer in accordance with (14.24), i.e. $G - \varepsilon = \partial T_y/\partial y$. In the log-law layer, on the other hand, we find that there is a close balance between generation and dissipation, with

$$G/\varepsilon \approx 0.91, \quad \text{and} \quad k = \frac{1}{2}\overline{(\mathbf{u}')^2} \approx 3.5V_*^2, \tag{14.70}$$

although k exhibits a weak logarithmic dependence on y, as we shall see.

Let us now consider the stream-wise structure function, which must take the form

$$\left\langle [\Delta v]^2 \right\rangle = \left\langle [u_x'(\mathbf{x} + r\hat{\mathbf{e}}_x) - u_x'(\mathbf{x})]^2 \right\rangle = \beta \varepsilon^{2/3} r^{2/3}, \tag{14.71}$$

at small scales. In the outer regions of a boundary layer, the 2/3rds law simply merges into the large-scale turbulence. However, in the log-law layer it is observed that a new regime appears wedged between the $r^{2/3}$ range and the largest scales (see Davidson et al., 2006). This is roughly of the form

$$\left\langle [\Delta v]^2 \right\rangle \sim V_*^2 \ln(r/y), \quad y/4 < r < 2\delta, \tag{14.72}$$

where δ is the boundary-layer thickness. The appearance of this new regime is explained in Davidson and Krogstad (2014), who employ asymptotic matching of the type used to deduce (14.61). They show that, in the log-law layer, we have

$$\boxed{\frac{\left\langle [\Delta v]^2 \right\rangle}{V_*^2} = A_2 + B_2 \ln\left(\frac{\varepsilon r}{V_*^3}\right), \quad r_* \ll r \ll \delta}, \tag{14.73}$$

where $r_* = V_*^3/\varepsilon$ and B_2 is (probably) a universal constant of approximate value 2.0. This merges smoothly into (14.71), rewritten as

$$\frac{\left\langle [\Delta v]^2 \right\rangle}{V_*^2} = \beta \left(\frac{\varepsilon r}{V_*^3} \right)^{2/3}, \quad \eta \ll r \ll r_*, \tag{14.74}$$

where η is the Kolmogorov scale. Moreover, in the log region we have $G = V_*^3/\kappa y$ and $G \approx \varepsilon$, which yield

$$r_* = V_*^3/\varepsilon = (G/\varepsilon)\,\kappa y \approx \kappa y,$$

and so these two laws merge at around $r \sim y$.

The same expression, $V_*^3/\varepsilon = (G/\varepsilon)\kappa y$, tells us that we can rewrite (14.73) as

$$\frac{\left\langle [\Delta v]^2 \right\rangle}{V_*^2} = A_2^* + B_2 \ln \left(\frac{r}{\kappa y} \right), \quad y \ll r \ll \delta, \tag{14.75}$$

where $A_2^* = A_2 + B_2 \ln (\varepsilon/G)$, which is consistent with the observations (14.72). Note that the structure function saturates at $2\overline{(u_x')^2}$ for $r \gg \delta$, and so (14.75) tentatively suggests

$$\frac{2\overline{(u_x')^2}}{V_*^2} \approx C + B_2 \ln \left(\frac{\delta}{\kappa y} \right), \tag{14.76}$$

for some constant C. This logarithmic variation of $\overline{(u_x')^2}$ with y is indeed observed and some authors associate it with the *inactive motion* introduced in §14.4.3, that is, the planar sweeping motion generated near the wall by the core eddies.

In summary, within the log-law region, we have the two structure function laws

$$\frac{\left\langle [\Delta v]^2 \right\rangle}{V_*^2} = \beta \left(\frac{\varepsilon r}{V_*^3} \right)^{2/3} \approx \beta \left(\frac{r}{\kappa y} \right)^{2/3}, \quad \eta \ll r < y/4, \tag{14.77}$$

$$\frac{\left\langle [\Delta v]^2 \right\rangle}{V_*^2} = A_2 + B_2 \ln \left(\frac{\varepsilon r}{V_*^3} \right) \approx A_2 + B_2 \ln \left(\frac{r}{\kappa y} \right), \quad y/4 < r < 2\delta, \tag{14.78}$$

as shown in Figure 14.13. The upper limit on r in (14.78) comes from observations, as does the approximate transition between the two regimes at $r \sim y/4$.

Note that expression (14.73) is more general than (14.75) in the sense that it applies equally to homogeneous shear flow, where there is no boundary and hence no y, and to flow over rough surfaces. One physical interpretation of (14.73) rests on the idea of the *signature function*, $V(r)$, introduced in Exercise 13.3. In particular, $rV(r)$ is a measure of the kinetic energy of eddies of scale r, and the fact that (14.73) yields

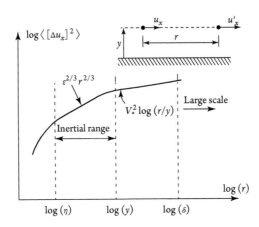

Figure 14.13 The shape of the longitudinal structure function $\left\langle [\Delta v]^2 \right\rangle$ near the wall.

$$rV\left(r\right) = -\frac{r^3}{2}\frac{\partial}{\partial r}\frac{1}{r}\frac{\partial}{\partial r}\left[\frac{3}{4}\left\langle [\Delta v]^2 \right\rangle\right] = \frac{3}{4}B_2 V_*^2 \qquad (14.79)$$

suggests that it represents a range of scales whose kinetic energy is of order V_*^2.

14.4.5 Coherent Structures and Near-wall Cycles

An important feature of boundary layers is the persistence of so-called *coherent structures*. This rather vague term is meant to suggest a class of vortices which reappear time and again and are relatively long-lived. The most commonly observed coherent structure is a *hairpin vortex*, which is a vortex loop which arches up from the boundary to span much of the boundary layer, as shown in Figure 14.14. Such vortices have a maximum length of the order of δ and a diameter which may be as small as $10\nu/V_*$. Experiments at modest Re suggest that turbulent boundary layers are heavily populated with hairpin vortices of a variety of sizes, many of them orientated at around $45°$ to the mean flow. As shown in Figure 14.5, such vortices are ideally suited for extracting energy from the mean flow, and so they contribute greatly to the Reynolds stress and to the production of turbulent energy.

Hairpin vortices develop as follows. The mean flow $\bar{u}_x(y)$ is associated with span-wise vorticity, $\bar{\omega} = (0,0,\bar{\omega}_z)$, which we might think of as an array of vortex filaments. Since Re is large, such vortex filaments are more or less frozen into the fluid and so a stream-wise turbulent gust will sweep out an axial component of vorticity, as shown in Figure 14.14. This represents the birth of a hairpin vortex. Next, it is readily confirmed that the curvature of a hairpin vortex induces a velocity field that tends to advect the tip of the hairpin upwards into the flow (see Exercise 14.2). As the vortex rotates, the tip of the loop finds itself in a region of high mean velocity relative to its base. The vortex loop is then stretched out by the mean shear, intensifying the vortex loop.

It is likely that hairpin vortices are primarily initiated in the buffer region close to the wall, since the mean vorticity is most intense there. It is also likely that hairpin vortices are

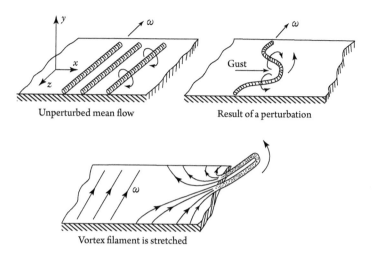

Figure 14.14 The formation of a hairpin vortex.

subsequently destroyed through their interaction with other boundary-layer structures. In summary, then, a typical sequence of events might be:

(i)   an axial gust close to the wall creates a small, horizontal vortex loop;
(ii) self-advection of the vortex loop gives rise to rotation of the loop;
(iii) an inclined loop sitting in a mean shear is stretched and intensified;
(iv) the hairpin vortex is destroyed through interactions with other vortices.

This is a somewhat idealized cartoon, however. For example, hairpin vortices are rarely symmetric. Rather, one leg is usually more pronounced than the other. Also, there is some disagreement as to what triggers the formation of a hairpin. One possibility is that a large eddy in the upper part of the boundary layer interacts with the buffer region. The mean vortex lines in the wall layer are then bent out of shape, initiating the sequence of events shown in Figure 14.14. In this picture, then, hairpin vortices are triggered by events which begin far from the wall. An alternative suggestion is that the initial disturbance is local to the wall, being part of a *near-wall cycle*, as discussed below. Yet another complication is that these vortices often appear in groups, with each hairpin initiating another in its wake (Figure 14.15).

There are yet other interpretations of the observed vortex loops, although most are little more than cartoons. In any event, the unifying theme across these various interpretations is the power of the mean shear to stretch out the vortex tubes to form inclined, elongated vortices. While authors argue about the precise details of the origins of hairpin vortices, most agree that hairpins are common in turbulent boundary layers.

Close to the boundary, a second type of structure is frequently observed. Specifically, pairs of stream-wise vortex tubes of opposite polarity are often observed for $y^+ < 50$ (see

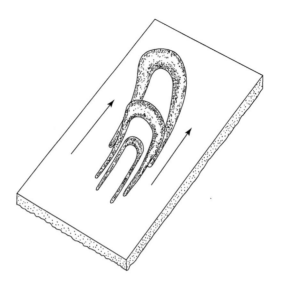

Figure 14.15 A packet of hairpin vortices.

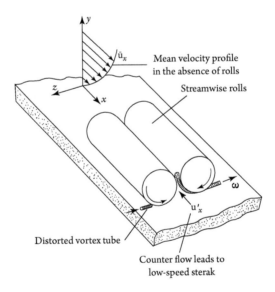

Figure 14.16 Deformation of the mean-flow vorticity by a pair of vortex rolls.

Figure 14.16). The rotation in these tubes is such that fluid near the wall is swept towards the gap between the tubes and then pushed upward, away from the boundary. There is therefore an interaction between these rolls and the mean span-wise vorticity, in which the span-wise vortex lines are deflected upward into the gap between the rolls. This induces a velocity perturbation, u'_x, which is opposite to the mean flow, and so the local stream-wise velocity is diminished. This, in turn, results in a so-called *low-speed streak*, such steaks being commonly observed for $y^+ < 20$. Low-speed streaks are eventually ejected from the lower buffer zone by the updraft between the rolls and this is often followed by a rapid destabilization of the flow, called a *burst*. Such near-wall bursts result in a local intensification of the turbulence and are thought to be one of the primary mechanisms of turbulent kinetic energy generation within a boundary layer.

In some cartoons the stream-wise rolls constitute the lower regions of a hairpin vortex, having become highly elongated as a result of the strong local shear. In this picture the stream-wise streaks are an inevitable consequence of the rolls, and the rolls themselves are the footprints of large-scale hairpin vortices. There is a second popular cartoon, however, in which the rolls are not attached to vortices in the outer part of the boundary layer, but rather participate in a near-wall cycle in which both the streaks and rolls are dynamically interacting structures. Specifically, some researchers advocate a regenerative cycle in which rolls produce streaks, the streaks become unstable, and the resulting non-linear instability generates new rolls.

14.4.6 Turbulent Heat Transfer near a Surface and the Log-law for Temperature

We now turn to turbulent heat transfer. Consider the flow shown in Figure 14.17, where the random eddying motion carries hot fluid away from the lower wall and cold fluid down from the upper surface. In the core of the flow the eddies span much of the channel and are very efficient at transporting and mixing heat. Very close to the walls, however, the eddies are small and the burden of transporting the heat falls to a combination of molecular diffusion and turbulent convection. Hence the wall region tends to be resistant to turbulent heat transfer, and so usually dominates the thermal resistance.

Let us start by revisiting the advection–diffusion equation for heat, which has its origins in energy conservation. Let $\dot{\mathbf{q}}$ be the heat flux density associated with molecular conduction in accordance with Fourier's law, $\dot{\mathbf{q}} = -k_c \nabla T$, where k_c is the thermal conductivity and T the temperature. Then conservation of energy applied to a *material* volume element δV gives

$$\frac{D}{Dt}(\rho c_p T \delta V) = -\oint_{\delta S} \dot{\mathbf{q}} \cdot d\mathbf{S} = -\int_{\delta V} \nabla \cdot \dot{\mathbf{q}} dV \approx -\nabla \cdot \dot{\mathbf{q}} \delta V, \qquad (14.80)$$

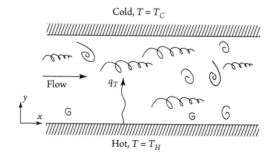

Figure 14.17 Turbulent heat transfer from a wall.

where c_p is the specific heat. This comes from equating the rate of loss of thermal energy from the fluid within δV to the rate at which heat diffuses out through the surface of δV. Dividing through by δV brings us back to (5.42),

$$\frac{D}{Dt}(\rho c_p T) = -\nabla \cdot \dot{\mathbf{q}}, \quad \dot{\mathbf{q}} = -k_c \nabla T, \tag{14.81}$$

and substituting for $\dot{\mathbf{q}}$ gives (5.43),

$$\frac{DT}{Dt} = \alpha \nabla^2 T, \quad \alpha = k_c / \rho c_p. \tag{14.82}$$

Now let us divide T into mean and fluctuating parts, $T = \bar{T} + T'$. If $\mathbf{u}$ and T are statistically steady, then averaging (14.81) yields

$$\bar{\mathbf{u}} \cdot \nabla (\rho c_p \bar{T}) + \nabla \cdot (\rho c_p \overline{T'\mathbf{u}'}) = -\nabla \cdot (-k_c \nabla \bar{T}), \tag{14.83}$$

which we rewrite as

$$\boxed{\bar{\mathbf{u}} \cdot \nabla (\rho c_p \bar{T}) = -\nabla \cdot \dot{\mathbf{q}}_T,} \qquad \boxed{\dot{\mathbf{q}}_T = -k_c \nabla \bar{T} + \rho c_p \overline{T'\mathbf{u}'}.} \tag{14.84}$$

Evidently, $\dot{\mathbf{q}}_T$ is the net turbulent heat flux density which includes both molecular conduction and turbulent mixing. The turbulent contribution, $\rho c_p \overline{T'\mathbf{u}'}$, is analogous to the Reynolds stresses that appear in the averaged momentum equation.

To make progress, we must now estimate $\overline{T'\mathbf{u}'}$. One common approach is to write $\overline{T'\mathbf{u}'} = -\alpha_t \nabla \bar{T}$, where α_t is a *turbulent diffusivity*, and so (14.84) becomes

$$\bar{\mathbf{u}} \cdot \nabla \bar{T} = -\nabla \cdot (\dot{\mathbf{q}}_T / \rho c_p) = \nabla \cdot \left[(\alpha + \alpha_t) \nabla \bar{T} \right]. \tag{14.85}$$

Such an estimate, which is known as a *gradient-diffusion* approximation, is similar to the eddy–viscosity estimate (14.12). The idea is that, on average, turbulent mixing will tend to eradicate gradients in mean temperature, just as molecular diffusion does.

We now return to the case shown in Figure 14.17, where the flow is statistically independent of x and z. Here, (14.84) tell us that $\nabla \cdot \dot{\mathbf{q}}_T = 0$, or $\partial \dot{q}_{Ty} / \partial y = 0$, and so

$$\frac{\dot{q}_{Ty}}{\rho c_p} = -\alpha \frac{d\bar{T}}{dy} + \overline{T'u_y'} = -(\alpha + \alpha_t) \frac{d\bar{T}}{dy} = \text{constant}. \tag{14.86}$$

If we now invoke a mixing-length approximation for α_t then, from (14.14), $\alpha_t = \ell_m V_t$. Here ℓ_m is the mixing length and we might expect $V_t \sim V_*$. This amounts to the assertion that

$$\overline{T'u_y'} \sim -\ell_m V_* \frac{d\bar{T}}{dy}, \tag{14.87}$$

which transfers the problem to one of determining the mixing length. Now, mixing-length theory applied to the log-law for *velocity* has $\ell_m = \kappa y$. On the assumption that the same eddies are transporting both momentum and heat, this suggests we take $\ell_m = \kappa y$ in (14.87), at least near the lower boundary in Figure 14.17. Outside the viscous sublayer, this yields

$$T^* = \frac{\dot{q}_T y}{\rho c_p V_*} \sim -\kappa y \frac{d\bar{T}}{dy},$$

which defines T^*. We now rewrite this as

$$T^* = \frac{\dot{q}_T y}{\rho c_p V_*} = -\kappa_T y \frac{d\bar{T}}{dy}, \tag{14.88}$$

which defines κ_T. Finally, this integrates to give a *log-law for temperature*,

$$\frac{\Delta T}{T^*} = \frac{T_H - \bar{T}(y)}{T^*} = \frac{1}{\kappa_T} \ln \left[\frac{V_* y}{\alpha} \right] + A_T, \tag{14.89}$$

where A_T is a dimensionless constant and T_H is the temperature at $y = 0$. We must now match (14.89) to a conduction-only region next to the wall, which is readily shown to take the form $\Delta T/T^* = V_* y/\alpha$. This is illustrated in Figure 14.18. We conclude that the *additive* constant A_T is a function of the dimensionless temperature drop across the conduction-only region. The details of this matching process are discussed in Davidson (2015), where it is shown that A_T is a function of the Prandtl number, $\Pr = \nu/\alpha$, and in particular $A_T \approx 5/3$ for $\Pr < 0.3$ while $A_T \approx (5/3)(3\Pr^{1/3} - 1)^2$ for $0.3 < \Pr < 10^4$.

The appearance of a log-law for temperature is interesting because of its similarity to the log-law for velocity. It turns out that the log-law for temperature is not simply an artefact

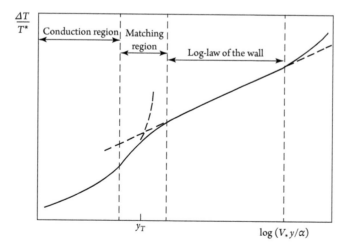

Figure 14.18 The log-law for temperature.

of the mixing-length approximation, but rather a genuine law (Landau and Lifshitz, 1959). Following arguments analogous to those leading up to (14.62), it may be shown that, when $V_* y/\alpha$ is large, yet y/W or y/δ is small, the expression

$$\boxed{\frac{\Delta T}{T^*} = \frac{1}{\kappa_T} \ln \left[\frac{V_* y}{\alpha} \right] + A_T(\mathrm{Pr})}$$

(14.90)

is exact. The coefficient κ_T is (probably) a universal constant, with a value of ~ 0.48.

14.5 The Influence of Surface Roughness and Stratification on Turbulent Shear Flow

14.5.1 The Log-law for Flow over a Rough Surface

So far we have focussed exclusively on flow over smooth surfaces. We now turn to turbulent flow over a rough boundary, which is particularly important in meteorology. Let us return to the channel flow shown in Figure 14.8, only now we take the boundaries to be rough. Let the rms roughness height be $\hat{k}$. If $\hat{k}$ is comparable with, or greater than, the thickness of the viscous sublayer in the equivalent smooth-walled flow, then $\hat{k}$ becomes an important parameter. The velocity profile in the inner region, (14.58), then takes the form

$$\bar{u}_x/V_* = f(y/\hat{k}, y^+), \quad y/W \ll 1.$$

(14.91)

For $\hat{k}^+ = V_* \hat{k}/\nu \gg 1$, viscous effects are negligible by comparison with the turbulence generated by the roughness, except very close to the wall. In such cases, ν ceases to be a relevant parameter outside the viscous sublayer and (14.91) simplifies to $\bar{u}_x/V_* = f(y/\hat{k})$. If we now repeat the arguments which lead up to (14.62), the law of the wall is modified to

$$\boxed{\bar{u}_x/V_* = \frac{1}{\kappa} \ln \left(y/\hat{k} \right) + \text{constant} = \frac{1}{\kappa} \ln \left(y/y_0 \right)}.$$

(14.92)

The parameter y_0, which is defined by (14.92) in terms of $\hat{k}$ and the additive constant, characterizes the height, shape, and sparseness of the roughness. Experiments show that, for sand-grain roughness, the additive constant has a value of ~ 8.5, which gives $y_0 \approx \hat{k}/30$. The same experiments suggest that (14.92) holds when $\hat{k}^+ > 60$, while roughness may be ignored provided $\hat{k}^+ < 4$. In the atmospheric boundary layer y_0 is found to lie in the range $y_0 \sim 0.02\,\mathrm{m} - 0.1\,\mathrm{m}$ for crops, $y_0 \sim 0.2\,\mathrm{m} - 0.5\,\mathrm{m}$ for woodland, and $y_0 \sim 0.5\,\mathrm{m}$ in the suburbs.

14.5.2 The Atmospheric Boundary Layer, Stratification, and the Flux Richardson Number

In §14.4.6 we considered the effect of turbulence on a temperature field, but not the back reaction of the buoyancy force on the turbulence. The effects of buoyancy are particularly

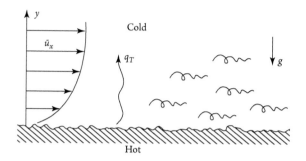

Figure 14.19 Turbulence generated by shear and buoyancy.

important in the atmospheric boundary layer (ABL), which is the lower part of the atmosphere where the effects of ground friction and surface heating or cooling are important. The ABL may be between 0.5 km and 2 km deep, with the Coriolis force playing an important role in the upper regions of the boundary layer, called the outer layer. However, in the lowest 10% of the ABL, the Coriolis force is relatively unimportant and the turbulent shear stress is roughly constant and equal to the surface stress. It is this inner layer that concerns us here. The effects of buoyancy on the inner ABL may be stabilizing, destabilizing, or neutral. Stable conditions (hot air over cold air) are common at night as the ground cools by radiation, whereas unstable conditions are more common during the day when the ground heats up by solar radiation. Neutral conditions tend to correspond to windy conditions and dense cloud cover.

To focus our thoughts, suppose that we have a turbulent shear flow over a rough surface, $\bar{u}_x(y)$, and that there is a uniform heat flux, $\dot{q}_T$, away from that surface, as shown in Figure 14.19. For simplicity, we adopt the Boussinesq approximation in which changes in density are sufficiently small for ρ to be treated as a constant, except to the extent that there is a buoyancy force of $\delta\rho\mathbf{g}$ per unit volume acting on the fluid. This force can be written as $-\rho\beta(T - T_0)\mathbf{g} = -\rho\beta\Theta\mathbf{g}$, where $\beta = -\rho^{-1}(d\rho/dT)$ is the expansion coefficient and $\Theta = T - T_0$ is the departure of T from a reference temperature T_0. For an ideal gas it may be shown that $\beta = T_0^{-1}$, and so the buoyancy force becomes $-\rho(\Theta/T_0)\mathbf{g}$.

Since $\dot{\mathbf{q}}_T = \rho c_p \overline{\Theta'\mathbf{u}'}$ (provided molecular conduction is ignored), the mean rate of working of this buoyancy force is

$$\left(\overline{\Theta'u_y'}\right)\rho g/T_0 = \dot{q}_T g/c_p T_0. \tag{14.93}$$

Hence, if all statistical averages are independent of x, the turbulent kinetic energy equation, (14.11), may be modified to incorporate buoyancy according to

$$0 = \tau_{xy}^R \frac{d\bar{u}_x}{dy} - 2\rho\nu\overline{\left(S_{ij}'\right)^2} - \frac{\partial}{\partial y}\left[\frac{1}{2}\overline{\rho(\mathbf{u}')^2 u_y'} + \overline{p'u_y'} - 2\rho\nu\overline{u_i' S_{iy}'}\right] + \left(\overline{\Theta'u_y'}\right)\rho g/T_0. \tag{14.94}$$

The third term on the right merely redistributes turbulent kinetic energy, and so we have two sources of turbulent energy, $\tau_{xy}^R d\bar{u}_x/dy$ and $\left(\overline{\Theta'u_y'}\right)\rho g/T_0$, the sum of which is absorbed by the small-scale viscous dissipation, $2\rho\nu\overline{\left(S_{ij}'\right)^2}$.

If the turbulent heat flux has the opposite sign (hot air over cold air), the buoyancy suppresses, rather than generates, turbulent kinetic energy and so $\overline{\Theta'u'} < 0$. Either way, the *flux Richardson number*, Ri_f, is defined as (minus) the ratio of the rate of production of turbulent energy by buoyancy and shear forces:

$$\mathrm{Ri}_f = -\frac{g\,\overline{\Theta'u_y'}/T_0}{(-\overline{u_x'u_y'})d\bar{u}_x/dy} = -\frac{\dot{q}_{Ty}g/c_pT_0}{\tau_{xy}^R d\bar{u}_x/dy}. \tag{14.95}$$

This might be compared with the *gradient Richardson number* (11.70). Note that it is conventional to define Ri_f so that it is negative for an upward transfer of heat and positive for a downward transfer. It follows that, when Ri_f is large and negative, the primary source of turbulence is the buoyancy force, while $0 < \mathrm{Ri}_f < O(1)$ corresponds to a stable stratification of temperature and hence to a partial suppression of the turbulence. Of course, a small value of $|\mathrm{Ri}_f|$ implies that the buoyancy is negligible.

This simple energy analysis, as represented by (14.94) and (14.95), forms the basis of most models of buoyant shear flow, with the flux Richardson number playing a central role in many theories of the inner ABL. This is particularly true of the popular model of Monin and Obukhov, which we shall discuss in §14.5.4. First, however, we describe an early model of the inner ABL which is usually attributed to Prandtl. This relates specifically to the case where the wind shear is relatively weak.

14.5.3 Prandtl's Weak-shear Model of the Atmospheric Boundary Layer

When Ri_f is large and negative, so the wind shear is relatively weak, the generation of turbulence by buoyancy is more or less matched by the viscous dissipation. We then have

$$\dot{q}_T g/c_p T_0 \sim \rho\varepsilon.$$

In addition, a mixing length estimate of $\dot{q}_T$ suggests

$$\frac{\dot{q}_T}{\rho c_p} = \overline{\Theta'u_y'} \sim -(u'\ell_m)\frac{d\bar{T}}{dy}, \quad u' = \sqrt{\overline{u_y'^2}},$$

where ℓ_m is the mixing length and u' a characteristic fluctuation associated with u_y. Now suppose that the buoyancy force does not significantly alter the conventional energy cascade. Then the dissipation per unit mass can be estimated as $\varepsilon \sim u'^3/\ell_m$ and we may combine these various estimates, to yield

$$\frac{\dot{q}_T}{\rho c_p} \sim \Theta'u' \sim u'\ell_m \left|\frac{d\bar{T}}{dy}\right| \sim \frac{T_0}{g}\frac{u'^3}{\ell_m}. \tag{14.96}$$

Near the ground we might expect the size of the eddies to grow as $\ell_m \sim \kappa y$, and so for the inner ABL, (14.96) yields

$$\sqrt{\overline{u_y'^2}} \sim \left(\frac{\dot{q}_T}{\rho c_p}\right)^{1/3} \left(\frac{T_0}{\kappa g}\right)^{-1/3} y^{1/3}, \tag{14.97}$$

$$\Theta' \sim \left(\frac{\dot{q}_T}{\rho c_p}\right)^{2/3} \left(\frac{T_0}{\kappa g}\right)^{1/3} y^{-1/3}, \tag{14.98}$$

$$\left|\frac{d\bar{T}}{dy}\right| \sim \left(\frac{\dot{q}_T}{\rho c_p}\right)^{2/3} \left(\frac{T_0}{\kappa g}\right)^{1/3} y^{-4/3}. \tag{14.99}$$

Although some doubts have been expressed as to the accuracy of (14.97)–(14.99), Monin and Yaglom (1971) provide considerable evidence in support of these expressions.

14.5.4 The Monin–Obukhov Theory of the Atmospheric Boundary Layer

We now consider a more general model of the inner ABL that allows for both stable and unstable stratification and is not restricted to weak wind shear. We adopt the convention that $\dot{q}_T > 0$ when the turbulent heat flux is upward and $\dot{q}_T < 0$ when it is downward, so that (14.93) holds in either case. Since we restrict attention to the inner layer, $\hat{k} \ll y \ll \delta$, we have $\tau_{xy}^R = \rho V_*^2$ and (14.95) becomes

$$\mathrm{Ri}_f = -\frac{\dot{q}_T g/\rho c_p T_0}{V_*^2 d\bar{u}_x/dy}. \tag{14.100}$$

If the mean velocity profile is approximately logarithmic, $d\bar{u}_x/dy \approx V_*/(\kappa y)$, as would be the case for weak buoyancy, then this reduces to

$$\mathrm{Ri}_f \approx \frac{y}{L}, \quad L = -\frac{(T_0/\kappa g)V_*^3}{(\dot{q}_T/\rho c_p)}, \tag{14.101}$$

which defines the *Monin–Obukhov length*, L. Note that L is effectively determined by just two parameters, $\dot{q}_T$ and V_*, with strong shear favouring a large value of $|L|$ and an intense heat flux a small value of $|L|$. Moreover, as with Ri_f, a positive value of L indicates stable stratification, which leads to the suppression of turbulence, whereas a negative value indicates unstable stratification. The magnitude of $|L|$ is seldom much less than a few metres, unless there is little or no wind, and it may be a few tens of metres on a windy day. By way of contrast, the inner (constant stress) region of the ABL may be around 100 m deep.

Equation (14.101) tells us that, close to the ground, and in particular when $y \ll |L|$, buoyancy plays little role in the generation of turbulence and conditions are close to neutral. Conversely, if $y \gg |L|$ and the stratification is unstable, (14.101) tentatively suggests that the generation of turbulence by shear is negligible and buoyancy is the dominant source of turbulent energy. If all of this is true (and it is), then there must be a transition from mechanically forced turbulence to free convection at a height of $y \sim |L|$.

Monin and Yaglom (1971) go further and suggest that, within the constant stress region of the ABL, y/L may be used to modify the log-laws for velocity and temperature, *i.e.* (14.92) and (14.89). In particular, they show that dimensional considerations require

$$\frac{d\bar{u}_x}{dy} = \frac{V_*}{\kappa y}\Phi_u\left(y/L\right), \quad \frac{d\bar{T}}{dy} = -\frac{T^*}{\kappa_T y}\Phi_T(y/L), \tag{14.102}$$

where $T^* = \dot{q}_T/\rho c_p V_*$ and Φ_u and Φ_T are the corrections to the log-laws arising from the influence of buoyancy. It turns out that both of these expressions provide a particularly powerful means of rationalizing measurements made in the ABL. We shall focus here on the suggested correction to the velocity log-law.

When $y \ll |L|$, so that buoyancy mildly perturbs the inner layer, it is found that the effects of buoyancy can be taken into account by modifying the usual form of the velocity log-law with a simple linear correction. This modified version, valid for positive or negative L, is

$$\frac{d\bar{u}_x}{dy} = \frac{V_*}{\kappa y}\left[1 + \gamma\frac{y}{L}\right], \quad y \ll |L|, \tag{14.103}$$

where γ is found to lie in the range $4 \to 7$. This integrates to give

$$\frac{\bar{u}_x}{V_*} = \frac{1}{\kappa}\left[\ln\left(\frac{y}{y_0}\right) + \gamma\frac{y}{L}\right], \quad y \ll |L|, \tag{14.104}$$

where y_0 is the surface roughness parameter. Thus $\bar{u}_x$ changes less rapidly with height for unstable conditions, due to increased mixing, and more rapidly for a stable stratification.

Let us now consider the case of unstable stratification, and heights for which $y \gg |L|$. Here observations tentatively suggest that

$$\frac{d\bar{u}_x}{dy} \sim \frac{V_*}{\kappa y}\left(\frac{y}{|L|}\right)^{-1/3}, \quad y \gg |L|, \tag{14.105}$$

which is nothing more or less than Prandtl's model for weak wind shear. That is to say, the mixing-length model would have

$$\frac{d\bar{u}_x}{dy} \sim \frac{V_*^2}{u'\ell_m} \sim \frac{V_*^2}{u'\kappa y}, \quad u' = \sqrt{\overline{u_y'^2}},$$

and substituting for u' using (14.97) brings us back to (14.105).

Both (14.103) and (14.105) are consistent with (14.102) and a common interpolation formula for $\Phi_u\left(y/L\right) = \Phi_u\left(\eta\right)$ is

$$\begin{aligned} \Phi_u(\eta) &\approx 1 + \gamma\eta, & 0 < \eta < 1, \\ \Phi_u(\eta) &\approx (1 + \beta|\eta|)^{-1/3}, & -2 < \eta < 0, \end{aligned} \tag{14.106}$$

where β lies in the range $\beta = 9 \to 20$. (Sometimes the exponent $-1/3$ is replaced by $-1/4$.)

14.6 Closure Models for Turbulent Shear Flows: the k-ε Model as an Example

We now consider a class of turbulent closure models that are commonly applied to shear flows in order to estimate the Reynolds stresses and so allow the mean flow to be calculated. These are called *one-point closure models* as they focus attention on quantities that relate to just one point in space, such as τ_{ij}^R. The most popular scheme is the k-ε *model* and so we shall focus on that, though it has many cousins that are in common use.

14.6.1 The Basis of the k-ε Closure Model

Our aim is to find a closed set of equations that allow the instantaneous distribution of τ_{ij}^R to be estimated from a knowledge of the mean flow, so that the evolution of the mean flow itself may be determined. Recall that the mixing-length model of Prandtl applied to a simple shear flow predicts

$$\tau_{xy}^R = \rho \ell_m^2 \left| \frac{\partial \bar{u}_x}{\partial y} \right| \frac{\partial \bar{u}_x}{\partial y} = \rho \nu_t \frac{\partial \bar{u}_x}{\partial y}. \tag{14.107}$$

It turns out that this works reasonably well for one-dimensional shear flows of the form $\bar{u}_x(y)$. The underlying reason for the comparative success of (14.107) is vortex dynamics. The vorticity associated with the mean flow is $|\partial \bar{u}_x / \partial y|$, while the vorticity of the energy-containing eddies comes from a distortion of the mean flow vortex lines, as illustrated in Figure 14.14. We would expect, therefore, the vorticity of the large eddies to be of the order of $|\partial \bar{u}_x / \partial y|$. Hence, if u' is a typical measure of $|\mathbf{u}'|$ and ℓ_m the size of the large, energy-containing eddies, then we expect

$$u'/\ell_m \sim |\bar{\omega}_z| = \left| \frac{\partial \bar{u}_x}{\partial y} \right|. \tag{14.108}$$

Now, in a neutrally buoyant turbulent shear flow, u_x' and u_y' are of the same order of magnitude and they are strongly correlated for the reasons discussed in section 14.1.2. It follows that

$$\left| \overline{u_x' u_y'} \right| \sim u'^2 \sim \ell_m^2 \left(\frac{\partial \bar{u}_x}{\partial y} \right)^2, \tag{14.109}$$

from which the mixing-length estimate follows.

This justification of the use of mixing length applies only to one-dimensional shear flows, and not to flows of greater complexity. Nevertheless, modellers interested in more complex flows tend to revert to the Boussinesq–Prandtl generalization of (14.107), which is

$$\tau_{ij}^R = 2\rho \nu_t \bar{S}_{ij} - (\rho/3) \overline{(u_k' u_k')} \delta_{ij}, \tag{14.110}$$

$$\nu_t = \ell_m V_t. \tag{14.111}$$

Equation (14.110) is sometimes justified on the grounds that momentum is exchanged through eddying motion in a manner analogous to the microscopic transport of momentum by molecular action. However, this is not very convincing as molecules are discrete and interact only intermittently, whereas turbulent eddies are distributed in space and interact continually. Moreover, the mean-free path length of the molecules is tiny by comparison with the macroscopic dimensions of the flow, which is not true of the large eddies.

An alternative rationalization of the Boussinesq–Prandtl equations might be to suggest that (14.110) merely defines ν_t, while (14.111) is a dimensional necessity. However, this too is a flawed argument, with three major weaknesses. First, we have chosen to relate τ_{ij}^R and $\bar{S}_{ij}$ through a scalar, ν_t, rather than through a tensor. This suggests that such eddy viscosity models will not work well when the turbulence is strongly anisotropic, as is the case when the flow is subject to a significant stratification or background rotation. Second, when $\bar{S}_{ij} = 0$, (14.110) predicts that freely decaying turbulence is isotropic in the sense that $\langle (u_x')^2 \rangle = \langle (u_y')^2 \rangle = \langle (u_z')^2 \rangle$. However, it is known from numerical simulations, and from grid turbulence experiments, that anisotropy in freely decaying turbulence can be stubbornly persistent. (See, for example, the discussion in Davidson et al., 2012.) Finally, (14.110) assumes that τ_{ij}^R is determined by the *local* strain rate, and not by the *history* of the straining of the turbulence. This, in turn, assumes that the turbulent eddies have relaxed to some sort of statistical equilibrium, governed by local conditions alone. But this cannot be true for turbulence that has a significant stream-wise development, since the magnitude of the Reynolds stresses at some particular location depends on the shape and intensity of the local vortices and this, in turn, depends on the history of the straining of those vortices. So we are not, in general, at liberty to assume the turbulence has relaxed to some form of local equilibrium. Evidently we must treat the Boussinesq–Prandtl equations with some care.

Despite these concerns, many one-point closure models adopt the Boussinesq–Prandtl equations as their starting point, and this is true of the k-ε model. It we accept (14.110), the next task is to determine ν_t. It seems natural to take V_t as $k^{1/2}$, where $k = \frac{1}{2}\overline{\mathbf{u}'^2}$, which gives us $\nu_t \sim k^{1/2}\ell_m$. The k-ε model uses $\varepsilon \sim u'^3/\ell_m$ to rewrite this as $\nu_t \sim k^2/\varepsilon$, and so we have

$$\nu_t = c_\mu k^2/\varepsilon \tag{14.112}$$

for some coefficient c_μ. The constant c_μ is given a value of 0.09, which allows the k-ε model to reproduce the observed relationship between shear stress and turbulent kinetic energy in the log-law region of a boundary layer, as discussed in §14.6.2.

The problem now is that we need to be able to estimate k and ε at each point in the flow. To that end, the k-ε model proposes empirical evolution equations for both k and ε. The k equation is based on the exact energy equation, (14.24). For unsteady flows, this may be generalized to read

$$\frac{\partial k}{\partial t} + \bar{\mathbf{u}} \cdot \nabla k = \left(\tau_{ij}^R/\rho\right)\bar{S}_{ij} - \varepsilon - \nabla \cdot \mathbf{T}, \tag{14.113}$$

provided that the timescale for changes in statistically averaged quantities is slow by comparison with the timescale for turbulent fluctuations. (We need a separation in these timescales in order to justify the continued use of a time average to distinguish between the mean flow and the turbulence.) There remains the problem of what to do with the unknowns $\overline{p'u_i'}$ and $\overline{u_i'u_j'u_j'}$ contained within $\mathbf{T}$. In the k-ε model it is assumed that fluctuations in pressure induced by the turbulent eddies act to spread the turbulent kinetic energy in a diffusive manner from regions of strong turbulence to those of weak turbulence. So $\mathbf{T}$ is simply modelled as $\mathbf{T} = -\nu_t \nabla k$, from which we have

$$\frac{\partial k}{\partial t} + \bar{\mathbf{u}} \cdot \nabla k = \nabla \cdot (\nu_t \nabla k) + \left(\tau_{ij}^R/\rho\right) \bar{S}_{ij} - \varepsilon. \tag{14.114}$$

Since the use of ν_t for the diffusivity in (14.114) is somewhat arbitrary, this is often rewritten in the more general form

$$\frac{\partial k}{\partial t} + (\bar{\mathbf{u}} \cdot \nabla)k = \nabla \cdot \left[\left(\nu + \frac{\nu_t}{\sigma_k}\right)\nabla k\right] + \left(\tau_{ij}^R/\rho\right) \bar{S}_{ij} - \varepsilon, \tag{14.115}$$

for some constant σ_k. In practice, however, σ_k is usually set equal to unity, while $\nu \ll \nu_t$.

While the evolution equation for k may seem plausible, the origins of the ε equation are more obscure. It is

$$\frac{\partial \varepsilon}{\partial t} + (\bar{\mathbf{u}} \cdot \nabla)\varepsilon = \nabla \cdot \left[\left(\nu + \frac{\nu_t}{\sigma_\varepsilon}\right)\nabla \varepsilon\right] + c_1 \frac{G\varepsilon}{k} - c_2 \frac{\varepsilon^2}{k}, \tag{14.116}$$

where $G = \left(\tau_{ij}^R/\rho\right) \bar{S}_{ij}$ is the rate of generation of turbulent kinetic energy and σ_ε, c_1, and c_2 are empirical coefficients which are chosen to reproduce certain features of simple turbulent flows. The standard values of these coefficients are: $\sigma_\varepsilon = 1.3$, $c_1 = 1.44$, and $c_2 = 1.92$.

One common interpretation of the ε equation is the following. The characteristic vorticity of a large-scale eddy is u'/ℓ_m, which can be rewritten as ε/k, since $\varepsilon \sim u'^3/\ell_m$. Now consider a one-dimensional shear flow of the form $\bar{u}_x(y)$. The vorticity of the energy-containing eddies, which we label ω', comes from a distortion of the mean-flow vortex lines, as illustrated in Figure 14.14, and so we would expect ω' to be of the order of $|\partial \bar{u}_x/\partial y|$. Thus we have

$$\omega' \sim \varepsilon/k \sim |\partial \bar{u}_x/\partial y|. \tag{14.117}$$

Consider, in particular, the case of a *homogeneous* shear flow, $\partial \bar{u}_x/\partial y = S = \text{constant}$. Then (14.117) becomes $\omega' = \lambda S$ for some constant λ. If the initial value of ω' were different to λS, then we might expect ω' to relax towards λS on a timescale of S^{-1}. Now the k-ε equations model this homogeneous flow, as

$$\frac{dk}{dt} = G - \varepsilon, \quad \frac{d\varepsilon}{dt} = c_1 \frac{G\varepsilon}{k} - c_2 \frac{\varepsilon^2}{k}, \tag{14.118}$$

where

$$G = \nu_t S^2 = c_\mu k^2 S^2 / \varepsilon. \tag{14.119}$$

These may be rearranged to give

$$\frac{d\omega'}{dt} = c_\mu (c_1 - 1) S^2 - (c_2 - 1) \omega'^2,$$

where $\omega' = \varepsilon / k$, or equivalently,

$$\frac{d\omega'}{dt} = (c_2 - 1) \left[(\lambda S)^2 - \omega'^2 \right], \tag{14.120}$$

where $\lambda^2 = c_\mu (c_1 - 1)/(c_2 - 1)$. This does indeed ensure that ω' relaxes to λS on a timescale of S^{-1}. So, provided c_1 and c_2 are both greater than unity, the ε equation ensures that the large-scale vorticity behaves in a plausible manner, at least in a homogeneous shear flow. Indeed, we can think of the ε equation as being an evolution equation for ω', continually pushing ω' towards the vorticity of the local mean flow.

In summary, then, the k-ε model adopts the Boussinesq–Prandtl equations:

$$\boxed{\tau_{ij}^R / \rho = 2\nu_t \bar{S}_{ij} - (2/3) k \delta_{ij}}, \tag{14.121}$$

$$\boxed{\nu_t = c_\mu k^2 / \varepsilon}. \tag{14.122}$$

These are then supplemented by two empirical evolution equations for k and ε,

$$\boxed{\frac{\partial k}{\partial t} + (\bar{\mathbf{u}} \cdot \nabla)k = \nabla \cdot \left[(\nu + \frac{\nu_t}{\sigma_k}) \nabla k \right] + G - \varepsilon}, \tag{14.123}$$

$$\boxed{\frac{\partial \varepsilon}{\partial t} + (\bar{\mathbf{u}} \cdot \nabla)\varepsilon = \nabla \cdot \left[(\nu + \frac{\nu_t}{\sigma_\varepsilon}) \nabla \varepsilon \right] + c_1 \frac{G\varepsilon}{k} - c_2 \frac{\varepsilon^2}{k}}, \tag{14.124}$$

where $G = \left(\tau_{ij}^R / \rho \right) \bar{S}_{ij}$ and

$$c_\mu = 0.09, \quad \sigma_k = 1, \quad \sigma_\varepsilon = 1.3, \quad c_1 = 1.44, \quad c_2 = 1.92.$$

As already noted, the values of these coefficients have been chosen to allow the k-ε model to reproduce certain well-documented observations in standard turbulent flows. So, in some sense, the k-ε model is a sophisticated exercise in interpolation.

14.6.2 The k-ε Model applied to Some Simple Turbulent Flows

Consider freely decaying, homogeneous turbulence in which there is no mean flow, $\bar{\mathbf{u}} = 0$. The k-ε model then predicts

$$\frac{dk}{dt} = -\varepsilon, \quad \frac{d\varepsilon}{dt} = -c_2 \frac{\varepsilon^2}{k}, \quad \frac{d\omega'}{dt} = -(c_2 - 1)\omega'^2,$$

where $\omega' = \varepsilon/k$. These may be integrated to give $k = k_0 \left(1 + t/\tau\right)^{-n}$, where $\tau = nk_0/\varepsilon_0$ is the initial turn-over time and the decay exponent is $n = (c_2 - 1)^{-1} = 1.09$. (Actually, grid turbulence experiments show that k decays somewhat faster than this, with $1.1 < n < 1.4$.) Note that the expression $n = (c_2 - 1)^{-1}$ illustrates how tuning the coefficients in the k-ε model allows it to capture certain experimental observations.

Next we turn to the log-law region of a boundary layer. Here $\tau_{xy}^R = \rho V_*^2$ and so the rate of generation of turbulent energy is

$$G = \frac{\tau_{xy}^R}{\rho} \frac{\partial \bar{u}_x}{\partial y} = \frac{V_*^3}{\kappa y}.$$

We have also seen that, throughout the log-law region, $k \approx 3.5V_*^2$ and $G/\varepsilon \approx 0.91$ (see (14.70)), although there is a weak logarithmic variation of k with y. Given that k is almost uniform across the log-law layer, the k equation in the k-ε model reduces to $G = \varepsilon$, requiring a local balance between the generation and dissipation of energy. The eddy viscosity estimate (14.122) then becomes

$$\nu_t = \frac{\tau_{xy}^R/\rho}{\partial \bar{u}_x/\partial y} = \frac{V_*^2}{V_*/\kappa y} = c_\mu \frac{k^2}{\varepsilon} = c_\mu \frac{k^2}{V_*^3/\kappa y}, \tag{14.125}$$

which yields $c_\mu = (V_*^2/k)^2$. Given the measured value of $k \approx 3.5V_*^2$, this fixes c_μ as ~ 0.08, which is close to the value of $c_\mu = 0.09$ adopted in the model. Once again, we see how tuning the coefficients in the model allows it to capture certain experimental observations.

Consider now the ε equation applied to the log-law layer. Given that $\varepsilon = G = V_*^3/\kappa y$ and $c_\mu = (V_*^2/k)^2$ in the k-ε model, (14.124) reduces to

$$\kappa^2 = \sigma_\varepsilon c_\mu^{1/2}(c_2 - c_1). \tag{14.126}$$

For specified values of κ, c_μ, c_2, and c_1, this determines σ_ε. Note that the model values of $\sigma_\varepsilon = 1.3$ and $c_2 - c_1 = 0.48$ correspond to $\kappa = 0.43$, which is close to the measured value of κ. It would seem, therefore, that the k-ε model works well in the log-law region of a boundary layer, largely because it has been tuned to do so.

Finally, consider the one-dimensional, homogeneous shear flow, $\bar{u}_x(y) = Sy$. Here it may be shown that the k-ε model predicts that the flow evolves towards a state in which $\tau_{xy}^R/(\rho k)$ and G/ε are both constant (see Exercise 14.4), with

$$\tau_{xy}^R/(\rho k) = \sqrt{c_\mu(c_2 - 1)/(c_1 - 1)} = 0.43,$$
$$G/\varepsilon = (c_2 - 1)/(c_1 - 1) = 2.1.$$

Wind tunnel data also suggests that, after a transient, $\tau_{xy}^R/(\rho k)$ and G/ε are both constant, but with the values $\tau_{xy}^R/(\rho k) \sim 0.28$ and $G/\varepsilon \sim 1.7$. It is reassuring that the k-ε model captures the correct physical trend, albeit with different numerical values.

In general, the k-ε model performs well in thin shear layers, such as a zero-pressure-gradient boundary layer, largely because the various coefficients have been chosen to ensure this. However, the model often performs poorly in more complex flows, particularly those involving a stagnation-point, a large mean rate-of-strain, a strong adverse pressure gradient, or strong anisotropy. Perhaps it is reasonable to characterize the k-ε model as a useful if imperfect tool, whose deficiencies are reasonably well understood. It has several closely related cousins, and each formulation has its own strengths and weaknesses. However, one should not lose sight of the fact that all of these schemes are ultimately sophisticated exercises in interpolating between experimental data sets.

○ ○ ○

This concludes our brief introduction to turbulent shear flows. Readers keen to learn more will find a measured introduction to the topic in Tennekes and Lumley (1972), while an advanced treatment may be found in Monin and Yaglom (1971) and Townsend (1976).

....

EXERCISES

14.1 *The log-law of the wall derived using Prandtl's mixing length theory.* Consider a one-dimensional shear layer $\bar{u}_x(y)$ adjacent to a wall. Show that, if $\partial \bar{p}/\partial x$ can be neglected, then the total shear stress, $\bar{\tau}_{xy}(y) + \tau_{xy}^R(y)$, is independent of y. Now find $d\bar{u}_x/dy$ using Prandtl's mixing-length equation, (14.16), where the mixing length is taken as $\ell_m = \kappa y$. You may assume that the viscous stress is negligible except in the immediate vicinity of the wall. Confirm that $\bar{u}_x/V_* = \kappa^{-1} \ln y + \text{const.}$ and hence deduce the log-law of the wall.

14.2 *Self-induced rotation of a hairpin vortex.* Use the Biot–Savart law to show that, if an incipient hairpin vortex is created by a stream-wise gust, as shown in Figure 14.14, then the vortex will rotate by self-induction, moving the tip of the vortex upwards into the flow.

14.3 *The evolution of homogeneous shear flow.* Consider a one-dimensional, homogeneous shear flow of the form $\bar{u}_x(y) = Sy$. Show that the energy equation (14.113) reduces to

$$\frac{dk}{dt} = (\tau_{xy}^R S/\rho) - \varepsilon = G - \varepsilon,$$

which may be rewritten as

$$\frac{d}{d(St)}(\ln k) = \frac{\tau_{xy}^R}{\rho k}(1 - \varepsilon/G).$$

Experiments show that the turbulence in such a flow tends to a state in which the ratio of τ_{xy}^R to ρk is constant, as is the ratio of G to ε. Show that the asymptotic state then takes the form

$$k = k_0 \exp\left(\hat{\lambda} St\right), \quad \varepsilon = \varepsilon_0 \exp\left(\hat{\lambda} St\right),$$

where $\hat{\lambda}$ is the constant,

$$\hat{\lambda} = \frac{\tau_{xy}^R}{\rho k} \left[1 - \varepsilon/G\right].$$

Experiments give $\tau_{xy}^R/\rho k \approx 0.28$ and $G/\varepsilon \approx 1.7$, from which $k/k_0 \approx \varepsilon/\varepsilon_0 \approx \exp\left(0.12\,St\right)$.

14.4 *The k-ε model applied to homogeneous shear flow.* Consider the homogeneous shear flow $\bar{u}_x(y) = Sy$. Integrate the k-ε model equation, (14.120), and show that, for $St \gg 1$, $\omega' = \varepsilon/k$ tends to λS, where $\lambda^2 = c_\mu(c_1 - 1)/(c_2 - 1)$. Now consider late times for which $\omega' = \lambda S$ and show that the k-ε model requires that $\tau_{xy}^R/(\rho k)$ and G/ε are both constant and equal to

$$\tau_{xy}^R/(\rho k) = c_\mu/\lambda = \sqrt{c_\mu(c_2 - 1)/(c_1 - 1)} = 0.43,$$
$$G/\varepsilon = (c_2 - 1)/(c_1 - 1) = 2.1.$$

Finally, show that the k-ε model predicts that k and ε grow as $k/k_0 = \varepsilon/\varepsilon_0 = \exp\left(0.23\,St\right)$.

14.5 *The vortex dynamics of homogeneous shear flow.* It is instructive to consider the generation of turbulent kinetic energy in a homogeneous shear flow $\bar{u}_x(y) = Sy$ in terms of vortex stretching. Writing $\boldsymbol{\omega} = \bar{\boldsymbol{\omega}} + \boldsymbol{\omega}'$, show that the vorticity equation for this flow is

$$\frac{D\boldsymbol{\omega}'}{Dt} = -S\frac{\partial \mathbf{u}'}{\partial z} + S\omega_y'\hat{\mathbf{e}}_x + \boldsymbol{\omega}' \cdot \nabla\mathbf{u}' + \nu\nabla^2\boldsymbol{\omega}'.$$

The first contribution on the right, $-S\partial\mathbf{u}'/\partial z$, comes from $\nabla \times (\mathbf{u}' \times \bar{\boldsymbol{\omega}})$ and so represents the advection of the mean vortex lines by the turbulent velocity field. This may be pictured as shown in Figure 14.20, where a turbulent gust distorts the mean vortex lines, producing some turbulent vorticity. For example, a vertical gust, u_y', induces vertical vorticity, ω_y'. The second contribution on the right, $S\omega_y'\hat{\mathbf{e}}_x$, derives from $\nabla \times (\bar{\mathbf{u}} \times \boldsymbol{\omega}')$ and so represents the action of the mean shear on the turbulent vorticity. As noted in §14.4.5, this process is one by which the mean flow intensifies the turbulent vorticity by vortex stretching, as shown in Figure 14.14. The reason why it appears as a source of stream-wise vorticity is because the mean velocity, $\bar{u}_x$, tilts

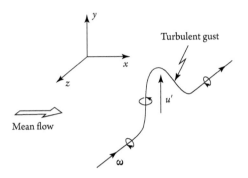

Figure 14.20 A turbulent gust creates turbulent vorticity by bending the mean vortex lines.

the vertical vorticity ω'_y, thus acting as a source of ω'_x. Now confirm that

$$\frac{D}{Dt}\left[\tfrac{1}{2}(\boldsymbol{\omega}')^2\right] = -S\nabla\cdot[u'_z\boldsymbol{\omega}'] + \omega'_x\omega'_y S + \omega'_i\omega'_j S'_{ij} + \nu\boldsymbol{\omega}'\cdot\nabla^2\boldsymbol{\omega}',$$

and use the fact that, in homogeneous turbulence, the ensemble average of a divergence is zero to show that

$$\frac{\partial}{\partial t}\left\langle\tfrac{1}{2}(\boldsymbol{\omega}')^2\right\rangle = \left\langle\omega'_x\omega'_y\right\rangle S + \left\langle\omega'_i\omega'_j S'_{ij}\right\rangle - \nu\left\langle(\nabla\times\boldsymbol{\omega}')^2\right\rangle.$$

Discuss the relative roles played by the various terms on the right of this equation in the generation and destruction of turbulent enstrophy.

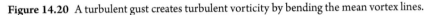

REFERENCES

Davidson, P.A., 2015, *Turbulence: an introduction for scientists and engineers,* 2nd Ed., Oxford University Press.

Davidson, P.A. & Krogstad, P.-A., 2014, A universal scaling for low-order structure functions in the log-law region of smooth and rough-wall boundary layers. *J. Fluid Mech.*, **752**, 140–56.

Davidson, P.A., Krogstad, P.-A., & Nickels, T.B., 2006, A refined interpretation of the logarithmic structure function law in wall layer turbulence. *Phys. Fluids*, **18**, 065112.

Davidson, P.A., Okamoto, N., & Kaneda, Y., 2012, On freely decaying, anisotropic, axisymmetric Saffman turbulence. *J. Fluid Mech.*, **706**, 150–72.

Landau, L.D. & Lifshitz, E.M., 1959, *Fluid mechanics*, 1st Ed., Pergamon Press.

Monin, A.S. & Yaglom, A.M., 1971, *Statistical fluid mechanics I.* MIT Press.

Schlichting, H., 1979, *Boundary-layer theory*, 7th Ed., McGraw-Hill.

Tennekes, H. & Lumley, J.L., 1972, *A first course in turbulence.* MIT Press.

Townsend, A.A., 1976, *The structure of turbulent shear flow*, 2nd Ed., Cambridge University Press.

Dimensional Analysis

Dimensional analysis is a powerful technique for reducing the number of independent variables in a problem by gathering all of the relevant physical parameters into dimensionless groups. It was championed by, amongst others, Fourier, Maxwell, Reynolds, and Rayleigh.

In fluid mechanics, most quantities have dimensions which can be constructed from some subset of mass (M), length (L), and time (T). For example, the dimensions of speed, density, and pressure are LT^{-1}, ML^{-3}, and $ML^{-1}T^{-2}$, respectively. Now, a physical parameter, say speed, has a particular numerical value in a given set of units, and of course if we change units, that numerical value must also change. For example, on moving from SI to CGS units, $V = 2$ m/s becomes $V = 200$ cm/s.

A physically meaningful equation must hold for all systems of units, and so each term in an equation like $A = B + C$ must have the same dimensions. If they did not, then the numerical values of A, B and C would change in different ways when we changed units, and the equation would no longer hold. This is the *principle of dimensional homogeneity*, and it is the starting point for much of dimensional analysis. For example, in the Navier–Stokes equation, (2.20), the inertial, pressure, gravitational and viscous terms all have the dimensions of acceleration. More generally, suppose we have an equation of the form

$$A = B + \frac{C + D}{E} \cos \theta.$$

Then dimensional homogeneity tells us that C and D have the same dimensions, as do A, B and C/E. Moreover, θ must be dimensionless because the cosine can be expanded as $\cos \theta = 1 - \theta^2/2 + \dots$.

A typical application of dimensional analysis is the following. Suppose we have laminar flow over a sphere, as shown in Figure A1.1, and we are interested in the drag force F_D acting on the sphere. It is clear that F_D depends in some way on ρ, ν, the radius of the sphere, R, and the far-field speed, V, and so we might write

$$F_D = f_1(\rho, \nu, R, V),\tag{A1.1}$$

for some function f_1. It is natural to ask if we can gain more insight into the functional relationship between F_D and the other parameters of the problem without having to solve the governing equations. This is the type of question for which dimensional analysis is particularly well suited.

One approach is the following. We start by writing down the governing equation and boundary conditions

$$(\mathbf{u} \cdot \nabla)\mathbf{u} = -\nabla(p/\rho) + \nu \nabla^2 \mathbf{u}, \quad \text{with} \quad \mathbf{u}_\infty = V \hat{\mathbf{e}}_z, \quad \mathbf{u}_{|\mathbf{x}|=R} = 0.\tag{A1.2}$$

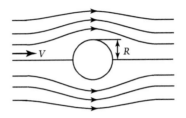

Figure A1.1 Flow over a sphere.

Note that density is *not* a free parameter in this equation because we have folded it in with the pressure to produce a rescaled pressure field. So the free parameters on which $\mathbf{u}$ depends are ν, R and V, the latter two appearing in the boundary conditions rather than the governing equation itself. Next, we introduce dimensionless versions of all of the dependent and independent variables in the equation, say

$$\mathbf{u}^* = \mathbf{u}/V, \quad (p/\rho)^* = (p/\rho)/V^2, \quad \mathbf{x}^* = \mathbf{x}/R,$$

and replace the dimensional variables in (A1.2) by their dimensionless counterparts,

$$(\mathbf{u}^* \cdot \nabla^*)\mathbf{u}^* = -\nabla^* (p/\rho)^* + \frac{\nu}{VR}\nabla^{*2}\mathbf{u}^*, \quad \text{with} \quad \mathbf{u}^*_\infty = \hat{\mathbf{e}}_z, \quad \mathbf{u}^*_{|\mathbf{x}^*|=1} = 0. \tag{A1.3}$$

Dimensional homogeneity tells us that ν/RV must be dimensionless because everything else in (A1.3) is dimensionless, and of course it is just the inverse of the Reynolds number, $\mathrm{Re} = RV/\nu$. Now, the key point is that in (A1.2) we have three independent parameters that can be varied in either the governing equation or the boundary conditions (*i.e.* ν, R, and V), whereas in (A1.3) there is only one free parameter, which is $\mathrm{Re} = RV/\nu$. This is an example of an *independent dimensionless group*, and it *uniquely determines the velocity field* when that field is expressed in dimensionless variables.

More generally, if we carry out this procedure on a more complicated set of equations we find various groups of parameters pre-multiplying the starred terms, and dimensional homogeneity tells us that each of these groups is dimensionless. In effect, we have a procedure for flushing out the various independent dimensionless groups that uniquely control the flow.

Returning to Figure A1.1, we note that the pressure and viscous stresses acting on the surface of the sphere are given by

$$p = \rho V^2 \left[(p/\rho)^*\right], \qquad \tau_{ij} = 2\rho\nu S_{ij} = \rho V^2 \left[2\mathrm{Re}^{-1}S_{ij}^*\right],$$

where the dimensionless starred quantities are themselves functions of Re (and of course position on the surface of the sphere). Since the drag force is a surface integral of these stresses, we deduce that

$$F_D = \rho V^2 R^2 f_2(\mathrm{Re}), \tag{A1.4}$$

for some function f_2. This is most conveniently expressed in dimensionless form. By convention, we introduce the drag coefficient C_D and write

$$C_D = \frac{F_D}{\frac{1}{2}\rho V^2 A} = f_3(\mathrm{Re}), \tag{A1.5}$$

where $A = \pi R^2$ is the frontal area of the sphere. If we compare this with (A1.1), it is clear that we have made a great deal of progress, and without having to solve any equations! This is the power of dimensional analysis. C_D is an example of a *dependent dimensionless group*, because it depends on the details of the flow field, and hence on Re. We read (A1.5) as saying that the dependent dimensionless group, C_D, is some function of the independent dimensionless group, Re.

Of course, a similar analysis may be performed for flow over a circular cylinder, only now F_D is the drag force per unit length of the cylinder and the frontal area is $A = 2R$. The variation of C_D with Re for a circular cylinder, determined experimentally, is shown in Figure A1.2. Expressions such as (A1.5) are of great help in the design experiments. If we insisted on measuring the dependence of F_D on ρ, ν, R, and V individually, in accordance with (A1.1), then many thousands of measurements would be required, whereas (A1.5) tells us that we need only vary one parameter in the experiment, *i.e.* Re.

Now, there is a faster, more efficient, way of getting from (A1.1) to (A1.5) which does not even require a detailed knowledge of the governing equations. There is a theorem, called the *Buckingham Pi theorem*, which tells us how many dimensionless groups to expect in any given situation. First, one counts the number of independent parameters which control the velocity field (in this case ν, R and V), to which we add the dependent parameter of interest (F_D in this case) and any other parameter which might influence the dependent parameter (ρ for this particular problem). This gives us a total of five parameters, say $P = 5$. Next, we count the number of relevant dimensions, which for the drag on a sphere is M, L and T, giving $D = 3$. The Buckingham Pi theorem then tells us that the number of dimensionless groups involved is (usually) $G = P - D$, and so we expect $G = 2$. By tradition, these dimensionless groups are labelled Π_1, Π_2, etc., which is the origin of the name 'Pi theorem'. For flow over a sphere, inspection of (A1.1) tells us that these groups are

$$\Pi_1 = C_D = \frac{F_D}{\frac{1}{2}\rho V^2 A} \,, \quad \Pi_2 = \text{Re} = \frac{VR}{\nu}. \tag{A1.6}$$

The final step is to note that dimensional homogeneity requires that any given dimensionless group can be a function *only* of another dimensionless group, so we have $\Pi_1 = f(\Pi_2)$, which is (A1.5). It is extraordinary that we can get so far without a detailed knowledge of the governing equations!

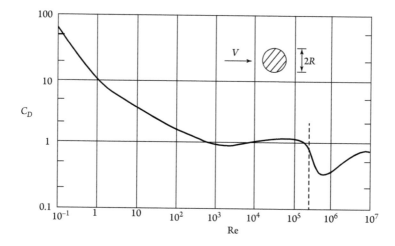

Figure A1.2 Drag coefficient for a cylinder, $C_D = F_D / \frac{1}{2}\rho V^2 A$, versus Re $= 2RV/\nu$.

Admittedly, there are occasions, albeit exceptional occasions, when $G \neq P - D$, but rather $G > P - D$. (See, for example, the discussion in Barenblatt, 2003.) However, these degenerate cases are rare and the rule $G = P - D$ almost invariably holds good. The Pi theorem is generally attributed to Buckingham (1914), although several other scientists had similar ideas, such as Rayleigh.

Perhaps some comments are in order. First, the Pi theorem relies on choosing the correct independent parameters, and this often takes considerable physical insight. A common mistake is to choose the dynamic viscosity, rather than the kinematic viscosity, as a control parameter for an incompressible velocity field, the former introducing mass as a dimension. (In the case above, this does not matter since ρ ends up being included in the list of parameters for different reasons, but that is not always the case.) Second, there are systematic procedures for calculating the dimensionless groups associated with a given list of dimensional parameters. However, in practice it is almost always faster to simply spot the dimensionless groups by inspection, as we shall see. Third, given two or more dimensionless groups, one can always construct additional ones by combining them. For example, in the case above $\Pi_1 \Pi_2$ is also dimensionless. The Pi theorem merely tells us what the minimum number of dimensionless groups is, but then that is all we need to know. Finally, the special case where there is only one dimensionless group, say Π_1, is of particular interest. Here there are no other dimensionless groups on which Π_1 can depend and so it must be a constant for the problem in hand. This then yields a *scaling law*, as discussed below.

Let us illustrate the Pi theorem with some simple examples involving surface gravity waves. Suppose we have waves of wavelength λ on the surface of an inviscid fluid of depth h, as shown in Figure A1.3. We are interested in the angular frequency of the waves and our physical intuition tells us that it should depend on λ, g, and h:

$$\varpi = f(\lambda, g, h). \tag{A1.7}$$

We have four parameters that between them contain only two dimensions, L and T. The Pi theorem therefore tells us that the number of dimensionless groups is $G = 2$. Clearly one of them is $\Pi_2 = h/\lambda$, and since time appears only in ϖ and g, the other group should be proportional to ϖ^2/g. To eliminate length from ϖ^2/g we must multiply by either λ or h, and so we conclude that $\Pi_1 = \varpi^2 \lambda/g$ is a legitimate choice for the other group. Dimensional homogeneity then yields

$$\varpi^2 \lambda/g = f(h/\lambda), \tag{A1.8}$$

for some function f. In terms of the wavenumber, $k = 2\pi/\lambda$, this reads

$$\varpi^2/gk = f(kh). \tag{A1.9}$$

This might be compared with the exact form of the dispersion relationship derived in Chapter 7:

$$\varpi^2/gk = \tanh kh. \tag{A1.10}$$

In the case of deep-water waves, h ceases to be a relevant parameter and (A1.7) becomes $\varpi = f(k, g)$, giving $P = 3$, $D = 2$, and $G = 1$. Our single dimensionless group is then $\Pi_1 = \varpi^2/gk$. Now, as noted above, one dimensionless group can depend only on another dimensionless group, and since there are no other dimensionless groups available to us, we conclude that ϖ^2/gk is a constant. This yields the *scaling law* $\varpi^2 \sim kg$ for deep-water waves, the exact law being $\varpi^2 = kg$.

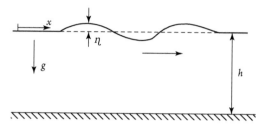

Figure A1.3 A surface gravity wave.

Returning to waves on water of finite depth, suppose that we mistakenly assumed that density is a relevant parameter. Then (A1.7) is amended to read

$$\varpi = f(\rho, k, g, h), \tag{A1.11}$$

which adds mass to the list of dimensions. We now have $P = 5$ and $D = 3$, which again yields $G = 2$. There is, however, a problem as M appears only in ρ and so there is no possibility of incorporating density into a dimensionless group. We conclude that (A1.11) is an illegitimate list of parameters and that we must revert to (A1.7).

Finally, suppose that the wavelength is sufficiently short for surface tension to be important but h to be irrelevant. Then we must include the surface tension coefficient γ in our list of parameters. Since this involves the dimension M, we are now free to include density as one of our parameters. In fact, we are *obliged* to include density because without it we cannot make γ dimensionless. We now have

$$\varpi = f(\rho, \gamma, k, g), \tag{A1.12}$$

which gives $P = 5$, $D = 3$ and $G = 2$. We now need to find a new dimensionless group involving surface tension and density, and since γ has the dimensions of MT^{-2}, we conclude that one possibility is $\gamma k^2 / \rho g$. The other group may be taken as ϖ^2 / gk and dimensional homogeneity then gives us

$$\varpi^2 / gk = f\left(\gamma k^2 / \rho g\right). \tag{A1.13}$$

This might be compared with the exact result (7.52):

$$\varpi^2 / gk = 1 + \gamma k^2 / g\rho. \tag{A1.14}$$

We close this appendix by listing, in Table A1.1, some of the more commonly used dimensionless groups in fluid dynamics. In Chapter 1 we introduced the Reynolds number, Re, and the free-surface Froude number, Fr:

$$\mathrm{Re} = V\ell/\nu, \quad \mathrm{Fr} = V\big/\sqrt{gh},$$

where h is the fluid depth. The former is indicative of the ratio of inertial to viscous forces and the latter is the ratio of the fluid velocity to the shallow-water wave speed. An alternative definition of the Froude number, appropriate for stratified flow, is

Table A1.1 Some common dimensionless groups in fluid mechanics.

Name of group	Defining equation	Definition	Interpretation of group
Reynolds number	-	$Re = \frac{V\ell}{\nu}$	Ratio of inertial to viscous forces. (V = typical speed, ℓ = typical length)
Froude number (free-surface flow)	-	$Fr = \frac{V}{\sqrt{gh}}$	Ratio of velocity to surface wave speed in shallow-water flow. (h = depth)
Froude number (for stratified flow)	(9.6)	$Fr = \frac{u}{N\ell}$	Ratio of velocity to internal wave speed (N = Väisälä–Brunt frequency)
Rossby number	(10.15)	$Ro = \frac{u}{\Omega\ell}$	Ratio of inertial to Coriolis forces. (Ω = background rotation rate)
Taylor number	(11.22)	$Ta = \frac{\Omega^2 R d^3}{\nu^2}$	Ratio of destabilizing angular momentum gradient to stabilizing viscous forces.
Raleigh number	(11.25)	$Ra = \frac{g\beta\Delta T d^3}{\nu\alpha}$	Ratio of destabilizing buoyancy forces to stabilizing diffusive processes.
Global Richardson number	(11.68)	$Ri = \frac{g(\Delta\rho/2d)}{\rho_0 (du_x/dz)^2}$	Ratio of stabilizing stratification to destabilizing shear.
Local (or gradient) Richardson number	(11.70)	$Ri = \frac{g\lvert d\rho_0/dz\rvert}{\bar{\rho}(du_x/dz)^2}$	Ratio of stabilizing stratification to destabilizing shear.
Drag coefficient	(4.25)	$C_D = \frac{F_D}{\frac{1}{2}\rho V^2 A}$	Dimensionless drag force. (A = frontal area)
Lift coefficient	(6.50)	$C_L = \frac{F_L}{\frac{1}{2}\rho V_\infty^2 L}$	Dimensionless lift force per unit span of wing. (L = chord)
Prandtl number	-	$Pr = \frac{\nu}{\alpha}$	Ratio of momentum diffusivity to thermal diffusivity.
Weber number	-	$We = \frac{\rho V^2 \ell}{\gamma}$	Ratio of inertial forces to surface tension forces. (γ = surface tension coefficient)
Mach number	-	$Ma = V/a$	Ratio of flow speed to speed of sound, a.

$$Fr = u/N\ell,$$

where N is the Väisälä–Brunt frequency. This represents the ratio of the fluid velocity to *internal* wave speed, while the Rossby number,

$$Ro = u/\Omega\ell,$$

is usually interpreted as the ratio of inertial to Coriolis forces in a rotating fluid.

In Chapter 11 we introduced the Taylor, Rayleigh, and Richardson numbers:

$$\text{Ta} = \frac{\Omega^2 R d^3}{\nu^2}, \quad \text{Ra} = \frac{g\beta\Delta T d^3}{\nu\alpha}, \quad \text{Ri} = \frac{g\,|d\rho_0/dz|}{\bar{\rho}\,(du_x/dz)^2},$$

which are all important in stability theory. The Richardson number is a measure of the ratio of the stabilizing influence of stratification to the destabilizing role of shear.

The lift and drag coefficients, C_L and C_D, are simply normalized measures of lift and drag, the Prandtl number the ratio of two diffusivities, and the Weber number,

$$\text{We} = \rho V^2 \ell/\gamma,$$

a measure of the ratio of inertial to surface tension forces. Finally, the Mach number is the ratio of the flow speed to the speed of sound. Note that many of these dimensionless groups may be thought of as the ratio of two competing processes.

....

REFERENCES

Barenblatt, G.I., 2003, *Scaling*, Cambridge University Press.

Buckingham, E., 1914, On physically similar systems: illustrations of the use of dimensional equations. *Phys. Rev.*, **4**(4), 345–76.

Vector Identities and Theorems

Triple products:

$$(\mathbf{a} \times \mathbf{b}) \cdot \mathbf{c} = (\mathbf{b} \times \mathbf{c}) \cdot \mathbf{a} = (\mathbf{c} \times \mathbf{a}) \cdot \mathbf{b}$$

$$\mathbf{a} \times (\mathbf{b} \times \mathbf{c}) = (\mathbf{a} \cdot \mathbf{c}) \mathbf{b} - (\mathbf{a} \cdot \mathbf{b}) \mathbf{c}$$

$$\mathbf{a} \times (\mathbf{b} \times \mathbf{c}) + \mathbf{b} \times (\mathbf{c} \times \mathbf{a}) + \mathbf{c} \times (\mathbf{a} \times \mathbf{b}) = 0$$

Grad, div, and curl in Cartesian coordinates:

$$\nabla \varphi = \frac{\partial \varphi}{\partial x} \mathbf{i} + \frac{\partial \varphi}{\partial y} \mathbf{j} + \frac{\partial \varphi}{\partial z} \mathbf{k}$$

$$\nabla \cdot \mathbf{F} = \frac{\partial F_x}{\partial x} + \frac{\partial F_y}{\partial y} + \frac{\partial F_z}{\partial z}$$

$$\nabla \times \mathbf{F} = \left[\frac{\partial F_z}{\partial y} - \frac{\partial F_y}{\partial z} \right] \mathbf{i} + \left[\frac{\partial F_x}{\partial z} - \frac{\partial F_z}{\partial x} \right] \mathbf{j} + \left[\frac{\partial F_y}{\partial x} - \frac{\partial F_x}{\partial y} \right] \mathbf{k}$$

Grad, div, and curl in cylindrical polar coordinates (r, θ, z):

$$\nabla \varphi = \frac{\partial \varphi}{\partial r} \hat{\mathbf{e}}_r + \frac{1}{r} \frac{\partial \varphi}{\partial \theta} \hat{\mathbf{e}}_\theta + \frac{\partial \varphi}{\partial z} \hat{\mathbf{e}}_z$$

$$\nabla \cdot \mathbf{F} = \frac{1}{r} \frac{\partial}{\partial r} (r F_r) + \frac{1}{r} \frac{\partial F_\theta}{\partial \theta} + \frac{\partial F_z}{\partial z}$$

$$\nabla \times \mathbf{F} = \left[\frac{1}{r} \frac{\partial F_z}{\partial \theta} - \frac{\partial F_\theta}{\partial z} \right] \hat{\mathbf{e}}_r + \left[\frac{\partial F_r}{\partial z} - \frac{\partial F_z}{\partial r} \right] \hat{\mathbf{e}}_\theta + \left[\frac{1}{r} \frac{\partial}{\partial r} (r F_\theta) - \frac{1}{r} \frac{\partial F_r}{\partial \theta} \right] \hat{\mathbf{e}}_z$$

$$\nabla^2 \varphi = \frac{1}{r} \frac{\partial}{\partial r} \left(r \frac{\partial \varphi}{\partial r} \right) + \frac{1}{r^2} \frac{\partial^2 \varphi}{\partial \theta^2} + \frac{\partial^2 \varphi}{\partial z^2}$$

$$\nabla^2 \mathbf{F} = \left[\nabla^2 F_r - \frac{1}{r^2} F_r - \frac{2}{r^2} \frac{\partial F_\theta}{\partial \theta} \right] \hat{\mathbf{e}}_r + \left[\nabla^2 F_\theta - \frac{1}{r^2} F_\theta + \frac{2}{r^2} \frac{\partial F_r}{\partial \theta} \right] \hat{\mathbf{e}}_\theta + (\nabla^2 F_z) \hat{\mathbf{e}}_z$$

Vector identities:

$$\nabla(\varphi\psi) = \varphi\nabla\psi + \psi\nabla\varphi$$
$$\nabla(\mathbf{F}\cdot\mathbf{G}) = (\mathbf{F}\cdot\nabla)\mathbf{G} + (\mathbf{G}\cdot\nabla)\mathbf{F} + \mathbf{F}\times(\nabla\times\mathbf{G}) + \mathbf{G}\times(\nabla\times\mathbf{F})$$
$$\nabla\cdot(\varphi\mathbf{F}) = \varphi\nabla\cdot\mathbf{F} + \mathbf{F}\cdot\nabla\varphi$$
$$\nabla\cdot(\mathbf{F}\times\mathbf{G}) = \mathbf{G}\cdot(\nabla\times\mathbf{F}) - \mathbf{F}\cdot(\nabla\times\mathbf{G})$$
$$\nabla\cdot(\nabla\times\mathbf{F}) = 0$$
$$\nabla\times(\varphi\mathbf{F}) = \varphi(\nabla\times\mathbf{F}) + (\nabla\varphi)\times\mathbf{F}$$
$$\nabla\times(\mathbf{F}\times\mathbf{G}) = \mathbf{F}(\nabla\cdot\mathbf{G}) - \mathbf{G}(\nabla\cdot\mathbf{F}) + (\mathbf{G}\cdot\nabla)\mathbf{F} - (\mathbf{F}\cdot\nabla)\mathbf{G}$$
$$\nabla\times(\nabla\varphi) = 0$$
$$\nabla^2\mathbf{F} = \nabla(\nabla\cdot\mathbf{F}) - \nabla\times(\nabla\times\mathbf{F})$$

Integral theorems:

$$\int_V \nabla\cdot\mathbf{F}\,dV = \oint_S \mathbf{F}\cdot d\mathbf{S}, \quad \int_V \nabla\varphi\,dV = \oint_S \varphi\,d\mathbf{S}$$
$$\int_V \nabla\times\mathbf{F}\,dV = -\oint_S \mathbf{F}\times d\mathbf{S}, \quad \int_S \nabla\times\mathbf{F}\cdot d\mathbf{S} = \oint_C \mathbf{F}\cdot d\mathbf{l}$$
$$\int_S \nabla\varphi\times d\mathbf{S} = -\oint_C \varphi\,d\mathbf{l}$$

Green's inversion of Poisson's equation:
Consider the Poisson equation

$$\nabla^2\varphi = -s(\mathbf{x}), \quad \varphi_\infty = 0, \tag{A2.1}$$

where the source $s(\mathbf{x})$ is a known, localized function of position. This may be inverted in an infinite domain to give

$$\varphi(\mathbf{x}) = \frac{1}{4\pi}\int \frac{s(\mathbf{x}')}{|\mathbf{x} - \mathbf{x}'|}\,d\mathbf{x}', \tag{A2.2}$$

where $\mathbf{x}'$ is a dummy variable that samples the source and $d\mathbf{x}' = dx'\,dy'\,dz'$ is a volume element. This is known as the Green's function solution of Poisson's equation. The origin of Green's solution becomes clear if one considers the case where the source is a unit delta function at the origin, $s(\mathbf{x}) = \delta(\mathbf{x})$. The solution of Poisson's equation is then

$$\varphi = \frac{1}{4\pi\,|\mathbf{x}|}\,,\quad\text{for } s(\mathbf{x}) = \delta(\mathbf{x}), \tag{A2.3}$$

in the sense that $\nabla^2\varphi = 0$ for $|\mathbf{x}| \neq 0$ and integrating over the sphere $|\mathbf{x}| \leq R$ gives

$$\int_V \nabla^2\varphi\,d\mathbf{x} = \oint_S \nabla\varphi\cdot d\mathbf{S} = -\oint_S \frac{1}{4\pi\,|\mathbf{x}|^2}dS = -1 = -\int \delta(\mathbf{x})d\mathbf{x}.$$

If the source is a delta function of strength s' located at $\mathbf{x}'$, then (A2.3) generalizes to $\varphi = s'/4\pi\,|\mathbf{x} - \mathbf{x}'|$. Superposition then yields the Green's function solution (A2.2).

Helmholtz's decomposition:

Any vector field $\mathbf{F}$ may be written as the sum of an irrotational and a solenoidal vector field. The irrotational field may be written in terms of a scalar potential, φ, and the solenoidal field in terms of a vector potential, $\mathbf{A}$:

$$\mathbf{F} = -\nabla\varphi + \nabla\times\mathbf{A}\,,\quad \nabla\cdot\mathbf{A} = 0.$$

These two potentials are then solutions of

$$\nabla^2\varphi = -\nabla\cdot\mathbf{F}\,,\qquad \nabla^2\mathbf{A} = -\nabla\times\mathbf{F},$$

which may be unambiguously inverted using the Green's function to give φ and $\mathbf{A}$.

· · **·** · ·

Navier–Stokes Equation in Cylindrical Polar Coordinates

The viscous contribution to the stress tensor in cylindrical polar coordinates:

$$\tau_{rr} = 2\rho\nu \frac{\partial u_r}{\partial r}, \qquad \tau_{\theta\theta} = 2\rho\nu \left[\frac{1}{r} \frac{\partial u_\theta}{\partial \theta} + \frac{u_r}{r} \right], \quad \tau_{zz} = 2\rho\nu \frac{\partial u_z}{\partial z}$$

$$\tau_{r\theta} = \rho\nu \left[r \frac{\partial}{\partial r} \left(\frac{u_\theta}{r} \right) + \frac{1}{r} \frac{\partial u_r}{\partial \theta} \right], \quad \tau_{\theta z} = \rho\nu \left[\frac{1}{r} \frac{\partial u_z}{\partial \theta} + \frac{\partial u_\theta}{\partial z} \right], \quad \tau_{zr} = \rho\nu \left[\frac{\partial u_r}{\partial z} + \frac{\partial u_z}{\partial r} \right]$$

Navier–Stokes equation in cylindrical polar coordinates:

$$\frac{\partial u_r}{\partial t} + \left[(\mathbf{u} \cdot \nabla) u_r - \frac{u_\theta^2}{r} \right] = -\frac{1}{\rho} \frac{\partial p}{\partial r} + \nu \left[\nabla^2 u_r - \frac{u_r}{r^2} - \frac{2}{r^2} \frac{\partial u_\theta}{\partial \theta} \right]$$

$$\frac{\partial u_\theta}{\partial t} + \left[(\mathbf{u} \cdot \nabla) u_\theta + \frac{u_r u_\theta}{r} \right] = -\frac{1}{\rho r} \frac{\partial p}{\partial \theta} + \nu \left[\nabla^2 u_\theta - \frac{u_\theta}{r^2} + \frac{2}{r^2} \frac{\partial u_r}{\partial \theta} \right]$$

$$\frac{\partial u_z}{\partial t} + (\mathbf{u} \cdot \nabla) u_z = -\frac{1}{\rho} \frac{\partial p}{\partial z} + \nu \left[\nabla^2 u_z \right]$$

The Reynolds-averaged Navier–Stokes equation for steady-on-average flow, viscous terms omitted:

$$\left[(\bar{\mathbf{u}} \cdot \nabla) \bar{u}_r - \frac{\bar{u}_\theta^2}{r} \right] = -\frac{1}{\rho} \frac{\partial \bar{p}}{\partial r} - \left[\frac{1}{r} \frac{\partial}{\partial r} \left(r \overline{u_r' u_r'} \right) + \frac{1}{r} \frac{\partial}{\partial \theta} \left(\overline{u_r' u_\theta'} \right) + \frac{\partial}{\partial z} \left(\overline{u_r' u_z'} \right) - \frac{\overline{u_\theta' u_\theta'}}{r} \right]$$

$$\left[(\bar{\mathbf{u}} \cdot \nabla) \bar{u}_\theta + \frac{\bar{u}_r \bar{u}_\theta}{r} \right] = -\frac{1}{\rho r} \frac{\partial \bar{p}}{\partial \theta} - \left[\frac{1}{r^2} \frac{\partial}{\partial r} \left(r^2 \overline{u_\theta' u_r'} \right) + \frac{1}{r} \frac{\partial}{\partial \theta} \left(\overline{u_\theta' u_\theta'} \right) + \frac{\partial}{\partial z} \left(\overline{u_\theta' u_z'} \right) \right]$$

$$\left[(\bar{\mathbf{u}} \cdot \nabla) \bar{u}_z \right] = -\frac{1}{\rho} \frac{\partial \bar{p}}{\partial z} - \left[\frac{1}{r} \frac{\partial}{\partial r} \left(r \overline{u_z' u_r'} \right) + \frac{1}{r} \frac{\partial}{\partial \theta} \left(\overline{u_z' u_\theta'} \right) + \frac{\partial}{\partial z} \left(\overline{u_z' u_z'} \right) \right]$$

APPENDIX 4

· · • · ·

The Fourier Transform

The Fourier transform of some function $f(x)$ is defined as the integral

$$F(k) = \frac{1}{2\pi} \int_{-\infty}^{\infty} f(x) e^{-jkx} dx.$$

(A4.1)

This integral exists, and $F(k)$ is well defined, provided that the integral of $|f(x)|$ from $-\infty$ to ∞ exists. Fourier's integral theorem then says that the inverse transform is

$$f(x) = \int_{-\infty}^{\infty} F(k) e^{jkx} dk.$$

(A4.2)

If $f(x)$ is an even function then

$$F(k) = \frac{1}{\pi} \int_{0}^{\infty} f(x) \cos(kx) dx,$$

(A4.3)

which implies that $F(k)$ is also even and so, from (A4.2),

$$f(x) = 2 \int_{0}^{\infty} F(k) \cos(kx) dk.$$

(A4.4)

The transform pair (A4.3) and (A4.4) is known as the cosine transform.

One reason why the Fourier transform is so useful is its ability to transform a linear differential equation into an algebraic equation. This arises from the property

$$\text{Transform}\, [df/dx] = jkF(k),$$

(A4.5)

which may be confirmed by differentiating (A4.2). Thus, for example,

$$f''(x) + \beta f(x) = g(x)$$

transforms to $(\beta - k^2)F = G$, or equivalently $F = G/(\beta - k^2)$, and if $G(k)$ is known this can (at least in principle) be inverse transformed to give $f(x)$. Additional properties of the Fourier transform are listed below, where † represents a complex conjugate.

(i) The power theorem (for real functions of f and g):

$$2\pi \int_{-\infty}^{\infty} F(k) G^{\dagger}(k) \, dk = \int_{-\infty}^{\infty} f(x) g(x) \, dx. \tag{A4.6}$$

(ii) The convolution theorem:

$$2\pi F(k) G(k) = \text{Transform} \left[\int_{-\infty}^{\infty} f(u) g(x-u) \, du \right]. \tag{A4.7}$$

(iii) The autocorrelation theorem (for a real function f):

$$2\pi |F(k)|^2 = \text{Transform} \left[\int_{-\infty}^{\infty} f(u) f(u+x) \, du \right]. \tag{A4.8}$$

Of these, property (ii) is the most fundamental. In fact, it is readily confirmed that theorems (i) and (iii) follow directly from (ii).

The Fourier transform is readily generalized to functions of more than one variable. Thus, for example,

$$F(k_x, k_y) = \frac{1}{(2\pi)^2} \int_{-\infty}^{\infty} \int_{-\infty}^{\infty} f(x,y) e^{-jk_x x} e^{-jk_y y} \, dx \, dy, \tag{A4.9}$$

$$f(x,y) = \int_{-\infty}^{\infty} \int_{-\infty}^{\infty} F(k_x, k_y) e^{jk_x x} e^{jk_y y} \, dk_x \, dk_y, \tag{A4.10}$$

represents the process of transforming first with respect to one variable, say x, and then the other variable, y. Such a transform pair may be written more compactly as

$$F(\mathbf{k}) = \frac{1}{(2\pi)^2} \int f(\mathbf{x}) e^{-j\mathbf{k}\cdot\mathbf{x}} \, d\mathbf{x}, \tag{A4.11}$$

$$f(\mathbf{x}) = \int F(\mathbf{k}) e^{j\mathbf{k}\cdot\mathbf{x}} \, d\mathbf{k}, \tag{A4.12}$$

where $\mathbf{k} = (k_x, k_y)$ is a vector. This generalizes in an obvious way to three dimensions.

The Fourier transform of a vector field is found by transforming each of its components, one by one. Thus, for example, the Fourier transform, $\hat{\mathbf{u}}(\mathbf{k})$, of a three-dimensional velocity field, $\mathbf{u}(\mathbf{x})$, satisfies the transform pair,

$$\hat{\mathbf{u}}(\mathbf{k}) = \frac{1}{(2\pi)^3} \int \mathbf{u}(\mathbf{x}) e^{-j\mathbf{k}\cdot\mathbf{x}} \, d\mathbf{x}, \tag{A4.13}$$

$$\mathbf{u}(\mathbf{x}) = \int \hat{\mathbf{u}}(\mathbf{k}) e^{j\mathbf{k}\cdot\mathbf{x}} \, d\mathbf{k}. \tag{A4.14}$$

Note that the continuity equation gives

$$\nabla \cdot \mathbf{u} = \int j\mathbf{k} \cdot \hat{\mathbf{u}} e^{j\mathbf{k}\cdot\mathbf{x}} \, d\mathbf{k} = 0, \tag{A4.15}$$

and since this must be true for all $\mathbf{x}$, we have $\mathbf{k} \cdot \hat{\mathbf{u}} = 0$. Moreover, the vorticity is related to $\hat{\mathbf{u}}$ by

$$\boldsymbol{\omega} = \nabla \times \mathbf{u} = \int (j\mathbf{k}) \times \hat{\mathbf{u}} e^{j\mathbf{k}\cdot\mathbf{x}} d\mathbf{k}, \tag{A4.16}$$

and so the transform of the vorticity field satisfies $\hat{\boldsymbol{\omega}} = j\mathbf{k} \times \hat{\mathbf{u}}$.

Now suppose we have a function $g(\mathbf{x})$ which is defined in three dimensions but which has spherical symmetry, i.e. $g = g(r)$ where $r = |\mathbf{x}|$. Then it is readily confirmed that $G(\mathbf{k})$ also has spherical symmetry in $\mathbf{k}$-space. Thus we have

$$G(k) = \frac{1}{(2\pi)^3} \int g(r) e^{-j\mathbf{k}\cdot\mathbf{x}} d\mathbf{x}, \quad k = |\mathbf{k}|,$$

and integrating over the polar and azimuthal angles we find

$$G(k) = \frac{1}{2\pi^2} \int_0^\infty r^2 g(r) \frac{\sin(kr)}{kr} dr \tag{A4.17}$$

(see Bracewell, 1986). The inverse transform turns out to be

$$g(r) = 4\pi \int_0^\infty k^2 G(k) \frac{\sin(kr)}{kr} dk. \tag{A4.18}$$

The reason why the Fourier transform is widely used in turbulence theory is that it may be thought of as a sort of filter that sifts through the various scales present in a random signal. The way this is achieved is best explained in one dimension. Consider a function $f(x)$ which is composed of fluctuations of different characteristic scale. For example, $f(x)$ might be the instantaneous velocity component u_x measured along the x-axis in a turbulent flow. We now introduce the so-called *box function*,

$$h(x) = \begin{cases} (2L)^{-1}, & |x| < L \\ 0, & |x| > L \end{cases} \tag{A4.19}$$

and consider the new function $f^L(x)$, defined by

$$f^L(x) = \int_{-\infty}^\infty h(x') f(x - x') dx'. \tag{A4.20}$$

Physically, f^L is constructed from f as follows. For each x, we evaluate the average value of f in the vicinity of x by integrating f from $x - L$ to $x + L$ and dividing by $2L$. Thus $f^L(x)$ is a smoothed version of $f(x)$, the smoothing operation being carried out over the scale L. The effect of this smoothing is to filter out those parts of $f(x)$ whose characteristic length scale is significantly smaller than L, as shown in Figure A4.1 (a).

Let us now take the Fourier transform of $f^L(x)$ and invoke the convolution theorem (A4.7). Since the Fourier transform of $h(x)$ is $\sin kL/(2\pi kL)$, this gives

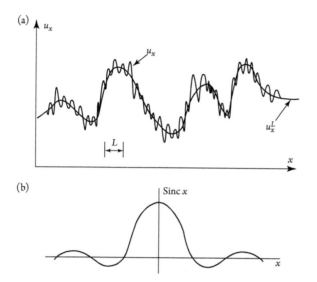

Figure A4.1 (a) The effect of filtering. (b) Shape of sinc(x). (From Davidson, 2015.)

$$F^L(k) = 2\pi H(k)F(k) = \frac{\sin(kL)}{kL}F(k).$$
(A4.21)

The function $\sin x / x$ (also known as sinc x) peaks at the origin, falls to zero at $|x| = \pi$, and oscillates with a diminishing amplitude thereafter (see Figure A4.1 (b)). So, if we ignore the weak oscillations in $H(k)$ for $|k| > \pi/L$, we may regard $F^L(k)$ as a clipped version of $F(k)$, with contributions to $F^L(k)$ suppressed for $|k| > \pi/L$.

We conclude that the operation of smoothing $f(x)$ over a length scale L, eliminating contributions of scale less than L, is roughly equivalent to suppressing the high-wavenumber contribution to $F(k)$, the cut-off being around $k \sim \pi/L$. The implication is that the short length-scale information contained in $f(x)$ is held in the high-wavenumber part of $F(k)$. However, the Fourier transform is a somewhat leaky filter, as there is no simple one-to-one relationship between k and L.

A simple toy problem helps reinforce these ideas. Suppose that $f(x)$ is given by

$$f(x) = \sum A_i \cos(k_i x),$$

where the k_i represent fluctuations of various different scales. Then, from (A4.20),

$$f^L(x) = \sum A_i \frac{1}{2L} \int_{-L}^{L} \cos(k_i(x - x')) \, dx',$$

which integrates to give

$$f^L(x) = \sum A_i \frac{\sin k_i L}{k_i L} \cos k_i x.$$

Evidently, all contributions to $f(x)$ for which $k_i > \pi/L$ are suppressed in f^L.

....

REFERENCES

Bracewell, R.N., 1986, *The Fourier transform and its applications*, McGraw-Hill.

Davidson, P.A., 2015, *Turbulence: an introduction for scientists and engineers*, 2nd Ed., Oxford University Press.

⋅ ⋅ ● ⋅ ⋅

The Physical Properties of Some Common Fluids

Table A5.1 Properties of pure water at various temperatures

Temperature (°C)	Density ρ (kg m^{-3})	Thermal diffusivity α (m^2 s^{-1})	Kinematic viscosity ν (m^2 s^{-1})	Prandtl number Pr
10	999.7	1.38×10^{-7}	1.30×10^{-6}	9.4
20	998.2	1.42×10^{-7}	1.00×10^{-6}	7.1
30	995.7	1.46×10^{-7}	0.802×10^{-6}	5.5
40	992.3	1.52×10^{-7}	0.659×10^{-6}	4.3

Table A5.2 Typical properties of some common liquids at 20° C

Liquid	Density ρ (kg m^{-3})	Surface tension coef. γ (N m^{-1})	Thermal diffusivity α (m^2 s^{-1})	Kinematic viscosity ν (m^2 s^{-1})
Glycerol	1260	6.3×10^{-2}	9.4×10^{-8}	1.2×10^{-3}
Mercury	13.5×10^3	0.47	4.6×10^{-6}	1.1×10^{-7}
Olive oil	920	3.2×10^{-2}	9.4×10^{-8}	9.1×10^{-5}
Engine oil	888	–	8.7×10^{-8}	9.0×10^{-4}

Table A5.3 Properties of air at atmospheric pressure at various temperatures

Temperature (°K)	Density ρ (kg m^{-3})	Thermal diffusivity α (m^2 s^{-1})	Kinematic viscosity ν (m^2 s^{-1})	Prandtl number Pr
250	1.41	1.57×10^{-5}	1.13×10^{-5}	0.72
300	1.18	2.22×10^{-5}	1.57×10^{-5}	0.71
350	0.998	2.98×10^{-5}	2.08×10^{-5}	0.70
400	0.883	3.76×10^{-5}	2.59×10^{-5}	0.69

Table A5.4 Properties of some common gasses at atmospheric pressure and 300° K

Gas	Density $\rho\,(\mathrm{kg\,m^{-3}})$	Thermal diffusivity $\alpha\,(\mathrm{m^2\,s^{-1}})$	Kinematic viscosity $\nu\,(\mathrm{m^2\,s^{-1}})$	Prandtl number Pr
Carbon dioxide	1.80	1.06×10^{-5}	8.32×10^{-6}	0.78
Hydrogen	0.0819	1.55×10^{-4}	1.10×10^{-4}	0.71
Nitrogen	1.14	2.20×10^{-5}	1.56×10^{-5}	0.71
Oxygen	1.30	2.24×10^{-5}	1.59×10^{-5}	0.71

INDEX

A

acceleration of fluid particle 9
advection 56
advection-diffusion equation
  for heat 56, 141, 470
  for vorticity 56
aerofoil
  angle of attack 178
  circulation around 175
  drag on 165
  Kutta condition 175
  Kutta-Joukowski lift
    theorem 22, 172
  lift force 22, 173, 177
  slender, theory of 178
angular impulse 256
angular momentum 29, 256, 303,
  437
anisotropy, in turbulence 417, 442
Arnold's variational
  principle 361–367
atmospheric boundary
  layer 472–476
attached vortices 236
autocorrelation function 403,
  409, 497
averages
  ensemble 392
  time 392
  volume 393
axisymmetric flows
  jets 147, 454
  stream function 11, 75
  viscous 73–84, 246, 249

B

Batchelor's k^4 spectrum 415, 436
Bearings
  journal (Sommerfeld) 109
  slipper (Reynolds) 107
  stepped (Rayleigh) 111
Bénard convection 341–347, 374

Bernoulli equation
  for inviscid flow 14–17
  for irrotational flow 156
bifurcation theory 384–388
biharmonic equation 98
Biot-Savart law 53, 62, 222
Bjerknes' theorem 286
blocking, in a stratified fluid 270
bluff-body flow 139, 165
Borda-Carnot head-loss 24, 401
bore 214
boundary layer
  accelerated 132–137
  adverse pressure gradient
    (decelerated) 50, 138
  approximation 49, 125
  Blasius solution 127
  converging channel flow 136
  flat plate 127
  instability of 381
  laminar 127, 132–137
  log-law layer
    (turbulent) 456–466
  mean velocity profile
    (turbulent) 461
  on circular cylinder 135
  separation 138–140
  similarity solution 127, 133,
    137
  thermal 141–144
  transition to turbulence 381
bound vorticity 176
Boussinesq approximation 141,
  266, 473
bubble, flow past 93, 120
Buckingham Pi theorem 487–489
Buoyancy
  force 141, 266, 473
  influence on the log-law 476
  instability 341–347
  internal gravity waves 271–284
Burgers' vortex tube 245–248

Burgers-like vortex sheet 262
β model of intermittency in
  turbulence 428

C

capillary
  length 202
  waves 202–204
cascade of blades 20–22
cascade of energy, in
  turbulence 398–400
Cauchy's equation of motion 41
Cauchy-Riemann equations 160
centrifugal instability 293,
  334–341
channel flow
  (turbulent) 457–460
chaos, deterministic 384–386
circular cylinder, flow past
  irrotational 162–164
  high Reynolds
    number 134–136
  low Reynolds
    number 118–120
  vortex street behind 51, 57,
    230, 238
  with circulation 171
circular flow 39, 78–83, 293,
  334–341
circulation
  Bjerknes' theorem 286
  definition of 22
  Kelvin's theorem 174, 225
  Kutta-Joukowski lift
    theorem 22, 172
  relationship to lift 22, 173
closure models of turbulence
  constant skewness model 423
  k-ε model 477–482
closure problem of
  turbulence 394, 421, 444
cnoidal waves 212, 218

coherent structures (in
turbulence) 466–469
columnar vortices
in atmosphere 239–241
in oceans 240
in rapidly-rotating flows 296,
301–312, 325–327
complex potential
definition 160
elementary examples 161
flow into duct 166
flow past plate 164
method of images 167–170
conservation of
angular impulse (angular
momentum) 256
linear impulse (linear
momentum) 254–256
Loitsyansky's integral 437
mass 10
mechanical energy (Bernoulli's
theorem) 14
Saffman's integral 438
constant-skewness model of
turbulence 423
constitutive equation, for
Newtonian fluid 41
continuity equation 10
continuum hypothesis 2
convection
Bénard 341–347
forced 141–143
free 143
from a heated plate 143–144
in the atmospheric boundary
layer 473–476
convective derivative 9
Coriolis force 294
corner eddies 98
correlation, velocity
auto 409, 497
lateral 403
longitudinal 403, 407
tensor 402, 407
triple 403, 407
two-point 402–408
Couette flow
between plates 66
between rotating cylinders 78,
337
creeping flow
in wedge-like cavity 98
Oseen's correction to 95, 119

past cylinder 94, 119
past sphere 91–96
critical Reynolds number
Bénard's experiment 346
pipe flow 372
Taylor's experiment 339
cylindrical polar
coordinates 73–75, 492, 495

D

D'Alembert's paradox 159
D'Alembert's solution of wave
equation 186
decay of homogeneous
turbulence 438
deformation of fluid
elements 36–38
delta wing 178, 237
diffusion (molecular)
of heat 56, 141–144
of vorticity 56–58, 69–73,
242–244
diffusion (turbulent)
of heat 469–472
Richardson diffusion (relative
diffusion) 439
diffusion length 70–71
diffusivity, thermal 56, 70, 141
dimensional analysis 70–71,
485–491
dimensionless groups 490
dispersion relationship for
waves 187, 190, 278, 298
dispersive waves 188, 194–198,
278–280, 298–311
dissipation length scale (in
turbulence) 395, 401, 418
dissipation of energy
in turbulent flow 25, 27,
400–402, 444
viscous 45
dissipation range (in
turbulence) 395–399, 401,
415
divergence theorem 493
diverging channel flow 137
downwash 179
drag
at high Reynolds
number 139–140, 165
at low Reynolds number 94–96
coefficient 96, 139–140, 487
crisis 139–140

due to waves 193–194
on circular cylinder 50, 140,
487
on sphere 94–96, 120–121,
139–140, 487

E

eddy
diffusivity 470
viscosity 445–447, 477–478
Ekman
layer 317–319
number 320, 490
pumping 319
spiral 319
energy, in laminar flow
dispersion due to
waves 195–203, 279,
301–311
dissipation due to viscosity 45
energy, in turbulent flow
cascade 398–400
decay law for isotropic
turbulence 438
dissipation 25, 27, 400–401
distribution across
scales 416–419
loss in hydraulic jump 27
loss in pipe expansion 25
production in shear
flows 448–450
spectrum 409–416
ensemble average 392
enstrophy 58, 396, 412, 484
entrainment
in laminar jet 146–148
in turbulent jet 452–456
equation of motion
averaged, for turbulent
flow 421
Cauchy 41
Euler 13
Navier-Stokes 43
equilibrium range, in
turbulence 417, 422
Ertel's potential 287
Euler equation 13

F

Falkner-Skan solutions for flow
over wedge 132
Feigenbaum sequence 384

filtering using the Fourier
transform 410–412, 499
five-thirds law of
Kolmogorov 417–419
Fjortoft's theorem 349
flat plate
boundary layer 126–127
drag 127, 165
inviscid flow over 164–165
flatness factor 429–433
fluid impulse, of localised vorticity
distribution
angular impulse 256–257
linear impulse 254–256
force
buoyancy 141, 266, 473
centrifugal 260, 294, 377
Coriolis 294
pressure 12, 41–43
viscous 39–43
four-fifths law of
Kolmogorov 419, 422
Fourier analysis 496–500
Fourier's law of heat
conduction 141, 470
free shear flows 450–456
free surface
boundary condition 189, 202
gravity waves 189–194,
199–219
Froude number
for free surface flow 26, 194,
213, 490
for stably stratified flow 267,
490

G

Gaussian distribution 429
generation of turbulent energy by
mean shear 448–450
geostrophic balance 296, 314
geostrophic flow 296, 314
gravity waves
deep water 190–194, 200
dispersion of 194–204
energy flux 199, 215
energy in 199–200
finite amplitude 204–214,
216–217
group velocity 195–203
internal 272–284
nonlinear steepening 206

refraction of 201
shallow water 191, 200,
204–214, 216–217
standing 214
Stokes drift 192
surface tension 202–204
grid turbulence 393–397
group velocity
concept of 195–198
for capillary waves 203
for inertial waves 298–300
for internal gravity waves 278
for Rossby waves 315–316
for surface gravity
waves 199–203

H

Hagen-Poiseuille flow 76–78
hairpin vortex 381, 467
Hamilton's principle 369
heat
conduction 141, 470
diffusion 56, 141–144, 470
equation 56, 141, 470
flux 470, 473–475
heat transfer from a wall
141–144
helicity 225–227, 299, 310–312
Helmholtz's decomposition of
vector fields 494
Helmholtz's laws of vortex
dynamics 223–225
Hill's spherical vortex 248
homogeneous turbulence
402–416, 420–424
hurricanes 321–324
hydraulic jump 25–28, 213
hydrostatics 4–6

I

ideal flow 7, 12–23
impulse, of a localised vorticity
distribution
angular impulse 256–257
linear impulse 254–256
impulsively started plate 69–72
inactive motion 462
induced drag 180
inertial subrange, in
turbulence 417–419
inertial waves 229, 293, 298–313,
325–328

inflection point theorem,
Rayleigh 347–349, 366
instability
Bérnard 341–347, 373–375
boundary layer 381
Kelvin-Arnold
theorem 361–367
Kelvin-Helmholtz 350–355
parallel shear flow 347–360
pipe flow 372
Rayleigh's centrifugal
criterion 292, 334–335
Rayleigh's inflection point
theorem 347–349
rotating flow 334–341
stratified shear flow 355–360
subcritical 387–388
supercritical 387
intermittency in
turbulence 424–428, 432
internal gravity waves 272–284
invariants (dynamic)
helicity 225–227
Loitsyansky integral 436–438
Saffman's integral 436–438
irrotational flow, *see under*
potential flow
isotropic turbulence
definition of 406
governing equation
for 406–415, 420–422
local 416–420

J

Jeffrey-Hamel flow 136–138
jets (laminar)
plane 144–147
round 147–148, 151
jets (turbulent)
plane 450–453
round 454–456
journal bearing 109–111

K

Karman constant 459, 463
Karman-Howarth
equation 420–424
Karman vortex street 51
Kelvin-Helmholtz
instability 350–355
Kelvin's circulation theorem 174,
225

Kelvin's variational principle in
   stability theory 361–367
kinematics
   Lagrangian versus Eulerian
      description 8
   material derivative 9
   relative motion near a
      point 36–38
   streamfunction 11
   vorticity and intrinsic
      rotation 38
Kolmogorov's
   constant, in inertial range 418,
      424–426
   decay law (for isotropic
      turbulence) 438
   equation for the equilibrium
      range 422
   five-thirds law 418–419
   four-fifths law 419, 422
   microscales 395, 401
   theory of the small
      scales 416–422, 424–428
   two-thirds law 417–419
Korteweg-de Vries
   equation 210–212, 218
Kutta condition 175
Kutta-Joukowski lift theorem 22,
   172–173
k-ε closure model
   examples of use 480–482, 483
   governing equations 477–480

L

Lagrangian
   description of motion 8
   displacement 367–368
laminar flow (definition) 7
Landau equation, in stability
   theory 386–388
Landau's interpretation of
   Loitsyansky's integral
   437
law of the wall
   for momentum 459
   for temperature 471
least action 369
lee waves 271–277
lift force
   on aerofoil 22–23, 172–179
   on cylinder with
      circulation 171

linear impulse (linear
   momentum) 253–256
linear stability theory
   Bérnard convection 341–347,
      373–375
   Kelvin-Helmholtz 350–355
   Rayleigh's inflection point
      theorem 347–349
   rotating flow 334–341
   shear flow 347–360
   stratified shear flow 355–360
local isotropy hypothesis 416
logistic equation 384–386
log-law of the wall
   for momentum 457–465
   for temperature 469–471
log-normal hypothesis for
   dissipation 427
Loitsyansky's integral 436–438
long-range correlations in
   turbulence 436–437
low speed streaks 467–469
lubrication theory
   Rayleigh's analysis (stepped
      bearing) 111–112
   Reynolds' analysis (slipper
      bearing) 107–108
   Sommerfeld's analysis (journal
      bearing) 109–111

M

marginal stability 339, 344–347,
   350, 355
mass conservation 10–11
material derivative 9
microorganisms, swimming
   of 102–105
microscales of turbulence
   Kolmogorov scale 395, 401
   Taylor scale 408
minimum dissipation
   theorem 97–98
mixing length theory 445–447
momentum equation, in integral
   form 18, 46–48
Monin-Obukhov theory of
   atmospheric flow 475

N

narrow-gap
   approximation 338–339
Navier-Stokes equation

derivation of 41–43
   in cylindrical polar
      coordinates 73–75
Newtonian fluid 6–7, 34, 42
Newton's law of viscosity 41–42
non-Newtonian fluids 6–7
normal distribution 429
no-slip condition 6, 35

O

one-dimensional energy
   spectrum 413
Orr-Sommerfeld equation 349
oscillating plate 72
Ossen's correction for Stokes
   flow 95, 119
outer layer in turbulent boundary
   layer 460–461
overlap region in turbulent
   boundary layer 458–461

P

parallel shear flows
   inviscid instability of 347–349
   stratified 355–360
particle paths in surface gravity
   waves 192, 215
Pascal's law 4
period doubling 384
phase speed 187–188
pipe flow
   stability of 371–272
   steady, laminar 75–77
   turbulent 77–78, 461
Pi theorem, Buckingham 71,
   487–489
Pitot tube 16
plane jet 144–147, 450–453
planetary waves 314
Poiseuille flow 75–77
polar coordinates 492, 495
potential flow
   complex potential 160–170
   D'Alembert's paradox 159
   in two dimensions 157–159,
      160–173
   into two-dimensional duct
      166
   past cylinder 163
   past plate 164–165
   past sphere 184
   velocity potential 155

Prandtl-Batchelor
theorem 242–244
Prandtl number 141–144, 471
Prandtl's theory of atmospheric
convection 474
principle of least action 369
probability density function
(pdf) 429
production of turbulent energy by
shear 448–450
puff (turbulent) 372

Q

quasi-equilibrium range 417–419
quasi-geostrophic flow 314

R

rapidly rotating flow 298–316
rate-of-strain tensor 36–38
Rayleigh's
centrifugal stability
criterion 293, 334–335
equation for parallel flow
348
inflection point theorem 349
stepped bearing 111
Rayleigh number 341, 490
Rayleigh-Bénard
convection 341–347,
373–375
relative diffusion of two
particles 439
Reynolds
lubrication theory 107–108
number 7, 35, 490
stresses 392, 442
Richardson's law for relative
dispersion 439
Richardson number
flux 474
global 354
local (or gradient) 356
ripples 202–204
Rossby number 295, 490
Rossby waves 314–316
rotating flow
between cylinders 78, 334–340
between discs 328
spin down of 86, 320
waves in 229, 293, 298–316,
325–327
rough wall turbulence 472–476

round jet 147, 151, 454–456
Russel, J.S. 185, 207, 209

S

Saffman's spectrum 436–438
scales in turbulence
dissipation scale 395, 401, 418
integral scale 395, 400, 403,
406
Taylor scale 408
secondary flow due to Ekman
pumping 317–320, 329
self-propulsion at low Reynolds
number 102–105
self-similar solution of
boundary layer 127, 133, 137,
142, 144
decay of line vortex 80
impulsively started plate 71
jets 145, 147
sensitivity to initial
conditions 383, 385, 392
separation of boundary layer 35,
50, 138–140, 236- 238
shallow-water flow 191, 204–214,
216–219
shear layers 66, 127, 132–138,
347–360, 450–456, 461
shear production of turbulent
energy 448–450
shear stress
viscous 3, 34, 37, 41–43
Reynolds 392, 442
signature function 435
skewness factor
definition of 429–431
relationship to enstrophy
production 433–434
slipper bearing 107–108
solitary wave 207–212, 216–218
Sommerfeld, theory of journal
bearing 109–111
spectrum of energy
inertial range 417–419
one-dimensional 413
sphere, flow over
at high Reynolds
number 139–140
at low Reynolds
number 91–96, 120–121
irrotational 184
spherical vortex, Hill's 248

spin-down 86, 320
spreading of a circular pool 114
Squire's theorem 347
stability, see under instability
stagnation point flow 68, 134, 161
starting vortex 23, 175–176, 238
steepening of surface gravity
waves 205–207
Stokes
drag law for a sphere 94–96
drift 192
flow 90–105, 118–120
streamfunction 11, 75
strain-rate tensor 36–38
stratified flow
blocking 270, 281
internal gravity waves 271–284
lee waves 271–277
shear flow instability 350–360
streamfunction
for axisymmetric flows 11, 75
in two dimensions 11, 98
stress tensor
Reynolds 392, 442
viscous, Newtonian
fluid 39–42
stretching of vortex lines 55, 246,
433
structure functions
constant skewness model
of 423
definition of 404–407
subcritical and supercritical
bifurcations 387–388
subcritical and supercritical
free-surface flow 25, 211,
213
subrange, inertial (in
turbulence) 416–419,
423–428
surface tension 30, 202–204, 490
suspensions 99–102
swimming of
microorganisms 102–105

T

Taylor
centrifugal instability 336–340
columns 295–297
microscale 408
number 339
vortices 339–341

Taylor-Goldstein equation 357
Taylor-Proudman theorem 296
teacup, spin-down in 320
temperature log-law of the
    wall 471
thermal
    convection 141–144
    diffusion (laminar) 56, 70,
        141–144
    diffusion (turbulent) 470–475
thin aerofoil theory 22–23,
    172–180
thin film flow
    on an incline 67, 116
    on a rotating disc 82, 118
tidal vortex 240
tip vortices 179, 233
tornado 239–240
trailing vortices 179, 233
transition to turbulence 379–381,
    386–389
tropical cyclone 320–324, 330
triple velocity correlation 403,
    407, 421
two-thirds law, in
    turbulence 417–419

U

uniqueness theorem 96
universal equilibrium range, in
    turbulence 416–428
universality of the small scales of
    turbulence 416–419

V

Väisälä-Brunt frequency 266
variational principles in stability
    theory 361–371

vector identities and
    theorems 492–494
velocity correlation
    tensor 402–407
velocity potential 155
viscosity
    eddy 445–447
    molecular 3, 7, 34
viscous
    dissipation 45
    length scale (in
        turbulence) 395, 401
    sublayer near wall (in
        turbulence) 460
vortex
    atoms 227
    bound (in aerofoil theory)
        176
    Burgers 245–248, 262
    coherent 466–469
    double roller 468
    hairpin 381, 467–469
    Hill's 248
    rings 234–236, 249–251,
        258–260
    shedding 23, 51, 176, 179, 238,
        395
    sheet (stability) 350–353
    starting 23, 175–176, 179,
        237–238
    street 51, 57, 238
    stretching 55, 246, 433
    trailing 179, 233
vorticity
    advection-diffusion equation
        for 56
    correlation tensor 407
    definition of 37

generation at boundaries 57,
    69–73
generation by buoyancy 266,
    285–286
Kelvin's theorem for 174, 225
stretching 55, 246, 433

W

wake 50–51, 140, 165, 381, 383
wall jet, thermal 143–144
wall turbulence 456–476
waves
    capillary 202–204
    cnoidal 212, 218
    deep water 191–194, 200–204
    dispersion 187, 195–199, 203,
        278–280, 298–316
    drag 193
    energy in 199
    equation 186
    finite amplitude 204–214,
        216–218
    group velocity 195–200, 274,
        278–280, 298–300
    inertial (in rotating
        fluid) 298–313
    internal gravity 277–284
    lee 271–277
    nonlinear steeping of 206
    packet 195–198, 279,
        302–311
    Rossby 314–316
    shallow-water 191, 200,
        204–214, 216–218
    solitary 207–212, 216–218
    surface gravity 189–194,
        199–214
wind-tunnel turbulence 393–398